PRINCIPLES OF

ALTERNATING-CURRENT MACHINERY

PRINCIPLES OF
ALTERNATING-CURRENT MACHINERY

RALPH R. LAWRENCE
Professor Emeritus of Electrical Machinery
Massachusetts Institute of Technology
Fellow of the American Institute of Electrical Engineers
Fellow of the American Academy of Arts and Sciences

Revised by

HENRY E. RICHARDS

FOURTH EDITION

NEW YORK   TORONTO   LONDON
McGRAW-HILL BOOK COMPANY, INC.
1953

PRINCIPLES OF ALTERNATING-CURRENT MACHINERY

Library of Congress Catalog Card Number: 52-10334

v

THE MAPLE PRESS COMPANY, YORK, PA.

PREFACE TO THE FOURTH EDITION

Although there have been many developments in a-c machinery since the third edition was published, the general principles underlying a-c machinery have not changed. In this edition the general scope and treatment of the subject have not been altered, but many suggested changes have been incorporated. The order of presenting certain parts of the subject has also been rearranged and new chapters have been added on polyphase autotransformer connections and the power mercury-arc rectifier. In addition, problems have been included as a part of the text.

The authors appreciate the courtesy of the Westinghouse Electric Corporation, the General Electric Company, and the Allis-Chalmers Manufacturing Company in furnishing photographs for some of the illustrations and machine data used in some of the problems.

Ralph R. Lawrence
Henry E. Richards

Cambridge, Mass.
Boston, Mass.
January, 1953

PREFACE TO THE FIRST EDITION

This book deals with the principles underlying the construction and operation of alternating-current machinery. It is in no sense a book on design. It is the result of a number of years' experience in teaching the subject of Alternating-current Machinery to senior students in Electrical Engineering and has been developed from a set of printed and neostyled notes used for several years by the author at the Massachusetts Institute of Technology.

The transformer is the simplest piece of alternating-current apparatus and logically perhaps should be considered first in discussing the principles of alternating-current machinery. Experience has shown, however, that students just beginning the subject grasp the principles of the alternator more readily than those of the transformer. For this reason the alternator is taken up first.

No attempt has been made to treat all types of alternating-current machines, only the most important being considered. Certain types have been developed in considerable detail where such development seemed to bring out important principles, while other types have been considered only briefly or omitted altogether. No new methods have been used, but it is believed that bringing together material which has been much scattered and making it available for students is sufficient reason for the publication of the book.

Mathematical and analytical treatment of the subject has been freely employed where such treatment offered any advantage. The symbolic notation has been used throughout the book.

The author wishes to express his sincere thanks to Professor W. V. Lyon of the Massachusetts Institute of Technology for many suggestions and especially to Professor H. E. Clifford, Gordon McKay Professor of Electrical Engineering at Harvard University and the Massachusetts Institute of Technology, who critically read the original manuscript and offered many suggestions. The author also wishes to express his thanks to Mr. N. S. Martson for his care in reading the proof, and to the Crocker-Wheeler Company, the General Electric Company and the Westinghouse Electric and Manufacturing Company who furnished photographs from which the drawings of machines were prepared.

Ralph R. Lawrence

Boston, Mass.
September, 1916

CONTENTS

PART TWO. SYNCHRONOUS GENERATORS

TABLE OF SYMBOLS

A = armature reaction, generally expressed in ampere-turns per pole

a = ratio of transformation

$\mathcal{B}$ = flux density

b = susceptance

E = induced or generated effective voltage. In general, E is a voltage rise and $-E$ is a voltage drop

F = impressed field of a synchronous generator or motor

$\mathfrak{F}$ = magnetomotive force

f = frequency

$\boldsymbol{f}$ = function

g = conductance

I = current

I_φ = magnetizing current of a transformer or of an induction motor

I_{h+e} = core-loss current of a transformer or of an induction motor

I_n = exciting current of a transformer or of an induction motor

$I_1{}'$ = load component of primary current of a transformer or of an induction motor

$\mathfrak{J}$ = moment of inertia

j = operating factor which rotates a vector counterclockwise through 90 deg

k_b = breadth factor

k_p = pitch factor

N = number of turns

n = speed; number of phases

P = power

p = number of poles; number of anodes

pf = power factor

q = order of harmonic

R = air-gap field of a synchronous generator or motor

$\mathcal{R}$ = reluctance

r = resistance

r_e = effective resistance or equivalent resistance of a transformer

s = slip; number of slots per phase

T = torque

V = terminal effective voltage.   In general V is a voltage rise and $-V$ is a voltage drop

x = reactance

x_a = leakage reactance of a generator or a motor

x_e = equivalent reactance of a transformer

x_s = synchronous reactance

y = admittance

Z = number of inductors

z = impedance

z_s = synchronous impedance

α = angular acceleration or a phase angle; angle of retard

δ = power angle of a synchronous machine

η = efficiency or hysteresis constant

θ = phase angle

μ = angle of overlap

ρ = coil pitch

Σ = summation

φ = flux

ω = angular velocity or $2\pi f$

Where the letters given in the preceding table are used with other significance than indicated, it is so stated in the text. Where other letters are used, their meaning is stated in the text.

PART ONE

STATIC TRANSFORMERS

Chapter 1

Transformer. Types of transformers. Cores. Windings. Insulation. Terminals. Cooling. Oil. Indoor installations. Breathers.

Transformer. A transformer consists of two or more insulated coils coupled by mutual induction. Its action depends upon the self-inductances of its coils and the mutual induction between them. In its simplest form, the transformer consists of two coils interlinked by a common magnetic circuit. When power is supplied to one coil at a definite frequency and voltage, power can be taken from the other at the same frequency but in general at a different voltage. The ratio of the two terminal voltages depends upon the resistances of the coils, their inductances, the closeness of their coupling, and the load. When the resistances are small and the coupling is close, as in a power system transformer, the ratio of voltages depends chiefly upon the ratio of the number of turns in the two coils.

Transformers are used for widely different purposes. In communication circuits, they are used to adjust the impedances of the input and output circuits in order to attain the condition of maximum power transmission between the transmitting circuit and the receiving circuit. For power transmission, transformers are used to step up the generated voltage in order that the transmission may be accomplished economically with a small current and therefore a small amount of copper. At the receiving end of the line, they are used to step down the voltage to a value which can be conveniently and safely used. For impedance matching the efficiency is essentially low. For power transmission high efficiency is essential. The fundamental equations are the same for all types of transformers, and the fundamental principles underlying their operation are also the same.

Transformers may be of the air-core or the iron-core type, but for power purposes they are always of the iron-core type. Power system transformers are given major consideration in this book. A power system transformer consists essentially of a magnetic core, usually of silicon steel, linked with two or more windings of insulated copper wire. The windings to which power is supplied are called the primaries; the others which deliver power to the receiving circuit are called the secondaries. Either windings serve equally well as primaries or as secondaries. If the primary

1

windings have more turns than the secondary windings, the voltage is lowered and the transformer is a step-down transformer. If the secondary windings have more turns than the primary windings, the voltage is raised and the transformer is a step-up transformer.

The development of low-loss, nonaging silicon steels and the better design of ducts for the oil, which is almost universally used as a cooling medium to carry the internal heat due to the losses to special cooling devices such as radiators, have greatly increased the efficiency and the permissible size of transformers. Transformers of 50,000-kva output have been built and efficiencies of 99 per cent or even better have been attained under good operating conditions.

The improvement of insulating oils and insulating materials in general and the use of shielding have greatly increased the permissible voltage for which transformers can be successfully designed. Although the highest voltage in use at the present time (1952) for power transmission is slightly under 300,000 volts, transformers for 10,000,000 volts have been built for laboratory testing purposes.

Types of Transformers. Transformers are often classified according to the purposes for which they are used. In a power system, transformers are used to increase the voltage of the generating station for transmission purposes and then to reduce or step down the voltage at a point nearer a load center for application to the distribution system. Such transformers are of the largest sizes both in terms of dimensions and of kva and voltage ratings and are known as power transformers. The distribution voltage is further reduced by transformers before connection to house circuits or distribution throughout a factory. Transformers for this purpose are smaller in size and rating and are known as distribution transformers. Other types are instrument transformers, constant-current transformers, and audio-frequency transformers.

Whatever the application, two distinct types of construction are used, differentiated by the relative positions occupied by the windings and core. In the *core* type, two groups of windings envelop a considerable part of the magnetic circuit. In the *shell* type, the magnetic circuit envelops a considerable portion of the windings. These are illustrated in Figs. 1 and 2. For a given output and a definite voltage rating, the core type contains less iron but more copper than the shell type. By proper design both types can be made to have essentially the same electrical characteristics.

Cores. Since the core usually serves as a frame to hold the windings, it must be mechanically strong. Under normal conditions the mechanical stresses between windings and between adjacent turns of transformers are low, but under short circuit the stresses may be very great. On short circuit the steady-state currents of a normal power system transformer usually are 10 to 25 times full-load current if full voltage is maintained on the primary. These currents produce stresses between windings which are from $(10)^2 = 100$ to $(25)^2 = 625$ times those at full load. The transient currents may be even greater. These stresses are highly important in large transformers.

There are two general methods of forming the cores of transformers. The traditional construction involves the use of laminations which are stamped from silicon-steel plates. These punchings are then stacked to form the core. Laminations are often insulated from one another by a thin coat of varnish although for small transformers the oxide scale which forms may provide sufficient insulation. With this construction there are

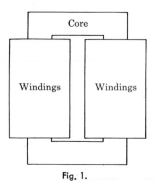

Fig. 1.

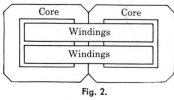

Fig. 2.

always certain places in the core where the flux goes across the direction of rolling of the metal, thus increasing the core loss. In a newer type of construction, the sections of the core are formed by winding a continuous steel ribbon into a tight coil. This provides flux paths which are more nearly parallel to the direction of rolling of the steel at all points and so reduces the losses. Nevertheless, the laminated construction is still widely used for power transformers and for distribution transformers of the larger sizes.

In the older of these two types of construction, the laminations are built up into a core within the finished coils with the joints of the successive layers reversed so as to overlap and make the reluctance of the core a minimum. The laminations are then firmly bolted together. Figure 3 shows common forms of stampings for a core-type

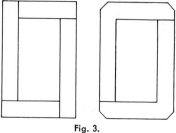

Fig. 3.

transformer. The thickness of the laminations used depends upon the kind of steel used and upon the frequency for which the transformer is designed. With ordinary transformer steel and 60 cycles, the thickness of the laminations may be as small as 0.014 in.

Some manufacturers have used butt joints in the magnetic circuits of large transformers to facilitate building up the cores and removing coils in case of breakdown. When such joints are used, it is customary to insulate each joint with a layer of thin tough paper about 0.005 in. thick to prevent a loss due to eddy currents at the joints. Reference to Fig. 4 shows that, unless the laminations of the two parts of the joints are exactly over each other, a path is provided by which the eddy currents can pass from one lamination to the next across the joint as shown by the wavy line in the figure. This permits the eddy currents to flow in portions of the core at the joints much the same as if the core were not laminated. It is sometimes claimed that the loss in butt joints which are without insulating paper is no greater than in ordinary lap joints. With lap joints, the greater part of the flux at the joints passes from one lamination to the next nearly perpendicularly to the plane of the laminations, and the planes of the paths of the eddy currents therefore lie in the laminations. The lamination of the core has no effect on these eddy currents.

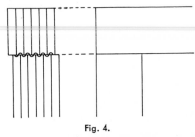

Fig. 4.

Lap joints are equivalent to small air gaps. They call for an increase in the magnetizing ampere-turns required for a core, the increase depending on the flux density. For flux densities at which transformers usually operate, this increase varies from 8 to 12 amp-turns per joint. Butt joints must be figured as air gaps equal in length to the thickness of the paper insulation in the joints.

When the laminations are assembled in core-type transformers, the resulting cross section of the core is rectangular in small and medium-sized transformers. This is therefore called the rectangular-core type of construction. In the larger sizes, laminations of different widths are used so as to form a cross section of cruciform shape. This gives the cruciform-core type and allows circular-shaped coils to be used which withstand better the forces due to high currents.

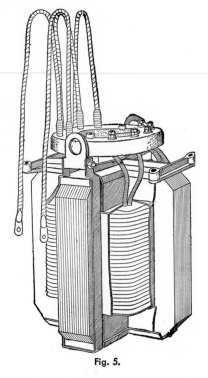

Fig. 5.

In the laminated shell-type transformers of the larger sizes, the core is assembled around form-wound coils and is shaped similar to that shown in Fig. 2, page 3. This is known as the simple shell type. Distribution transformers often have the core surrounding the windings on three or four sides as in Fig. 5, a type of construction known as the distributed-shell type.

In the newer wound-core construction, several ingenious methods of assembly are used. In the Spirakore transformer manufactured by the General Electric Company, the coils are first assembled. Two continuous strips of steel ribbon are then threaded through the coils and wound

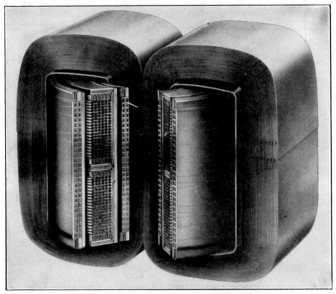

Fig. 6. (*Courtesy of General Electric Co.*)

around opposite sides of the coils to form a shell type of transformer, as illustrated in Fig. 6. In another method, used by the Line Material Company, the core is made up first by winding continuous strips of steel to the desired shape. The forms for the coils are then fitted loosely around the two legs of the core, and the coils are wound onto the forms by rotating the forms by a suitable driving mechanism. This results in a core type of construction. The Hypersil cores used in many of their transformers by the Westinghouse Electric Corporation necessitate a construction such that the path of the flux is nearly parallel to the direction of rolling of the steel. Hypersil is a silicon steel having, at normal excitations, about one-third higher permeability than that of the usual silicon steels. Except in the larger sizes, each of two cores is formed by winding the steel around a rectangular mandrel. Each core is then cut into two U-shaped parts, which are then fitted together around the

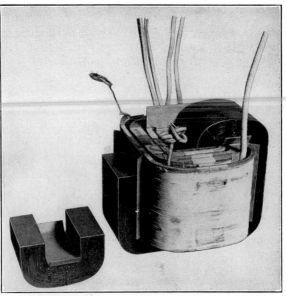

Fig. 7. (*Courtesy of Westinghouse Electric Corp.*)

windings so as to form a shell type of construction, as shown in Fig. 7. The Bent-Iron construction is used by the Kuhlman Electric Company.

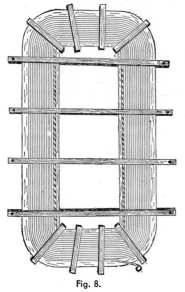

Fig. 8.

Steel strips are bent in a folding machine and interleaved to form two D-shaped cores. After annealing, the flat surfaces of the two are tied together and the interleaved portions opened up. The coils are then wound around the butting portions of the two cores and the D sections reformed, giving a shell-type construction.

Windings. The windings of all transformers of any substantial size and voltage are subdivided to increase the insulation and to diminish the leakage flux between primary and secondary. Each winding is made up of coils which are either pancake shaped, such as are commonly used on large shell-type transformers, or cylindrical, such as are used on core-type transformers. The two shapes or types of coils are illustrated in Figs. 8 and 9.

The coils are of cotton-covered wire of either round or rectangular cross section and are machine-wound. After being formed, the coils are taped to hold them together and then are thoroughly impregnated with insulating compound. Where proper insulation can be provided, wire of rectangular cross section is desirable since it has a greater space factor than round wire. The coils are sometimes formed of flat strip copper wound edgewise with strips of varnished paper, cambric, or mica paper between successive turns. The conductors of large transformers are laminated to decrease the eddy-current loss in them.

The primary and secondary windings of small core-type transformers are made in two sections, and one primary and one secondary section are placed on each of the two upright sides of the iron core. One winding is placed outside the other to minimize the magnetic leakage.

Insulation. The separate turns of the coils are insulated with cotton, cambric, or mica paper. To prevent breakdown between successive layers it is necessary, especially in high-voltage coils, to provide extra insulation in the form of pressboard or mica paper. When this is required, the extra insulation is extended slightly beyond the ends of the windings to prevent creepage. It is usual to limit the voltage of a single coil to about 5,000 volts. When the voltage of a winding exceeds this value, the

Fig. 9. (Courtesy of Westinghouse Electric Corp.)

winding should be subdivided into a number of coils with insulation between adjacent coils. Special insulation is required between the high-voltage and low-voltage coils and the core. High-voltage transformers, for 50,000 volts and over, require insulation of the greatest dielectric strength between the coils and the core. For this purpose built-up mica and shields of pressboard are used. The different sections of the windings are held apart by insulating barriers, and oil in the spaces left between the sections not only cools the coils but also provides the necessary insulation.

The voltage stresses in the insulation of the end turns of a transformer winding caused by abnormal operating conditions, such as are produced by lightning, may be very great. The breakdown of the insulation of many transformers, especially high-voltage transformers, has been due to such stresses.

A transformer not only has self-inductance and mutual inductance but there is capacitance between turns of the windings, between each winding and the core, between the windings themselves, and also between the windings and the case. At ordinary frequencies the effect of these capacitances is negligible, but their effect is far from negligible when there are high-frequency oscillations caused by abnormal operating conditions or by surges.

Before these capacitance effects were clearly understood, it was common practice to provide extra insulation on the end turns of the high-voltage windings of transformers, but the decrease in the capacitance between these turns due to the greater separation necessitated by the extra insulation often more than balanced the increase in insulation strength.

A transformer winding is equivalent to a series inductance with shunt capacitances between turns and distributed capacitance between the windings and the core and between the windings and the case. There is also mutual inductance and distributed capacitance between each pair of windings.

If the capacitance between coils, between the coils and the core, and between the coils and the case can be made negligible and the capacitance between turns can be made uniform, the voltage stress due to abnormal conditions is uniformly distributed throughout the windings.

This condition can be attained approximately by properly designing the windings and by using electrostatic shields. A transformer designed with electrostatic shields and with its windings arranged so that the voltage drop through them due to a surge is approximately uniformly distributed is known as a nonresonating transformer. A cylindrical coil can be successfully shielded by winding it so that one terminal is on the inside and the other is on the outside. Each terminal is then connected to a cylindrical metallic shield, one shield inside the coil, the other outside the coil. If the capacitance between turns is uniform, the voltage stress due to a surge is nearly uniformly distributed throughout the winding. The arrangement of shields is determined by the type of winding used.

Transformers have also been protected by spark gaps.

Figure 10 shows a high-voltage core-type transformer without its case. The low-voltage windings are next to the core. The high-voltage windings are on the outside and are sectionalized for better insulation. They are insulated from the low-voltage windings by cylindrical barriers which are shown in the figure.

Terminals. The insulation of the high-voltage terminals for 50,000 volts and over is a difficult problem which has been solved by the use of two quite different types of terminals, known as the oil-insulated terminal and the condenser terminal. The oil-insulated terminal consists of segments of porcelain or other molded material built up about the conducting rod to form an enclosure for oil. The oil space is subdivided vertically by insulating cylinders to prevent lining up of particles in the oil which would concentrate the dielectric stress and tend to cause breakdown. The porcelain segments are shaped on the outside to form a series of petticoats so as to increase the creepage distance. The condenser-type terminal is built up of alternate layers of metal foil and micarta paper wound on the conducting rod under heat and pressure. The purpose of the metal foil is to distribute the dielectric stress uniformly throughout the insulation. In order to accomplish this it is necessary that the lengths of the successive layers of metal foil and micarta paper decrease by such amounts that the condensers formed by the layers of metal foil and the paper all have the same capacitance. The decrease in the lengths of the metal foil and paper causes the two ends of the terminal to taper. For indoor service, the terminal is completed by the addition of a flat disk of insulating material at the top and an

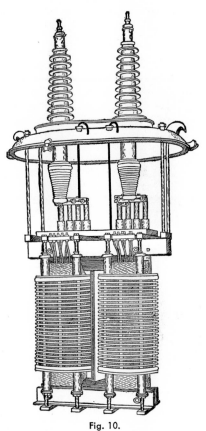

Fig. 10.

external insulating tube which extends the length of the terminal. The space between the insulating tube and the terminal is filled with insulating compound to prevent corona. For outdoor service, the portion of the terminal outside the transformer case is enclosed in a one-piece porcelain cover with petticoats to increase the creepage surface. The space between the terminal and its porcelain cover is filled with oil.

A and B of Fig. 11 show a complete Westinghouse high-voltage outdoor condenser-type terminal and also its cross section. C and D of this figure show a complete General Electric high-voltage outdoor oil-filled-type terminal and its cross section.

For moderate voltages, porcelain bushings are used. To prevent moisture and water from entering the transformer case, the leads of transformers for not over a few thousand volts leave the containing case

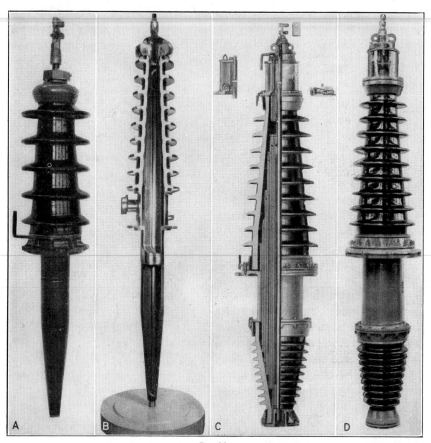

Fig. 11.

under projecting ledges on each side near the top and the high-voltage and low-voltage leads are placed on opposite sides.

Cooling. Since the output of any piece of electrical apparatus is limited by the rise in temperature produced by its losses, one of the most important problems in design is to provide some satisfactory means of cooling. Since the losses in a transformer vary approximately as the volume and since the amount of heat that can be dissipated depends upon the surface exposed, it is evident that the problem of cooling a transformer becomes more difficult as its size is increased.

The coils of most transformers are placed in oil contained in welded sheet-steel tanks. To ensure effective cooling, the coils and core are so arranged that the heated oil can rise readily to the top through ducts provided between coils and between coils and core. The oil then passes down along the cooler walls of the transformer case.

As only about 1 watt can be radiated for each degree centigrade rise in temperature from each 150 sq in. of dry surface of ordinary transformer cases with smooth sides, special means for cooling must be provided except for small transformers up to about 35-kva rating. For larger

Fig. 12. General Electric distribution transformer, 10 kva, 60 cps, 6,900 :11,950 volts.

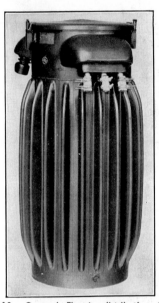

Fig. 13. General Electric distribution transformer, 75 kva, 60 cps, 2,400 :$^{240}_{120}$ volts.

transformers up to about 75- to 100-kva capacity, cases with corrugated sides can be used. Corrugating the sides of the transformer case increases by about 50 per cent the amount of heat that can be radiated.

When cooling water is available, a common method for keeping large transformers cool is to circulate water through coils of pipe placed in the tops of the transformer cases in the oil above the windings. When this is done, corrugated containing cases are not necessary. Occasionally the oil is circulated through coolers which are external to the transformer cases. Transformers for relatively low voltages and in sizes up to about 5,000 kva may be forced-air cooled. Such transformers are known as air-blast transformers. No oil is used for air-blast transformers. The cooling is effected by circulating air, by means of external blowers, through ducts provided in the windings and in the cores. In such cases, the cooling air must be filtered to remove the dust and dirt as otherwise the ducts would quickly become covered with a film of dust which would considerably decrease the heat transfer from the windings and cores to the cooling air. Comparatively few air-blast transformers are used except where fire hazard prevents the use of oil. In many cases, for example in unattended substations and in places where no cooling water is available, it is not possible to use artificial cooling. Under these con-

Fig. 14. General Electric oil-immersed single-phase power transformer, 1,000 kva, 60 cps, 13,800 to 2,300/4,000 Y.

ditions special transformer cases with large radiating surfaces are used. The required increase in surface can be obtained by welding vertical tubes to the cases. For very large self-cooled transformers, tanks with radiators through which the cooling oil can circulate are attached to the sides of the tanks. Fans mounted on the radiators in such a way as to blow air through them are frequently used to increase the cooling. These fans in some cases are operated only for heavy loads or for overloads.

Typical transformer cases are illustrated in Figs. 12, 13, 14, and 15. These show a case with smooth sides, a case with corrugated sides, a tubular case, and a radiator type of case. A transformer with a cooling coil for water is shown without its case in Fig. 16.

Oil. The oil in an oil-cooled transformer not only carries the heat by convection from the windings and core to the transformer case and also to the cooling coils, provided they are used, but also has the even more important function of insulation. The selection of a suitable oil is of the greatest importance. Transformer oil is obtained by fractional distillation of petroleum. It must be free from alkalies, sulfur, and moisture. Moisture has a very marked effect on the dielectric strength of transformer oil. The presence of as small an amount as 1 part in 20,000 by volume decreases the puncturing voltage to nearly one-third its normal value. Minute particles mechanically suspended in the oil greatly decrease the puncturing voltage by localizing the dielectric stresses in the oil. The

Fig. 15.   General Electric power transformer, 31,250 kva, 39,500/68,400 Y to 13,200 volt, 60 cps.

methods for cleaning and for removing moisture from oil can be found in any of the standard handbooks for electrical engineers.

The ordinary specifications for transformer oil are given in Table 1.

TABLE 1

	Medium (Transil—6)	Light (Transil—10c)
Flash point, °C...............	180 to 190	133
Burning point, °C............	205 to 215	148
Freezing point, °C...........	−10 to −15	−40
Density....................	0.865 to 0.870 (at 13.5°C)	0.87 (at 15°C)
Viscosity (Saybolt, sec).......	100 to 110 (at 40°C)	40 to 57 (at 37.8°C)
Acid, alkali, sulfur, moisture...	None	None

Medium oil is used for self-cooled transformers. For water-cooled transformers lighter oil is used. The dielectric strength of transformer oil, measured between brass spheres 0.5 in. in diameter and placed 0.15 in. apart, should not be less than 30,000 volts.

Indoor Installations.   For indoor installations several types of transformers are used.   Because of fire hazard, Fire Underwriters' Rules require that oil-insulated transformers be installed in special fire-resistive vaults.   Exceptions are small transformers of not over 600 volts.   Expensive vault construction may often be avoided for voltage ratings up to 15,000 volts by the use of either dry-type or Askarel-insulated transformers.   The dry-type either may be self-cooled, employing no fan for ventilation, or may be an air-blast transformer of the type discussed on page 11.   The self-cooled transformer should be insulated with class B materials such as porcelain, asbestos, mica, and spun glass which are noncombustible and which, because of the higher operating temperatures allowed, reduce the dimensions for a given kva capacity.   Askarel-insulated transformers use a nonflammable liquid in place of oil.   This liquid is available under the following trade names: Chlorextol, Abestol, Pyranol, Arachlor, Noflamol, Inerteen.   Because they are powerful solvents for the insulating materials commonly employed, they can be used only in transformers specially designed for their use.

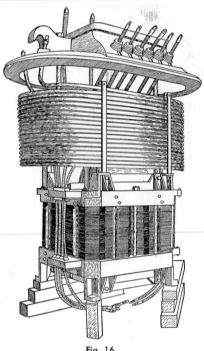

Fig. 16.

Breathers.   Since a small amount of moisture causes a great decrease in the puncturing voltage of transformer oil, it is necessary to take special precautions to prevent moisture from entering transformer cases.   In small transformers this is accomplished by making the cases airtight by sealing the leads where they pass through the cases with some compound such as asphaltum and by making the joints between the covers and cases airtight by clamping or bolting the covers down with gaskets between them and the cases.

It is practically impossible to make the cases of large transformers airtight.   For this reason it is customary to provide them with definite openings or vents through which the difference in pressure inside and outside the cases, caused by changes in temperature of the transformer or by changes in atmospheric conditions, can equalize.   When provision is made for the equalization of the pressure, devices technically known as breathers are used to prevent moisture from entering the cases through the vents. In its simplest form a breather is merely a chamber with baffle plates in it, connected to the top of the transformer by means of a small pipe. The baffles are arranged to prevent water or snow from entering the transformer when it is used out of doors or in an exposed place.

Unless the air in a transformer is dry, a sudden drop in the temperature of the surroundings may cool the air in the transformer below the dew point and cause moisture to precipitate. This is not likely to occur unless the transformer is lightly loaded and is in an exposed place. To prevent this precipitation, breathers with provision for drying the air which passes through them are used on transformers installed out of doors. Such breathers contain a considerable quantity of calcium chloride over which the air must pass before entering the transformer.

The size of transformers above which it is desirable to use breathers depends upon the conditions of service, but in general it is customary to install breathers on outdoor transformers of 500 kva or over and above 22,000 volts.

Moisture in transformer oil not only reduces the dielectric strength of the oil but aids in the formation of a sludge. Oxygen combines chemically with the oil also to form a sludge which in extreme cases shows as a darkening of the oil. This sludge tends to form a deposit on the transformer windings and on the core and to act as a heat insulator to retard the transfer of heat from the transformer windings and the core to the cooling oil. This formation of sludge is prevented by one type of breather which, by chemicals, removes the moisture from the air which enters the transformer through the breather and also removes the oxygen. Before a transformer with this type of breather is put in service, the air above the transformer oil is replaced by nitrogen. A special valve allows the transformer to breathe in only when the pressure in the tank falls a few pounds below atmospheric pressure and prevents breathing out except when the pressure in the gas space is about 5 lb above atmospheric pressure. The limiting of the breathing greatly increases the time between renewals of the chemicals. The gas space above the oil is made as large as possible, and when sufficient space cannot be provided above the oil, auxiliary chambers are used. With nothing but nitrogen in contact with the oil, no secondary explosion can take place in the tank. A relief diaphragm is provided which relieves any excessive pressure which might be developed in the tank due to a failure in the windings or a short circuit of the windings.

Another device which is used to prevent sludge reaching the transformer coils and the oil ducts is an expansion tank which is mounted on the top of the transformer case. When an expansion tank is used, the main tank is completely filled with oil and all the change in oil level caused by the expansion and contraction of the oil, due to changes in operating temperatures, takes place in the expansion tank.   No air is in contact with the oil in the main tank.   A breather is provided and also a sump from which any condensation which collects can be drawn off by a suitable valve.   A relief diaphragm is also provided for protection against excessive pressure which might be generated by a short circuit in the windings of the transformer.   This arrangement is shown in Fig. 14.

Chapter 2

Induced voltage.   Transformer on open circuit.   Reactor.

Induced Voltage.   The voltage induced in any winding depends only upon the number of turns in the winding and the rate of change of the flux linking it.   It makes no difference how the change in flux is produced.

Let φ_m be the maximum value of the flux linking a transformer coil and assume this flux to vary sinusoidally with respect to time.

$$\varphi = \varphi_m \sin \omega t$$

If N_1 is the number of turns in the coil, the voltage rise[1] induced in the coil at any instant by the flux φ is

$$e = -N_1 \frac{d\varphi}{dt}$$
$$= -\omega N_1 \varphi_m \cos \omega t$$

The maximum voltage is

$$e_m = \omega N_1 \varphi_m$$
$$= 2\pi f N_1 \varphi_m$$

The effective rms voltage in volts, when φ_m is expressed in webers, is

$$E_1 = \frac{2\pi f}{\sqrt{2}} N_1 \varphi_m$$
$$= 4.44 f N_1 \varphi_m \tag{1}$$

If the voltage is not a sine wave, Eq. (1) becomes

$$E_1 = 4(\text{form factor}) f N_1 \varphi_m \tag{2}$$

Transformer on Open Circuit.   When an alternating potential is impressed on an inductive circuit, the current increases until the total voltage drop around the circuit is zero.   Under this condition the total voltage drop due to induction plus (vectorially) the resistance drop in the circuit is equal to the impressed voltage drop.

$$-V = -E + Ir$$

---

[1] In this book, voltage rise in general is considered positive.   A voltage drop is therefore negative.   For the significance of voltage rise and fall, see R. R. Lawrence, "Principles of Alternating Currents."

$$\phi = 0.4 \pi NI \frac{\mu A}{?}$$

$-V$ and $-E$ are the impressed voltage drop and the total voltage drop induced in the circuit by the flux linking it.   The diagram of connections for such a circuit containing iron and its vector diagram are shown in Figs. 17 and 18.   The conditions shown in these figures correspond to those existing in a transformer with the secondary circuit open.

Referring to Fig. 18, E_1 is the voltage rise induced by the flux linking the winding.   To induce this voltage, a flux φ is required which is 90 deg ahead of the voltage.   This flux causes hysteresis and eddy-current losses in the iron core, and for this reason the current I_1 producing the flux must have an active component with respect to the voltage drop $-E_1$. I_1 can therefore be resolved into two components, one in phase with $-E_1$ and one in phase with the flux φ.   The component I_φ which is in phase

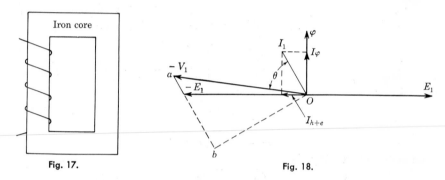

Fig. 17.   Fig. 18.

with the flux is the current which would be required to produce the flux if there were no core loss, and for this reason it is called the magnetizing component or simply the magnetizing current.   The effect of the active component of I_1, I_{h+e}, is to balance the effects of the hysteresis and eddy currents in the core, so far as the production of flux is concerned.   In reality the currents I_1, I_φ, and I_{h+e} should not be drawn as vectors, as will be shown later, since they are not sine waves even though the impressed voltage is a sine wave.   These currents must be considered to be replaced, in Fig. 18, by their equivalent sine waves.   The voltage $-V_1$ impressed across the coil must balance $-E_1$ and in addition must have a component equal to the resistance drop in the circuit.   It is equal to $-E_1$ plus the resistance drop, in a vector sense.   The current I_{h+e} depends upon the flux density in the core, upon the thickness of the laminations, and upon the amount and quality of the iron.   The magnetizing current I_φ depends upon the flux density, the kind of steel, the number of turns in the primary winding, and the magnetic circuit.   Its value may be calculated by usual magnetic-circuit methods.   The relationship between the magnetizing force in ampere-turns per unit length of magnetic path and the maximum flux density is generally available in the form of magnetization curves for the grades of steel of which transformer cores are made. In a transformer it is desirable that the magnetizing current be small in comparison with the rated current.

Reactor. A reactor consists essentially of a winding on a laminated iron core which is designed so that the winding takes current at a low power factor. The magnetizing current should therefore be large. The conditions, so far as the magnetic circuit is concerned, are therefore quite different from those required for a transformer.

If the winding has low resistance, $-E_1$ and $-V_1$ are nearly equal even for large variations in the current I_1. Therefore if $-V_1$ is constant, $-E_1$ and φ_m are nearly constant and the current I_{h+e}, which supplies the core losses, is also nearly constant. The magnetizing current I_φ can be increased without appreciably affecting either φ_m or I_{h+e} by introducing an air gap in the magnetic circuit. By adjusting the length of the air gap, the reactive component I_φ of the current I_1 can be made large compared with I_{h+e}. The power factor can be varied therefore by varying the length of the air gap. In this case, the permeability of the magnetic circuit is nearly constant, and I_φ is nearly a sinusoid when the impressed voltage is a sinusoid. To keep the losses down and consequently make I_{h+e} small, it is necessary to design reactors to operate at low flux density. The resistance of the winding should also be made small as the resistance drop introduces an energy component in the voltage impressed across the coil and raises the power factor.

When considering the iron-core transformer under load, conditions are simplified by dividing the total flux linking a winding into two components, a leakage component and a mutual component. The mutual component for any winding is the portion of the total flux linking the winding which also links a second winding that is associated with it. The leakage flux is the portion of the total flux which links only the winding considered. In general the leakage fluxes of an iron-core transformer are small compared with the mutual fluxes. For a reactor, nothing is gained by dividing the flux into the foregoing components. Reactors with iron cores such as have just been described are used for many purposes.

The line ba on the vector diagram shown in Fig. 18, page 18, is drawn parallel to the current I_1 and is the active component of the voltage impressed on the circuit. It is therefore the apparent resistance drop through the winding. The line ob is the reactive component of the voltage drop in the circuit and represents the apparent reactance drop in the circuit.

Air-core reactors are an extremely important adjunct to the large modern central station to limit the current at times of accidental short circuit of any part of the system. For this purpose they are placed in series with the part of the system to be protected.   They may be placed in series with the synchronous generators, in series with the feeders, or in the bus bars, or in any two or all three of these places.   Iron cores are not used in such current-limiting reactors for two reasons.   First, at times of short circuit the amount of iron which would be necessary to prevent saturation would far exceed that required for normal conditions, and consequently the coils would be prohibitively expensive, bulky, and

Fig. 19.

heavy.   Moreover, the loss in the core would be appreciable at times of normal operation.   Second, the losses in the iron core, *i.e.*, hysteresis and eddy-current losses, would so retard the change of flux through the core during the initial rush of current at times of short circuit as to reduce its usefulness very greatly.   The initial rush of current from a synchronous generator at times of short circuit may be ten times the sustained short-circuit current.   The current reaches its maximum in a fraction of a cycle.

A Westinghouse three-phase current-limiting reactor is shown in Fig. 19.

Chapter 3

Determination of the shape of the flux curve which corresponds to a given voltage curve. Determination of the voltage curve from the flux curve. Determination of the magnetizing current and the current supplying the hysteresis loss from the hysteresis curve and the curve of induced voltage. Current inrush.

Determination of the Shape of the Flux Curve Which Corresponds to a Given Voltage Curve. If the wave shape of the voltage induced by the flux linking the winding of a reactor or the primary winding of a transformer is known, the shape of the flux curve corresponding to this can be found.

In a reactor it has been shown that the impressed voltage is nearly equal to the voltage induced by the flux. The impressed and induced voltages of a transformer of ordinary design are also nearly equal, especially at no load. At no load, a transformer is exactly the same as a reactor which has a good magnetic circuit. A good magnetic circuit is necessary since it is desirable to make the magnetizing current of a transformer as small as possible, as its presence increases the primary copper loss and lowers the power factor. In determining the shape of the flux curve of a transformer, it is sufficiently accurate to assume that the induced and impressed voltages of the primary winding are equal and in phase when considered in the same sense, *i.e.*, both as voltage rises or voltage drops.

The voltage rise induced in a transformer is equal to minus the rate of change of flux through the winding multiplied by the number of turns contained in the winding. If e is the instantaneous value of the induced voltage rise and N_1 and φ are the number of turns and the flux linking the winding,

$$ e = -N_1 \frac{d\varphi}{dt} $$

$$ d\varphi = -\frac{1}{N_1} e\, dt $$

$$ \varphi = -\frac{1}{N_1} \int_{t \text{ for } \varphi=0}^{t \text{ for } \varphi=\varphi'} e\, dt $$

21

The integral represents the area under the voltage curve between an ordinate drawn through the value of t at which the flux is zero and an ordinate through the value of t at which the flux φ is desired. When φ is a maximum, $d\varphi/dt$ is zero. Therefore the maximum value of the flux, either positive or negative, occurs when the voltage is zero. Moreover, except when even harmonics are present, the maximum positive and negative values of flux have equal magnitudes. Since, in this case, between points of zero flux and points of maximum flux on each side, equal quantities of flux must be added to the winding and subtracted from it, ordinates drawn through the zero points of the flux curve must

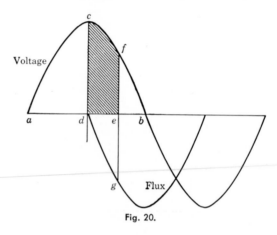

Fig. 20.

divide the areas enclosed by the positive and negative loops of the voltage curve into two equal parts.

Let $acfb$, Fig. 20, be the curve of the voltage induced in a transformer and let the ordinate dc divide the area under the positive loop of this curve into two equal parts. The point d is a zero point of the flux curve.

The ordinate eg of the flux curve at the point e is negative and is equal to $1/N_1$ times the area enclosed by the voltage curve between ordinates drawn through d and e. This area is shown crosshatched. The other points on the flux curve can be found in a similar way.

The maximum points of the flux curve lie on the ordinates drawn through the points of zero voltage. The flux and voltage curves therefore are 90 deg apart, and the voltage curve lags the curve of flux.

The wave forms of the voltage and the flux are not the same except when the voltage is sinusoidal. This can be shown by expressing the voltage in a Fourier series. Let the induced voltage be given by

$$e = E_1 \sin \omega t + E_3 \sin 3\omega t + E_5 \sin 5\omega t + \ldots$$

The flux corresponding to this is

$$\varphi = -\frac{1}{N_1}\int e\,dt = \frac{1}{N_1\omega}[E_1\cos\omega t + \tfrac{1}{3}E_3\cos 3\omega t$$
$$+ \tfrac{1}{5}E_5\cos 5\omega t + \dots]$$
$$= \frac{1}{N_1\omega}\left[E_1\sin\left(\omega t + \frac{\pi}{2}\right) + \tfrac{1}{3}E_3\sin\left(3\omega t + \frac{\pi}{2}\right)\right.$$
$$\left.+ \tfrac{1}{5}E_5\sin\left(5\omega t + \frac{\pi}{2}\right) + \dots\right]$$

If the voltage is sinusoidal, all the terms beyond the first in the expressions for voltage and flux drop out, leaving both waves sinusoidal. For any other wave form, any terms beyond the first may be present. Under this condition, the voltage and flux waves contain the same harmonics, but the relative magnitudes are different for the different harmonics and some harmonics may be reversed in phase. The two waves are therefore of different form. With respect to the fundamental, the third harmonic in the flux wave is only one-third as great, the fifth only one-fifth as great, the seventh only one-seventh as great, etc., as the corresponding harmonics in the voltage wave.

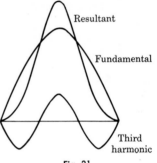

Fig. 21.

A peaked voltage wave contains a third harmonic which has a positive peak approximately coincident in time with the positive maximum of the fundamental as in Fig. 21.

If the amplitude of the third harmonic in the flux wave is only one-third as great with respect to the fundamental as it is in the voltage wave, the flux wave corresponding to the peaked voltage wave in Fig. 21 is flatter than that wave. This is particularly true because, for the peaked voltage wave shown, the value of E_3 is negative in the voltage and flux equations. Consequently, when the fundamental component of the flux is a maximum as at $\omega t = 0$, the third-harmonic component substracts from it. In a flat voltage wave the third harmonic is in opposite phase to that shown in Fig. 21. In this case the diminution of the third harmonic in the flux wave makes the flux wave less flat than the voltage wave. Also, E_3 is positive and the fundamental and third-harmonic components of flux aid one another when the fundamental is a maximum. In general, a flat voltage wave gives rise to a flux wave which is less flat, and a peaked voltage wave gives rise to a flux wave which is less peaked.

Figures 22 and 23 show flat and peaked voltage waves, of approximately the same rms value, and the corresponding flux waves.

Scale of Flux. The flux is given in webers by $1/N_1$ times the area enclosed by the voltage curve between the ordinate which divides each loop of the curve into two equal areas and an ordinate drawn through the

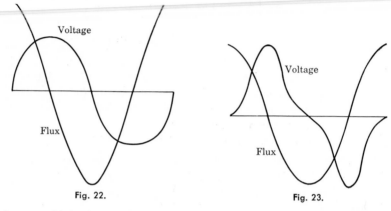

Fig. 22.   Fig. 23.

point at which the flux is desired. To get the numerical value of the flux, this area expressed in square inches must be multiplied by $1/N_1$, by the scale of the voltage in volts to the inch, and by the scale of time in seconds to the inch.

Determination of the Voltage Curve from the Flux Curve. The voltage induced in a coil is

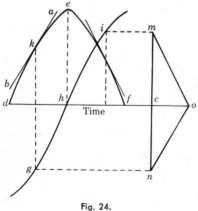

Fig. 24.

$$e = -N_1 \frac{d\varphi}{dt}$$

It is proportional to the rate of change of flux. Therefore if the flux curve is plotted with flux as ordinates and time as abscissas, the slope of a line drawn tangent to the curve at any point is proportional to the voltage at that point.

The voltage curve corresponding to any flux curve can be obtained graphically by the construction shown in Fig. 24.

The flux curve is *def* and *ghi* is a portion of the corresponding voltage curve. A point such as *g* on the voltage curve is obtained in the following manner. Draw a tangent *ab* at the point *k* of the flux curve. Select any point, as *o*, at the right of the diagram and draw a vertical line *mn* at a distance _oc_ to the left of _o_. From the point _o_ draw a line parallel to the tangent _ab_, and from the point of intersection _n_ of this line and _mn_ draw a horizontal line _gn_. The point of intersection _g_ of _gn_ and a vertical dropped from _k_ is the point on the voltage curve which corresponds to the point _k_ on the flux curve.

If _oc_ is, to scale, 1 sec, _cn_ measured to the scale of flux in webers and multiplied by N_1 is the voltage in volts. If, for example, _oc_ is made $\frac{1}{100}$ sec, _cn_ must be multiplied by $100N_1$.

Determination of the Magnetizing Current and the Current Supplying the Hysteresis Loss from the Hysteresis Curve and the Curve of Induced Voltage. To determine the magnetizing and hysteresis currents, it is necessary first to find the flux curve from the curve of impressed voltage

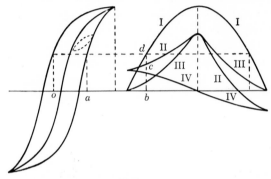

Fig. 25.

by the method just described. It is then necessary to obtain, by measurement, the hysteresis loop for the iron core and to plot this loop with total flux as ordinates and current as abscissas. The maximum value of the flux density for the hysteresis loop and for the flux curve must be the same. These two curves are plotted in Fig. 25.

Curves _I, II, III_, and _IV_ are the curves of flux, the combined magnetizing and hysteresis currents, the magnetizing current, and the hysteresis current.

The curve _II_ of combined magnetizing and hysteresis currents is obtained from the hysteresis loop and the flux curve in the following manner. For any point, as _d_, on the rising part of the flux curve, a current _oa_ is required. The current _oa_ is laid off on _bd_ giving a point _c_, where _bc = oa_, on the curve of the exciting current. The other points on this curve are obtained in a similar manner.

If $d\varphi/dt$ changes sign between the maximum points of positive and negative flux, this construction does not hold, since the hysteresis loop then contains a small loop shown dotted on Fig. 25. The area of this small loop represents an additional energy loss. If however the complete loop, *i.e.*, the full line plus the dotted portion of the hysteresis curve, is used, the correct result is obtained.

The wave form of the so-called magnetizing or reactive component of the exciting current can be obtained in a similar manner by making use of the magnetization curve for the iron core. If there were no hysteresis loss, the hysteresis curve would contract into a single line which would be the magnetization curve for the core. Both this line and the hysteresis curve for the core are slightly different from the corresponding curves for the iron of which the core is built on account of the extra ampere-turns required to overcome the reluctance introduced into the magnetic circuit by any joint or air gap that is present.

Subtracting the ordinates of curve *III* from those of curve *II* gives curve *IV*, which is the curve of the component current required to supply the hysteresis loss. This last curve leads and is in quadrature with the flux curve *I*. The flux curve is 90 deg ahead of the curve (not shown) of induced voltage rise and therefore 90 deg behind the curve of $-E_1$, the component of the impressed voltage which is required to balance the voltage drop induced by the flux. The component current which supplies the hysteresis loss is therefore in time phase with the voltage drop $-E_1$. This is the phase relation used on the vector diagram of the reactance coil.

The component current required for the eddy-current loss is of the same wave shape as $-E_1$ and in time phase with it, provided the local fluxes set up by the eddy currents can be neglected as on page 65.

The mmf due to the component current which supplies the eddy-current loss is just balanced by the equal and opposite mmf caused by the eddy currents themselves. These two mmfs act like the primary and secondary mmfs of a transformer.

The component of the current taken by a reactor or the component of the no-load current of a transformer which is in quadrature with the induced voltage has been called the magnetizing current, although it is not the actual current causing the flux. The current causing the flux is that shown by curve *II* on Fig. 25 plus the current required to supply the eddy-current loss. The current which has been called the magnetizing current is the current which would be required to produce the flux if there were neither hysteresis nor eddy-current losses. The component currents I_h and I_e are required to supply the losses due to hysteresis and to eddy currents in the core, and it is because of those two components that the current required to produce the flux in a transformer or in a reactor leads the flux by a small angle known as the angle of core-loss advance (see Fig. 18, page 18).

It is evident from Fig. 25 that the wave form of the current producing the flux in a reactor or in a transformer is different from the wave form of the flux. If the flux follows a sine wave, the current causing it is not sinusoidal but contains harmonics, among which the third is large unless the magnetization curve of the core is nearly a straight line, as it is if the core contains a long air gap. The magnetic circuit of the ordinary transformer has no air gap and the magnetizing current for such a magnetic circuit contains a large third harmonic. If the component currents taken by a reactor or by a transformer at no load contain harmonics, they cannot be combined as vectors as was done in Fig. 18, and the vector diagrams must be considered as an approximation.

Current Inrush. When a transformer is connected to a circuit, the current and also the flux do not immediately assume their final wave forms and magnitudes. The initial inrush of current and the number of cycles before the current wave assumes the final form depend upon the point of the voltage wave at which the circuit is closed and upon the residual magnetism in the iron core and its direction with respect to the instantaneous value of the initial mmf. Under certain conditions, the current at the instant the circuit is closed may be several times the full-load current of the transformer. The current inrush really depends upon the apparent instantaneous impedance of the circuit at the instant of closing the circuit.

When a transformer is disconnected from the mains, the exciting current becomes zero but the flux is not zero unless the circuit happens to be opened at the point on the impressed voltage wave which corresponds to zero flux under steady operating conditions. The flux remaining in the core after the circuit is opened may be either plus or minus φ_R where φ_R is the residual flux remaining in the core after the circuit is opened.

The resistance and leakage-reactance drops[1] in the windings of a transformer are small and even at full load are seldom more than a few per cent of the rated voltages of the windings. Even when the current in a winding far exceeds its full-load value, the impedance drop due to the current is still small. As an approximation, neglect these drops and assume that the instantaneous voltage impressed on a winding is equal to the voltage induced in the winding by the core flux. With this assumption, the instantaneous voltage impressed on the primary winding at any instant t is

$$-v = N \frac{d\varphi}{dt} \tag{3}$$

where φ is the flux linking the primary winding at the instant considered and N is the number of primary turns.

[1] See page 36 for the significance of leakage reactance.

$$d\varphi = \frac{-v}{N} dt$$

$$\varphi = \frac{1}{N} \int -v \, dt$$

Assume that the secondary winding is open and that the impressed voltage is sinusoidal and given by

$$-v = V_m \sin (\omega t + \alpha) \tag{4}$$

where V_m is the maximum value of the impressed voltage and α is the angle between the time when t is zero and the time when the instantaneous impressed voltage $-v$ passes through zero increasing in the positive direction.   Let t be zero at the instant when the voltage is impressed on the winding.   Assume no residual flux in the core when the circuit is closed.

The flux at any instant of time t' after the circuit is closed is

$$\varphi' = \frac{1}{N} \int_{=0}^{t=t'} V_m \sin (\omega t + \alpha) dt \tag{5}$$

The maximum value of the flux must occur when $d\varphi/dt = 0$.   From Eqs. (3) and (4)

$$\frac{d\varphi}{dt} = \frac{V_m}{N} \sin (\omega t + \alpha) = 0 \tag{6}$$

$d\varphi/dt$ is zero when $\sin (\omega t + \alpha)$ is zero or when $(\omega t + \alpha) = n\pi$, where n is an integer.   Therefore for maximum flux $t' = (n\pi - \alpha)/\omega$ and the maximum flux is

$$\varphi_m = \frac{V_m}{N} \int_{t=0}^{t=\frac{n\pi - \alpha}{\omega}} \sin (\omega t + \alpha) dt$$

$$= -\frac{V_m}{\omega N} (\cos n\pi - \cos \alpha) \tag{7}$$

Equation (7) assumes that the flux is zero when the time t is zero, or, in other words, that there is no residual flux.   The phase angle α may have any value depending upon the point on the voltage wave at which the circuit is closed.

Under steady-state conditions, the flux and the voltage induced by the flux are in quadrature and if the voltage is sinusoidal the flux is sinusoidal.   Assume that for steady-state conditions

$$\varphi = \varphi'_m \sin \omega t$$

where φ'_m is the maximum value of the flux corresponding to a sinusoidal impressed voltage of maximum value V_m.

$$-v = N \frac{d\varphi}{dt}$$

$$= N \frac{d}{dt} (\varphi'_m \sin \omega t)$$

$$= N\omega\varphi'_m \cos \omega t \tag{8}$$

Under steady-state conditions, the voltage $-v$ is a maximum when $\cos \omega t$ is a maximum, *i.e.*, when $\cos \omega t = 1$.

$$V_m = N\omega\varphi'_m$$

$$\varphi'_m = \frac{V_m}{\omega N} \tag{9}$$

Substituting φ'_m for $V_m/\omega N$ in Eq. (7),

$$\varphi_m = -\varphi'_m(\cos n\pi - \cos \alpha) \tag{10}$$

Equation (10) gives the maximum transient flux φ_m in terms of the maximum steady-state flux for the same effective impressed voltage when a transformer is switched on with no load across its secondary. The point on the voltage wave at which the transformer is connected in circuit is determined by the value of the phase angle α. Equation (10) assumes that the residual flux in the core is zero at the time of switching. If the residual flux is $\pm\varphi_R$, the maximum transient-flux equation becomes

$$\varphi_m = \pm\varphi_R - \varphi'_m(\cos n\pi - \cos \alpha) \tag{11}$$

The greatest maximum values of the flux occur when $\cos n\pi$ has the same sign as $-\cos \alpha$. If α is greater than 0 but less than $\pi/2$, $-\cos \alpha$ is negative and the successive maximum values of the transient flux occur when n equals 1, 3, 5, etc. If α is greater than $\pi/2$ but less than $3\pi/2$, $-\cos \alpha$ is positive and the successive maximum values of the transient flux occur when n equals 2, 4, 6, etc. The greatest possible maximum value of the transient flux occurs when $-\cos \alpha = \pm1$. Cos α has this value when the voltage is impressed at the time its instantaneous value is zero.

This greatest possible value of the transient flux is

$$\varphi_R + 2\varphi_m$$

It is twice the maximum value of the flux under steady-state conditions plus the residual core flux at the time the transformer is switched in.

The minimum value of the maximum flux occurs when $\cos \alpha$ equals zero, *i.e.*, when the voltage is impressed at the time its instantaneous value is a maximum. Under this condition and if there is no residual flux, the maximum flux is the same as under steady-state conditions for the same effective impressed voltage. In this case, both the flux and the exciting current immediately assume their steady-state values.

Curves of the transient flux, the steady-state flux, and the impressed voltage, when there is no residual flux and the voltage is impressed at the instant its value is zero, are shown in Fig. 26.

Actually, the maximum flux is somewhat less than the value given by Eq. (11), on account of the resistance and leakage-reactance drops which are neglected in Eq. (11).

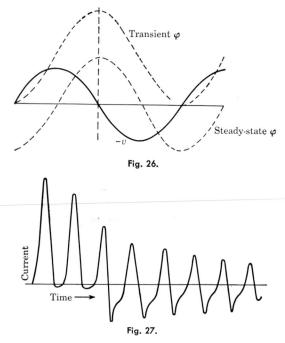

Fig. 26.

Fig. 27.

The curve of the transient magnetizing current corresponding to the transient flux curve can be obtained from the magnetization curve for the core by the use of the equation

$$h = \frac{N i_\varphi}{l}$$

where h is the magnetizing force in ampere-turns per unit length corresponding to any transient flux density and l is the mean length of the magnetic core. The magnetizing current is i_φ. N is the number of turns in the winding.

On account of the high saturation which may be produced by the transient flux when a transformer is connected to a line, the actual current inrush may be very large and may be equal to several times the normal full-load current of the transformer, in spite of the limiting effect of the resistance and leakage-reactance drops which have been neglected in the equations.

A curve of the exciting current when a transformer is switched on the line is shown in Fig. 27.

From Eq. (11) for the maximum transient flux it appears that the successive maximums do not decrease in magnitude with time. This is because the resistance loss and the core loss have been neglected. If the condition of no losses could actually be realized, the axis of the flux curve when a transformer is switched on a line would be permanently displaced from the axis of time. Actually, the successive maximums decrease with time and after a few cycles become equal to the maximum flux under steady-state conditions.

Chapter 4

Conventions for positive direction of voltage, current, and flux. Fluxes concerned in the operation of air-core transformers. Fluxes concerned in the operation of an iron-core transformer. Fundamental equations of the iron-core transformer. Ratio of transformation. Tap changing under load. Reaction of secondary current. Reduction factors. Relative values of resistances. Relative values of leakage reactances. Calculation of leakage reactance. Load vector diagram. Analysis of vector diagram. Solution of vector diagram and calculation of regulation. An example of the calculation of the regulation of a transformer from its complete vector diagram. The vector diagram of a transformer using the reduction factors and also using the primary terminal voltage and the voltage induced in the primary by the mutual flux as voltage rises.

Conventions for Positive Direction of Voltage, Current, and Flux. In order to interpret properly the signs in the voltage equations of a transformer and to determine the phase angles of the voltage, current, and flux vectors of a vector diagram, it is necessary to assign in advance the assumed positive directions of these quantities. Any reasonable set of conventions may be adopted, but the signs in the equations and the phase angles of the vectors will differ according to the assumptions made. In this text the following assumptions are used:

1. Any direction of flux may be chosen as the positive direction, but it must be made definite and all components of flux must have the same assumed positive direction through the windings.

2. The positive direction of current is that which, when acting alone, produces a positive flux.

3. The positive direction of a voltage rise is that produced by a decreasing positive flux ($d\varphi/dt$ negative). It is as a result of this assumption that the negative sign appears in $e = -N \, (d\varphi/dt)$.

4. Either v or e, when positive, is used to represent a voltage rise, *i.e.*, an increase in potential in the assumed positive direction.

To illustrate the use of the above conventions, refer to Fig. 28. Consider the flux to be positive when it is acting upward through the right-hand coil and downward through the left-hand coil, *i.e.*, in a counterclockwise direction around the magnetic circuit. Then the positive directions of the currents i_1 and i_2 must be taken as indicated by the arrows in order to satisfy rule (2). Moreover, when the positive flux is decreasing, it produces an emf in each coil which is a voltage rise acting upward through each coil. The upper terminals of the coils therefore have the same relative polarities as indicated by the dots and also by the plus and minus signs and e_1 and e_2 are acting upward as indicated by the arrows.

Fluxes Concerned in the Operation of Air-core Transformers.
In Fig. 29, the flux produced by the primary current i_1 acting alone may be conveniently divided into two components φ_{11} and φ_{12} where φ_{12} is the component of the total flux produced by i_1 which is common to or links both windings and φ_{11} is the component of the flux produced by i_1

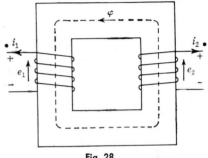

Fig. 28.

which links only the primary winding. This latter component is often called the primary leakage flux. In a like manner the total flux produced by i_2 acting alone, as in Fig. 30, may be divided into two components φ_{22} and φ_{21} which have the same significance in regard to i_2 that φ_{11} and φ_{12} have in regard to i_1. The total flux φ_1 linking the primary winding due to the combined action of primary and secondary currents, i_1 and i_2, is

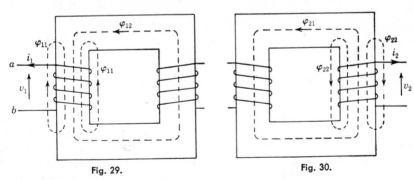

Fig. 29.   Fig. 30.

therefore $\varphi_1 = \varphi_{11} + \varphi_{12} + \varphi_{21}$ and the total flux φ_2 linking the secondary winding due to the combined action of i_1 and i_2 is $\varphi_2 = \varphi_{22} + \varphi_{21} + \varphi_{12}$.

The voltage of self-induction in the primary winding, $-L_1(di_1/dt)$, is the voltage induced by the total flux produced by the current i_1 acting alone, $i.e.$, by the flux $\varphi_{11} + \varphi_{12}$. The voltage of mutual induction $-M(di_2/dt)$ induced in the primary winding by the current i_2 is the voltage produced by the flux φ_{21}. The total voltage rise from b to a,

Fig. 29, is $v_1 = -i_1r_1 - L_1(di_1/dt) - M(di_2/dt)$, or the voltage rise from a to b is

$$-v_1 = i_1r_1 + L_1\frac{di_1}{dt} + M\frac{di_2}{dt} \tag{12}$$

Likewise for the secondary,

$$-v_2 = i_2r_2 + L_2\frac{di_2}{dt} + M\frac{di_1}{dt} \tag{13}$$

The positive directions of v_1 and v_2 are taken in accordance with rule 3 and are indicated by the arrows in the figures. $-v_1$ is positive when v_1 is negative and so $-v_1$ represents a voltage rise in the direction from a to b. $-v_1$ may also be considered as the voltage drop from b to a and is the sum of the drops due to resistance, self-induction, and mutual induction, which is in accordance with Eq. (12).

If the inductances are constant, as in an air-core transformer, and the currents and voltages are sinusoidal, Eqs. (12) and (13) may be written in terms of effective currents and voltages as follows:

$$-V_1 = I_1r_1 + j\omega L_1I_1 + j\omega MI_2 \tag{14}$$

and

$$-V_2 = I_2r_2 + j\omega L_2I_2 + j\omega MI_1 \tag{15}$$

The above equations are most useful in problems involving the air-core transformer. When an iron core is used, however, the inductances are not constant and vary with the degree of saturation of the iron. In some communication circuits however, when there is direct as well as alternating current in one of the transformer windings, the alternating component of the flux may be small in comparison with the flux due to the direct current. Under these conditions, if the inductances are determined at the saturation due to the direct current, the above equations give useful approximate results.

Fluxes Concerned in the Operation of an Iron-core Transformer. In most problems involving the iron-core transformer, where the inductances are not constant, and particularly in the case of transformers such as are used for power purposes, a different method of analysis gives better and more dependable results. This method involves the concept of mutual and leakage fluxes. Because of the impossibility of having the primary and secondary windings occupy the same position on the core of a transformer, there is a certain amount of magnetic leakage between them. All the flux which links the primary winding does not link the secondary winding, and all the flux which links the secondary winding does not link the primary winding. The reluctances for the leakage fluxes are almost entirely due to air or other nonmagnetic materials. Consequently these reluctances are nearly constant and the leakage fluxes are nearly proportional to the currents producing them. The voltages produced in the windings by the leakage fluxes may therefore be replaced by reactance drops due to constant leakage reactances.

For the present, only the two-winding transformer is considered. The primary leakage flux is that part of the total primary flux which does not link the secondary winding, and similarly the secondary leakage flux is that part of the total secondary flux which does not link the primary winding. That part of the total primary and secondary flux which is common to or links both windings is called the mutual flux. The mutual flux is produced by the combined action of the primary and secondary currents, but since for the iron-core transformer its path is almost entirely in iron, the mutual flux is not proportional to the total mmf producing it.

Referring to Figs. 29 and 30, the primary leakage flux is φ_{11}, the secondary leakage flux is φ_{22}, and the mutual flux is $\varphi_{12} + \varphi_{21}$. The total flux linking the primary winding is therefore $\varphi_1 = \varphi_{11} + \varphi_{12} + \varphi_{21} = \varphi_{11} + \varphi_M$ where φ_M is the mutual flux.

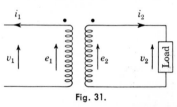

Fig. 31.

The voltage produced in the primary winding by the flux φ_1 may be found by combining the voltages due to φ_{11} and to φ_M. The voltage due to the flux φ_{11} is $-N_1(d\varphi_{11}/dt) = -S_1(di_1/dt)$ where S_1 is the primary leakage inductance. The voltage due to φ_M is $-N_1(d\varphi_M/dt) = e_1$ where e_1 is the voltage induced in the primary winding by the mutual flux. Likewise, for the secondary, $\varphi_2 = \varphi_{22} + \varphi_M$ where φ_M is the same as the mutual flux linking the primary. The voltages produced in the secondary winding by these component fluxes are $-S_2(di_2/dt)$ and e_2, respectively. Because the same flux is involved $e_2 = e_1(N_2/N_1)$.

Consider the schematic diagram of a transformer as shown in Fig. 31. The voltage $v_1 = -i_1r_1 - S_1(di_1/dt) + e_1$ and

$$-v_1 = +i_1r_1 + S_1 \frac{di_1}{dt} - e_1 \tag{16}$$

Likewise

$$v_2 = -i_2r_2 - S_2 \frac{di_2}{dt} + e_2$$

and

$$e_2 = v_2 + i_2r_2 + S_2 \frac{di_2}{dt} \tag{17}$$

If sinusoidal quantities are assumed, these equations may be written as

$$-V_1 = I_1r_1 + j\omega S_1 I_1 - E_1$$
$$= I_1r_1 + jI_1x_1 - E_1 \tag{18}$$

and

$$E_2 = V_2 + I_2r_2 + j\omega S_2I_2$$
$$= V_2 + I_2r_2 + jI_2x_2 \tag{19}$$

where x_1 and x_2 are the leakage reactances of the primary and secondary windings, respectively.

Equations (18) and (19) may also be derived directly from Eqs. (14) and (15), respectively. Add and substract $j\omega MI_1(N_1/N_2)$ from the right-hand member of Eq. (14). This does not change the value of the equation. N_1 and N_2 are the number of turns in the primary and secondary windings. Then

$$-V_1 = I_1r_1 + j\omega I_1 \left(L_1 - M \frac{N_1}{N_2}\right) + j\omega \frac{M}{N_2}(I_1N_1 + I_2N_2) \tag{20}$$

By the definitions of self-inductance and mutual inductance, $j\omega I_1L_1$ and $j\omega I_1M$ are the voltages induced in the primary and secondary windings by the flux linkages with these windings due to the current I_1 acting alone. Since N_1/N_2 is the ratio of primary and secondary turns, $j\omega I_1M(N_1/N_2)$ must be the voltage induced in the primary winding by the flux linkages with the primary winding due to that part of the flux, produced by the primary current I_1 acting alone, which links the secondary winding, i.e., produced by the primary mutual flux. The term $j\omega I_1[L_1 - M(N_1/N_2)]$ must therefore be the voltage induced in the primary winding by the primary leakage flux. $[L_1 - M(N_1/N_2)]$ is the primary leakage inductance S_1 and $\omega[L_1 - M(N_1/N_2)]$ is the primary leakage reactance. The last term of Eq. (20), $j\omega(M/N_2)(I_1N_1 + I_2N_2)$, is the voltage induced in the primary winding by the resultant mutual flux linkages with the primary winding when both primary and secondary currents act. It is the voltage induced in the primary winding by the resultant mutual flux caused by both currents and is usually called the primary induced voltage. Indicating this voltage by E_1, Eq. (20) becomes

$$-V_1 = I_1r_1 + jI_1x_1 + E_1$$

where x_1 is the primary leakage reactance.

$$x_1 = 2\pi f \left(L_1 - M \frac{N_1}{N_2}\right) = 2\pi f S_1$$

Since E_1 and E_2 are produced by the same flux, the resultant mutual flux, their ratio is equal to the ratio of primary to secondary turns. The ratio of turns is the ratio of transformation. The vector sum of I_1N_1 and I_2N_2 is the total number of ampere-turns acting to produce mutual flux. For the air-core transformer, all the inductances are constant. For the iron-core transformer, the inductances are not constant, but the effect of those parts of the leakage-flux paths due to air is so great compared with the effect of the iron parts of the paths that the reluctances of these paths are nearly constant and may be assumed constant with negligible error. Therefore the leakage reactances of the iron-core transformer may be assumed constant.

In an iron-core transformer, the mutual flux is determined at each instant by the instantaneous resultant mmf $(i_1N_1 + i_2N_2)$ produced by the combined action of the primary and secondary currents. The mutual flux lies almost wholly in the iron core. The reluctance of its path is far from constant under normal operating conditions, and the mutual flux is not proportional at each instant to the resultant mmf acting to produce it. This causes little difficulty in considering the iron-core transformer. It calls for a small nonsinusoidal component in the primary current, the effect of which is so small under operating conditions that usually it can be neglected.

In an air-core transformer, the leakage fluxes may be large compared with the mutual flux. In an iron-core transformer as ordinarily designed, the leakage fluxes are small compared with the mutual flux and lie largely in air. They are usually not over a few per cent of the mutual flux. The mutual flux of the iron-core transformer lies almost wholly in the iron core and under full-load

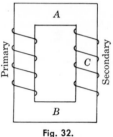

Fig. 32.

conditions is 97 to 98 per cent of the total flux linking either winding. The percentage at no load is still greater.

That the leakage flux of a transformer should be nearly proportional to the current can be seen from what follows. Consider a simple case where the primary and secondary coils of a core-type transformer are on opposite sides of the iron core. Refer to Fig. 32.

At no load, the magnetic potential between A and B produced by the primary coil is that necessary to force the flux through the iron circuit ACB. When the secondary circuit is loaded, the magnetic potential between A and B produced by the primary winding must not only overcome the reluctance of the iron path ACB, but in addition it must balance the opposing mmf caused by the current in the secondary winding. Since the increase in mmf between A and B is proportional to the secondary current, the leakage between A and B should be nearly proportional to the load current carried by the transformer.

In what follows, the voltages which are induced in the primary and secondary windings by the mutual flux are called the induced voltages and the voltages induced by the leakage fluxes are replaced by primary and secondary leakage-reactance drops.

Since the leakage flux of the usual transformer is small and is largely in air, its effect on the apparent resistance is small.

The following notation is used:

N_1 = turns in primary winding
N_2 = turns in secondary winding
a = ratio of transformation
V_1 = primary impressed or terminal voltage
V_2 = secondary terminal voltage
E_1 = voltage induced in the primary by the mutual flux
E_2 = voltage induced in the secondary by the mutual flux
I_1 = total primary current
I_2 = secondary current
I_φ = magnetizing component of the primary current
I_{h+e} = component of the primary current to supply the core loss
$I_n = I_\varphi + I_{h+e}$ (vectorially) and is the exciting current
I_1' = load component of the primary current, *i.e.*, the component caused by the load on the secondary
r_1 = primary effective resistance
r_2 = secondary effective resistance
x_1 = primary effective leakage reactance
x_2 = secondary effective leakage reactance

V and E without signs are voltage rises. Minus signs before them are used to indicate voltage drops.

Figure 18, page 18, is the vector diagram of a reactor or of a transformer with the secondary circuit open. In a transformer, the magnetic circuit is made as good as possible in order to make the magnetizing component of the current small. The resistance and the leakage reactance are also made small. In the diagram shown in Fig. 18, φ is the total flux linking the winding and E_1 is the voltage rise induced by this flux. In the transformer, the total flux linking the primary winding is divided into a mutual flux and a leakage flux and the effect of the leakage flux is replaced by a reactance drop. This leakage flux is in phase with the total primary current, and the reactance voltage drop, which replaces the effect of the leakage flux, leads the current by 90 deg.

Figure 33 gives the vector diagram of a transformer with the secondary circuit open and with the total primary flux replaced by the mutual flux and a leakage reactance.

Referring to Fig. 33, φ is the mutual flux and I_φ is the magnetizing component of the primary current producing this flux. I_{h+e} is the component of the current to supply the core loss due to the mutual flux. I_n, the exciting current, is equal to the primary current I_1 at no load. E_1 is the voltage rise induced in the primary winding by φ. $I_1 x_1$ and

I_1r_1 are the leakage-reactance and resistance drops. The vector sum of $-E_1$ and I_1x_1 corresponds to what is marked $-E_1$ in Fig. 18, page 18, and is the voltage drop which must be impressed across the primary winding to balance the voltage drop induced in that winding by the total primary flux, *i.e.*, by the mutual flux plus the primary leakage flux.

The voltage drop $-V_1$ impressed across the primary winding of a transformer is

$$-V_1 = -E_1 + I_1(r_1 + jx_1) \tag{21}$$

In a properly designed transformer, the resistance and leakage reactance are small. The drop in voltage in the primary due to these is small, when compared with the impressed voltage, especially at no load. Therefore $-E_1$ and $-V_1$ are nearly equal. They seldom differ by more than

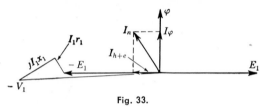

Fig. 33.

1 or 2 per cent at full load, and at no load they do not differ by more than a small fraction of 1 per cent. Therefore, as an approximation,

$$V_1 = E_1 = 4.44fN_1\varphi_m \tag{22}$$

It is evident from Eq. (22) that the mutual flux in a transformer is determined by the frequency, the number of turns in the winding, and the impressed voltage. The magnetizing current I_φ is determined by the mutual flux and the reluctance of the magnetic circuit. In any given transformer the voltage impressed fixes the flux, and the magnetizing current must adjust itself to produce this flux. A modification of the dimensions of the iron core of a transformer is not accompanied by a change in the flux but by a change in the magnetizing current required to produce the flux.

Equation (22) shows that for a fixed frequency, the voltage per turn, *i.e.*, E_1/N_1, is proportional to the flux. The voltage per turn multiplied by the number of turns must always be equal to the impressed voltage to within 1 or 2 per cent. The two would be exactly equal if the primary had neither resistance nor leakage reactance. If the turns are doubled, the voltage induced per turn and the flux are halved. If the turns are halved, the voltage induced per turn and the flux are doubled, provided the increase in the saturation of the core is not sufficient to increase the magnetic leakage beyond the point where the relation $-V_1 = -E_1$ is still approximately true.

Fundamental Equations of the Iron-core Transformer. The following six fundamental equations of the transformer are important.

$$-V_1 = -E_1 + I_1 z_1 \tag{23}$$

$$E_1 = 4.44 N_1 \varphi_m f \tag{24}$$

$$N_1 I_{m\varphi} = \sum H_m l = \sum \frac{\beta_m}{\mu} l \tag{25}$$

$$\text{Maximum flux density} = \beta_m = \frac{\varphi_m}{\text{cross section of core}} = \frac{\varphi_m}{A} \tag{26}$$

$$\text{Core loss} = P_{h+e} = \text{volume} \ (k_e t^2 f^2 \beta_m{}^2 + k_h f \beta_m{}^{1.6}) \tag{27}$$

$$\text{Core-loss current} = I_{h+e} = \frac{P_{h+e}}{E_1}$$

$$= \frac{P_{h+e}}{V_1} \ (\text{approximately}) \tag{28}$$

All the above equations are arithmetical equations with one exception, the first, which is a vector equation.

Equation (23). Since $I_1 z_1$ is very small, E_1 is nearly equal to V_1. For ordinary purposes, E_1 may be assumed equal to V_1, especially at no load.

Equation (24). With E_1 fixed, the product $N_1 \varphi_m f$ is also fixed. For example, if N_1 and f are fixed and E_1 is decreased, φ_m must decrease by the same percentage as the decrease in E_1 in order to balance the change in E_1.

Equation (25). $I_{m\varphi}$ is the maximum value of the magnetizing component of the current, H_m the maximum value of the magnetizing force, l the length of the magnetic path, and μ its permeability. With the maximum flux density fixed, the maximum value of the magnetizing component of the primary current is determined by the number of primary turns and the permeability of the magnetic circuit. It is the flux that fixes the current and not the current that fixes the flux.

Equation (26). This equation needs no comment.

Equation (27). The core loss is fixed by this equation if the volume of the core, the frequency, the maximum flux density, the character of the magnetic material, and the thickness of its laminations are known. For more complete explanation of this formula, see Chap. 6.

Equation (28). This equation fixes the core-loss component of the exciting current.

The following example illustrates the use of these equations. Assume that the number of turns in the primary winding of a transformer is doubled and that the cross section of the core is halved without changing the mean length. If V_1 and f remain the same and the primary impedance drop is negligible, the primary flux is halved, Eq. 24. Under these conditions the flux density is not affected since the flux and the cross section of the core are each halved. Since the maximum flux density in the core remains the same, the permeability of the core is unchanged. The magnetizing component of the primary current is therefore halved. Since β_m and f are unaltered, the core loss is halved because of the decrease in the volume of the core, Eq. (27). The core-loss component of the primary current is reduced to one-half its original value because of the change in the core loss, Eq. (28).

Ratio of Transformation. Both the primary and the secondary windings of a transformer are on the same iron core and are subjected to the same variation in the mutual flux. Therefore the voltages E_1 and E_2 induced in the two windings by the mutual flux must be in time phase. The voltage induced per turn in each winding must be the same.

$$\frac{E_1}{N_1} = \frac{E_2}{N_2}$$
$$\frac{E_1}{E_2} = \frac{N_1}{N_2} = a \tag{29}$$

The ratio of the two induced voltages is called the ratio of transformation. It is fixed by the ratio of turns in the primary and secondary windings. This is the true ratio of transformation and is constant. Commercially, the ratio of the terminal voltages is often called the ratio of transformation. This ratio is not constant but varies with the load and its power factor. Under ordinary conditions this variation is small and lies within 1 to 5 per cent.

Although the ratio of transformation of a transformer is always given as a number greater than unity, in all equations in this text a is the ratio of the number of primary turns to the number of secondary turns and is greater or less than unity according as the transformer is used as a step-down transformer or as a step-up transformer.

Tap Changing under Load. With the increase in size and complexity of power systems and their interconnection, it is often necessary to change the voltage ratio of a transformer under load. This can be accomplished by changing the number of turns between the terminals of one of the windings by the use of taps brought out from the windings. The tap-changing device must be arranged in such a way that the change from one tap to the next can be made without either opening the main circuit or short circuiting any portion of the transformer windings. Tap changing under load can be done in several ways. One of the simplest is shown in Fig. 34.

The transformer winding is a-f and 1, 2, 3, 4, 5 are switches connected to the taps b, c, d, e, f. SC is a short-circuiting switch. R is a reactor which has its middle point connected to the line terminal X_2 of the transformer. X_1 is the other terminal

To obtain full voltage, switch 5 and short-circuiting switch SC are closed. This short-circuits the reactor and connects the line terminal X_2 through the two halves of the reactor in parallel to the tap f of the transformer winding. The voltage between the terminals X_1 and X_2 is now that of the entire transformer winding. The currents in the two halves of the reactor under this condition are equal and produce equal and opposite magnetic effects on the reactor core. The resultant reactance between the tap f and the terminal X_2 is therefore sensibly zero.

To reduce the voltage, short-circuit switch SC is opened and switch 4 is closed. This puts the reactor across the transformer taps e and f, and since the transformer terminal X_2 is connected to the middle point of the reactor, its potential is equal to that midway between the transformer taps e and f.

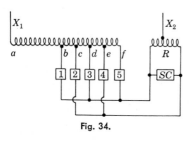

Fig. 34.

To connect the transformer terminal X_2 to tap e, switch f is opened and short-circuit switch SC is closed. During the transition period, half of the reactor carries the full current but the reactor is designed to carry this current during this period. The reactance drop due to half the reactor is small compared with the transformer voltage.

Reaction of Secondary Current. If the secondary circuit of a transformer is closed, a certain current I_2 flows through the N_2 secondary turns producing an mmf equal to $I_2 N_2$ acting to modify the flux in the core. From Eq. (21), page 39, the primary current is equal to

$$I_1 = \frac{-V_1 + E_1}{r_1 + jx_1} \tag{30}$$

Since E_1 is proportional to the mutual flux, the current I_2 in the secondary will cause the current I_1 in the primary to differ from its no-load value. Since, in accordance with rule (2), page 32, the positive directions of primary and secondary currents are such that each independently produces positive flux, the total mmf producing the mutual flux under load conditions is $N_1 I_1 + N_2 I_2$. It is convenient to resolve the primary current into a load component I_1' whose mmf exactly balances the mmf of the secondary current, i.e., such that $N_1 I_1' + N_2 I_2 = 0$, an exciting or magnetizing component I_φ whose mmf by itself is sufficient to produce the mutual flux required to satisfy Eq. (30), and a core-loss component I_{h+e} which as at no load is required because of hysteresis and eddy-current effects. If r_1 and x_1 are small, E_1 and therefore φ_M, I_φ, and I_{h+e} are changed only slightly from their no-load values. The maximum variation with load of the mutual flux of power and distribution transformers does not as a rule exceed 3 per cent. The additional primary current resulting from current I_2 in the secondary is therefore nearly the same as the component I'_1. Since

$$I'_1 N_1 + I_2 N_2 = 0 \tag{31}$$

then

$$-\frac{I'_1}{I_2} = \frac{N_2}{N_1} = \frac{1}{a} \tag{32}$$

The load component of the primary current is opposite in phase to the secondary current and is equal to that current multiplied by the inverse ratio of transformation.

Reduction Factors. When making transformer calculations and when drawing vector diagrams, it is convenient and often necessary to reduce all currents, voltages, resistances, and reactances to the corresponding currents, voltages, resistances, and reactances of an equivalent transformer having a ratio of transformation equal to unity. An equivalent transformer is one which has the same kva rating, the same losses, and the same regulation as the transformer it replaces. Usually one of the windings of this equivalent transformer is the same as the corresponding winding on the actual transformer. To make the substitution, it is necessary only to replace all currents, voltages, resistances, and reactances of either of the windings of the actual transformer by their equivalent values in terms of the other winding. This is accomplished by multiplying each by its proper reduction factor.

Induced voltages are proportional to the number of turns in the windings of a transformer. Load currents are inversely proportional to the number of turns. Therefore if an equivalent winding is to be substituted for one of the actual windings of a transformer, the ratio of the voltage induced in the equivalent winding to the voltage induced in the actual winding is the same as the ratio of turns in the two windings. Since the equivalent winding must give the same regulation as the actual winding, all component voltages induced in it must be in the same ratios to one another as they are in the actual winding. Not only are the induced voltages in the two windings in the ratio of turns, but all component voltages are also in this same ratio. For a similar reason, all corresponding currents or component currents must be in the inverse ratio of turns. If a transformer is to be replaced by an equivalent transformer having a ratio of transformation of unity, the ratio of the turns on the equivalent winding to the turns on the winding it replaces is equal to the inverse ratio of transformation of the actual transformer. It follows that, to reduce resistances and reactances to their equivalent values, the inverse square of the ratio of transformation must be used.

To refer primary voltages, currents, resistances, and reactances to their equivalents in terms of the secondary, multiply

$$\text{Primary voltages by } \frac{1}{a}$$

$$\text{Primary currents by } a$$

$$\text{Primary resistances by } \left(\frac{1}{a}\right)^2$$

$$\text{Primary reactances by } \left(\frac{1}{a}\right)^2$$

The reciprocals of these reduction factors are used to reduce from secondary to primary.

Relative Values of Resistances.   The temperature reached under load conditions is an important factor in determining the limiting output of any piece of electrical apparatus.   For best economy of material, all parts should reach their ultimate safe temperatures at the same time under the limiting condition of load.   In a transformer, the amounts of copper used in the primary and secondary windings should be so proportioned that each winding reaches its limiting temperature at the same time under the maximum load to be carried.   If one winding is still cool when the other has reached its ultimate safe temperature, an unnecessarily large amount of copper has been used in the cool winding.

If the conditions for the radiation and conduction of heat from the two windings of a transformer are the same, the copper loss in the primary and secondary windings of a properly designed transformer should be equal at full load.

For this condition,

$$I_1^2 r_1 = I_2^2 r_2$$
$$\frac{I_2^2}{I_1^2} = \frac{r_1}{r_2}$$

Since I_1 and I_1' are nearly equal at full load, the following approximate relation should hold:

$$\frac{I_2^2}{I_1'^2} = \frac{r_1}{r_2} = a^2 \tag{33}$$

The actual ratio of r_1 to r_2 is a matter of design.   It may differ somewhat from the ratio given by Eq. (33) chiefly because of the differences in the exposures of the two windings for cooling.

Relative Values of Leakage Reactances.   Reactance is proportional to the flux linkages per ampere with a winding and hence is proportional to the number of turns in a winding multiplied by the flux per ampere which links these turns.

Figure 35 is a section through a core-type transformer having a primary and a secondary winding on each leg of the core. The spaces between the windings and between the secondary winding and the core are occupied by insulation. The primary and secondary windings carry currents which are nearly opposite in phase and which produce mmfs which are nearly equal and opposite in phase. The mmfs would be equal if it were not for the small components of primary current which supply the flux and the core loss.

Consider the two windings in Fig. 35 that are on the left-hand leg of the core. Assume that at the instant considered the secondary current when viewed from the top of the winding is directed in a right-hand or clockwise direction about the core. The flux produced by this current is

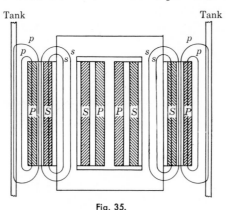

Fig. 35.

downward through the winding and upward outside the winding. All of the flux that lies in the annular space between the two windings and is due to the secondary current is secondary leakage flux since it does not link any part of the primary winding. The secondary current also produces flux which is directed upward within the secondary winding itself. This flux links some of the secondary turns but does not link any of the primary turns and is also secondary leakage flux. The return paths for these secondary leakage fluxes lie mainly in the iron core.

According to the assumed direction of the secondary current, the primary current as viewed from the top of the winding is directed in a left-hand or counterclockwise direction about the core, since the primary and secondary currents are nearly opposite in phase. They are opposite in phase if the exciting current is neglected. The current in the primary winding produces flux which is directed upward inside the winding. All of the flux that lies in the space between the two windings and is due to the primary current is leakage flux for the primary winding, since it does not link any part of the secondary winding. The primary current also produces flux which passes upward through the primary winding itself. This flux produces some linkages with the primary winding but does not link any of the secondary turns and is also primary leakage flux. The return paths for the primary leakage fluxes are chiefly in the space between the winding and the case and in the sides of the case. The reluctance of the return paths for all of the leakage fluxes is small compared with the reluctance of the portions of the leakage paths which lie between the ends of the windings. The reluctance of these return portions is usually neglected when computing leakage reactance, or allowance is made for them by the use of an empirical factor.

Typical leakage flux lines are shown in Fig. 35. These lines are marked with the small letters s and p to indicate secondary and primary. There are also leakage-flux lines (not shown in the figure) which do not pass completely through the windings but pass outward through them, as in Fig. 37, page 49.

If the two windings had the same length and the reluctances of the return paths for the leakage fluxes were neglected, the leakage inductances of the two windings would be proportional to the square of the turns in the two windings if the two windings had the same mean diameter and the same thickness. These last conditions are not possible because of the greater amount of insulation required for the high-voltage winding, always the outside winding, and the necessary difference in the diameters of the windings. In spite of these differences and the differences in the return paths of the leakage fluxes, the ratio of the leakage inductances and hence the ratio of the leakage reactances do not differ greatly in many cases from the ratio of the squares of the number of turns in the two windings.

$$\frac{x_1}{x_2} = \left(\frac{N_1}{N_2}\right)^2 \text{ approximately} \tag{34}$$

In many cases when it is necessary to know the ratio of the leakage reactances of a transformer, it is sufficiently close to assume that the relation given in Eq. (34) holds.

Calculation of Leakage Reactance. The following method of calculating the leakage reactance of a transformer makes certain simplifying assumptions in regard to the leakage fluxes, which are justified by the fact that the leakage reactances of transformers calculated in accordance with these assumptions check closely with the measured values. The leakage reactances of a core-type transformer with concentric cylindrical coils, shown in Figs. 35 and 36, will now be found.

The approximate leakage-flux paths are shown in Fig. 35. For the purpose of calculation, assume that all of the leakage flux is parallel to the axis of the coils and that the reluctance of its path is due to that portion of the path which lies between the ends of the coils. When the leakage flux passes beyond the ends of the coils, it spreads rapidly and its density is reduced to a low value. Most of the leakage flux which lies without the coils takes the shortest path to the iron core and returns with negligible reluctance drop. Therefore the reluctance of the portion of the path which lies outside the ends of the coils can be neglected. Because of the simplifying assumptions, it is not necessary to take into account the change in the length of a turn when passing outward through the windings. The mean length of the turns in both windings is used for the length of each turn in finding the cross sections of the leakage-flux paths. This mean length is $2\pi r$ where r (Fig. 36) is the mean radius of both windings. Let N_3 and N_2 be the number of turns in the outer and inner windings.

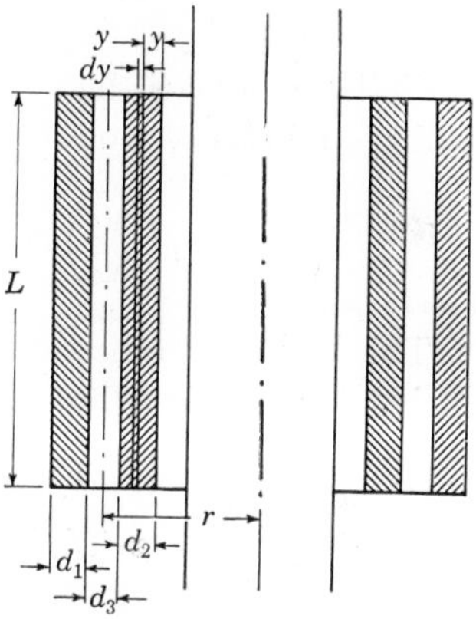

Refer to Fig. 36. When a transformer is in operation, the primary and secondary ampere-turns are equal and opposite in phase except for the effect of the exciting current which is small and is neglected. The magnetizing force H, acting in the space d_3, per ampere in the inner winding is N_2/L. The flux density is therefore $\mu_0 N_2/L$ where μ_0 is the permeability of free space, and the resulting leakage flux in the space d_3 between coils is $\mu_0 N_2 2\pi r d_3/L = K d_3$ where $K = \mu_0 N_2 2\pi r/L$, a constant.

Fig. 36.

Let the fraction of this flux which links the inner winding alone be p. The leakage-flux linkages with the inner winding due to the part p of the leakage flux are

$$p\left(\frac{\mu_0 N_2{}^2 2\pi r d_3}{L}\right)$$

The magnetizing force per ampere acting to produce flux in the element dy of the inner winding is due to y/d_2 times as many turns as act to produce flux in the space d_3. The flux per ampere in the element dy is therefore

$$\frac{\mu_0 N_2 \dfrac{y}{d_2} 2\pi r \, dy}{L} = \frac{\mu_0 N_2 2\pi r y \dfrac{dy}{d_2}}{L} = K y \frac{dy}{d_2}$$

This links $N_2(y/d_2)$ turns. Hence the flux linkages per ampere for the inner winding due to the flux within the winding itself are

$$KN_2 \int_0^{d_2} \frac{y^2}{d_2{}^2} \, dy = KN_2 \frac{d_2}{3}$$

The total leakage-flux linkages with the inner winding per ampere are

$$\frac{2\pi r\mu_0 N_2^2}{L}\left(pd_3 + \frac{d_2}{3}\right)$$

and the leakage reactance of the inner winding is

$$x_2 = 2\pi f\,\frac{2\pi r\mu_0 N_2^2}{L}\left(pd_3 + \frac{d_2}{3}\right)$$

The leakage reactance of the outer winding is

$$x_1 = 2\pi f\,\frac{2\pi r\mu_0 N_1^2}{L}\left[(1-p)d_3 + \frac{d_1}{3}\right]$$

If the outer coil had the same number of turns as the inner coil, its leakage reactance would be

$$x_1\left(\frac{N_2}{N_1}\right)^2 = 2\pi f\,\frac{2\pi r\mu_0 N_2^2}{L}\left[(1-p)d_3 + \frac{d_1}{3}\right]$$

This is the reactance of the outer winding referred to the inner winding. It is the reactance the outer winding would have if it were replaced by a winding of unchanged dimensions but the same number of turns as the inner coil.

The expression $x_1(N_2/N_1)^2 + x_2 = x_e$ is the equivalent leakage reactance of the transformer referred to the inner winding (see page 58). It is the reactance drop in the transformer per ampere in the inner winding as seen from the terminals of that winding.

$$x_e = 2\pi f\,\frac{m\mu_0 N_2^2}{L}\left(d_3 + \frac{d_1}{3} + \frac{d_2}{3}\right) \tag{35}$$

where $m = 2\pi r$ is the mean length of a turn.

Any consistent rationalized system of units may be used. For instance in the mks system, dimensions are in meters and $\mu_0 = 1.256 \times 10^{-6}$. If dimensions are in inches, $\mu_0 = 3.192$ and

$$x_e = 20.1f\,\frac{N_2^2 m}{L}\left(d_3 + \frac{d_1}{3} + \frac{d_2}{3}\right) \times 10^{-8} \tag{36}$$

If there are two primary windings and two secondary windings, the expression, Eq. (35) or (36), must be multiplied by 2 if the two secondary windings are in series. If the two secondary windings are in parallel, the expression, Eq. (35) or (36), must be divided by 2.

The fractional parts of the leakage flux, p and $(1-p)$, which link the two windings are not important in many transformer problems, since these problems involve only the combined leakage reactance of both windings, *i.e.*, the equivalent leakage reactance. The equivalent leakage reactance depends only on the total flux linkages with the windings.

In a transformer with more than two windings, the leakage reactance of a winding is not definite. It depends on the winding with which it is associated. For example, consider a three-winding transformer. The leakage reactance of winding 1 with respect to the leakage reactance of winding 2 is not the same as the leakage reactance of winding 1 with respect to the leakage reactance of winding 3.

The calculation of the leakage reactance of a transformer by the method just outlined takes no account of the leakage between turns of the windings. All of the leakage flux was assumed to be parallel to the axis of the coils. In reality a portion of the leakage flux passes between turns, as shown in Fig. 37.

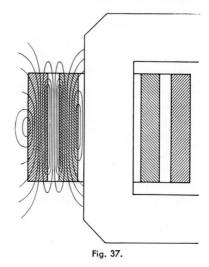

Fig. 37.

Load Vector Diagram. The complete vector diagram of a transformer is shown in Fig. 38.[1] The vectors representing the resistance and reactance drops and the core-loss components of the primary current are exaggerated in order to make the diagram clearer.

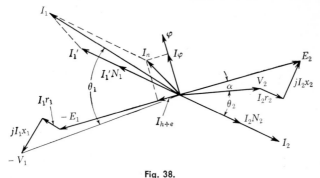

Fig. 38.

When drawing vector diagrams it is customary to reduce all vectors to their equivalents in terms of either the primary or the secondary winding. In Fig. 38 all vectors may be considered with respect to the secondary.

[1] For equations and vector diagrams of the air-core transformer, see R. R. Lawrence, "Principles of Alternating Currents," 2d ed., p. 201.

All primary quantities are then referred to the secondary side of the transformer by multiplying currents by $N_1/N_2 = a$, voltages by $1/a$, and resistance and reactance by $(1/a)^2$.

The mmf N_2I_2 of the secondary winding is balanced by the opposite and equal mmf in the primary winding due to the load component of the primary current. Since these two mmfs neutralize each other, they usually are omitted from the diagram. The resultant mmf which produces the mutual flux φ is N_1I_φ. The voltage E_2 is the voltage induced in the secondary by the mutual flux φ. This voltage is in phase with the voltage induced in the primary by the same flux. E_1 is this voltage referred to the secondary by multiplying the actual voltage drop in the primary by $1/a$. This statement in regard to the relative phase of the primary and secondary voltages is in accordance with rule 3, page 32, since both voltages are produced by the same flux.

The magnetizing current I_φ cannot properly be used on the vector diagram of an iron-core transformer since I_φ is far from sinusoidal. However, it is so small compared with the total primary current under load conditions that using either its fundamental component or its equivalent sine-wave value introduces no appreciable error. The omission of I_φ altogether when considering an iron-core transformer under load conditions generally introduces negligible error.

Analysis of Vector Diagram. The primary power P_1 is

$$P_1 = V_1I_1 \cos \theta_1 = V_1I_1 \cos \theta_{-I_1}^{-V_1}$$

Resolving $-V_1$ into its components gives

$$P_1 = E_1I_1 \cos \theta_{-I_1}^{-E_1} + (I_1r_1)I_1 \cos 0 + (I_1x_1)I_1 \cos \frac{\pi}{2} \qquad (37)$$

If I_1 in the first term of Eq. (37) is replaced by its components, the equation can be further developed to give

$$P_1 = E_1I_1' \cos \theta_{-I_{1'}}^{-E_1} + E_1I_{h+e} \cos 0 + E_1I_\varphi \cos \frac{\pi}{2}$$

$$+ (I_1r_1)I_1 \cos 0 + (I_1x_1)I_1 \cos \frac{\pi}{2}$$

$$= \begin{Bmatrix} \text{power trans-} \\ \text{ferred to sec-} \\ \text{ondary by} \\ \text{magnetic in-} \\ \text{duction} \end{Bmatrix} + \begin{Bmatrix} \text{core} \\ \text{loss} \end{Bmatrix} + \{\text{zero}\} + \begin{Bmatrix} \text{primary} \\ \text{copper loss} \end{Bmatrix} + \{\text{zero}\}$$

$$E_1I_1' \cos \theta_{-I_{1'}}^{-E_1} = E_2I_2 \cos \theta_{I_2}^{E_2} = P_2'$$

where P_2' is the total secondary power.

Expressing E_2 in terms of its components

$$P'_2 = (I_2 r_2) I_2 \cos 0 + (I_2 x_2) I_2 \cos \frac{\pi}{2} + V_2 I_2 \cos \theta - \frac{V_2}{I_2}$$

$$= \left\{ \begin{matrix} \text{secondary} \\ \text{copper loss} \end{matrix} \right\} + \left\{ \text{zero} \right\} + \left\{ \begin{matrix} \text{secondary} \\ \text{output} \end{matrix} \right\}$$

Solution of Vector Diagram and Calculation of Regulation.   Take V_2 as axis of reference.

$$V_2 = V_2 + j0$$
$$I_2 = I_2(\cos \theta_2 - j \sin \theta_2)$$
$$E_2 = V_2 + I_2 z_2$$
$$= V_2 + I_2(\cos \theta_2 - j \sin \theta_2)(r_2 + jx_2)$$
$$= V_2 + I_2(r_2 \cos \theta_2 + x_2 \sin \theta_2) + jI_2(x_2 \cos \theta_2 - r_2 \sin \theta_2)$$
$$= A + jB \tag{38}$$

where A and B are the real and j terms of Eq. (38).

$$-E_1 = -E_2 a$$
$$= -(A + jB)a$$

The load component I'_1 of the primary current is

$$I'_1 = -\frac{1}{a} I_2$$
$$= -\frac{1}{a} I_2 (\cos \theta_2 - j \sin \theta_2)$$

Let $I_n = -I_{h+e} + jI_\varphi$ referred to E_2 as an axis of reference be the exciting component of the primary current measured at voltage

$$E_1 = aE_2 = a(A + jB)$$

and let P_{h+e} be the core loss also measured at voltage E_1.   For practical purposes it is sufficiently accurate to measure the core loss at the voltage V_1 or V_2.

Then

$$I_{h+e} = \frac{P_{h+e}}{E_1}$$

and

$$I_\varphi = \sqrt{I_n{}^2 - I_{h+e}{}^2}$$

I_n referred to E_2 as an axis is

$$I_n = -I_{h+e} + jI_\varphi$$

I_n referred to V_2 as an axis is

$$I_n = -I_{h+e}(\cos \alpha + j \sin \alpha) + jI_\varphi(\cos \alpha + j \sin \alpha)$$

where

$$\sin \alpha = \frac{B}{\sqrt{A^2 + B^2}}$$

and

$$\cos \alpha = \frac{A}{\sqrt{A^2 + B^2}}$$

$$I_1 = I_1' + I_n$$

$$= -\frac{1}{a} I_2 + I_n$$

$$= -\frac{1}{a} I_2(\cos \theta_2 - j \sin \theta_2)$$

$$\quad -I_{h+e}(\cos \alpha + j \sin \alpha) + jI_\varphi(\cos \alpha + j \sin \alpha)$$

$$= -\left(\frac{I_2}{a} \cos \theta_2 + I_{h+e} \cos \alpha + I_\varphi \sin \alpha\right)$$

$$\quad +j\left(\frac{I_2}{a} \sin \theta_2 - I_{h+e} \sin \alpha + I_\varphi \cos \alpha\right)$$

$$= C + jD \tag{39}$$

C and D are the real and j parts of Eq. (39).

$$-V_1 = -E_1 + I_1(r_1 + jx_1)$$

$$= -a(A + jB) + (C + jD)(r_1 + jx_1)$$

$$= F + jG$$

At no load $-E_1$ and $-V_1$ are nearly equal since the no-load current of a transformer is small. The no-load current is 3 to 6 per cent of the full-load current, and the impedance drop in the primary due to this current is not greater than 0.1 or 0.2 per cent of the rated primary voltage. Since $E_2 = V_2$ at no load, the no-load secondary voltage is

$$V_2' = \frac{E_1'}{a}$$

where E_1' is $V_1 + I_n z_1$ and the addition is made vectorially.

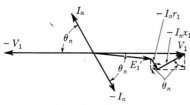

Fig. 39.

Refer to Fig. 39, which represents the no-load condition with the vector for the leakage impedance drop exaggerated. The voltage V_1 is taken as axis of reference. θ_n is the no-load power-factor angle.

$$E_1' = V_1(1 + j0) + I_n(-\cos \theta_n + j \sin \theta_n)(r_1 + jx_1)$$

$$= V_1(1 + j0) - I_n(r_1 \cos \theta_n + x_1 \sin \theta_n) - jI_n(x_1 \cos \theta_n - r_1 \sin \theta_n)$$

The term $jI_n(x_1 \cos \theta_n - r_1 \sin \theta_n)$ is small and is in quadrature with the terms $V_1(1 + j0)$ and $I_n(r_1 \cos \theta_n + x_1 \sin \theta_n)$. It has negligible effect on the magnitude of E'_1. Therefore

$$V'_2 = \frac{V_1 - I_n(r_1 \cos \theta_n + x_1 \sin \theta_n)}{a} \tag{40}$$

where the expression is in an arithmetical sense.

The regulation is

$$\frac{V'_2 - V_2}{V_2} 100 \tag{41}$$

In the actual solution of the transformer diagram, the angle between E_2 and V_2 under load conditions may be neglected so far as the phase of I_n is concerned, and the magnitudes of I_n and P_{h+e} at no load may be found for a voltage equal to the terminal voltage instead of for the voltage E_1 without producing any appreciable error.

An Example of the Calculation of the Regulation of a Transformer from Its Complete Vector Diagram. The following are the constants of a 300 kva 11,000:2,300-volt 60-cycle power transformer.

$$r_1 = 1.28 \text{ ohms} \qquad r_2 = 0.0467 \text{ ohm}$$
$$x_1 = 4.24 \text{ ohms} \qquad x_2 = 0.162 \text{ ohm}$$

The core loss and the no-load current when 2,300 volts at 60 cycles are impressed on the low-voltage winding are

$$P_n = 2,140 \text{ watts} \qquad I_n = 3.57 \text{ amp}$$

Assume that an inductive load of 300 kva at 0.8 power factor is connected to the low-voltage terminals. Calculate the voltage which it is necessary to impress on the high-voltage terminals in order to maintain 2,300 volts across the load. Refer to the vector diagram of Fig. 38, page 49. Also refer to the equations given on pages 51 and 52.

$$I_2 = \frac{300,000}{2,300} = 130.4 \text{ amp}$$

Take the secondary voltage as reference axis.

$$V_2 = 2,300(1 + j0) \text{ volts}$$
$$I_2 = 130.4(0.8 - j0.6) = 104.3 - j78.2 \text{ amp}$$
$$E_2 = 2,300(1 + j0) + (104.3 - j78.2)(0.0467 + j0.162)$$
$$= 2,317.5 + j13.45 \text{ volts}$$
$$a = \frac{11,000}{2,300} = 4.783$$
$$I'_1 = -\frac{1}{a}I_2 = -\left(\frac{104.3 - j78.2}{4.783}\right) = -21.81 + j16.35 \text{ amp}$$

As an approximation assume that I_n is fixed by the terminal voltage instead of by E_1. Referred to the high-voltage side of the transformer the exciting current is

$$I_n = \frac{3.57}{4.783} = 0.746 \text{ amp}$$

$$I_{h+e} = \frac{2,140}{2,300} \times \frac{1}{a} = \frac{2,140}{11,000} = 0.195 \text{ amp}$$

As an approximation assume that I_φ is sinusoidal.

$$I_\varphi = \sqrt{(0.746)^2 - (0.195)^2} = 0.720 \text{ amp}$$
$$I_1 = I_1' + I_n = I_1' + I_{h+e}(-\cos\alpha - j\sin\alpha) + jI_\varphi(\cos\alpha + j\sin\alpha)$$
$$\cos\alpha = \frac{2,318}{\sqrt{(2,318)^2 + (13.4)^2}} = 1.000$$
$$\sin\alpha = \frac{13.4}{\sqrt{(2,318)^2 + (13.4)^2}} = 0.0058$$
$$I_1 = (-21.81 + j16.35) + 0.195(-1.000 - j0.0058)$$
$$+ j0.720(1.000 + j0.0058)$$
$$= (-21.81 + j16.35) + (-0.195 - j0.001) + (j0.720 - 0.004)$$
$$= -22.01 + j17.07 \text{ amp}$$
$$-V_1 = -4.783(2,318 + j13.45) + (-22.01 + j17.07)(1.28 + j4.24)$$
$$= (-11,087 - j64.3) + (-100.6 - j71.5)$$
$$= -11,188 - j135.8 \text{ volts}$$
$$V_1 = \sqrt{(11,188)^2 + (135.8)^2} = 11,189 \text{ volts}$$

The open-circuit secondary voltage corresponding to a given primary impressed voltage is given by Eq. (40), page 53. The no-load primary copper loss is negligible and is omitted when finding the cosine and sine of the no-load power-factor angle.

$$\cos\theta_n = \frac{2,140}{2,300 \times 3.57} = 0.261$$
$$\sin\theta_n = 0.966$$
$$I_n \text{ (on the high-voltage side)} = \frac{3.57}{4.783} = 0.746 \text{ amp}$$
$$V_2' = \frac{11,189 - 0.746(1.28 \times 0.261 + 4.24 \times 0.966)}{4.783}$$
$$= \frac{11,186}{4.783} = 2,338.7 \text{ volts}$$
$$\text{Regulation} = \frac{V_2' - V_2}{V_2} \times 100 = 1.68 \text{ per cent}$$

It is seldom necessary to include the exciting current when finding the regulation of a transformer. If it is omitted, both in finding the primary voltage which is necessary to maintain a given secondary voltage under certain load conditions and in finding the open-circuit secondary voltage corresponding to the calculated primary voltage, the error nearly cancels. If the exciting current is omitted the solution is simplified.

The Vector Diagram of a Transformer Using the Reduction Factors and Also Using the Primary Terminal Voltage and the Voltage Induced in the Primary by the Mutual Flux as Voltage Rises.   It is often convenient to draw the vector diagram of a transformer with both the terminal voltages and the voltages induced in the windings by the mutual flux as voltage rises.   Such a diagram is given in Fig. 40.   The primary voltages V_1 and E_1, primary current I_1 and its components, the primary resistance r_1 and reactance x_1 are all actual values.   They are referred

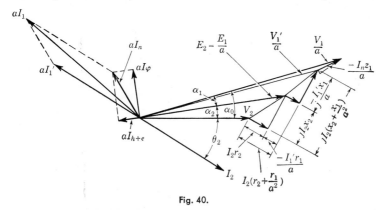

Fig. 40.

to the secondary by the use of the proper reduction factors.   In Fig. 38, the reduction factors are omitted and V_1, E_1, I_1, r_1, x_1 are assumed to have been multiplied by the proper factors to reduce them to the secondary side.   The magnitudes of the resistance and reactance drops and also of the exciting current and its components have been much exaggerated on the vector diagram in Fig. 40 in order to make the diagram clearer.

It is evident by referring to the vector diagram that the primary terminal voltage referred to the secondary side of the transformer may be found by adding the equivalent impedance drop $I_2\left[\left(r_2 + \dfrac{r_1}{a^2}\right) + j\left(x_2 + \dfrac{x_1}{a^2}\right)\right]$ vectorially to V_2 and then adding to V_1'/a vectorially the impedance drop $-I_n z_1/a$ which is due to the exciting component of the primary current.   This simplifies the solution of the vector diagram. The impedance drop $-I_n z_1/a$ is small and is nearly in phase with V_1'/a so that little error is made in the magnitude of V_1/a if this drop is added arithmetically to the magnitude of V_1'/a. Using this simplification,

$$\frac{V_1}{a} = \sqrt{(V_2 \cos \theta_2 + I_2 r_e)^2 + (V_2 \sin \theta_2 + I_2 x_e)^2} + \frac{I_n \sqrt{r_1^2 + x_1^2}}{a} \quad (42)$$

where $r_e = r_2 + (r_1/a^2)$ and $x_e = x_2 + (x_1/a^2)$. The resistance r_e and the reactance x_e as given are the equivalent resistance and the equivalent reactance of the transformer referred to its secondary side. The ratio of transformation a as used in all of the equations is the ratio from primary to secondary and is greater or less than unity according to whether the primary is the high-voltage or the low-voltage side. The equivalent resistance and the equivalent reactance are discussed more fully in the next chapter.

Chapter 5

The equivalent circuit of a transformer. Vector representation of the approximate equivalent circuit. Calculation of the regulation from the approximate equivalent circuit. An example of the calculation of the regulation of a transformer from its approximate equivalent circuit. Per-unit values. Calculation of regulation by the per-unit method.

The Equivalent Circuit of a Transformer. If the resistances and the reactances of a transformer are assumed constant, the transformer can be exactly represented by the T-connected circuit of Fig. 41. In this cir-

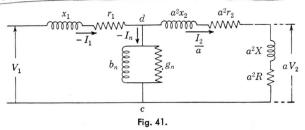

Fig. 41.

cuit, R and X are the resistance and the reactance, respectively, of the load connected to the secondary terminals of the transformer.

$$g_n = \frac{I_{h+e}}{E_1}$$

and carries the current $-I_{h+e}$, $b_n = (I_\varphi/E_1)$ and carries the current $-I_\varphi$. g_n and b_n are the conductance and susceptance, respectively, which make the branch current equal to the exciting current I_n. The negative signs associated with these primary currents are the result of the conventions adopted with regard to the positive directions of the currents in the transformer. It should be noted that, as in the vector diagrams of Chap. 4, the primary power-factor angle θ_1 is the angle between $-V_1$ and I_1 or between V_1 and $-I_1$.

Equation (18), page 35, applies to the primary winding of a transformer and may be written in the form of Eq. (43).

$$-V_1 = I_1 r_1 + j I_1 x_1 - E_1$$

Therefore

$$V_1 = -I_1(r_1 + j x_1) + E_1 \tag{43}$$

57

Likewise Eq. (19), page 36, can be converted to the form of Eq. (44).

$$E_2 = V_2 + I_2 r_2 + j I_2 x_2$$

and

$$aE_2 = aV_2 + \frac{I_2}{a}(a^2 r_2) + \frac{I_2}{a}(ja^2 x_2)$$

$$= aV_2 + \frac{I_2}{a}(a^2 r_2 + ja^2 x_2) \tag{44}$$

Then since $E_1 = aE_2$

$$V_1 = -I_1(r_1 + jx_1) + \frac{I_2}{a}(a^2 r_2 + ja^2 x_2) + aV_2 \tag{45}$$

Also, $I_1 = I_1' + I_n$ as discussed on page 42 where I_n is the sum of the magnetizing component of the current I_φ and the core-loss component I_{h+e} and $I_1' = -(I_2/a)$. Therefore

$$-I_1 = -I_n + \frac{I_2}{a} \tag{46}$$

These equations, (45) and (46), are exactly represented by the equivalent circuit of Fig. 41. The voltage rise from c to d is equal to aE_2 and is equal to E_1. Therefore $-I_n = E_1(g_n - jb_n)$.

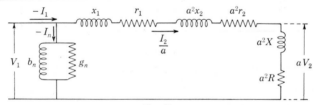

Fig. 42.

Both $I_1 r_1$ and $I_1 x_1$ are small compared with V_1, and since I_n is also small, as a rule only a few per cent of I_1, the error of using $-I_n$ corresponding to the voltage V_1 instead of to the voltage E_1 is negligible. For the same reason the error in V_1 produced by neglecting the component of current $-I_n$ in r_1 and x_1 and using I_2/a in place of $-I_1$ is also negligible. Therefore the real equivalent diagram may be replaced by the approximate form of Fig. 42.

By combining x_1 and $a^2 x_2$ and r_1 with $a^2 r_2$, the diagram can be modified still further to give that of Fig. 43.

$x_1 + a^2 x_2 = x_e$ and $r_1 + a^2 r_2 = r_e$ are called the equivalent reactance and equivalent resistance of the transformer. In Fig. 43, x_e and r_e are referred to the primary side of the transformer and must be used with the secondary current referred to the primary side.

If everything on the diagrams of Figs. 41, 42, and 43 is referred to the secondary side, the equivalent resistance and the equivalent reactance are referred to the secondary side. Referred to the secondary side, x_e and r_e are

$$x_e = \frac{x_1}{a^2} + x_2$$

$$r_e = \frac{r_1}{a^2} + r_2$$

The secondary current must be used with the equivalent reactance and equivalent resistance if they are referred to the secondary winding.

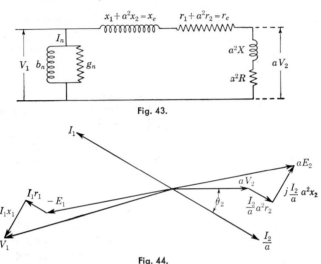

Fig. 43.

Fig. 44.

Vector Representation of the Approximate Equivalent Circuit. The vector diagram of a transformer with the exciting current omitted is shown in Fig. 44. Everything in this diagram is referred to the primary side.

If voltage rises are used for the primary impressed voltage and the voltage induced in the primary winding by the mutual flux, the vector diagram becomes that of Fig. 45. This diagram is the same as that of Fig. 40, page 55, with the voltage drop $-I_n z_1/a$ in the primary omitted and all vectors referred to the primary side of the transformer instead of to the secondary side as in Fig. 40.

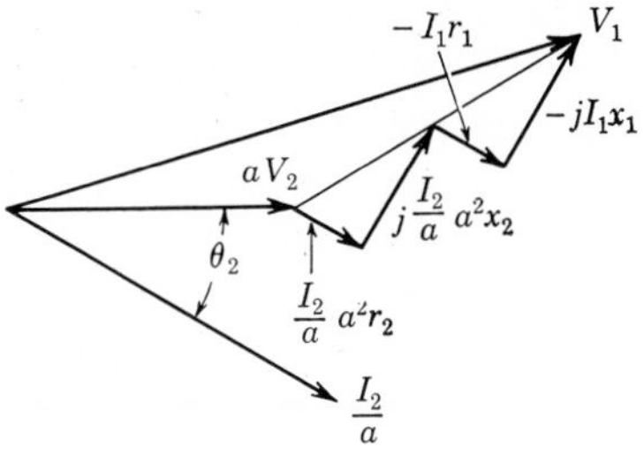

Fig. 45.

If the two resistance drops and the two reactance drops are replaced by the equivalent resistance drop and the equivalent reactance drop, the diagram shown in Fig. 45 is still further simplified and becomes that of Fig. 46.

The vector diagram of Fig. 46 represents the conditions existing in the approximate equivalent circuit of a transformer shown in Fig. 43. The exciting current has been omitted from Figs. 44, 45, and 46 since in the approximate diagrams it does not influence the secondary voltage.

Calculation of the Regulation from the Approximate Equivalent Circuit.

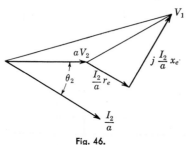

Fig. 46.

$$V_1 = aV_2 + \frac{I_2}{a}(\cos\theta_2$$
$$- j\sin\theta_2)(r_e + jx_e) \quad (47)$$

and the per cent regulation is

$$\frac{V_1 - aV_2}{aV_2}100$$

If it is more convenient, all vectors may be referred to the secondary side. In this case

$$\frac{V_1}{a} = V_2 + I_2(\cos\theta_2 - j\sin\theta_2)(r_e + jx_e) \qquad (48)$$

$\dfrac{V_1/a - V_2}{V_2}100$ is the per cent regulation[1]

The values of r_e and of x_e in Eq. (47) are not the same as in Eq. (48). In Eq. (48) both r_e and x_e are referred to the secondary side and are $1/a^2$ times their values in Eq. (47), where they are both referred to the primary side.

In so far as voltage regulation is concerned, a transformer may be replaced by a series impedance with a resistance equal to r_e and a reactance equal to x_e.

The approximate method of calculating the regulation of a transformer gives results which are nearly enough correct except when applied to transformers with large exciting currents or very large resistance and leakage-reactance drops in the primary winding. The method should not be used for such transformers.

An Example of the Calculation of the Regulation of a Transformer from Its Approximate Equivalent Circuit.

The 300-kva 11,000:2,300-volt 60-cycle transformer for which the regulation is found in Chap. 4, page 53, is used. Refer to the approximate equivalent circuit of a transformer in Fig. 43, page 59, and to Eq. (48). The resistances and the ratio of transformation of the transformer are (page 53)

[1] Since the exciting current is neglected in the approximate equivalent circuit, V_1 should not be corrected for the no-load drop.

$$r_1 = 1.28 \text{ ohms} \qquad r_2 = 0.0467 \text{ ohm}$$
$$x_1 = 4.24 \text{ ohms} \qquad x_2 = 0.162 \text{ ohm}$$
$$a = \frac{11,000}{2,300} = 4.783$$

The low-voltage side is used as the secondary. As in the example on page 53, the regulation is found for full kva inductive 0.8-power-factor load.

$$r_e \text{ (referred to the secondary)} = \frac{r_1}{a^2} + r_2 = \frac{1.28}{(4.783)^2} + 0.0467$$
$$= 0.05595 + 0.0467 = 0.103 \text{ ohm}$$

$$x_e \text{ (referred to the secondary)} = \frac{x_1}{a^2} + x_2 = \frac{4.24}{(4.783)^2} + 0.162$$
$$= 0.1853 + 0.162 = 0.347 \text{ ohm}$$

The equivalent resistance and the equivalent reactance would ordinarily be found from the short-circuit test described in Chap. 7, pages 77 and 79.

$$I_2 \text{ (full-load secondary current)} = \frac{300,000}{2,300} = 130.4 \text{ amp}$$

$$\frac{V_1}{a} = V_2(1 + j0) + I_2(\cos \theta_2 - j \sin \theta)(r_e + jx_e)$$
$$= 2,300(1 + j0) + 130.4(0.8 - j0.6)(0.103 + j0.347)$$
$$= (2,300 + j0) + (37.9 + j28.1)$$
$$= 2,337.9 + j28.1$$

$$\frac{V_1}{a} = \sqrt{(2,337.9)^2 + (28.1)^2} = 2,338.2 \text{ volts}$$

$$\text{Regulation} = \frac{2,338.2 - 2,300}{2,300} \times 100 = 1.66 \text{ per cent}$$

The regulation found on page 54, Chap. 4, was 1.68. In order to show any difference between the regulation as determined from the full transformer diagram and from the approximate equivalent circuit neglecting the exciting current, more significant figures are retained in the computations than would be warranted ordinarily by the precision of the data as usually obtained.

Per-unit Values. A method which is commonly used in making calculations involving electrical machinery and power circuits is the per-unit method. In using this method all quantities are expressed in decimal form as a ratio of the actual value to some base value. They are therefore the same as a percentage of the base value with the multiplier of 100 omitted. For instance, if 100 volts were taken as the base value of voltage, then 125 volts would be 125 per cent of the base and would have a per-unit value of $125/100 = 1.25$. Some of the more important per-unit values are as follows:

$$\text{Per-unit voltage} = \frac{\text{voltage in volts}}{\text{base volts}}$$

$$\text{Per-unit current} = \frac{\text{current in amperes}}{\text{base amperes}}$$

$$\text{Per-unit impedance} = \frac{\text{impedance in ohms}}{\text{base ohms}}$$

$$\text{Per-unit resistance} = \frac{\text{resistance in ohms}}{\text{base ohms}}$$

$$\text{Per-unit reactance} = \frac{\text{reactance in ohms}}{\text{base ohms}}$$

$$\text{Per-unit power} = \frac{\text{power in kilowatts}}{\text{base kilovolt-amperes}}$$

Rated voltage is generally taken as the base volts. The base value used for kilovolt-amperes is often the rated kilovolt-amperes. In dealing with a single machine such as a single transformer, the rated value is usually selected as a base, but in the analysis of power systems involving the interconnection of a number of machines a common base value of kilovolt-amperes is often selected to be applied to all parts of the system, which base therefore would not be the rated kilovolt-amperes for all of the machines involved.

Base amperes is computed from the base values of voltage and kilovolt-amperes and is the rated current if rated values are used as the base for voltage and kilovolt-amperes. The base value of ohms is the base voltage divided by the base current.

If per-unit values are used, the reduction factors of pages 43 and 44 may be omitted in all formulas and calculations. This is because the per-unit values for both primary and secondary of an equivalent 1:1 ratio transformer are equal to the per-unit values of the actual transformer which the equivalent transformer replaces.

This method of expressing the size of the various quantities often conveys more information than to express them in the more usual units. Thus in the example on page 53, the no-load current $I_n = 3.57$ amp. The rated current of the low-voltage winding is 130.4 amp. The per-unit value of $I_n = 3.57/130.4 = 0.0274$ which shows immediately that this no-load current is less than 3 per cent of the rated current. If the impedance of a transformer is given in ohms, this conveys little idea as to whether the impedance is high or low. To determine whether it is high or low, the impedance drop at full-load current must be found. This drop expressed as a fraction of rated voltage shows whether the transformer is one of relatively high or low impedance, and this drop when so expressed is equal to the per-unit impedance since the base value of ohms equals rated voltage divided by rated current. One transformer may have an impedance in ohms which is twice as large as the impedance of another, and yet in its effect on the voltage characteristics, the former may be the low-impedance transformer if its impedance drop at full-load current expressed as a fraction of rated voltage is smaller than the impedance drop of the other machine similarly expressed, *i.e.*, if it has the smaller per-unit impedance.

Calculation of Regulation by the per-unit Method. For the transformer whose regulation was calculated on pages 60 and 61, if we neglect the no-load component of the current and omit the reduction factors, then

$$V_1 = V_2(1 + j0) + I_2(\cos \theta_2 - j \sin \theta_2)(r_e + jx_e)$$

Using rated values as the base values for voltage and kilovolt-amperes, the base value of V_2 is 2,300 and the per unit value is $2,300/2,300 = 1$. The base value of V_1 is 11,000. The base value of current is computed from the base kilovolt-amperes and for the secondary is

$$\frac{(300 \times 1,000)}{2,300} = 130.4$$

The base amperes for the primary is similarly

$$\frac{(300 \times 1,000)}{11,000} = 27.3$$

I_2 has a per-unit value of $130.4/130.4 = 1$. The base ohms for the primary is $11,000/27.3 = 402$ and for the secondary is $2,300/130.4 = 17.64$. The transformer resistances and reactances have the following per-unit values:

$$r_1 = \frac{1.28}{402} = 0.00318 \qquad x_1 = \frac{4.24}{402} = 0.01054$$

$$r_2 = \frac{0.0467}{17.64} = 0.00265 \qquad x_2 = \frac{0.162}{17.64} = 0.00920$$

$$r_e = r_1 + r_2 = 0.00583 \qquad x_e = x_1 + x_2 = 0.01974$$

Therefore

$$V_1 = 1.0(1 + j0) + 1.0(0.8 - j0.6)(0.00583 + j0.01974)$$
$$= 1.0 + j0 + 0.01652 + j0.01228$$
$$= 1.01652 + j0.01228$$

The magnitude of V_1 is 1.0166. The primary voltage required is therefore

$$1.0166 \times 11,000 = 11,183 \text{ volts}$$

and the regulation is

$$\frac{V_1 - V_2}{V_2} = \frac{1.0166 - 1.0}{1.0} = 0.0166 \text{ or } 1.66 \text{ per cent}$$

Chapter 6

Losses in a transformer. Eddy-current loss. Hysteresis loss. Screening
effect of eddy currents. Efficiency. All-day efficiency.

Losses in a Transformer. The losses in a transformer are (1) core loss,
(2) primary copper loss, (3) secondary copper loss. The core loss is
nearly constant and nearly independent of the load. The primary and
secondary copper losses vary as the squares of the primary and secondary
currents. The core loss is caused by the variation of the flux in the iron
core and depends upon the frequency, the maximum value of the flux

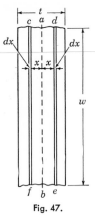

Fig. 47.

density in the core, the wave form of the time variation of the flux in the core, the quality of the iron,
the thickness of the laminations, and the volume or
weight of the core. The iron loss can be separated
into the loss due to eddy currents and the loss due
to hysteresis. These components follow different laws.

Eddy-current Loss. Let Fig. 47 represent a section,
taken perpendicularly to the flux, through one of the
laminations which form the iron core. Let t be the
thickness of the lamination and w its width measured
perpendicular to the flux, which is assumed to be perpendicular to the paper. Consider two elements, one
on each side of a line ab drawn through the middle of
the lamination parallel to its sides. Let the distance
of each element from the line ab be x and let the width be dx.

The flux which is enclosed between these two elements causes an eddy
current to circulate in the circuit $cdefc$ of which the elements form two
sides. If the thickness of the plate is small compared with its width,
the length of the path for the eddy current may be considered equal to
twice the length of an element, i.e., to $2w$.

Assume that the flux density is uniform throughout the cross section of
the laminations. With the usual thickness of lamination used in practice
for power transformers, this assumption is nearly true at commercial
frequencies. Let E_1 be the effective voltage induced in the primary winding of N_1 turns. The effective voltage induced per turn is $E = E_1/N_1$.
The voltage induced in the eddy-current path $cdefc$, Fig. 47, is equal to

64

the induced voltage per turn in the primary winding multiplied by the ratio of the flux enclosed by the area *cdefc* to the flux linked by the primary winding. Since the flux density in the lamination is assumed to be uniform, this ratio is equal to the ratio of the area enclosed within *cdefc* to the area of the cross section of the core. Therefore

$$E_e = \frac{E_1}{N_1} \frac{2xw}{A}$$

where E_e is the effective voltage induced in the eddy-current path *cdefc* and A is the cross section of the iron core.

Assume that the flux produced by the eddy currents is so small that the time lag of the eddy currents behind the voltage producing them can be neglected. This assumption follows from the assumption that the flux density in the lamination is uniform. Under these conditions, the eddy current in the elementary circuit *cdefc* can be found by dividing the voltage induced in this elementary circuit by its resistance. The eddy-current loss in the element is equal to the square of the voltage induced in the element divided by the resistance r of the element. Let ρ be the resistivity of the lamination. The eddy-current loss in the elementary circuit *cdefc* per unit length of lamination is

$$\frac{E_e^2}{r} = \left(\frac{2xw}{A} \frac{E_1}{N_1}\right)^2 \frac{dx}{2w\rho}$$

The loss per unit length of lamination is

$$P_e = \int_0^{t/2} \left(\frac{2xw}{A} \frac{E_1}{N_1}\right)^2 \frac{dx}{2w\rho} = \left(\frac{wE_1}{AN_1}\right)^2 \frac{1}{12w\rho} t^3$$

If E_1 is expressed in volts, the loss P_e is in watts. The volume corresponding to this loss is wt times 1 since unit length is considered. To find the loss P_e' per unit volume of core, divide by $wt1$.

$$P_e' = \left(\frac{E_1}{AN_1}\right)^2 \frac{1}{12\rho} t^2 \tag{49}$$

If E_1 is sinusoidal,

$$E_1 = \frac{2\pi f}{\sqrt{2}} N_1 \varphi_m = \frac{2\pi f}{\sqrt{2}} N_1 A \beta_m \tag{50}$$

where φ_m and β_m are the maximum flux and the maximum flux density in the core in webers and webers per unit area.

Putting the value of E_1, Eq. (50), in Eq. (49),

$$P_e' = \frac{t^2 \pi^2 f^2 \beta_m^2}{6\rho} \tag{51}$$

It should be noted that the eddy-current loss varies as the square of the thickness of the laminations. It also varies as the square of the effective voltage induced per turn in the windings of the transformer, Eq. (49). Any system of units may be used in Eqs. (49) and (51) as long as the dimensional units correspond to the unit of resistivity. For example, if ρ is in terms of ohms per meter cube and the dimensions are expressed in meters, P'_e gives the eddy-current loss in watts per cubic meter. Also if ρ is given in ohms per inch cube and the dimensions are in inches, then P'_e gives the loss in watts per cubic inch.

In the calculation of P'_e it is assumed that the magnetic effect of the eddy currents is negligible, that the flux density is uniform over the cross section of the laminations, and that the flux is parallel to the sides of the laminations. These assumptions are deviated from in practice. The value of ρ for transformer silicon steel lies between 50×10^{-8} and 62×10^{-8} ohm per meter cube.

Hysteresis Loss. Let N_1 be the number of turns in the winding producing the flux in the iron core and let the flux density at any instant be equal to b. If A is the cross-sectional area of the iron core, measured perpendicularly to the flux, the voltage induced in the winding is

$$e = -N_1 A \frac{db}{dt} \text{ volts}$$

If i is the instantaneous value of the current corresponding to e, the corresponding power is

$$p = ei$$

The energy in joules in a time dt is

$$p \, dt = ei \, dt$$

If it is assumed that the reluctance of the magnetic circuit per meter length is constant, the mmf per meter length of magnetic circuit also is constant. The mmf producing the instantaneous flux b is

$$4\pi N_1 i = hl$$

where h is the magnetizing force in praoersteds and l is the length of the magnetic circuit in meters.

Solving for i,

$$i = \frac{hl}{4\pi N_1}$$

If the values of e and i are substituted in the equation for the energy in joules expended in time dt,

$$p \, dt = \frac{Al}{4\pi} h \, db$$

$$= \frac{V}{4\pi} h \, db$$

where V is the volume in cubic meters.

The expenditure of energy in joules per unit volume during a magnetic cycle is

$$\frac{1}{4\pi} \int h \, db$$

If the iron is carried through f magnetic cycles per second, the loss per second per unit volume, or the rate at which energy is expended, is

$$\frac{f}{4\pi} \int_{+\beta}^{+\beta} h \, db \text{ watts}$$

A complete cycle of variation of flux density is from a density of $+\beta$ through a density of $-\beta$ back to a density of $+\beta$, which is the reason that the limits of integration are identical.

The integral $\int h \, db$ represents the area enclosed by the hysteresis loop of the core, plotted with flux density as ordinates and mmf per unit length as abscissas.

As a result of experiment, Steinmetz showed that the hysteresis loss in iron can be expressed empirically by

$$\eta f V \beta_m{}^x \text{ watts} \tag{52}$$

$$\eta f V \beta_m{}^x = \frac{fV}{4\pi} \int_{+\beta}^{+\beta} h \, db \tag{53}$$

Both η and x are constant. η is the hysteresis coefficient. Steinmetz showed that x is about 1.6. Although later experiments have shown that the hysteresis exponent may vary from 1.4 to 2.5, the value 1.6 is occasionally used.

The hysteresis loss depends upon the maximum value of the flux and is independent of how this maximum is reached, provided the change of flux between the limits zero and maximum is continuous and without reversal, *i.e.*, provided there are no small loops in the hysteresis curve.

Since the hysteresis component of the core loss is the larger part and depends upon the maximum value of the flux, the core loss corresponding to voltages impressed on the windings of an iron core varies with the wave form of the impressed voltage even if the rms value of the voltage remains constant. A flat voltage wave gives rise to a flux wave which is less flat and a peaked voltage wave gives rise to a flux wave which is less peaked (page 23). Hence the core loss corresponding to a flat voltage wave is greater than the core loss corresponding to a peaked voltage wave having the same rms value.

For an iron core, the expression for the core loss can be written

$$P_{h+e} = V\left[k_h f \beta_m{}^x + \left(\frac{E_1}{AN_1}\right)^2 \frac{1}{12\rho} t^2 \right] \tag{54}$$

For a sinusoidal voltage this becomes

$$P_{h+e} = V(k_h f \beta_m{}^x + k_e t^2 f^2 \beta_m{}^2) \tag{55}$$

The effective voltage in a transformer winding is given by

$$E_1 = 4(\text{form factor})f\varphi_m N_1$$

If φ_m is the maximum value of an alternating flux linking a coil, f is the frequency, and N_1 is the number of turns in the winding, the average voltage induced in the winding is

$$E_{av} = \frac{\varphi_m}{\frac{1}{4}f} N_1 = 4f\varphi_m N_1 \text{ volts}$$

provided the voltage wave is symmetrical in its positive and negative loops, i.e., provided it does not contain even harmonics.

If the thickness of laminations is fixed, Eq. (54) can be expressed by

$$P_{h+e} = V(K_h E_{av}{}^x f^{1-x} + K_e E_1{}^2) \tag{56}$$

where K_h and K_e are constant and E_{av} is the average voltage induced in the winding by the core flux.

The eddy-current loss in a transformer is fixed by the effective voltage induced in the winding by the core flux; the hysteresis loss is fixed by the average voltage induced in the winding by the core flux. The effective voltage can be measured by a voltmeter which gives the rms or effective voltage. Special voltmeters are available for measuring the average voltage of a circuit.

Table 2 gives the core loss per pound for 4.25 per cent silicon steel with No. 29 gauge (14 mils) laminations.

TABLE 2

Core Loss for 4.25 Per Cent Silicon Steel with No. 29 Gauge (14 mils) Laminations

Flux density, kilolines per square inch	Frequency, cycles per second	Core loss, watts per pound
70	60	0.70
65	60	0.61
70	25	0.27
65	25	0.24

Screening Effect of Eddy Currents. In deducing the expressions for the eddy-current and hysteresis losses, the flux density is assumed to be uniform throughout the laminations. Although this assumption is approximately correct for thin laminations, it is far from true for thick laminations.

The effect of eddy currents on the flux is similar to the effect of the secondary current in a transformer. They tend to demagnetize the core. This demagnetizing action is greatest at the center of each lamination and is zero at its surface, since all the eddy currents in any lamination flow in concentric paths about its center and produce the greatest effect at the center. Any point in a lamination is subjected to a demagnetizing action which is due to the eddy currents in that portion of the lamination which lies outside this point.

On account of this action of the eddy currents there is a diminution in the resultant mmf in passing from the surface to the center of a lamination. The flux density at the center of laminations of different thickness in per cent of the flux density at the perimeter is given in Table 3 for ordinary transformer iron, at 60 and 25 cycles.

From Table 3 it is evident that thicker laminations can be used for 25-cycle transformers than for 60-cycle transformers. With laminations less than 0.5 mm (0.196 in.) thick, the decrease in flux density from perimeter to center is only a few per cent for either 60 or 25 cycles. As the laminations used for transformer cores are seldom more than 0.014 in. (14 mils) thick, the flux density may be assumed to be uniform with negligible error.

TABLE 3

Maximum Flux Densities at the Center of Any Lamination in Terms of the Maximum Flux Density at Its Surface

Thickness............	2mm	1mm	0.5 mm
60 cycles..............	0.18	0.61	0.96
25 cycles..............	0.30	0.89	0.99

Table 3 is calculated for ordinary annealed transformer iron. For silicon iron, such as is almost universally used for transformer cores, the variation in flux is much less than is indicated in the table, owing to the high resistivity of silicon steel.

The effect of the diminution in flux density toward the center of each lamination is to increase the impressed mmf required for a given total flux in the core. The effect is the same, so far as the magnetizing current is concerned, as decreasing the cross section of the core. Since the hysteresis loss varies as the x (greater than 1) power of the maximum flux density, this loss is greater for a given total flux than if the density were uniform. The eddy-current loss is less, since for a given total flux the eddy currents are the same at the surface and at the center of the laminations whether the flux density is uniform or not, and at every other point in the laminations they are less. For these reasons, Eq. (55) gives an eddy-current loss which is greater and a hysteresis loss which is less than the actual losses. However, the difference between the actual core loss and that calculated by Eq. (55) is small for laminations of thicknesses ordinarily used, especially for silicon iron, and the error resulting from disregarding this difference is negligible.

Efficiency. The efficiency of a transformer is given by

$$\frac{\text{Output}}{\text{Input}} = \frac{V_2 I_2 \cos \theta_2}{V_2 I_2 \cos \theta_2 + P_{h+e} + I_1{}^2 r_1 + I_2{}^2 r_2} \tag{57}$$

The core loss of a transformer depends upon the flux density in the core and the frequency. Not all of the mutual flux of a transformer is in the iron core. A small part of the mutual flux is in the air outside the core, but because of the high reluctance of the air path as compared with the iron path, all but a small percentage of the mutual flux of a transformer is in the iron core. Hence it may be assumed that the core loss is fixed by the mutual flux and the frequency. Since at any frequency the mutual flux is proportional to the voltages it induces in the windings, these voltages may be taken as a measure of the core loss. Strictly speaking the value of the core loss to be used when calculating the efficiency of a transformer is that corresponding to the induced voltage and not the impressed voltage. The difference between the impressed voltage and the corresponding voltage induced in a winding by the mutual flux is small. Little error is introduced by using the impressed voltage instead of the induced voltage for finding the core loss when calculating the efficiency of a transformer. Because of the slight variation of the induced voltages of a transformer with change in load, the core loss of a transformer which operates with constant impressed voltage and frequency but with variable load is not constant. However, the variation in the induced voltage with load is so small that the core loss of a transformer which operates with constant impressed voltage and frequency is usually assumed constant.

Since the no-load or exciting current I_n of the primary current is small, the primary current can be replaced by its load component I_1'. If this is done, the primary and secondary copper losses can be combined by replacing the primary and secondary resistances by the equivalent resistance of the transformer. Making this substitution,

$$\text{Efficiency} = \frac{V_2 I_2 \cos \theta_2}{V_2 I_2 \cos \theta_2 + P_{h+e} + I_2{}^2 r_e} \tag{58}$$

The resistance r_e is referred to the secondary winding and is equal to $(r_1/a^2) + r_2$. A still further approximation can be made by replacing the load terminal voltage by the no-load or rated voltage. The net effect of the three approximations is slight and negligible for ordinary commercial work. The error introduced by the approximations should not exceed 0.1 per cent in the efficiency except at low power factors. The chief reason why the error is so small is that the total losses of a transformer are small. The efficiency of commercial transformers varies from 96 to 99 per cent. With an efficiency of 96 per cent, an error as great as 10 per cent in the losses would produce an error of about 0.5 per cent in the full-load efficiency.

The efficiency of a transformer can also be found from the solution of the transformer diagram on pages 51 and 52. The power in a circuit when both current and voltage are expressed as complex quantities is equal to the sum of the products of the real and the j parts of the current and voltage. Applying this to the results given on pages 51 and 52,

$$\text{Efficiency} = \frac{V_2(I_2 \cos \theta_2) + 0(I_2 \sin \theta_2)}{FC + GD} \tag{59}$$

It is usually more convenient and more satisfactory to calculate the efficiency of a transformer by Eq. (58).

The hysteresis component of the core loss of a transformer is the larger of the two components, especially in transformers having silicon-steel cores. It depends upon the x power of the maximum flux density. Therefore for the same rms induced voltage, the maximum flux density and consequently the core loss vary considerably for impressed voltages of different wave forms. According to the American Institute of Electrical Engineers rules, transformer core losses must be given for sinusoidal impressed voltages. The efficiency of a transformer operated on a peaked voltage wave is higher than when measured on a flat voltage wave. The difference may amount to as much as 0.2 per cent. All efficiency determinations are now based on sinusoidal conditions.

The efficiency of a transformer which is zero at no load increases with the current output and is a maximum when the core and copper losses are equal. This can be proved by making use of Eq. (58). The core loss and the secondary voltage V_2 are assumed constant. The maximum efficiency occurs for that value of the secondary current which makes the differential coefficient of the efficiency with respect to the secondary current zero. If η is the efficiency, the maximum efficiency occurs when $d\eta/dI_2 = 0$. Differentiating Eq. (58) with respect to I_2 and equating the differential coefficient to zero,

$$(V_2 I_2 \cos \theta_2 + P_{h+e} + I_2{}^2 r_e) V_2 \cos \theta_2 = V_2 I_2 \cos \theta_2 (V_2 \cos \theta_2 + 2 I_2 r_e)$$

or

$$P_{h+e} = I_2{}^2 r_e$$

This relation for maximum efficiency assumes that the core loss and V_2 are constant.

Transformers which are to be operated continuously under load should be designed to have equal core and copper losses at the average load at which they are to be used. Many transformers are connected permanently to the mains and operate under no load or under very small loads for a large part of the time. In such cases it is obviously impossible to reduce the amount of iron sufficiently to make the maximum efficiency occur for the average load, both on account of the large amount of copper required and the poor voltage regulation which would result. The full-load copper losses of such transformers are usually made somewhat greater than the core losses.

All-day Efficiency. The efficiencies of transformers discussed up to this point are power efficiencies, *i.e.*, they are the ratio of the power output to the power input under definite load conditions. Most transformers do not operate under constant load. As a rule all except those in central stations or in substations are permanently connected to the power lines on their high-voltage sides and consume power during 24 hr of each day corresponding to their core losses whether or not the transformers are loaded. For transformers which operate on definite load cycles, it is the energy efficiency that is important. The energy efficiency is the ratio of the total kilowatthour output to the total kilowatthour input during the load cycle, which is usually a period of 24 hr. This efficiency is called the all-day efficiency.

The all-day efficiency of a transformer depends not only on the watt losses at rated load but also on the division of those losses between core loss and copper loss.

The all-day efficiency is numerically equal to the total kilowatthour output for 24 hr, divided by the kilowatthour output for 24 hr plus the core loss for 24 hr plus the copper loss for 24 hr.

In algebraic form this is

$$\frac{\Sigma t V_2 I_2 \cos \theta_2}{\Sigma t V_2 I_2 \cos \theta_2 + \Sigma t I_2{}^2 r_e + 24 P_{h+e}} \tag{60}$$

where t is time in hours.

Reducing the number of turns on the windings of a transformer decreases the copper loss but increases the flux density in the core and consequently the core loss. Usually if the turns are decreased, the cross section of the core must be increased in order to keep the flux density down. Increasing the number of turns has the opposite effect. In general, decreasing the turns in both primary and secondary in the same proportion decreases the all-day efficiency, and increasing the turns in the same proportion increases the all-day efficiency.

The flux density used in the design of 60-cycle transformers varies from about 65 kilolines per square inch for distribution transformers to as high as 90 kilolines per square inch for large power transformers. For 25-cycle transformers, the flux densities used are from about 75 kilolines per square inch for distribution transformers to about 80 kilolines for large power transformers. The flux density may be higher in large transformers with artificial cooling.

The distribution of the losses in a transformer is often an important factor in determining the type of transformer which is most suitable for a given service. When transformers are connected only when loaded, it makes little difference, so far as the efficiency of operation is concerned, how the losses are distributed between core and copper, provided the total losses are not changed. When, however, transformers remain permanently connected to the mains, as is usually the case except in central stations or substations, the distribution of the losses may have a substantial influence on the economy of operation. For example, consider a 25-kva transformer having a total full-load loss of 750 watts, two-fifths of which is core loss. Assume the transformer to operate under the following conditions during 24 hr.

Full load..................	1 hr
One-half load.............	2 hr
One-quarter load.........	3 hr
No load..................	18 hr

The all-day efficiency of this transformer for the specified load conditions is 89.6 per cent, Eq. (60), page 72. If three-fifths of the total loss had been core loss instead of two-fifths, the all-day efficiency would have been 85.9 per cent. With power at 1 cent a kilowatthour, the operating cost of this transformer for 1 year with the assumed load conditions would be nearly $12 more when three-fifths of the total loss was core loss than when only two-fifths was core loss. In a large station having many distributing transformers connected to its feeders, a slight change in the distribution of the losses in those transformers may make a difference of several thousand dollars in the annual operating cost, even if the cost of power at the switchboard is low. Under certain conditions the greater first cost of the transformers which give the most desirable distribution of losses may more than offset the saving in the cost of operation. The most efficient piece of apparatus is not always the most economical one to install, as in some cases the increase in the interest on the investment necessary to obtain the increase in efficiency may more than balance the saving in the cost of power effected by the use of the more efficient apparatus.

Chapter 7

Measurement of core loss. Separation of eddy-current and hysteresis losses. Measurement of equivalent resistance. Calculation of the efficiency of a transformer from its losses as determined from open-circuit and short-circuit tests. Measurement of equivalent reactance, short-circuit method. Measurement of equivalent reactance, highly-inductive-load method. Opposition method for testing transformers.

Measurement of Core Loss. The power input to a transformer with its secondary open is equal to its core loss plus a small copper loss in its primary winding which under ordinary conditions is negligible. The value of the core loss obtained in this way corresponds to a voltage which is equal to the voltage induced in the transformer coils by the core flux. At no load this voltage does not differ appreciably from the voltage impressed across the terminals of the transformer.

As an example of the error introduced by neglecting the no-load copper loss when determining the core loss of a transformer, consider a 500-kva 12,700:2,300-volt 25-cycle transformer. This transformer has a relatively large ratio of full-load copper loss to core loss and a large no-load current. The full-load copper and core losses of this transformer are 4,680 and 3,300 watts. The no-load current is 5.2 per cent of the full-load current. If the neglect of the no-load copper loss of this transformer, when determining its core loss, produces no appreciable error, it is fair to assume that in general the no-load copper loss may be neglected. As a rule, the full-load copper loss is divided about equally between primary and secondary windings. If this assumption is made, the no-load copper loss of this transformer is

$$\frac{4,680}{2} (0.052)^2 = 6.3 \text{ watts}$$

or

$$\frac{6.3}{3,300} \, 100 = 0.19 \text{ per cent}$$

of the core loss of the transformer.

When measuring the core loss of a transformer, the copper losses in the measuring instruments must not be overlooked. Some of these losses are always included in the power indicated by the wattmeter. In small transformers of 5 to 15 kw, neglecting to make proper correction for the instrument losses may introduce an error of 5 to 10 per cent.

The hysteresis loss, which is the larger part of the core loss, depends on the average value of the voltage, as noted on page 68. Consequently, in measuring the core loss, care must be taken to use a voltage of sinusoidal wave form. A recommended method is to use a voltmeter which reads the average voltage. If the average value of the test voltage is adjusted to be the same as the average value of the desired sine wave of voltage, and the proper frequency held, the hysteresis loss will be the desired sine-wave value.

Separation of Eddy-current and Hysteresis Losses. From Eq. (55), page 68, the core loss for a sinusoidal impressed voltage and a core of fixed dimensions can be written

$$P_{h+e} = K'_k f B_m{}^x + K'_e f^2 B_m{}^2 \tag{61}$$

If the values of the two constants K'_h and K'_e can be determined, the total core loss P_{h+e} can be separated into its two components. These constants can be found for any given iron core by solving two simultaneous equations obtained by measuring the core loss, either at two different maximum flux densities and the same frequency or at two different frequencies and the same maximum flux density. Whichever method for obtaining the two equations is adopted, the wave form of the impressed voltage must be sinusoidal during both determinations of the core loss.

The objection to the separation of the losses by measuring the core loss at two different maximum flux densities and the same frequency is that the hysteresis exponent is not known. Although 1.6 is frequently used for the value of this exponent, it may actually differ considerably from this value. Unless the actual value of the exponent is known, the separation of the eddy-current and hysteresis losses by making use of two different maximum flux densities and the same frequency gives results which are far from reliable. When the losses are separated by making use of two frequencies but the same maximum flux density, no assumption is made in regard to the magnitude of the exponent x.

Since the wave form must be sinusoidal in both determinations, the maximum flux density in the equation can be replaced by its value in terms of the frequency and the rms voltage induced in the winding by the flux.

$$E_1 = 4.44 N_1 \varphi_m f$$

For any fixed wave form and number of turns this can be written

$$E_1 = k_1 \beta_m f$$

$$\beta_m = \frac{1}{k_1} \frac{E_1}{f}$$

Substituting this value of β_m in Eq. (61) for the core loss,

$$P_{h+e} = K_h'' f^{(1-x)} E_1^x + K_e'' E_1^2 \tag{62}$$

It must be remembered that Eq. (62) holds only for a sinusoidal wave form and a definite iron core. It also assumes that the flux density is uniform over the cross section of the core. It therefore applies only when the core has a uniform cross section. The constants K_h'' and K_e'' are different for different cores.

Whichever method of separating the losses is adopted, the voltage impressed across the terminals of the transformer must be changed. This cannot be done by putting resistance or reactance in series with the transformer since any harmonics in the current taken by the transformer would appear in the drop in potential through the resistance or reactance and would consequently be present in the voltage impressed across the transformer. It has been shown that the current producing the flux in a transformer contains marked harmonics, especially the third, even when the impressed voltage is sinusoidal.

The voltage impressed across the terminals of a transformer can be varied without changing the wave form by using a transformer which has large current rating compared with the current to be taken from it or by varying the excitation of an alternator which is large enough for its voltage not to be influenced by armature reaction or by armature resistance and leakage reactance drops caused by the current taken by the transformer under test.

If the core loss is to be separated into its components by measuring the loss at two different frequencies and the same maximum flux density, it is necessary to vary the voltage directly in proportion to the frequency in order to keep the maximum flux density constant. If the method using two different maximum flux densities is adopted, it is necessary merely to measure the core loss at two different voltages at the same frequency. It has been pointed out that the second method of separating the losses is unsatisfactory because of the uncertainty as to the magnitude of the exponent x for the hysteresis loss.

If the separation of losses is to be carried out at constant maximum flux density, a graphical method is preferable provided a sufficient range of frequency is available. Let Eq. (61) be divided by the frequency. This gives Eq. (62a), which is the expression for the core loss per cycle.

$$\frac{P_{h+e}}{f} = K_h' \beta_m^x + K_e' f \beta_m^2 \tag{62a}$$

Equation (62a) is an equation of the first degree with respect to f. If P_{h+e}/f is taken as one variable and is plotted as ordinates with f as abscissas, the result is a straight line as shown in Fig. 48. The intercept oa of this line on the axis of ordinates is $K_h'\beta_m^x$. $K_e'f\beta_m^2$ is equal to the ordinate minus oa at a point on the line for a frequency f. Equation (62a) is plotted in Fig. 48.

Referring to Fig. 48, oa and cd multiplied by the frequency f are the hysteresis loss and the eddy-current loss corresponding to that frequency.

Measurement of Equivalent Resistance. The equivalent resistance of a transformer can be calculated from the ohmic resistance of its primary and secondary windings, but it is sometimes better to measure it directly in order to include all local eddy-current and hysteresis losses which are produced in the conductors or in the iron core by the currents in the primary and secondary windings. The equivalent resistance, including these losses, can be obtained from measurements made with the transformer short-circuited.

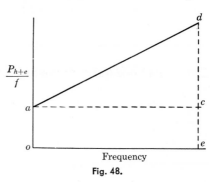

Fig. 48.

The vector diagram of a short-circuited transformer is given in Fig. 49. The flux in a short-circuited transformer is that required to produce a voltage E_2 equal to the impedance drop in the secondary winding (Fig. 49). The secondary impedance drop is approximately one-half the total impedance drop in the transformer referred to the secondary winding.

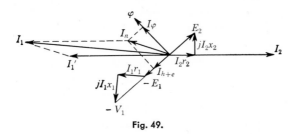

Fig. 49.

The total impedance drop of a transformer at rated current is 5 to 6 per cent of its rated voltage. The secondary impedance drop which is equal to the voltage induced in the secondary winding on short circuit is $2\frac{1}{2}$ to 3 per cent of the secondary rated voltage. Since at a fixed frequency the flux is proportional to the induced voltage and since the core loss produced by the flux varies between the 1.6 and 2 power of the flux density, the core loss in a short-circuited transformer is negligible in comparison with the copper loss. The input to a short-circuited transformer therefore is equal to the copper loss corresponding to the short-circuit current plus all other losses occurring under short-circuit. If P and I are the input and the short-circuit current, both measured on the side of the transformer to which the power is supplied, the equivalent resistance of the transformer referred to that side is P/I^2.

If the equivalent resistance is to be used for determining the copper loss in efficiency calculations, it should be determined at the operating temperature which is usually taken as 75°C. The total loss due to the currents in the windings as measured in the short-circuit test is often called the load loss. The difference between this loss and the copper loss determined from the d-c resistances is known as the stray-load loss. If $r_{e(dc)}$ is determined from d-c measurements at a temperature θ°C, the $r_{e(dc)}$ at 75°C equals $r_{e(dc)}$ at θ°C multiplied by $(234.5 + 75)/(234.5 + \theta)$. The stray-load loss, however, decreases with temperature. It includes eddy-current losses produced in the conductors themselves as well as eddy-current losses produced by the leakage fluxes in the iron and steel structural parts. If the operating temperature of these parts were 75°C and the temperature coefficient were the same as that for copper, the component of r_e due to eddy currents should be multiplied by $(234.5 + \theta)/(234.5 + 75)$, where θ is again the temperature at which r_e is measured. Since the stray-load loss is a relatively small part of the load loss, this temperature correction may be applied to the stray-load loss without introducing appreciable error. If therefore $r_{e(sc)}$ is the equivalent resistance determined from short-circuit measurements, and both $r_{e(sc)}$ and $r_{e(dc)}$ are determined at θ°C, then

$$r_e \text{ at } 75°C = r_{e(dc)} \frac{234.5 + 75}{234.5 + \theta} + (r_{e(sc)} - r_{e(dc)}) \frac{234.5 + \theta}{234.5 + 75} \quad (63)$$

and the load loss at 75°C due to a current I equals $I^2 r_e$ (at 75°C) where r_e and I are both referred to the same side of the transformer.

Calculation of the Efficiency of a Transformer from Its Losses as Determined from Open-circuit and Short-circuit Tests. The 300-kva 11,000:2,300-volt 60-cycle transformer for which the regulation was found in Chap. 4, page 53, and Chap. 5, page 60, is used in this calculation. When this transformer is operated at no load, with 2,300 volts at 60 cycles impressed on its low-voltage winding, the current and power are $I_n = 3.57$ amp and $P_n = 2,140$ watts. When the high-voltage winding is short-circuited and 48.4 volts at 60 cycles are impressed on the low-voltage winding, the imput current and power are $I_{sc} = 132.6$ amp and $P_{sc} = 1,934$ watts. The d-c resistances r_1 and r_2, from Chap. 5, page 61, were measured at a temperature of 25°C and the temperature during the short-circuit test was also 25°C

$$r_e = \text{equivalent resistance} = \frac{1{,}934}{(132.6)^2} = 0.110 \text{ ohm}$$

$$z_e = \text{equivalent impedance} = \frac{48.4}{132.6} = 0.365 \text{ ohm}$$

$$x_e = \text{equivalent reactance} = \sqrt{(0.365)^2 - (0.110)^2} = 0.347 \text{ ohm}$$

All three of the above constants are referred to the low-voltage side since the voltage was impressed on the low-voltage side when making the short-circuit test. From Chap. 5, page 61, $r_{e(dc)} = 0.103$ ohm and

$$r_e \text{ at } 75°C = 0.103 \frac{234.5 + 75}{234.5 + 25} + (0.110 - 0.103) \frac{234.5 + 25}{234.5 + 75}$$
$$= 0.122 + 0.006 = 0.128$$

The no-load copper loss is negligible. Therefore the power imput when the secondary is open is the core loss. The core loss at 60 cycles is therefore 2,140 watts. The full-load current on the low-voltage side was found in the previous examples to be 130.5 amp. The efficiency for a full kva load at 0.8 power factor is

$$\text{Efficiency} = \frac{300{,}000 \times 0.8}{300{,}000 \times 0.8 + 2{,}140 + (130.5)^2 \times 0.128}$$
$$= \frac{240{,}000}{240{,}000 + 4{,}320} = 1 - \frac{4{,}320}{240{,}000 + 4{,}320}$$
$$= 0.982 \text{ or } 98.2 \text{ per cent}$$

When the core loss of a transformer is considered as fixed by the terminal voltage instead of by the voltage induced by the core flux, the efficiency of a transformer for given terminal voltage, frequency, kva load, and power factor is independent of whether the current lags or leads the terminal voltage. Actually the efficiencies with equal leading and lagging currents are slightly different because the voltage which is induced by the core flux and which fixes the core flux is not quite the same with leading and lagging currents for a fixed terminal voltage. The difference is negligible in ordinary transformers.

Measurement of Equivalent Reactance, Short-circuit Method. When a transformer is short-circuited

$$-V_1 = I_1 z_1 - aI_2 z_2 \text{ vectorially}$$
$$= I_1 z_1 + a^2 I_1' z_2 \text{ vectorially}$$

Since on short circuit the currents I_1 and I_1' are nearly equal in magnitude and in phase,

$$V_1 = I_1 \sqrt{(r_1 + a^2 r_2)^2 + (x_1 + a^2 x_2)^2} \text{ arithmetically}$$
$$= I_1 z_e$$

z_e is the equivalent impedance and is referred to the primary side since I_1 is the primary current.

$$z_e = \frac{V_1}{I_1} \tag{64}$$

and

$$x_e = \sqrt{z_e{}^2 - r_e{}^2} \tag{65}$$

If the primary and secondary leakage reactances x_1 and x_2 are assumed to be proportional to the square of the number of turns in the two windings, the equivalent reactance may be divided into its two component parts.

Although the short-circuit method for determining the leakage reactance of a transformer involves low saturation, the value of the reactance

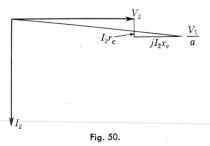

Fig. 50.

given by it differs only slightly from the value corresponding to normal saturation since the reluctance of the path of the leakage flux in most transformers is nearly independent of the saturation of the iron core.

Measurement of Equivalent Reactance, Highly-inductive-load Method. The simplified vector diagram for a transformer delivering a highly inductive load is shown in Fig. 50. Everything on the diagram is referred to the secondary winding.

The equivalent reactance may be calculated from the following equation which is approximately true when applied in an arithmetical sense to a transformer which carries a highly inductive load (Fig. 50)

$$\frac{V_1/a - V_2}{I_2} = x_e$$

The advantage of this method is that it gives a value of reactance which corresponds to nearly normal saturation of the transformer core. The disadvantage is that it necessitates the subtraction of two voltages, V_1/a and V_2, which are nearly equal, and any percentage error in the determination of either is greatly increased in their difference.

Opposition Method for Testing Transformers. The limit of the output of a transformer is determined by the rise in temperature of its parts and by its regulation. Of the two, the temperature rise is by far the more important in most cases.

The methods for determining the regulation of a transformer have already been given. In order to obtain the increase in temperature of a transformer under load, it is necessary to operate it under conditions which produce normal full-load heating for a sufficient length of time for the temperature of its parts to become constant. This requires 2 to 3 hr for small transformers and much longer for large transformers. When merely the ultimate temperatures are desired, the time required to make a heat run can be reduced considerably by accelerating the heating during the first part of the run by operating at overload.

Small transformers can be tested by applying an actual load, but when large transformers have to be tested, the cost of the power required for loading is prohibitive. In such cases, the opposition method may be used, provided two similar transformers are available. A modification of this method may be applied to a single transformer if it has two primary and two secondary windings. The opposition method is equally applicable to small and large transformers and is in general use. It requires merely enough power to supply the core and copper losses of the two transformers which are being tested.

For the opposition method, the primary windings of the two transformers are connected in parallel to mains of the proper voltage and frequency. The secondary windings are connected in series with their voltages opposing. Figure 51 gives the proper connections.

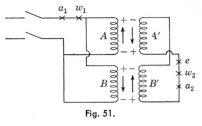

Fig. 51.

A and A' are the primary and secondary windings of one transformer; B and B' are the primary and secondary windings of the other transformer.

If the secondary windings are opposed with respect to the series circuit, they are virtually on open circuit so far as their primaries are concerned, and no current flows in them when the primaries are excited. So far as the secondaries are concerned, the primaries are virtually short-circuited with respect to any current which is sent through the secondaries.

The correctness of these two statements is made clear by referring to Fig. 51. The plus and minus signs on the figure indicate merely the polarity of the transformer windings at some particular instant. The arrows show the direction of the current which would be produced at some instant by inserting an alternating voltage anywhere in the secondary circuit, as at e. Following through the circuit in the direction of the arrows shows that the transformers are short-circuited so far as the voltage inserted at e is concerned. The secondary voltages are in opposition when considered with respect to the voltage impressed on the primaries. Therefore if the rated voltage is applied to the primary windings, the transformers operate under normal conditions so far as core loss is concerned. If at the same time the voltage inserted at e is adjusted so that full-load current exists in the secondary windings, full-load current also exists by induction in the primary windings and the transformers are operating under conditions of full load so far as the copper loss is concerned.

The only power required under these conditions is that necessary to supply the core loss, which is measured by a wattmeter placed in the primary circuit at w_1, and the power required to supply the copper loss, which is measured by a wattmeter at w_2 with its potential coil connected about the source of voltage at e. One-half the reading of the wattmeter at w_2 divided by the square of the current measured by an ammeter in series with w_2 is the equivalent resistance of one transformer. A voltmeter connected about the source of voltage at e records twice the equivalent impedance drop in one transformer. The reading of this voltmeter divided by twice the current in the circuit, as given by an ammeter placed at a_2, is the equivalent impedance of one transformer. An ammeter placed at a_1, in the primary circuit, records twice the no-load current of one transformer.

The temperature rise may be obtained by thermometers and from resistance measurements. The resistances for the calculation of the temperature rise can be obtained either from measurements made by any suitable method at the beginning and end of the run, or from the readings of the wattmeter and the ammeter placed at w_2 and a_2.

The best way to obtain the voltage required at e is to insert the secondary of a suitable transformer at e. The voltage may be varied by a resistance in series with the primary of this auxiliary transformer.

If the core losses are supplied on the low-voltage side of the transformers and the voltage at e is obtained from a third transformer, all necessity for handling high-voltage circuits when adjusting for load conditions is avoided.

The resistances obtained from the wattmeter and ammeter readings are not quite correct for determining the temperature rise in the transformers under load, as they are the effective resistances and include the effect on the resistance of the local losses due to the leakage fluxes. For a given current, these local losses do not vary in the same way with a change in temperature as the loss in the copper, but since the local losses in a properly designed transformer are small, the error in using the effective resistance for determining the temperature rise is not great.

The two transformers used in an opposition test do not operate under identical conditions. There is a transfer of power between them. As a result, the primary currents are nearly opposite in phase when considered with respect to the common terminal voltage. This makes the phase angle between the primary current of one transformer and the exciting current of this transformer differ by nearly 180 deg from the corresponding phase angle for the other transformer. Consequently the primary current of one transformer is slightly larger than the primary current of the other transformer.

Chapter 8

Current transformer. Potential transformer. Constant-current transformer. Autotransformer. Induction regulator.

Current Transformer. Current transformers are used with a-c instruments and serve the same purpose as shunts with d-c instruments. When a current transformer is used, its primary winding is placed in the line carrying the current to be measured and its secondary is short-circuited through the instrument which is to measure the current. Current transformers serve the double purpose of increasing the current range of an instrument and insulating it from the line.

The ratio of the secondary current in any transformer to the load component of the primary current is constant and is fixed by the ratio of turns on the primary and secondary windings. The two currents are opposite in phase. The total primary current and the secondary current are not opposite in phase, neither is their ratio constant. Both their phase relation and their ratio vary on account of the magnetizing current in the primary and the component current in the primary which is required to supply the core loss.

When the secondary winding is closed through a low impedance, such as an ammeter or the current coil of a wattmeter or both, the secondary induced voltage becomes small and is equal to the impedance drop in the instrument plus the impedance drop in the secondary of the transformer. The mutual flux required to produce this small induced voltage is correspondingly small, and since it is the mutual flux which determines the magnetizing current and the component current supplying the core loss, these two components of the primary current are small. Under normal conditions, with the secondary winding short-circuited through an instrument, neither of these two components of the primary current should be more than a fraction of 1 per cent of the rated current of the transformer. The voltage drop across the primary winding is merely the equivalent impedance drop in the transformer plus the impedance drop in the instrument, both referred to the primary winding.

Although the induced voltage in the current transformer and therefore the mutual flux are both directly proportional to the secondary current, assuming constant impedance for the transformer and the instrument, the small exciting current is not exactly proportional to the induced voltage and does not make a constant angle with it, since neither component of the exciting current varies as the first power of the mutual flux.

The magnitudes of both components of the exciting current depend upon the degree of saturation of the iron core of the transformer. For this reason direct current should not be put through a current transformer unless the precaution is taken afterward to demagnetize the core thoroughly. For the same reason the secondary winding should not be opened while the primary carries current. Passing direct current through the windings of a current transformer or opening its secondary circuit while its primary winding carries current may change its ratio of transformation because of the residual flux which may be left in the core. The winding with the fewer turns is placed in the line; therefore if the second-

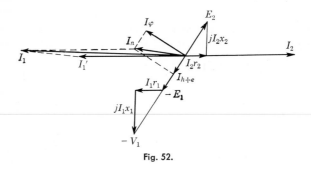

Fig. 52.

ary circuit is opened, the current transformer becomes a step-up transformer and a voltage dangerous to the insulation of the transformer and to life may be induced in the transformer windings. This voltage is limited by the saturation of the core. It is much less than the voltage of the circuit in which the transformer is placed multiplied by the ratio of turns. If the secondary is accidentally opened, the core must be completely demagnetized before putting the transformer in service. A current transformer should be insulated for the full voltage of the line with which it is to be used and should be operated with its secondary winding and also its case solidly grounded.

On account of the effect of the exciting component of the primary current upon the ratio of the primary and secondary currents and upon the phase angle between them, the exciting current of current transformers must be made small by designing such transformers to operate at relatively low flux density. The secondary winding must also be arranged for minimum leakage since any increase in the leakage reactance increases the mutual flux and hence increases both components of the exciting current.

From what precedes, it is obvious that current transformers should be calibrated for current ratio and phase angle with the instruments with which they are to be used in circuit, as well as at the currents to be measured. The phase angle of a current transformer is the angle between the secondary current and the reversed primary current. For power measurements where accuracy is essential, it is often necessary to apply corrections to the phase displacement between the primary and secondary currents because of the exciting current.

Figure 52 applies to a current transformer if x_2 and r_2 include the reactance and resistance of the instrument with which the transformer is used.

Current transformers are made for two classes of work, for use with instruments and for operating protective and regulating devices such as automatic oil switches. For the second class of service great accuracy and constancy of transformation ratio with change in load are not required, but great reliability is of prime importance.

In high-voltage power stations, current transformers are an important part of the auxiliary apparatus and require considerable space. They vary in weight from 40 to 50 lb for low voltages to 4,000 lb for 110,000 volts, and in height from 6 or 8 in. to 8 ft and a diameter of 3 ft. A current transformer for a 66,000-volt circuit is shown in Fig. 53.

Fig. 53. Outdoor-type high-voltage current transformer. (*Westinghouse Electric and Manufacturing Co.*)

Potential Transformer. Potential transformers are used to increase the range of a-c voltmeters and wattmeters and at the same time to insulate them from the line voltage. They do not differ from ordinary transformers except in details of design.

The ratio of the terminal voltages of an ordinary transformer does not change by more than a few per cent from no load to full load, and the voltages would be in phase when considered in the same sense[1] if it were not for the resistance and reactance drops. By designing a potential transformer with low resistance and reactance, the change in phase and in magnitude of the terminal voltages can be made small. The angle α_o, Fig. 40, page 55, is the phase angle of the transformer. The phase

[1] If both are voltage rises or both are voltage drops, they are in phase. On the vector diagram of a transformer in Fig. 38, page 49, the primary voltage is drawn as a voltage drop, the secondary as a voltage rise. In Fig. 40, page 55, both voltages are drawn as voltage rises.

relation is of importance only when potential transformers are used in connection with wattmeters. Since the magnetizing current and the current supplying the core loss are important components of the primary current, these component currents should be kept small. The influence of the resistance and leakage reactance of the windings is far more important in a potential transformer than in a current transformer, since these factors directly affect the ratio of transformation and the phase relation between the primary and secondary terminal voltages. The exciting current of a properly designed potential transformer should have little influence on either the ratio of transformation or the phase relation between the terminal voltages. When potential transformers are used for accurate power measurements, correction should be applied for the phase angle α_o, Fig. 40, page 55, between the primary and secondary voltages caused by the resistances and leakage reactances. Potential transformers as well as current transformers should always be calibrated for ratio of transformation and phase angle. The effect of the phase angle in a current transformer or a potential transformer on the reading of a wattmeter with which the transformer is used depends upon the power factor of the load. Because of the way the cosine of an angle varies with the angle, the effect is greatest for low power factors and least for high power factors. A 34,500-volt potential transformer is shown in Fig. 54.

Fig. 54. Outdoor-type potential transformer, 34,000:115 volts. (*Westinghouse Electric and Manufacturing Co.*)

Constant-current Transformer. When arc or incandescent lamps are operated in series, as is generally done when they are used for street lighting, they must all have the same current rating and must be operated in a circuit which carries constant current and has a voltage which varies with the number of lamps in circuit. Constant-current or "tub" transformers are now nearly always employed for such circuits. When constant-current transformers are employed for circuits requiring unidirectional current, they are used in conjunction with mercury-arc rectifiers.

Constant-current transformers are an important part of the auxiliary apparatus of central stations supplying power for street lighting.

If a transformer of the ordinary type is designed with high leakage reactance, it has a dropping voltage characteristic and may even be short-circuited without producing excessive current. A core-type transformer which has its primary and secondary windings on opposite sides of the core and is designed to give excessive leakage has a characteristic of this kind. A transformer designed in this way, if operated on the drooping part of its characteristic, gives a considerable range of voltage at nearly constant current. The characteristic of a transformer which has great magnetic leakage is shown in Fig. 55.

Between a and b on the characteristic there is a large change in voltage with a comparatively small change in current. If the leakage reactance can be increased automatically as the current tends to increase, the

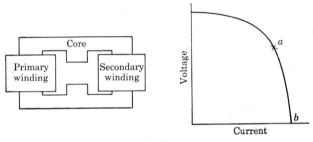

Fig. 55.

transformer can be made to regulate for constant current throughout any desired range of load.

The necessary automatic increase in the reactance is obtained in the constant-current transformer by arranging the primary and secondary windings so that they can move relatively to each other. The increase in the repulsion between the two windings, produced by an increase in the current, causes them to move apart, to increase the cross section of the path for magnetic leakage and thus to increase the leakage reactance.

The simple arrangement by which this is usually accomplished is shown in Fig. 56. CCC is the iron core which should be long and the central leg of which should operate at relatively high density in order to increase the magnetic leakage between the windings as they move apart. A and B are the primary and secondary windings. The secondary winding B is movable and is supported from an arm pivoted at D. A weight W, which is hung from the sector S attached to the swinging arm, partially counterbalances the combined weights of the secondary winding and the arm.

Due to the force of repulsion between the two windings caused by

the primary and secondary currents, the winding B moves away from
the winding A until the force of repulsion is just equal to the unbalanced
weight of the arm and the winding.   If the impedance of the external
circuit is diminished, the current increases and the winding B moves
farther away from A, increasing the leakage reactance and diminishing

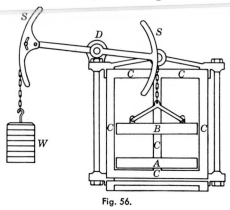

Fig. 56.

the current.   By properly adjusting the counterweight W, the shape of
the sectors, and the angle at which they are set, the transformer can be
made to regulate for nearly constant current over any desired range of
load, provided the core is long enough to allow the windings to get
sufficiently far apart at no load, $i.e.$, at short circuit on the secondary.
The maximum load is that for which the windings come in contact.

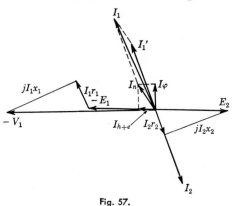

Fig. 57.

Since the secondary current is constant in a properly adjusted constant-current transformer, the load component of the primary current
is also constant.   If it were not for the variation in the exciting current
with change in the relative positions of the windings on change in load,
the primary current would be constant.   The primary winding operates

at constant voltage and nearly constant current, and the change in input is caused by a change in the primary power factor. The secondary winding delivers power at constant current and variable voltage and at a power factor determined by the constants of the load.

The method by which a constant-current transformer regulates for constant current should be made clear by inspecting Figs. 57 and 58. Figure 57 is for no load, *i.e.*, short circuit; Fig. 58 is for a large inductive

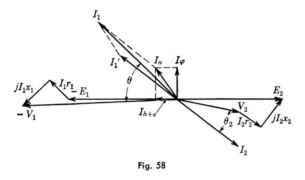

Fig. 58

load. The regulation is secured by the change in leakage-reactance drop with change in load.

A constant-current transformer is started with the secondary winding lifted to its highest position. After the primary circuit has been closed, the secondary winding is released and allowed to take the position corresponding to the load.

Autotransformer. In addition to the type of transformer in which the primary and secondary windings are independent, there is a type known as the autotransformer which has a single continuous winding, a portion of which may be considered to serve as both primary and secondary. The size of wire used for the continuous winding is not the same throughout unless the ratio of transformation is such that its two parts carry equal currents. The arrangement of the autotransformer

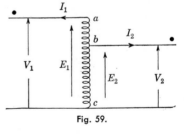

Fig. 59.

should be made clear by Fig. 59, which shows schematically a step-down transformer. If, however, the source were applied to terminals b and c, the same transformer could be used equally well to step up the voltage, in which case primary and secondary nomenclature would be interchanged. The portion of the winding between a and b is frequently called the series winding and the portion between b and c, which is common to both primary and secondary, the common winding.

Since part of the winding serves for both primary and secondary, an autotransformer requires less material and therefore is cheaper than a two-winding transformer of the same output and efficiency. The saving, however, is substantial only when the ratio of transformation is small. Since the primary and secondary windings of an autotransformer are in electrical connection, the use of autotransformers is limited to those places where the electrical connection between the low-voltage winding and a high-potential circuit is not objectionable.

In the analysis which follows, the same conventions with regard to positive directions of the fluxes, currents, and voltages will be adopted as were used for two-winding transformers. The directions indicated in Fig. 59 are in accordance with these conventions.

Since all the turns of the autotransformer are linked by the same mutual flux, the voltage induced per turn is the same throughout the winding. If N_1 and N_2 are the number of turns on the winding between a and c and between b and c, respectively, then

$$\frac{E_1}{E_2} = \frac{N_1}{N_2} = a \tag{66}$$

If the secondary circuit is closed, a current I_2 flows to the load. Since, as in the two-winding transformer, the mutual flux is considered to be produced by the magnetizing component of the primary current, then if this exciting current is neglected, the resultant magnetizing effect of the current is zero. Therefore, $I_sN_s + I_cN_c = 0$, where the subscripts s and c are used to indicate the series and the common windings, respectively. From Fig. 59, $I_s = I_1$, $I_c = I_1 + I_2$, $N_s = N_1 - N_2$. and $N_c = N_2$. Therefore $I_1(N_1 - N_2) + (I_1 + I_2)N_2 = 0$, and

$$\frac{I_1}{I_2} = -\frac{N_2}{N_1} = -\frac{1}{a} \tag{67}$$

Again referring to Fig. 59, we may write the following equations:

$$V_1 = E_1 - I_1z_s - (I_1 + I_2)z_c \tag{68}$$

and

$$E_2 = V_2 + (I_1 + I_2)z_c \tag{69}$$

where z_s and z_c are the leakage impedances of the series and the common windings, respectively.

From Eqs. (66) and (69)

$$E_1 = aE_2 = aV_2 + (I_1 + I_2)az_c$$

Substituting this value of E_1 in Eq. (68) and simplifying

$$V_1 = aV_2 + (I_1 + I_2)az_c - I_1z_s - (I_1 + I_2)z_c$$
$$= aV_2 - I_1z_s + (I_1 + I_2)(a - 1)z_c$$

Since from Eq. (67),

$$I_1 = -\frac{1}{a}I_2 \quad \text{and} \quad I_1 + I_2 = -\frac{1}{a}I_2 + I_2 = I_2\frac{a-1}{a},$$

then

$$V_1 = aV_2 + \frac{I_2}{a}z_s + \frac{I_2}{a}(a-1)^2z_c$$

$$= aV_2 + \frac{I_2}{a}[z_s + (a-1)^2z_c] \tag{70}$$

Thus the autotransformer has an equivalent impedance as indicated by the brackets in Eq. (70), and the equivalent resistance r_e and the equivalent reactance x_e, when referred to the primary side, are given by the following equations:

$$r_e = r_s + (a-1)^2r_c \tag{71}$$

and

$$x_e = x_s + (a-1)^2x_c \tag{72}$$

The equivalent impedance may be measured, as in the case of the two-winding transformer, by shorting the secondary terminals and applying a reduced voltage to the primary terminals. The voltage could as well be applied to the series winding since terminals b and c (Fig. 59) are at the same potential when the secondary terminals are shorted. If the exciting component of the primary current is neglected, the vector diagram of the autotransformer is the same as Fig. 46, page 60, which was drawn for a two-winding transformer.

Relative Efficiencies of an Autotransformer and a Two-winding Transformer. In many cases either an autotransformer or a two-winding transformer may be used to accomplish the same transformation. In such a case they would have the same primary and secondary voltage ratings and equal kva ratings, which would mean also the same rated values of primary and secondary currents. For purposes of comparison, the following assumptions will be made with respect to the design of the autotransformer.

1. Its core is identical with that of the corresponding two-winding transformer.

2. The number of turns used in its winding equals that of the high-voltage winding of the two-winding transformer. The two transformers would then operate at the same flux density and their core losses would be equal.

3. The wire size for the series winding is the same as that used in the high-voltage winding of the two-winding transformer. Since, at rated load, these windings carry equal currents, this is likely to be the case.

4. The wire size for the common winding is so chosen that at rated load

the current density in this winding equals the current density in the low-voltage winding of the two-winding transformer at the same load.

5. The mean length of a turn for the autotransformer is assumed to be the same as that for the two-winding transformer. Because of the fewer turns required it is likely that the autotransformer would actually have a somewhat smaller value.

On the basis of the above assumptions, and referring to Fig. 59, page 89, and the nomenclature previously used, the copper loss at rated load in the series winding equals $I_s{}^2r_s$. The copper loss in the primary of the two-winding transformer equals $I_1{}^2r_1$. Since

$$r_s = r_1 \frac{N_1 - N_2}{N_1} = r_1 \frac{a - 1}{a}$$

then

$$I_s{}^2r_s = I_1{}^2r_1 \frac{a - 1}{a} \tag{73}$$

The current in the common winding is $I_1 + I_2$. If we neglect the exciting component of the primary current, then

$$I_1 = -I_2 \frac{N_2}{N_1} = -\frac{I_2}{a}$$

and

$$I_c = I_1 + I_2 = -\frac{I_2}{a} + I_2 = I_2 \frac{a - 1}{a}$$

If A_c and A_2 are, respectively, the cross-section area of copper used in the common winding and in the secondary winding of the two-winding transformer, then for equal current densities

$$A_c = A_2 \frac{I_c}{I_2}$$

and since both windings consist of N_2 turns with the same mean length of turn

$$r_c = r_2 \frac{A_2}{A_c} = r_2 \frac{I_2}{I_c} = r_2 \frac{a}{a - 1}$$

The copper loss in the common winding equals $I_c{}^2r_c$ and that of the secondary of the two-winding transformer equals $I_2{}^2r_2$.

$$I_c{}^2r_c = I_2{}^2 \left(\frac{a - 1}{a}\right)^2 r_2 \frac{a}{a - 1} = I_2{}^2r_2 \frac{a - 1}{a} \tag{74}$$

Since the copper loss in the series winding of the autotransformer equals that of the primary of the two-winding transformer times $(a - 1)/a$ and the same relationship holds between the common winding and the secondary of the ordinary transformer, then the total copper loss of the autotransformer equals the total copper loss of the two-winding transformer times $(a - 1)/a$ or

$$I_s{}^2r_s + I_c{}^2r_c = (I_1{}^2r_1 + I_2{}^2r_2) \frac{a - 1}{a} \tag{75}$$

Moreover it can be shown that the total amount of copper used in the autotransformer is related to that of the two-winding transformer through the same ratio, $(a - 1)/a$. We find, therefore, that the autotransformer not only is the more efficient but is likely to cost less because of the saving in copper. An exactly similar saving is obtained if a step-up transformer is used. It will be noted, however, that the reduction in copper loss and in the amount of material used decreases as the ratio of the transformation is increased.

Although the assumptions used may depart somewhat from practice, good design is likely to result in a comparison even more favorable to the autotransformer. For example, because of its reduced losses, the kva rating of the autotransformer as designed above might be increased, thereby still further increasing its full-load efficiency, or the size of wire or even the size of the core could be reduced and still keep the losses and heating within the limits of those used for the two-winding transformer, resulting in smaller size and a saving in material and cost.

Briefly, the advantages of the autotransformer are better regulation and efficiency and lower cost for the same output. Its disadvantages are that its low-voltage winding is in electrical connection with the high-voltage winding and forms part of it and that, because of its reduced impedances, larger currents result from a short-circuit on its secondary side. Moreover, the advantages of the autotransformer decrease rapidly as the ratio of transformation is increased and for practical purposes disappear for ratios of transformation greater than 5.

One of the principal uses of autotransformers is to obtain reduced voltages for starting polyphase motors. Autotransformers are also used to raise or lower the voltage of a three-phase system or to tie together two three-phase systems of different voltages. In the smaller sizes a variable ratio autotransformer, known as a variac, is often used to obtain an adjustable voltage. In this application a sliding contact at connection b, Fig. 59, page 89, is used, thereby allowing the output voltage to be varied.

Induction Regulator. When several circuits are fed from one central station or from a single generator, it is often necessary to have some means of regulating independently the voltages of the different circuits. The induction regulator is commonly used for this purpose.

The induction regulator is essentially a step-down transformer with one of its windings mounted in such a way that it can be rotated into different positions with respect to the other winding. When the axes of the two windings are coincident, the maximum voltage is induced in the secondary winding. When they are at right angles, the secondary voltage is zero. Any intermediate voltage can be obtained. The primary winding of the regulator is placed across the circuit to be regulated. The secondary is placed in series with the circuit beyond the point at which the primary is connected. The regulator either adds to or subtracts from the voltage of the circuit according to the relative position of the two windings of the regulator. A diagram of connections of a single-phase induction regulator is given in Fig. 60. *PP* and *SS* are the primary and secondary windings. *CC* is a short-circuited wind-

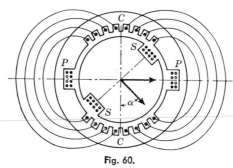

Fig. 60.

ing. The secondary winding *SS* can be placed in different positions with respect to the primary winding *PP* by rotating about its axis the core carrying the secondary winding *SS*. So far as the operation of the induction regulator is concerned, either the primary winding or the secondary winding can be stationary.

When the axes of the two windings of a single-phase induction regulator are at right angles, there can be no mutual induction between them. Under this condition the secondary acts like an impedance in series with the line. To prevent this, it is necessary to provide the single-phase regulator with a short-circuited winding on the core which carries the primary winding, and to have the axis of the short-circuited winding placed at right angles to the axis of the primary winding. When the primary and secondary windings are at right angles, this short-circuited or compensating winding acts, with respect to the secondary of the regulator, like the secondary of a short-circuited transformer and neutralizes the reactance of the secondary winding of the regulator. The only voltage drops across the secondary of the regulator are the equivalent resistance and the equivalent leakage-reactance drops of the secondary and compensating windings.

The compensating winding cannot have any effect on the primary winding since the axes of the two windings are at right angles. When the axes of the primary and secondary windings make any angle α with each other, the effect is the same as if the actual secondary winding were replaced by two windings in series, one (A) with its axis coincident with the axis of the primary winding and the other (B) with its axis perpendicular to the axis of the primary winding, hence coincident with the axis of the compensating winding. Let the effective turns in the primary, secondary, and compensating windings be N_1, N_2, and N_c. Then the number of turns in the two windings which replace the actual secondary winding are $N_2 \cos \omega$ for winding A and $N_2 \sin \alpha$ for winding B. The winding A and the primary winding together act like a loaded transformer with N_1 primary turns and $N_2 \cos \omega$ secondary turns and a ratio of

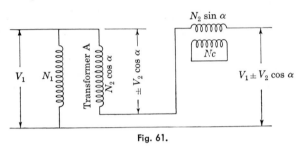

Fig. 61.

transformation $N_1/(N_2 \cos \alpha)$. This transformer has its primary across a circuit of which the voltage is to be regulated and its secondary in series with this circuit. The winding B and the compensating winding together act like a short-circuited transformer with $N_2 \sin \alpha$ primary turns and N_c secondary turns and a ratio of transformation $(N_2 \sin \alpha)/N_c$. This short-circuited transformer has its primary in series with the circuit to be regulated. The second of the two transformers, which replace the actual induction regulator, inserts a small impedance in series with the line to be regulated. This impedance is the equivalent impedance of a short-circuited transformer. Without the compensating winding to act as a short-circuited secondary, the impedance inserted in the line by the second transformer would be the large primary impedance of an open-circuited transformer. The equivalent circuit showing the windings of the two transformers which replace the induction regulator is given in Fig. 61.

Refer to Fig. 61. When $\alpha = 0$, $N_2 \cos \alpha = N_2$ and the voltage inserted in the line by transformer A is a maximum. Under this condition, $N_2 \sin \alpha$ is zero and the short-circuited transformer has no effect. The voltage drop across its terminals is zero since the number of turns in its primary winding is zero. There is no current in the compensating winding.   When $\alpha = 90$ deg, $N_2 \cos \alpha$ is zero and transformer A is simply an impedance shunted across the primary side of the induction regulator.   Under this condition, $N_2 \sin \alpha = N_2$ and the short-circuited transformer B has its maximum effect.   Even in this case the impedance drop across its primary is small since it is only the equivalent impedance drop of a short-circuited transformer.   Any voltage from zero to the maximum voltage of transformer A can be inserted in series with the circuit by giving the angle α the proper value.   If V_1 is the voltage impressed on the primary winding of the compensator and V_2 is the voltage of the secondary winding when the angle α is zero, the voltage range of the induction regulator is $V_1 \pm V_2$.   To give a voltage range of 20 per cent, 10 per cent above and 10 per cent below the normal voltage of the circuit in which the induction regulator is used, the kva rating of the regulator needs to be only 10 per cent of the kva rating of the circuit.

The polyphase regulator is similar in principle to the single-phase regulator.   It is a polyphase induction motor which has its armature blocked but capable of being placed in different positions with respect to its field winding.   The polyphase induction regulator requires no compensating winding.   The voltages given by its secondary windings are constant, but their phase relations with respect to the line voltages can be varied by moving the secondary windings with respect to the primary. The secondary voltages can be made to add vectorially at any desired phase angle to the corresponding line voltages.   The action of a polyphase induction regulator depends upon the revolving magnetic field of a polyphase induction motor and will be understood more clearly after considering that type of motor (see page 398).

Chapter 9

Three-winding transformers. Parallel operation of single-phase transformers.

Three-winding Transformers. Two-winding transformers have been considered up to this point. Such a transformer has a single primary winding and a single secondary winding or their equivalents. Many transformers have two identical primary windings and two identical secondary windings. If the two primary windings operate either in series or in parallel and the two secondary windings also operate either in series or in parallel, the transformer is equivalent to a two-winding transformer so far as its operation is concerned. When two identical windings on a transformer are connected in series,they are equivalent to a single winding of twice the number of turns and twice the voltage rating, with twice the resistance and twice the leakage reactance of each of the two identical windings. When two identical windings are connected in parallel, they are equivalent to a single winding with the same number of turns as each of the two windings, the same voltage rating, and with half the resistance and half the leakage reactance of each winding. These statements are true only when the assumption holds which is made in the consideration of two-winding transformers, that the effect on the core flux of a given number of ampere-turns of a winding is independent of the position of the winding on the core. This assumption is nearly true for commercial iron-core transformers. According to this assumption, if the exciting current is neglected, the arithmetical sum of the ampere-turns in all windings at any instant is zero, and if the currents are sinusoidal, the vector sum of the ampere-turns in all windings is zero.

When two secondary windings of a single-phase transformer are connected in series to feed a three-wire system, the two secondaries in general are unequally loaded and carry currents which are unequal in magnitude and which may be out of phase. The operation of the transformer is in no way influenced by the common connection between the two secondaries. The two secondary windings are independently loaded and the transformer operates as a three-winding transformer. A transformer may have any number of independent secondary windings. In such a case it is a multiwinding transformer. There are many cases where multiwinding transformers are used. Transformers with two or more secondaries are used to supply two or more distribution circuits of the same voltage or of different voltages from the same source of power. In certain polyphase connections, as the T connection discussed in Chap. 10, parts of the same winding carry different currents. Single-phase transformers used for certain three-phase transformations have three independent windings: a primary winding, a secondary winding, and a tertiary winding. The tertiary windings are connected in Δ. If the primary windings and the secondary windings are each connected in Y, the Δ-connected tertiaries provide a short-circuited path for the third-harmonic components in the exciting currents which are eliminated by the Y-connected primaries and which are necessary for a sinusoidal variation of the core fluxes. The Δ-connected tertiaries can also be used to feed local circuits in a power station. If an unbalanced load is connected between the line terminals and the neutral of the Y-connected secondaries, the tertiary windings carry a zero-sequence current. Transformers operating under these conditions must be treated as three-winding transformers. Because of the many uses of three-winding transformers, such transformers are considered somewhat in detail.

Although the following discussion is limited to three-winding transformers, it can be extended to transformers with more than three windings. In what follows it is assumed, just as in the discussion of two-winding transformers, that the effect on the core flux of a given number of ampere-turns in any winding is independent of the position of that winding on the core.

The fundamental equations of a three-winding transformer involve three self-inductances and three mutual inductances: the self-inductance of each of the three windings and the mutual inductance between each pair of the three windings. It must be remembered that the mutual inductance between any two windings such as A and B is the same whether it is considered as the mutual inductance of A on B or of B on A. The small letters v and i are used for instantaneous voltage and instantaneous current. Subscripts attached to letters indicate the circuit considered. The L's are the self-inductances and the M's are the mutual inductances of the windings. $M_{12} = M_{21}$ is the mutual inductance of winding 1 on winding 2 or of 2 on 1. $M_{23} = M_{32}$ is the mutual inductance of winding 2 on winding 3 or of 3 on 2. $M_{31} = M_{13}$ is the mutual inductance of winding 3 on winding 1 or of 1 on 3.

The fundamental equations for a three-winding transformer are

$$-v_1 = i_1 r_1 + L_1 \frac{di_1}{dt} + M_{21} \frac{di_2}{dt} + M_{31} \frac{di_3}{dt} \tag{76}$$

$$-v_2 = i_2 r_2 + L_2 \frac{di_2}{dt} + M_{32} \frac{di_3}{dt} + M_{12} \frac{di_1}{dt} \tag{77}$$

$$-v_3 = i_3 r_3 + L_3 \frac{di_3}{dt} + M_{13} \frac{di_1}{dt} + M_{23} \frac{di_2}{dt} \qquad (78)$$

If the effect of saturation is neglected and the current and voltage waves are assumed to be sinusoidal, Eqs. (76), (77), (78) can be written in vector form.

$$-V_1 = I_1 r_1 + jI_1 \omega L_1 + jI_2 \omega M_{21} + jI_3 \omega M_{31} \qquad (79)$$
$$-V_2 = I_2 r_2 + jI_2 \omega L_2 + jI_3 \omega M_{32} + jI_1 \omega M_{12} \qquad (80)$$
$$-V_3 = I_3 r_3 + jI_3 \omega L_3 + jI_1 \omega M_{13} + jI_2 \omega M_{23} \qquad (81)$$

Owing to the presence of the iron core, these equations cannot be used directly for the iron-core transformer because the self-inductances and mutual inductances are not constant and are not a function of any one current alone but depend on the currents in all the windings since the saturation of the core at any instant depends on the combined action of all the currents. However, a simple transformation similar to that used in the discussion of the two-winding iron-core transformer (page 34) makes the equation applicable to an iron-core transformer.

In order to simplify the discussion of the three-winding transformer, assume that all the constants, the voltages, and the currents are referred to a common base, *i.e.*, all are referred to the same winding by the use of proper reduction factors. For the mutual inductances, the factor is the inverse ratio of the number of turns in the winding considered to the number of turns in the winding to which the constants are referred. For the self-inductances and for the resistances, the factor is the square of the ratio of turns. For the voltages, the factor is the ratio of turns. For the currents, the factor is the inverse ratio of turns. Referring the constants, the currents, and the voltages to a common base is equivalent to replacing the actual transformer by an equivalent transformer with the same number of turns in all three windings.

Let M_c be that part of the mutual inductance due to the flux which is wholly in the iron core. M_c is the same for all windings since it is referred to a common base. In Eqs. (79), (80), (81) add to and subtract from each term involving inductance the term $j\omega M_c$ multiplied by the current appearing in the term. This does not change the values of the equations and gives the following results:

$$-V_1 = I_1 r_1 + jI_1 \omega(L_1 - M_c) + jI_2 \omega(M_{21} - M_c) + jI_3 \omega(M_{31} - M_c)$$
$$+ j(I_1 + I_2 + I_3)\omega M_c \qquad (82)$$
$$-V_2 = I_2 r_2 + jI_2 \omega(L_2 - M_c) + jI_3 \omega(M_{32} - M_c) + jI_1 \omega(M_{12} - M_c)$$
$$+ j(I_1 + I_2 + I_3)\omega M_c \qquad (83)$$
$$-V_3 = I_3 r_3 + jI_3 \omega(L_3 - M_c) + jI_1 \omega(M_{13} - M_c) + jI_2 \omega(M_{23} - M_c)$$
$$+ j(I_1 + I_2 + I_3)\omega M_c \qquad (84)$$

The last term in each equation is the voltage induced by the core flux, *i.e.*, by the flux which lies entirely in the iron core. This flux is produced by the simultaneous action of the three currents and induces equal voltages in the three windings since the windings are all referred to a common base. The actual voltages are proportional to the numbers of turns in the windings. Call this core-flux voltage E_c.

The voltage drop $j\omega I_1(L_1 - M_c)$ is the voltage induced in winding 1 by that part of the flux due to I_1 which lies entirely or partially outside the iron core. It does not include the voltage produced in winding 1 by any part of the flux due to I_1 which lies entirely within the iron core, since this voltage is included in E_c. The quantity $\omega(L_1 - M_c)$ is the core-leakage reactance of winding 1. The flux corresponding to the core-leakage reactance lies so largely in air that it can be considered to be proportional to the current producing it, and the core-leakage reactance of a winding can therefore be assumed to be constant. The core-leakage reactance of a winding must not be confused with the leakage reactance of the winding with respect to another winding. The leakage reactance of winding 1 with respect to winding 2 is caused by that part of the flux due to I_1 which does not link winding 2. It may produce linkages with winding 3 or with other windings if there are more than three. The leakage reactance of one winding with respect to another is the reactance used in the treatment of the two-winding transformer. The flux causing the core-leakage reactance of a winding may give linkages with all windings but it must not lie entirely within the core. The flux which lies entirely within the core gives rise to the induced voltage E_c.

The quantity $\omega(M_{21} - M_c)$ is the core-mutual-leakage reactance of winding 1 with respect to winding 2 or of winding 2 with respect to winding 1. It differs from the mutual reactance between windings 1 and 2 by being caused by only that portion of the mutual flux of those two windings that does not lie entirely within the core. The flux corresponding to the core-mutual-leakage reactance between any two windings lies so largely in air that this reactance can be assumed constant. Let

$$x_{11} = \omega(L_1 - M_c) = \text{core-self-leakage reactance of winding 1}$$
$$x_{22} = \omega(L_2 - M_c) = \text{core-self-leakage reactance of winding 2}$$
$$x_{33} = \omega(L_3 - M_c) = \text{core-self-leakage reactance of winding 3}$$

$x_{12} = x_{21} = \omega(M_{12} - M_c) = $ core-mutual-leakage reactance between windings 1 and 2 or 2 and 1

$x_{23} = x_{32} = \omega(M_{23} - M_c) = $ core-mutual-leakage reactance between windings 2 and 3 or 3 and 2

$x_{13} = x_{31} = \omega(M_{13} - M_c) = $ core-mutual-leakage reactance between windings 1 and 3 or 3 and 1

The equations can now be written in the following form by making use of the above relations:

$$-V_1 = I_1 r_1 + jI_1 x_{11} + jI_2 x_{12} + jI_3 x_{13} + E_c \qquad (85)$$
$$-V_2 = I_2 r_2 + jI_2 x_{22} + jI_3 x_{23} + jI_1 x_{12} + E_c \qquad (86)$$
$$-V_3 = I_3 r_3 + jI_3 x_{33} + jI_1 x_{31} + jI_2 x_{23} + E_c \qquad (87)$$

If the exciting current is neglected, the vector sum of the three currents I_1, I_2, I_3 is zero.

$$I_1 + I_2 + I_3 = 0 \qquad (88)$$

Equation (88) assumes that all three currents are referred to the same winding by the use of the proper reduction factors. This is the assumption made in all the preceding equations.

The operation of the three-winding transformer can be computed from Eqs. (85), (86), (87), (88), provided the constants are known.

Eliminate I_3 from Eqs. (85) and (86) by the use of Eq. (88) and then subtract $-V_1$ from $-V_2$.

$$V_1 - V_2 = -I_1[r_1 + j(x_{11} - x_{13} - x_{12} + x_{23})]$$
$$+ I_2[r_2 + j(x_{22} - x_{23} - x_{12} + x_{13})] \quad (89)$$

Eliminate I_1 from Eqs. (86) and (87) by the use of Eq. (88) and then subtract V_3 from V_2.

$$V_2 - V_3 = -I_2[r_2 + j(x_{22} - x_{12} - x_{23} + x_{13})]$$
$$+ I_3[r_3 + j(x_{33} - x_{13} - x_{23} + x_{12})] \quad (90)$$

Eliminate I_2 from Eqs. (85) and (87) by the use of Eq. (88) and then subtract V_1 from V_3.

$$V_3 - V_1 = -I_3[r_3 + j(x_{33} - x_{23} - x_{13} + x_{12})]$$
$$+ I_1[r_1 + j(x_{11} - x_{12} - x_{13} + x_{23})] \quad (91)$$

The reactance x_{11} is ω times that portion of the flux linkages with winding 1 per ampere in that winding that are due to flux which does not lie entirely within the core (see page 100). The reactance x_{12} is ω times the mutual-flux linkages between windings 1 and 2 per ampere in winding 1 that are due to flux which does not lie entirely within the core. The difference between x_{11} and x_{12} is caused by the leakage flux of winding 1 with respect to winding 2 and is the true leakage reactance of winding 1 with respect to winding 2. A similar line of reasoning shows that the true leakage reactance of any winding A with respect to another winding B is the self-core-leakage reactance of winding A minus the mutual-core-leakage reactance of winding A with respect to winding B. Let the true leakage reactance of winding 1 with respect to winding 2 be represented by $x_{1(2)}$. Let the other true leakage reactances be similarly represented. Then,

$x_{1(2)} = x_{11} - x_{12} =$ true leakage reactance of winding 1 with respect to winding 2

$x_{1(3)} = x_{11} - x_{13} =$ true leakage reactance of winding 1 with respect to winding 3

$x_{2(1)} = x_{22} - x_{12} =$ true leakage reactance of winding 2 with respect to winding 1

$x_{2(3)} = x_{22} - x_{23} =$ true leakage reactance of winding 2 with respect to winding 3

$x_{3(1)} = x_{33} - x_{13} =$ true leakage reactance of winding 3 with respect to winding 1

$x_{3(2)} = x_{33} - x_{23} =$ true leakage reactance of winding 3 with respect to winding 2

By making use of the relations given, the reactance parts of Eqs. (89), (90), (91) can be written in terms of the true leakage reactances of the windings.

$$
\begin{aligned}
[(x_{11} - x_{13}) - x_{12} + x_{23}] &= [(x_{11} - x_{12}) - x_{13} + x_{23}] \\
&= x_{1(3)} + x_{2(1)} - x_{2(3)} \\
&= x_{1(2)} + x_{3(1)} - x_{3(2)} \\
&= X_1 \\
[(x_{22} - x_{23}) - x_{12} + x_{13}] &= [(x_{22} - x_{12}) - x_{23} + x_{13}] \\
&= x_{2(3)} + x_{1(2)} - x_{1(3)} \\
&= x_{2(1)} + x_{3(2)} - x_{3(1)} \\
&= X_2 \\
[(x_{33} - x_{13}) - x_{23} + x_{12}] &= [(x_{33} - x_{23}) - x_{13} + x_{12}] \\
&= x_{3(1)} + x_{2(3)} - x_{2(1)} \\
&= x_{3(2)} + x_{1(3)} - x_{1(2)} \\
&= X_3
\end{aligned}
$$

It should be noted that the reactance X_1 used with the current in winding 1, or assigned to winding 1, is equal to the true leakage reactance of winding 1 with respect to winding 3 plus the difference between the true leakage reactances of winding 2 with respect to winding 1 and with respect to winding 3; $i.e.$, it is equal to the true leakage reactance of winding 1 with respect to winding 2 plus the difference between the true leakage reactances of winding 3 with respect to winding 1 and with respect to winding 2. Similar relations exist between X_2 and X_3 and the true leakage reactances.

Let

$$r_1 + jX_1 = Z_1$$
$$r_2 + jX_2 = Z_2$$
$$r_3 + jX_3 = Z_3$$

These are the impedances used with the currents 1, 2, 3. By using these symbols, Eqs. (89), (90), (91) become

$$V_1 - V_2 = -I_1(r_1 + jX_1) + I_2(r_2 + jX_2) = -I_1Z_1 + I_2Z_2 \quad (92)$$
$$V_2 - V_3 = -I_2(r_2 + jX_2) + I_3(r_3 + jX_3) = -I_2Z_2 + I_3Z_3 \quad (93)$$
$$V_3 - V_1 = -I_3(r_3 + jX_3) + I_1(r_1 + jX_1) = -I_3Z_3 + I_1Z_1 \quad (94)$$

If one winding of the three-winding transformer is open, the equations for the three-winding transformer reduce to those for a two-winding

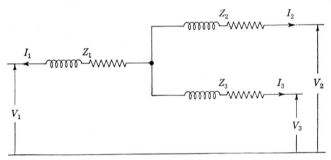

Fig. 62.

transformer. For example, let winding 3 be open. Under this condition the current in winding 3 is zero. Since the exciting current is neglected and all currents are referred to the same winding,

$$I_1 + I_2 + I_3 = I_1 + I_2 + 0$$
$$-I_1 = +I_2$$

From Eq. (92)

$$V_1 = V_2 - I_1Z_1 + I_2Z_2$$
$$= V_2 + I_2(Z_1 + Z_2)$$
$$= V_2 + I_2[(r_1 + r_2) + j(x_{1(2)} + x_{3(1)} - x_{3(2)})$$
$$\qquad\qquad\qquad\qquad + j(x_{2(1)} + x_{3(2)} - x_{3(1)})]$$
$$= V_2 + I_2[(r_1 + r_2) + j(x_{1(2)} + x_{2(1)})] \quad (95)$$

The same result is obtained by eliminating V_3 between Eqs. (93) and (94). In Eq. (95), $(r_1 + r_2)$ is the equivalent resistance of windings 1 and 2 and $(x_{1(2)} + x_{2(1)})$ is the equivalent reactance of these windings.

Equations (92), (93), (94), and (95) are exactly represented by the equivalent circuit of Fig. 62. This figure, therefore, represents the equivalent circuit of the equivalent three-winding transformer where all values are referred to a common winding. If it is desired to take into account the effect of the exciting current I_n, a branch circuit connected between the junction of the impedances and the return circuit could be inserted to provide a path for this current as in the equivalent circuit of the two-winding transformer (Fig. 41, page 57).

The impedances Z_1, Z_2, Z_3, which appear in Fig. 62, can be found from short-circuit measurements similar to those used for determining the equivalent resistance and the equivalent leakage reactance of a two-winding transformer. With winding 3 open and winding 2 short-circuited, measure the voltage impressed on winding 1, the current in winding 1, and the power taken by winding 1, and call these V_1, I_1, and P_1.
Then

$$Z_1 + Z_2 = \frac{V_1}{I_1} = Z_{12}$$

$$R_1 + R_2 = \frac{P_1}{I_1{}^2} = R_{12}$$

$$X_{12} = \sqrt{(Z_{12})^2 - (R_{12})^2}$$
$$Z_1 + Z_2 = Z_{12} = R_{12} + jX_{12} \tag{96}$$

In a similar manner,

$$Z_2 + Z_3 = Z_{23} = R_{23} + jX_{23} \tag{97}$$
$$Z_3 + Z_1 = Z_{31} = R_{31} + jX_{31} \tag{98}$$

From Eqs. (96), (97), (98),

$$\begin{aligned}
r_1 + jX_1 = Z_1 &= 0.5(Z_{12} + Z_{13} - Z_{23}) \\
&= 0.5[(R_{12} + R_{13} - R_{23}) + j(X_{12} + X_{13} - X_{23})] \quad (99)
\end{aligned}$$
$$\begin{aligned}
r_2 + jX_2 = Z_2 &= 0.5(Z_{12} + Z_{23} - Z_{13}) \\
&= 0.5[(R_{12} + R_{23} - R_{13}) + j(X_{12} + X_{23} - X_{13})] \quad (100)
\end{aligned}$$
$$\begin{aligned}
r_3 + jX_3 = Z_3 &= 0.5(Z_{13} + Z_{23} - Z_{12}) \\
&= 0.5[(R_{13} + R_{23} - R_{12}) + j(X_{13} + X_{23} - X_{12})] \quad (101)
\end{aligned}$$

The constants are all referred to winding 1.

The equations (92), (93), (94) can be solved as simultaneous vector equations provided the terminal conditions are known. For example, if V_1 is known and the impedances Z_{L2} and Z_{L3} of the loads across the terminals of windings 2 and 3 are known, V_2 and V_3 can be replaced in Eqs. (92), (93), (94), by $I_2 Z_{L2}$ and $I_3 Z_{L3}$. There remain only three unknown quantities with three vector equations to determine them. The load constants must be referred to the same winding as the other constants.

Parallel Operation of Single-phase Transformers. The conditions which must be fulfilled for the satisfactory parallel operation of transformers are the following.

1. The secondary currents must all be zero when the load on the system is zero.

2. The secondary current carried by each transformer must be proportional to its rating.

3. The secondary currents must be in phase with one another and therefore must be in phase with the current taken by the load on the system.

Whether or not the conditions for the parallel operation of transformers are fulfilled depends upon the ratios of transformation and the constants of the transformers. Transformers cannot be paralleled indiscriminately even if their ratios of transformation are the same.

The same voltage is impressed on the primaries of all transformers operating in parallel, and all primary terminal voltages must be equal and in phase. Similarly, since all the secondaries are connected in parallel, the secondary terminal voltages must be equal and in phase. If the ratios of transformation are the same, the primary voltages referred to the secondary side are equal and in phase, and if the exciting currents are neglected, the equivalent impedance drops of all transformers are equal and in phase since they must form the closing side of a voltage triangle which has for its other two sides the common impressed primary voltage referred to the secondary windings and the common secondary terminal voltage.

Transformers with Equal Ratios of Transformation. The current a transformer delivers when in parallel with others depends upon its equivalent impedance and not upon the way in which resistance and reactance are distributed between its primary and secondary windings.

Let the secondary currents delivered by any number of transformers connected in parallel be $I_2', I_2'', I_2''', \ldots$, and let the equivalent impedances and the equivalent admittances all referred to the secondary windings be $z_e', z_e'', z_e''', \ldots$ and $y_e', y_e'', y_e''', \ldots$. Then vectorially, if the exciting component of the primary current is neglected,

$$\frac{V_1}{a} = V_2 + I_2 z_e$$

as in Eq. (48), page 60. This applies to all transformers in parallel, and since all transformers must have the same primary voltage V_1 and the same secondary voltage V_2,

$$\frac{V_1}{a} - V_2 = I_2' z_e' = I_2'' z_e'' = I_2''' z_e''' = \ldots \tag{102}$$

and

$$I_2' : I_2'' : I_2''' : \ldots = \frac{1}{z_e'} : \frac{1}{z_e''} : \frac{1}{z_e'''} : \ldots$$
$$= y_e' : y_e'' : y_e''' : \ldots \tag{103}$$

Therefore the currents delivered by the transformers to the load are inversely proportional to their equivalent impedances or directly proportional to their equivalent admittances. The total current I_0 delivered by the system is equal to the vector sum of the component currents delivered by the separate transformers.

$$I_0 = I_2' + I_2'' + I_2''' + \ldots$$

A vector diagram for two transformers having equal ratios of transformation is given in Fig. 63. The no-load current is neglected. The

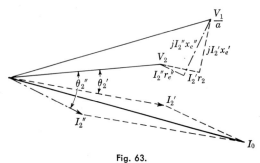

Fig. 63.

impedances on the diagram are in the ratio 2:1. Therefore

$$\frac{I_2'}{I_2''} = \frac{z_e''}{z_e'} = \frac{y_e'}{y_e''} = 2$$

If the two impedances are equal, the currents are also equal. The currents are in phase only when the ratio of the equivalent resistance of each transformer to its equivalent reactance is the same for both transformers.

The ratios of equivalent resistance to equivalent reactance for two transformers can be quite different without causing the vector sum of the output currents to differ much from their arithmetical sum, provided the ratios are small. The angle between the output currents is equal to the difference of the impedance angles. Suppose that the ratios of equivalent resistance to equivalent reactance for the two transformers are

$$\frac{r_e'}{x_e'} = \frac{1}{5} \quad \text{and} \quad \frac{r_e''}{x_e''} = \frac{1}{3}$$

The corresponding impedance angles are $\theta_z' = \tan^{-1} 0.20 = 11.3$ deg and $\theta_z'' = \tan^{-1} 0.33 = 18.4$ deg, and the output currents differ in phase by $18.4 - 11.3 = 7.1$ deg. In this example, where the impedances are in the ratio of 2:1, an angle of 7.1 deg makes the vector sum of the output currents differ by only about 0.8 per cent from their arithmetical sum.

The proper division of load among any number of transformers which operate in parallel is that which causes all the transformers to reach their maximum safe temperatures at the same time. This does not mean necessarily that the current supplied by each is in proportion to the manufacturer's rating, since this rating may be more conservative for some transformers than for others. The maximum safe temperatures of different types may not be the same.

Transformers having the same ratio of transformation can be represented by the equivalent circuit shown in Fig. 64.

The division of load among any number of transformers which have equal ratios of transformation can be found in the following manner. Referring to Fig. 64, let v be the voltage across the parallel admittances,

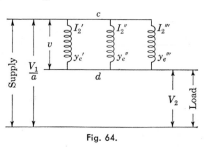

Fig. 64.

$$y'_e = g'_e - jb'_e, \ y''_e = g''_e - jb''_e, \ y'''_e = g'''_e - jb'''_e \ \ldots$$

These represent the equivalent admittances of the transformers. The resultant admittance between the points c and d of the equivalent circuit is

$$y_0 = g_0 - jb_0 = \Sigma g_e - j\Sigma b_e \qquad (104)$$

where $\Sigma g_e = g'_e + g''_e + g'''_e + \ldots$, and $\Sigma b_e = b'_e + b''_e + b'''_e + \ldots$

$$\begin{aligned} I'_2 &= vy'_e = v(g'_e - jb'_e) \qquad (105) \\ I''_2 &= vy''_e = v(g''_e - jb''_e) \\ I'''_2 &= vy'''_e = v(g'''_e - jb'''_e) \end{aligned}$$

$$\cdots \qquad \cdots$$

Since the total load current is

$$I_0 = I'_2 + I''_2 + I'''_2 + \ldots \qquad (106)$$

$$= vy'_e + vy''_e + vy'''_e + \ldots = vy_0$$

$$v = \frac{I_0}{y_0} \qquad (107)$$

$$I'_2 = vy_e = \frac{I_0}{y_0} y'_e \qquad (108)$$

$$I''_2 = \frac{I_0}{y_0} y''_e$$

$$I'''_2 = \frac{I_0}{y_0} y'''_e$$

$$\cdots \qquad \cdots$$

If only numerical values of the currents are desired, the y_e's should be expressed numerically.

It should be noticed that the distribution of the load among transformers with equal ratios of transformation is independent of the load on the system and its power factor.

The equivalent conductance and equivalent susceptance of a transformer can be obtained from the power, current, and voltage measured with the transformer short-circuited. Let the power input, the impressed voltage, and the current be P, V, and I. Then

$$y_e = \frac{I}{V}$$

$$g_e = \frac{P}{V^2}$$

$$b_e = \sqrt{y_e^2 - g_e^2}$$

All three constants are referred to the side of the transformer to which I and V are referred. If the equivalent resistance and equivalent leakage reactance are known, the equivalent conductance and equivalent susceptance can readily be found.

When the ratio of equivalent resistance to equivalent leakage reactance for each transformer is small, little error is made in the magnitude of y_0 [Eq. (106)] in assuming that it is equal to the sum of the magnitudes of the individual equivalent admittances. In such a case negligible error is made in the magnitude of the current delivered to the load by any transformer in assuming that the load current is equal to the magnitude of the total current taken by the load multiplied by the ratio of the magnitude of the equivalent admittance of the transformer to the sum of the magnitudes of the equivalent admittances of all the transformers in parallel.

When transformers do not have impedances in the correct ratio for sharing the load properly, it is possible to adjust the impedances by putting reactance in series with any transformer which has too low an impedance. Adding reactance may change the ratio of equivalent resistance to equivalent reactance, but this ratio has no effect on the relative magnitudes of the currents delivered to the load by transformers in parallel.

It is important to have the equivalent impedances in the correct ratio for transformers which are to operate in parallel in order that the transformers may share the load currents in proportion to the transformer ratings. Any transformer which has too small an equivalent impedance takes more than its proper share of the load current and may overheat before the other transformers are fully loaded. If the ratios of equivalent resistance to equivalent leakage reactance differ, the only effect is to make the currents delivered by the transformers out of phase with one another and with the load current. As a result, the permissible current output of the bank of transformers is less than if the currents were in phase.

If a transformer which is operating in parallel with others has an equivalent impedance which is much too low, it takes much more than its proper share of the load current, and as a bank of transformers which are operating in parallel cannot be safely loaded beyond the point where any transformer delivers much more than its rated current, the low-impedance transformer limits the permissible output of the bank. In

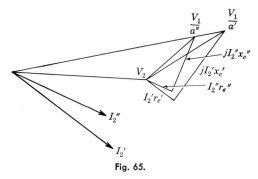

Fig. 65.

an extreme case, a larger output probably could be obtained safely by removing the low-impedance transformer.

Transformers Having Different Ratios of Transformation. Transformers of different designs but having the same nominal ratios of transformation may have actual ratios of transformation which differ slightly.

Differences in the ratios of transformation also occur when transformers which are provided with tap-changing devices are operated in parallel. These devices can be used to adjust the division of load current among the transformers by changing slightly their ratios of transformation by means of the tap-changing devices.

The vector relations which exist when two transformers having different ratios of transformation are put in parallel are shown in Fig. 65. The V's on the diagram are voltage rises.

Let a', a'', a''', . . . be the ratios of transformation of any number of transformers in parallel.

$$I_2' z_e' = \frac{I_2'}{y_e'} = \frac{V_1}{a'} - V_2$$

$$I_2'' z_e'' = \frac{I_2''}{y_e''} = \frac{V_1}{a''} - V_2$$

$$I_2''' z_e''' = \frac{I_2'''}{y_e'''} = \frac{V_1}{a'''} - V_2$$

$$\cdot \quad \cdot \quad \cdot \qquad \cdot \quad \cdot \quad \cdot$$

Solving for the currents,

$$I_2' = \frac{V_1}{a'} y_e' - V_2 y_e' \tag{109}$$

$$I_2'' = \frac{V_1}{a''} y_e'' - V_2 y_e''$$

$$I_2''' = \frac{V_1}{a'''} y_e''' - V_2 y_e'''$$

$$\cdots \qquad \cdots$$

Since the total current I_0 delivered by the system is equal to the vector sum of the component currents delivered by the separate transformers,

$$I_0 = \sum \frac{V_1}{a} y_e - \Sigma V_2 y_e$$

$$= V_1 \sum \frac{y_e}{a} - V_2 y_0 \tag{110}$$

Solving Eq. (110) for V_1,

$$V_1 = \frac{I_0 + V_2 y_0}{\sum \dfrac{y_e}{a}} \tag{111}$$

Substituting V_1 from Eq. (111) in Eq. (109),

$$I_2' = \frac{y_e'}{a'} \frac{I_0 + V_2 y_0}{\sum \dfrac{y_e}{a}} - V_2 y_e'$$

$$= \frac{y_e' I_0}{a' \sum \dfrac{y_e}{a}} + \frac{y_e' V_2}{a'} \left(\frac{y_0}{\sum \dfrac{y_e}{a}} - a' \right) \tag{112}$$

Equation (110) may be solved for V_2, and if this value of V_2 is substituted in Eq. (109), Eq. (113) results. Equation (113) gives I_2' in terms of V_1 and I_0.

$$I_2' = \frac{y_e'}{y_0} I_0 + y_e' V_1 \left(\frac{1}{a'} - \frac{\sum \dfrac{y_e}{a}}{y_0} \right) \tag{113}$$

Equation (112) is convenient for calculations inasmuch as the phase relation between V_2 and I_0 is usually known, it being determined by the power factor of the load. In Eq. (113), on the other hand, the phase angle between V_1 and I_0 is frequently not known. Equation (113), however, is useful for analyzing the effects of unequal ratios of transformation. In this equation, the first term is the current which would be carried if the ratios of transformation were all equal; the second term is the component current caused by differences among the ratios of transformation. If all values of y_e, in polar form, have nearly the same angle, then $\dfrac{\Sigma(y_e/a)}{y_0}$ has an angle close to zero and $\left[\dfrac{1}{a'} - \dfrac{\Sigma(y_e/a)}{y_0}\right]$ is mostly a real term with a very small j component. The entire second term, if positive, therefore represents a current which lags V_1 by nearly the angle of y_e'. If the term is negative, it represents a current which lags $-V_1$ by this same angle. If the equivalent conductance g_e' of the transformer is small compared with its equivalent susceptance b_e', the entire second term is essentially reactive with respect to V_1. The effects of unequal ratios are therefore more pronounced for lagging power-factor loads than for loads having a power factor near unity. It is undesirable to operate transformers in parallel when the component current caused by differences among the ratios of transformation exceeds, in any transformer, 10 per cent of its rated current.

From what precedes, it should be clear that transformers which are to be operated in parallel under ideal conditions should have

1. Equal voltage ratings
2. Equal ratios of transformation
3. Equivalent impedances which are inversely proportional to their current ratings
4. Ratios of equivalent resistance to equivalent reactance which are equal.

These four conditions are stated in the order of their relative importance.

That the transformers should have the same voltage rating needs no explanation. If their voltage ratings are not the same, some are operating on a higher and some on a lower voltage than that for which they are designed.

If the ratios of transformation are not the same, there are currents in the transformers, in addition to the exciting currents, when the load on the system is zero. The magnitudes of these currents depend upon the differences among the ratios of transformation, and the currents cannot be eliminated without redesigning the transformers.

If the impedances are not inversely proportional to the current outputs which produce the maximum safe temperature rises in the transformers, the transformers do not divide the load properly and some become overheated while others are below their safe temperatures, unless the system is operated at less than its total rated capacity.

If the ratios of equivalent resistance to equivalent reactance are not the same for all the transformers, the currents delivered by them are not in phase with one another or with the load current, and the transformers are carrying kilowatt loads which are not proportional to their current loads. As a result, the copper loss in all the transformers and in the system as a whole, for a given load on the system, is greater than it would be if all the currents were in phase. In other words, the maximum safe kilowatt output of the system is diminished. That this last condition is of minor importance when the ratio of equivalent resistance to equivalent leakage reactance is small is shown on page 106.

The last two faults, *i.e.*, impedances not in the proper ratio and unequal ratios of resistance to reactance, can be corrected by inserting the proper amount of resistance or reactance or both on either the primary or the secondary side of the transformers. Unless the ratios of equivalent resistance to equivalent leakage reactance are larger than those ordinarily found in practice, it is sufficient to add enough reactance to the low-impedance transformers to make their equivalent impedances inversely proportional to their kva ratings. If the transformers have tap-changing devices, these can sometimes be used to equalize the currents at the normal operating load or to make them nearly equal.

Chapter 10

Transformer connections for three-phase circuits using three identical transformers. The effect of the suppression of the third-harmonic components in the exciting currents of Y-Y-connected transformers. Unbalanced loads on transformers connected for three-phase transformation. An example of an unbalanced load on a bank of Y-Y-connected transformers with Δ-connected tertiary windings and an isolated primary neutral. Three-phase transformation with two transformers. Three-phase to four-phase transformation and vice versa. Three-phase to six-phase transformation. Two-phase or four-phase to six-phase transformation. Interconnected-star or zigzag connections. Three-phase zigzag connection. Three-phase to twelve-phase transformation.

Transformer Connections for Three-phase Circuits using Three Identical Transformers. *Δ and Y Connections.* When three identical single-phase transformers are used with three-phase circuits, they can be grouped in the following ways: (1) primaries in Δ, secondaries in Δ; (2) primaries in Y, secondaries in Y; (3) primaries in Δ, secondaries in Y; and (4) primaries in Y, secondaries in Δ.

Table 4 shows the open-circuit voltages resulting from the four three-phase transformer connections just stated, if the primary voltages are balanced. In this table, the primary line voltages are taken as unity.

All these arrangements are symmetrical and can give balanced secondary voltages for balanced loads, if the primary impressed voltages are balanced and the transformers are identical, *i.e.*, have equal ratios of transformation, equal equivalent resistances, and equal equivalent leakage reactances.

If the transformers are connected with primaries in Δ and secondaries in Y, any unbalanced Y-connected or Δ-connected load can be applied to the secondaries without unbalancing the secondary voltages by more than can be accounted for by the differences in the small impedance drops in the transformers. This is obvious since each primary can receive power directly from the line. When the transformers are connected Y-Y or Y-Δ, an unbalanced Δ-connected load can be carried without any serious unbalancing of the secondary voltages. Any unbalanced Δ-connected load, *i.e.*, a load without a neutral connection, can be resolved into two balanced three-phase loads of opposite phase rotation.[1] Each of these balanced loads alone does not produce unbalanced secondary voltages, but together they may cause slight unbalancing of the secondary voltages because of the way the two groups of impedance drops combine.

TABLE 4

Connection		Primary voltage		Secondary voltage	
Primary	Secondary	Between lines	To neutral	Between lines	To neutral
Δ	Δ	1	....	$\dfrac{1}{a}$	
Δ	Y	1	....	$\dfrac{\sqrt{3}}{a}$	$\dfrac{1}{a}$
Y	Y	1	$\dfrac{1}{\sqrt{3}}$	$\dfrac{1}{a}$	$\dfrac{1}{a\sqrt{3}}$
Y	Δ	1	$\dfrac{1}{\sqrt{3}}$	$\dfrac{1}{a\sqrt{3}}$	

If a single-phase load is applied between the line and neutral of a group of transformers which are connected Y-Y and which have no neutral connection on the primary side, only a small load current can be obtained even if the impedance of the load is reduced to zero. All the current on the primary side of the loaded transformer must come through the primaries of the other two transformers which are on open circuit. Since these transformers are on open circuit, the currents on their primary sides are exciting currents. Hence the only current that can be obtained from the loaded transformer, assuming a ratio of transformation of unity, is a current equal to the vector sum of the exciting currents of the other two transformers. If the impedance of the load is reduced to zero, the only voltage across the primary of the loaded transformer is the equivalent impedance drop in that transformer for a current which is much smaller than full-load current. As a result, the free neutral of the transformers on the primary side shifts until it almost coincides with the line to which the loaded transformer is connected. This puts the other two transformers across nearly line voltage, a voltage which is nearly $\sqrt{3}$ times the voltage for which the transformers are designed. This substantially increases their exciting currents, but even a large increase in the exciting currents allows less than full-load current to flow in the loaded transformer. Even a slight unbalancing of a Y-connected load produces a bad unbalancing of the secondary Y voltages. If the normal exciting currents of three transformers which are connected in Y on both

[1] See R. R. Lawrence, "Principles of Alternating Currents," 2d ed., p. 374.

primary and secondary sides are unequal, the secondary voltages to neutral are unbalanced at no load as well as under load.

The effect of a single-phase load applied between line and neutral of a Y-Y connected group of transformers, which have no neutral connection, is shown by the approximate vector diagrams of Fig. 66. The upper diagrams are for no-load conditions, while the lower diagrams show the conditions existing with a short circuit across the secondary terminals of one transformer, *i.e.*, from line to neutral. The subscripts 1, 2, 3 indicate phases, and single and double primes indicate primary and secondary

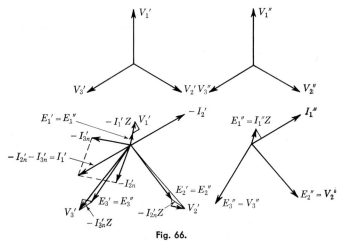

Fig. 66.

vectors, respectively. A ratio of transformation of unity is assumed, and the leakage impedances of all windings are assumed equal. Because of the low voltage impressed on the primary of the loaded transformer, its exciting current is much smaller than normal and is neglected. All impedance drops are greatly exaggerated.

The ratios of transformation for transformers used in any of the four three-phase connections must be identical, as otherwise the secondary voltages cannot be balanced even if the primary voltages and the load on the system are balanced. Moreover, if the ratios of transformation are not equal and if the secondaries are connected in Δ, there are currents in the transformers at no load, caused by the unbalanced voltages acting in the closed circuit formed by the secondaries. These currents increase the copper loss in the system without contributing to the useful output of the system.

For all four of the three-phase connections, not only must the ratios of transformation be equal but the transformers must have equal current and voltage ratings and their equivalent resistances and equivalent reactances must be equal. Under these conditions, all the given three-phase connections are symmetrical with respect to their line terminals and give balanced secondary voltages with balanced loads as well as at no load, if the primary voltages are balanced.

When the transformers are identical and the primary voltages and the load are balanced, the currents of the individual transformers are equal and form a balanced system. Under this condition the permissible three-phase output is three times the permissible output of one transformer.

When the ratios of transformation are equal and the primary voltages and load are balanced but the equivalent impedances are not identical,

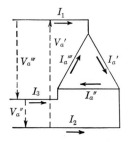 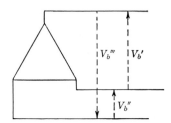

Fig. 67.

the transformers are not equally loaded if a Δ-Δ connection is used. Their currents are not equal and they do not form a balanced system. Under this condition, the permissible three-phase output is less than three times the permissible output of one transformer.

Consider the Δ-Δ connection shown in Fig. 67, which gives a diagram of connections.

The arrows on the figure indicate the directions taken as positive for currents and voltages. Subscripts 1, 2, 3 designate the three line currents. Single, double, and triple primes are used with the Δ currents and voltages. The subscripts a and b are used to distinguish primary and secondary currents and voltages. The k's denote ratios of transformation, and these ratios are assumed equal. Everything is referred to the primary side.

$$V'_a - I'_a z'_e = kV'_b$$
$$V''_a - I''_a z''_e = kV''_b$$
$$V'''_a - I'''_a z'''_e = kV'''_b$$

Since

$$V'_a + V''_a + V'''_a = 0 \tag{114}$$

and

$$V'_b + V''_b + V'''_b = 0$$
$$I'_a z'_e + I''_a z''_e + I'''_a z'''_e = 0 \tag{115}$$
$$I_1 = I'_a - I'''_a \tag{116}$$
$$I_2 = I''_a - I'_a \tag{117}$$

Substituting I'' and I''' from Eq. (116) and Eq. (117) in Eq. (115),

$$I'_a = \frac{I_1 z'''_e - I_2 z''_e}{z'_e + z''_e + z'''_e} \tag{118}$$

From the symmetry of connections, the equations for the other two transformer currents can be written.

$$I''_a = \frac{I_2 z'_e - I_3 z'''_e}{z'_e + z''_e + z'''_e} \tag{119}$$

$$I'''_a = \frac{I_3 z''_e - I_1 z'_e}{z'_e + z''_e + z'''_e} \tag{120}$$

From Eqs. (118), (119), (120) it is obvious that the currents in the transformers cannot be equal even when I_1, I_2, I_3 form a balanced system unless the three equivalent impedances of the transformers are identical. If the impedance of one transformer is smaller than that of the other two, this transformer takes more current than the others and fixes the permissible three-phase output of the transformers at a value less than three times the permissible output of one transformer.

If the equivalent impedances of the transformers are identical, Eq. (120) reduces to

$$I'''_a = \frac{I_3 - I_1}{3}$$

For balanced load, I_3 and I_1 are equal and differ in phase by 120 deg. Hence the magnitude of their differences is $\sqrt{3}$ times the magnitude of either. The transformer currents under this condition are equal to the line current divided by $\sqrt{3}$. This is the relation which exists between the magnitudes of line and phase currents in a balanced Δ-connected system.

The Effect of the Suppression of the Third-harmonic Components in the Exciting Currents of Y-Y-connected Transformers. Thus far in the consideration of three-phase transformer connections, the exciting currents have been neglected. With the exception of the Y-Y connection, the effect of the exciting currents is negligible. In a balanced three-phase system, the fundamentals and all harmonics of like order, with the exception of harmonics of triple frequency and multiples of triple frequency, differ in phase by 120 deg. If the phase order of the fundamentals is positive, the phase order of harmonics which differ in phase by 120 deg alternates with respect to the phase order of the fundamentals. For example, if the phase order of the fundamentals is positive, that of the fifth harmonics is negative, that of the seventh harmonics is positive, etc. The third harmonics and all harmonics of frequencies that are multiplies of triple frequency are in phase.[1] These statements neglect all even harmonics, as even harmonics do not occur ordinarily in power systems.

The magnetizing component of the exciting current of an iron-core transformer contains harmonics even when the voltage impressed on the transformer is sinusoidal. If the voltage impressed on a transformer is sinusoidal, the flux is nearly sinusoidal but the magnetizing current producing this flux cannot be sinusoidal because of the cyclic variation of the permeability of the core. The exciting current contains harmonics, among which the third harmonic is prominent. It can be as large as 30 to 40 per cent of the fundamental, depending on the degree of saturation at which the core is operated.

Since the vector sum of the currents at any junction point in a circuit must be zero, third-harmonic currents and all harmonic currents of frequencies that are multiples of triple frequency cannot exist in either the primaries or the secondaries of Y-Y-connected transformers, when balanced voltages are impressed, unless the common junction of the primary windings is connected to the neutral of the source of power. All currents of these frequencies are eliminated. The fundamentals and harmonic currents other than those of triple frequency and multiples of triple frequency are not eliminated, since in a balanced system their vector sum is zero.

The fundamental and third harmonic in the magnetizing current of a transformer are shown in Fig. 68. With a sinusoidal impressed voltage, the flux wave of a transformer is nearly sinusoidal. If the third-harmonic current is eliminated, the flux wave becomes flattened and contains a fundamental and a pronounced third harmonic. There are other harmonics, but these are small and are neglected in this discussion. Figure 69 shows the flux wave and its fundamental and third-harmonic components when the third harmonic in the magnetizing current is suppressed. The resultant flux and its fundamental and third-harmonic components are shown by full lines. Voltage drops induced in the winding by the components of flux are 90 deg ahead of the components of flux producing them and are proportional to the product of flux and frequency. If the elimination of the third harmonic in the magnetizing current produces a 35 per cent third harmonic in the flux, this flux harmonic causes a $3 \times 35 = 105$ per cent third harmonic in the voltage. The fundamental and third-harmonic voltages induced by the fundamental and third-harmonic components of flux are shown in Fig. 69 as dotted lines. It should be noted that the resultant voltage wave, also shown in the figure, is strongly peaked.

[1] R. R. Lawrence, "Principles of Alternating Currents," 2d ed., p. 319.

Since the third harmonics in the magnetizing currents of Y-Y-connected transformers are eliminated, large third-harmonic voltages are induced in the windings of each transformer. These voltages are in phase and cancel between line terminals of the Y-Y-connected transformers, but they appear in the voltages between line terminals and neutral. They cause resultant voltages between line terminals and neutral having high peak values as in Fig. 69, below. The peak values of these voltages may be 175 per cent or more of the peak voltage of the fundamentals of the voltages to neutral which are substantially equal to the line voltage divided by the square root of 3. These high

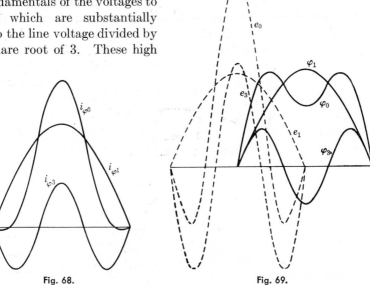

Fig. 68.   Fig. 69.

peak voltages between line terminals and neutral may put a prohibitively great strain on the insulation of the transformers and must not be permitted. The Y-Y connection of transformers is not safe from the standpoint of insulation stress unless the precaution is taken to eliminate the high peak voltages. Fortunately they can be eliminated either by connecting the neutral of the source of power to the primary neutral of the transformers by a low impedance conductor or by using a tertiary winding on each transformer and connecting these tertiary windings in Δ. The first method allows the necessary third harmonics in the exciting currents to come in from the source of power over the neutral connection. However the neutral of the source of power is seldom available at the transformer location. Moreover the neutral connection is objectionable because of the possible interference on telephone lines due to the third-harmonic currents and their multiples. Third-harmonic currents and their multiples may exist in the three line conductors to the transformers and in the neutral connection due to like harmonics in the line-to-neutral voltages of the source of power. The neutral connection and line conductors must carry also the components of triple frequency and multiples of this frequency in the exciting currents of the transformers. These components are confined to the transformers if the transformers have Δ-connected windings and there is no neutral connection. The use of Δ-connected tertiary windings is the common method of eliminating the third-harmonic voltages. Tertiary windings are generally used when transformers are connected Y-Y for three-phase transformation.

The third-harmonic voltages are in phase and are therefore short-circuited by the Δ-connected tertiary windings. As there cannot be any third-harmonic currents in either the primary or the secondary Y-connected windings, the tertiary windings act for the third-harmonic currents like primaries of transformers with secondaries open. The third-harmonic currents are limited in magnitude by the triple-frequency self-impedances of the windings and are the third-harmonic components of the magnetizing currents which are necessary for sinusoidal fluxes. With Δ-connected tertiary windings, the third-harmonic components in the flux and therefore in the voltages induced in the primary and secondary windings of the transformers are nearly eliminated. If the number of turns in the tertiary windings is equal to the number in the primary windings, the currents in the tertiary windings are equal to the third-harmonic component currents which would exist in the primary wind ings if not suppressed by the connections.

Any three-phase connection of single-phase transformers which has a group of windings connected in Δ operates without trouble from third-harmonic voltages. In the Y-Δ connection, the third-harmonic components of the exciting currents circulate in the Δ-connected secondaries. When the primaries are in Δ and the secondaries are in Y, the third-harmonic components circulate in the Δ-connected primaries. Under this Δ-Y condition, there are no third-harmonic currents in the three-phase line conductors when the primary line voltages are balanced, since the third-harmonic currents in any two windings of the Δ-connected primaries which are connected to a common line terminal are equal in magnitude and opposite in phase with respect to the line terminal and therefore cancel. When the primary and secondary windings are both connected in Δ, the third-harmonic components in the magnetizing ampere-turns divide between primary and secondary windings inversely as their triple-frequency leakage impedances.

Tertiary windings on Y-Y-connected transformers, so far as the third-harmonic exciting currents are concerned, carry only small currents. However, if an unbalanced load is connected to the secondary windings of the transformers between the line terminals and the common junction of the secondaries, the tertiary windings must carry zero-sequence components of fundamental frequency because of the unbalanced Y-connected load. The current-carrying capacity of the tertiary windings must be large enough to carry without overheating any zero-sequence component current that may occur because of an unbalanced Y-connected load. In some cases the tertiary windings are designed for voltages which may be useful to supply local circuits in a power station.

Unbalanced Loads on Transformers Connected for Three-phase Transformation. Any of the three-phase connections listed in Table 4, page 114, can be used to supply any unbalanced Δ-connected load. The Δ-Y connection, as well as the Y-Y connection with Δ-connected tertiary windings, can supply unbalanced Y-connected loads which have their neutral points connected to the common junction of the secondary windings. A Y-connected load which does not have its neutral point connected to the common junction of the secondary windings is the same, so far as its effect on the transformers is concerned, as a Δ-connected load, since any Y-connected load having no neutral connection which carries current can always be replaced by an equivalent Δ-connected load. Unbalanced loads do not cause serious unbalancing of the secondary voltages. Any unbalancing of these voltages that occurs is caused by the differences between the impedance drops produced in the windings by the unbalanced loads. As the impedance drop in a transformer is small compared with its terminal voltage, even with full-load current, the effect of the unbalanced impedance drops on the degree of balance of the secondary terminal voltages of a bank of transformers which carries an unbalanced load is not great.

Neglecting the exciting currents, the vector sum of the ampere-turns in all windings of a transformer must be zero. Therefore, in order that the secondary windings of a bank of transformers may carry a given three-phase load, there must be other windings on each transformer capable of carrying component currents similar to those in the secondary windings in order that the vector sum of the ampere-turns on each transformer shall be zero. That Δ-Δ-connected transformers and Δ-Y-connected transformers are able to supply a load of any character is obvious since each primary winding in both of these connections is tied directly to two terminals of the three-phase line, can receive power independently of the other two primary windings, and can carry any current necessary to balance the ampere-turns of the current in its secondary winding. That the other connections of Table 4 can also carry unbalanced loads is best understood when the actual load currents are replaced by their symmetrical-phase components.

An unbalanced three-phase system of currents can be resolved into three groups of components: the positive-phase components, the negative-phase components, and the zero-phase components.[1] The positive-phase components form a balanced system of currents with the same phase order as the actual currents. The negative-phase components also form a balanced system of currents but with a phase order opposite to that of the actual currents. The zero-phase components are identical. They are equal in magnitude and in phase with one another. The actual current in any phase is equal to the vector sum of the components of positive, negative, and zero sequence in that phase. Zero-sequence components cannot exist in the line currents of a three-phase load which does not have its neutral point connected to the neutral of the source of power. Therefore in either the Y-Δ connection or the Y-Y connection without the common junction of its secondary windings connected to the neutral of the load, there can be only positive-sequence and negative-sequence component currents in the transformer windings. As the vector sum of each of these groups of component currents is zero, they can be carried by transformer windings which are connected in Y, and the vector sum of the ampere-turns in the windings of each transformer can be zero. The ampere-turns due to the positive-sequence component currents in the secondaries are balanced by the ampere-turns due to similar component currents in the primaries. The ampere-turns due to the negative-sequence component currents in the secondaries are balanced by the ampere-turns due to similar component currents in the primaries.

No zero-sequence component currents can exist in primary windings of a bank of transformers with an isolated primary neutral since the vector sum of the primary currents under such a condition must be zero. As the zero-sequence component currents are equal in magnitude and are in phase, their vector sum can be zero only when each component is zero. Y-Y-connected transformers with an isolated primary neutral cannot carry a load which requires zero-sequence component currents in the primary windings to balance the ampere-turns of similar component currents in the secondary windings. Only Δ-connected loads can be carried, as these loads do not call for any zero-sequence component currents in the secondary windings. If the primary windings of a Y-Y-connected bank of transformers have their common junction connected to the neutral of the source of power, any zero-sequence component currents that may be required in the primaries can come over the neutral connection. Under this condition any kind of secondary load can be carried. If the primary neutral of a Y-Y-connected group of transformers is isolated but the transformers are provided with tertiary windings connected in Δ, the tertiary windings can carry the zero-sequence

[1] R. R. Lawrence, "Principles of Alternating Currents," 2d ed., Chap. 13.

currents necessary to balance the ampere-turns due to corresponding component currents in the secondary windings. Therefore Y-Y-connected transformers which are provided with Δ-connected tertiary windings can carry a secondary load of any character.

An Example of an Unbalanced Load on a Bank of Y-Y-connected Transformers with Δ-connected Tertiary Windings and an Isolated Primary Neutral. Consider an extreme case of unbalancing. Assume that a single-phase load is connected across one secondary winding between one line terminal and the neutral. Let the load current be $A\,\underline{/0°}$. Indicate the positive-sequence, the negative-sequence, and the zero-sequence components of the currents by the use of superscripts

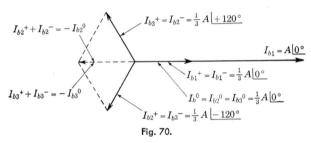

$$I_{b2}^+ + I_{b2}^- = -I_{b2}^0$$
$$I_{b3}^+ = I_{b2}^- = \tfrac{1}{3}A\,\underline{/+120°}$$
$$I_{b1} = A\,\underline{/0°}$$
$$I_{b1}^+ = I_{b1}^- = \tfrac{1}{3}A\,\underline{/0°}$$
$$I_{b3}^+ + I_{b3}^- = -I_{b3}^0$$
$$I_b^0 = I_{b2}^0 = I_{b3}^0 = \tfrac{1}{3}A\,\underline{/0°}$$
$$I_{b2}^+ = I_{b3}^- = \tfrac{1}{3}A\,\underline{/-120°}$$

Fig. 70.

$+$, $-$, and 0. Let the currents in the phases be distinguished by the use of subscripts 1, 2, 3. The subscripts a, b, t are used to distinguish the primary, secondary, and tertiary currents.

The actual secondary phase currents, *i.e.*, the currents in the secondary windings, are

$$I_{b1} = A\,\underline{/0°} \qquad I_{b2} = 0 \qquad I_{b3} = 0$$

The resolution[1] of these currents gives for the three component currents in each of the secondary phases

$$
\begin{array}{lll}
I_{b1}^+ = \tfrac{1}{3}A\,\underline{/0°} & I_{b2}^+ = \tfrac{1}{3}A\,\underline{/-120°} & I_{b3}^+ = \tfrac{1}{3}A\,\underline{/+120°} \\
I_{b1}^- = \tfrac{1}{3}A\,\underline{/0°} & I_{b2}^- = \tfrac{1}{3}A\,\underline{/+120°} & I_{b3}^- = \tfrac{1}{3}A\,\underline{/-120°} \\
I_{b1}^0 = \tfrac{1}{3}A\,\underline{/0°} & I_{b2}^0 = \tfrac{1}{3}A\,\underline{/0°} & I_{b3}^0 = \tfrac{1}{3}A\,\underline{/0°}
\end{array}
$$

The vectors representing these currents are shown in Fig. 70.

Each of the secondary windings carries all three components, but the primary windings carry only positive-sequence and negative-sequence components. No zero-sequence components can exist in the primaries with an isolated neutral. Only zero-sequence components can exist in the tertiary windings. Let N with the subscripts a, b, t represent the numbers of turns in the three windings of each transformer. If the

[1] For the method of resolution, see R. R. Lawrence, "Principles of Alternating Currents," 2d ed.

exciting current is neglected, since the vector sum of all the ampere-turns on any transformer must be equal to zero, the conditions in transformer 1 are represented by the following equations.

$$N_a I_{a1}{}^+ + N_b I_{b1}{}^+ = N_a \frac{N_b}{N_a}(-I_{b1}{}^+) + N_b I_{b1}{}^+ = 0$$

$$N_a I_{a1}{}^- + N_b I_{b1}{}^- = N_a \frac{N_b}{N_a}(-I_{b1}{}^-) + N_b I_{b1}{}^- = 0$$

$$N_t I_{t1}{}^0 + N_b I_{b1}{}^0 = N_t \frac{N_b}{N_t}(-I_{b1}{}^0) + N_b I_{b1}{}^0 = 0$$

The total number of ampere-turns acting on the transformer of phase 1, due to the load, is zero. Similarly the total number of ampere-turns on each of the other two transformers of phases 2 and 3, due to the load, is zero.

The transformers used for the Y-Y connection with Δ-connected tertiary windings are an example of three-winding transformers. In the case just considered, the primary current in the transformer across which the load is placed is $I_{a1} = I_{a1}{}^+ + I_{a1}{}^- = -(N_b/N_a)(I_{b1}{}^+ + I_{b1}{}^-)$ vectorially. Likewise, $I_{a2} = -(N_b/N_a)(I_{b2}{}^+ + I_{b2}{}^-)$ and

$$I_{a3} = -\frac{N_b}{N_a}(I_{b3}{}^+ + I_{b3}{}^-)$$

The tertiary current $I_{t1} = I_{t1}{}^0$ is equal to $-(N_b/N_t)I_{b1}{}^0$. The actual primary current in the primary of phase 1 is the current I_{a1} plus the excitation current. The three primary currents are different in magnitude and are not 120 deg different in phase. The three tertiary currents are identical since they are zero-sequence currents.

Summary of Symmetrical Three-phase Connections. Some of the advantages of the different connections given in Table 4, page 114, are the following.

Δ-Δ. If one transformer is damaged, the system can still be operated at about 58 per cent of its normal capacity with the remaining two transformers connected in open Δ or V. By reconnecting the secondary windings in Y, the line voltage can be increased if it becomes desirable to raise the transmission voltage. If the secondary voltage with the Δ-Δ connection is 66,000 volts, reconnecting the secondary windings in Y gives a secondary line voltage of approximately 115,000 volts. The connection is advantageous when the line currents are large, as the currents in the transformers are less than the line currents. Disadvantages are the lack of a neutral for grounding and other purposes and the dependence upon the relative equivalent impedances of the individual transformers for proper load division.

Δ-Y. This gives a higher secondary line voltage for transmission purposes than the connections with Δ secondaries without increasing the strain on the insulation of the transformers. It is the connection commonly used at the generating end of transmission lines.

Y-Y. This permits grounding the neutral points of both primary and secondary three-phase circuits. When the primary neutral is not connected to the source of power, it is necessary to use Δ-connected tertiary windings.

Y-Δ. This permits the primary neutral to be grounded. If the Y-Δ connection is used for transmission purposes, the secondaries can be reconnected in Y, if at any time it becomes desirable to raise the transmission voltage to increase the capacity of the line, provided the primary neutral is connected to the source of power or tertiary windings are available which can be connected in Δ. Without the primary neutral connection or Δ-connected tertiary windings, the necessary triple frequency components in the exciting currents are suppressed. This results in a dangerously high peak voltage in the transformer windings, as has been explained. The Y-Δ connection is commonly used at the receiving ends of high-voltage transmission lines.

Y connection of secondaries permits the use of a four-wire distributing system. This is sometimes desirable for lighting.

Method of Testing for Proper Connections. When three single-phase transformers are to be connected three phase, their primary windings can be connected at random since the transformers have entirely independent magnetic circuits and the phase relations between voltages depend merely upon the line and the way in which the transformers are connected. After the primary windings have been connected, the secondary voltages are fixed. The proper connections for the secondaries must be tested with a voltmeter or by other means.

To connect the secondary windings in Y, connect one terminal of each of two secondaries together and then put a voltmeter across the remaining free terminals. The voltage across these is equal either to the voltage of one secondary or to $\sqrt{3}$ times that voltage. It should be $\sqrt{3}$ times that voltage. If it is not, reverse the connections of either of the two secondaries. When the two secondaries have been connected properly, connect one end of the remaining secondary to the common junction of the other two. The voltage between the free terminal of this last secondary and the free terminal of either of the other two secondaries should be $\sqrt{3}$ times the voltage of one secondary. If it is not, reverse the connections of the last secondary.

The method of testing the proper connections for putting the secondaries in Δ is similar to the method of testing for putting them in Y. For the Δ connection, connect one end of each of two secondaries together.

The voltage across the free ends should be the same as the voltage of one secondary. If it is not, reverse one of the secondaries. Then connect one end of the remaining secondary to one of the free ends of the other two. The voltage across the remaining gap is either zero or twice the voltage of one winding. If it is double the voltage, reverse the connections. When it is zero, the remaining gap can be closed and the second-

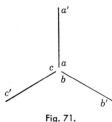

Fig. 71.

aries are in Δ. If this gap is closed when the last secondary is connected reversed, the transformers are virtually short-circuited. Twice the voltage of one winding acts on an impedance which is equal to three times the impedance of a single winding. The current under this condition is two-thirds of what would flow if a single transformer were short-circuited.

Let Fig. 71 represent the secondary windings and also a vector diagram of the secondary voltages.

If a and b are connected, the voltage across the free ends or across $a'b'$ is $V_{a'a} + V_{bb'}$ and is equal in magnitude to the magnitude of either $V_{a'a}$ or $V_{b'b}$ multiplied by $\sqrt{3}$ and lags the voltage $V_{bb'}$ by 30 deg. This is the correct connection of the windings aa' and bb' for Y. If b' is connected to a, the voltage across the free ends or across $a'b$ is $V_{a'a} + V_{b'b}$. This is equal in magnitude to the magnitude of either $V_{aa'}$ or $V_{bb'}$ and leads $V_{aa'}$ by 120 deg. This is the correct connection for Δ. The third winding should have c' connected to b if Δ connection is desired. The vector diagram for Δ connection is shown in Fig. 72. The connections are a to b', b to c', and c to a'.

The vector sum of the three voltages $V_{b'b}$, $V_{c'c}$, $V_{a'a}$ which act around the closed Δ is zero. If c is connected to b, the resultant voltage in the three windings is $(V_{b'b} + V_{a'a}) + V_{cc'} = 2V_{cc'}$.

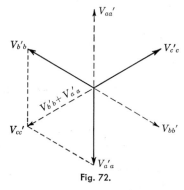

Fig. 72.

The statements which have just been made in regard to the magnitudes of the voltages to be expected assume that the primary windings are connected in such a way that they can receive exciting currents which contain the normal triple-frequency components. They must be connected either in Δ, or in Y with the neutral connected to the neutral of the source of power.

If the primary windings are connected in Y and their neutral is not connected to the neutral of the source of power, the suppression of the triple-frequency components in the exciting currents by the Y-connected primaries may cause difficulty when testing for zero voltage across the final gap before closing the Δ formed by the secondaries. If the secondary windings are properly connected for Δ connection, the fundamental voltage across the gap is zero but the third-harmonic voltage across the gap is three times the third-harmonic voltage induced in each secondary winding. This third-harmonic voltage across the gap may be higher than the normal secondary voltage of one secondary winding. If the secondary windings are properly connected for Δ connection, this high triple-frequency voltage causes no trouble when the final gap is closed which puts the secondary windings in Δ. The triple-frequency current which flows in the closed Δ is limited by three times the triple-frequency self-impedance of a single secondary winding. This impedance is very high. Only a small current flows in the secondaries, *viz.*, the triple-frequency component of the exciting current, which is ordinarily in the primaries. In most commercial installations, in place of the above tests, lead markings on the transformer terminals are used to achieve the correct connections.

Three-phase Transformation with Two Transformers. Three-phase transformation can be obtained with two single-phase transformers by connecting them either in open Δ, which is also called V, or in T. Both these connections are unsymmetrical and give slightly unbalanced voltages under load. The amount of this unbalancing is small and is negligible under ordinary conditions, especially with T connection.

Open-Δ Connection. The open-Δ or V connection is the same as the Δ connection with one transformer removed. When similar transformers are used, the voltages given by the Δ and V connections are the same and their outputs are proportional to their line currents. Let I be the maximum current output per transformer. The current output per line of the Δ connection is $\sqrt{3}I$. The current output per terminal of the V connection is equal to the current output of one transformer, *i.e.*, equal to I. Therefore the output of the open Δ is $1/\sqrt{3}$ or 58 per cent of the output of the closed Δ. The actual transformer capacity of the open Δ is two-thirds that of the Δ, but all of this capacity cannot be utilized on account of the power factors at which the transformers of the open Δ operate as compared with the power factor of the load. With a noninductive balanced load, each transformer of the Δ system carries one-third of the total load at unity power factor. Under the same conditions, each transformer of the open-Δ system carries one-half of the load at a power factor of $\sqrt{3}/2 = 0.866$. Multiplying 0.866 by $\frac{2}{3}$ gives 0.58, which is the capacity of the open Δ as compared with the Δ. The transformers of the open Δ do not carry equal watt loads except when the power factor of the three-phase load is unity. The current loads however are equal whenever the three-phase load is balanced.

The output of a system made up of two groups of transformers in parallel, one group of two transformers in open Δ or V and the other of three transformers in Δ, is only 33⅓ per cent greater than the output of the Δ-connected group alone and not 58 per cent greater as might be expected (see page 156).

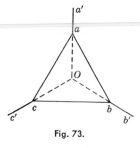

Fig. 73.

Figure 74 is a vector diagram for two transformers connected in open Δ or V. The lettering on this diagram corresponds to the lettering on the diagram of connections shown in Fig. 73. Equivalent resistances and equivalent leakage reactances are used. Single and double primes indicate primary and secondary values.

The transformers of the open Δ are ca and bc. Transformer ca carries the current $I_{aa'}$. Transformer bc carries the current $I_{bb'}$. The voltage across ab, the open part of the Δ, is equal to the vector sum of V_{ac} and V_{cb}. If θ is the angle of lag for the load, the current in the lines lags behind the Y voltage of the system by an angle θ. To simplify the construction of the vector diagram, let θ be the angle of lag of the secondary current with respect to the primary voltage referred to the secondary, and assume the current load to be balanced with respect to the primary voltage.

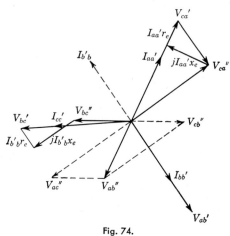

Fig. 74.

The current $I_{aa'}$, Fig. 74, lags behind V'_{ca} by an angle $(\theta - 30)$ deg. On Fig. 74, θ is 30 deg. V'_{ca}, V'_{ab}, V'_{bc} are the three primary voltages referred to the secondary windings. V''_{ca}, V''_{ab}, V''_{bc} are the three corresponding secondary voltages. V''_{ab} is the voltage across the open side of the Δ and is the vector sum of the voltages produced by the two transformers ac and cb.

$$V''_{ab} = V''_{ac} + V''_{cb}$$

It is evident from Fig. 74 that the secondary voltages cannot be balanced even for a balanced load. The unbalancing on the diagram is very much greater than is found in practice on account of the exaggerated impedance drops in the diagram.

T Connection. Two transformers with the same current ratings but with different voltage ratings are used. One transformer, which is called the teaser, is connected to the middle of the other as in Fig. 75. Both primary and secondary windings are connected in the same way. Figure 75 serves either for the diagram of connections or for the vector diagram of voltages. The teaser transformer is indicated by ad and the second transformer by cb. Three-phase voltages are impressed across the terminals a, b, c. If the secondaries are similarly connected, they supply three-phase power at a voltage which, except for the impedance drops, is equal to the impressed voltage divided by the ratio of transformation.

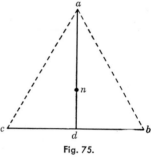

Fig. 75.

If the impressed voltages are balanced, the primary voltages V_{ab}, V_{bc}, V_{ca} are each equal to $2V_{cd}$. The voltages V_{da}, V_{dc} and also V_{da}, V_{db} are in quadrature.

$$V_{ca} = \sqrt{V_{da}^2 + V_{cd}^2}$$

The angle acd is 60 deg and

$$\frac{V_{da}}{V_{ca}} = \sin 60 = \frac{\sqrt{3}}{2} = 0.866$$

The teaser transformer therefore should be wound for a voltage which is 86.6 per cent of the voltage of the line or of the main transformer. Usually the teaser transformer is wound for the same voltage as the main transformer but is provided with a tap for 86.6 per cent of full voltage.

A neutral point n can be obtained from the T connection by bringing out a tap from the teaser transformer at a distance from a equal to two-thirds of the distance between a and d.

$$\frac{V_{na}}{V_{da}} = \frac{2}{3}$$

Since n, Fig. 75, is the neutral point of the three-phase system,

$$\frac{V_{na}}{V_{ca}} = \frac{1}{\sqrt{3}}$$

$$V_{na} = \frac{1}{\sqrt{3}} V_{ca}$$

As

$$V_{da} = 0.866 V_{ca} = \frac{\sqrt{3}}{2} V_{ca}$$

$$V_{na} = \frac{1}{\sqrt{3}} \frac{2}{\sqrt{3}} V_{da} = \frac{2}{3} V_{da}$$

The T system is unsymmetrical and cannot give balanced secondary voltages under load conditions, but the amount of unbalance is small.

Two identical transformers can be used for the T connection with fair results, but this is not advisable except for temporary work or in an emergency. If the two transformers are identical, the one which is used for the teaser has more turns than it should for the voltage impressed upon it and the leakage impedance drop is unnecessarily large. In such a case, it is better to connect the transformers in open Δ.

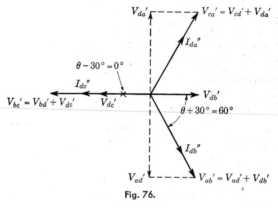

Fig. 76.

Figure 76 is a vector diagram for the T connection. The load is assumed to be balanced with respect to the primary voltage. The angle of lag θ is assumed to be 30 deg with respect to the primary voltage. All vectors are referred to the primary.

The voltage V'_{da} is in phase with the Y voltage of the system. The transformer da carries line current. Therefore the power factor for the transformer da is the same as the power factor of the three-phase load.

The halves of the secondary coil of the main transformer carry currents which are out of phase. In order to find the voltage across the secondary, the transformer bc must be treated as a transformer with two secondary windings which are independently loaded, i.e., like a three-winding transformer. By inspection of Fig. 76, it is evident that the T system is not symmetrical and cannot give balanced secondary voltages under load.

The capacity of the T system for three-phase transformation is somewhat less than the sum of the capacities of the two transformers.

Consider a load at unity power factor and assume that the teaser transformer da, Fig. 75, is wound for the correct voltage. Let the line current and line voltage of the three-phase system be I and V.

The transformer da has a voltage equal to $0.866V$ and operates at the power factor of the load. Its output is $0.866VI$. The halves of the secondary of transformer bc carry the current I at a power factor $\cos 30° = 0.866$. Its output is

$$VI \cos 30° = 0.866VI$$

The total output of the system is

$$2(0.866VI)$$

The total rated capacity of the two transformers is

$$0.866VI + VI = 1.866VI$$

Comparing the actual output with the rated capacity gives

$$\frac{1.732VI}{1.866VI} = 0.928$$

as the fraction of the total transformer rating which is available for three-phase output.

If the transformer da is wound for the same voltage as the transformer bc but has a voltage tap for 86.6 per cent of full voltage, 86.6 per cent of the rating of this transformer is utilized. In this case the output of the T system is 86.6 per cent of the total transformer rating, or is the same as the three-phase output of the same two transformers when connected in V or open Δ.

Three-phase to Four-phase Transformation and Vice Versa. Transformation from three-phase to four-phase or vice versa is easily accomplished by means of the Scott-transformer or T-transformer connections. Referring to Fig. 75 it is evident that the voltages across the primary terminals of the two transformers are in quadrature and are in the ratio of 1:0.866. The secondary voltages are also in quadrature and in the same ratio.

A symmetrical four-phase system can be obtained on the secondary side by connecting the secondary windings at their middle points and adjusting the turns on the two secondary windings so that these voltages are equal. This can be accomplished by making the ratios of transformation of the two transformers ad and bc equal to $1/0.866a$ and $1/a$. In order to have the two transformers interchangeable, they are usually provided with taps on the primary sides for 0.866 per cent of full voltage, though the tap on only one transformer is used.

The Scott connection for three-phase to four-phase transformation is shown in Fig. 77. The point n of the common connection is the neutral point of the four-phase side. The secondaries may be considered to give either a four-phase or a two-phase system. The four-phase voltages are na', nb', nc', nd' with n as a neutral point. The two-phase voltages are $a'd'$ and $b'c'$. To transform from two phase to three phase, it is necessary merely to consider a', b', d', c', Fig. 77, as the primary terminals and a, b, c as the secondary terminals.

If the Scott connection is used to transform from two phase to three phase, one of the three-phase voltages, V_{cb}, is derived directly from one transformer. Each of the other two voltages is equal to the vector sum

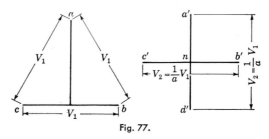

Fig. 77.

of two quadrature voltages, derived from the two transformers. Neglecting the effect of the small leakage impedance drops in the transformers, the wave form of the voltage V_{bc} is the same as the wave form of the voltage impressed on the two-phase side. The other two voltages, V_{ca}, V_{ab}, however, are not of the same wave form as the two-phase voltage and do not have the same wave form except when the two-phase voltage is sinusoidal.

If power is supplied on the three-phase side, one of the two-phase voltages, $c'b'$, has the same wave form as the three-phase voltages. The other, except when the impressed voltage is sinusoidal, is either more or less peaked than the impressed voltage. Whether it is more peaked or less peaked depends upon the harmonics present and their phase relations.

Let the voltage drops across the two-phase side of the Scott-connected transformers (Fig. 77) be

$$e_{b'c'} = E_1 \sin(\omega t + \alpha_1) + E_3 \sin(3\omega t + \alpha_3)$$
$$+ E_5 \sin(5\omega t + \alpha_5) + E_7 \sin(7\omega t + \alpha_7)$$
$$e_{d'a'} = E_1 \sin(\omega t + \alpha_1 - 90°) + E_3 \sin(3\omega t + \alpha_3 + 90°)$$
$$+ E_5 \sin(5\omega t + \alpha_5 - 90°) + E_7 \sin(7\omega t + \alpha_7 + 90°)$$

These two voltages have the same wave form but differ in phase by 90 deg.

Assume a ratio of transformation of unity between the three-phase and four-phase voltages. The voltage drops across the three-phase side of the transformers are

$$e_{ab} = 0.866e_{a'd'} + 0.5e_{c'b'}$$
$$e_{ca} = 0.866e_{d'a'} + 0.5e_{c'b'}$$

Referred to bc as an axis, the three-phase voltages are

$$
\begin{aligned}
e_{bc} &= E_1 \sin (\omega t + \alpha_1) + E_3 \sin (3\omega t + \alpha_3) \\
&\quad + E_5 \sin (5\omega t + \alpha_5) + E_7 \sin (7\omega t + \alpha_7) \quad &(121)
\end{aligned}
$$

$$
\begin{aligned}
e_{ca} &= E_1 \sin (\omega t + \alpha_1 - 120°) + E_3 \sin (3\omega t + \alpha_3 - 240°) \\
&\quad + E_5 \sin (5\omega t + \alpha_5 - 120°) + E_7 \sin (7\omega t + \alpha_7 - 240°) \quad &(122)
\end{aligned}
$$

$$
\begin{aligned}
e_{ab} &= E_1 \sin (\omega t + \alpha_1 - 240°) + E_3 \sin (3\omega t + \alpha_3 - 120°) \\
&\quad + E_5 \sin (5\omega t + \alpha_5 - 240°) + E_7 \sin (7\omega t + \alpha_7 - 120°) \quad &(123)
\end{aligned}
$$

It is evident from Eqs. (121), (122), (123) that the wave forms of the voltages V_{ab} and V_{ca} are different from the wave form of the voltage V_{bc}. All three of the three-phase voltages contain third harmonics which differ in phase by 120 deg. Except when three-phase voltages are obtained

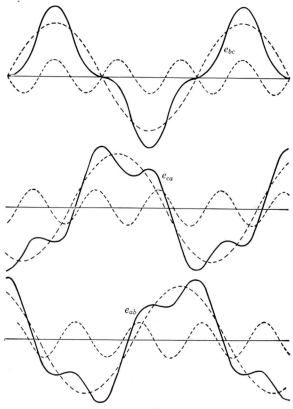

Fig. 78.

from Scott-connected transformers or from some other unsymmetrical system, the voltages cannot contain third harmonics.

The wave forms of the three-phase voltages are plotted in Fig. 78 for the case where the two-phase voltages contain 30 per cent third harmonics. The angles α_1 and α_3 are assumed to be 0 and 180 deg. The fundamentals and the third harmonics of each wave are shown dotted.

Three-phase to Six-phase Transformation. *Double Δ and Double Y.* A six-phase system can be derived from any three-phase system by the use of three single-phase transformers, each provided with two independent secondary windings. The primaries can be connected for three phase in either Y or Δ. They should not be connected in Y unless the

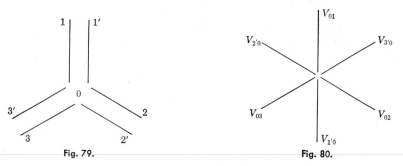

Fig. 79.   Fig. 80.

secondaries are connected in Δ to give a closed path for the third harmonic of the exciting current which is suppressed in the primaries by Y, connection. The two sets of secondaries are connected to form two independent three-phase systems with the connections of one set of secondaries reversed with respect to the connections of the other set.

The phase relations of the six secondary voltages are shown in Fig. 79. Reversing one group of secondaries gives the phase relations of Fig. 80.

The two groups of secondaries can be connected in Δ or in Y, giving what is known as the double-Δ or the double-Y connection. In either the double-Δ or the double-Y connection, one-half the power delivered by the transformers is supplied by each group of secondaries at three-phase voltage. The connections with the secondaries in double Δ and the primaries in Y are shown in Fig. 81. Figure 82 shows the connections for the secondaries in double Y and the primaries in Δ.

The two Δ's forming the double-Δ secondaries in Fig. 81 have no electrical connection and cannot be considered to form a true six-phase system. However when the secondaries are connected to the armature of a motor or a synchronous converter, the electrical connection between the two Δ's is established and the effect is the same as if six-phase power were fed to the machine. The two Y's forming the double Y can be interconnected at their neutral points n and n' and form under this condition a true six-phase star system.

Diametrical Connection. Three single-phase transformers with single secondaries can be used to supply six-phase power to a synchronous converter or a motor by making use of what is known as the diametrical connection for the secondaries. The diametrical connection is commonly

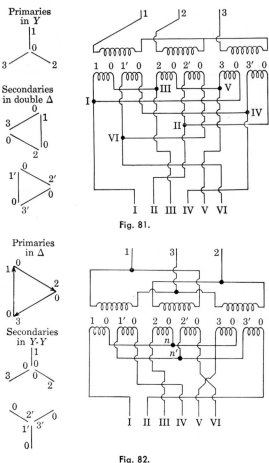

Fig. 81.

Fig. 82.

employed when synchronous converters are used to supply power for electrical railways. The double-Y connection is always used when a neutral is desired for grounding or for the neutral wire of a three-wire d-c system which receives power from a six-phase synchronous converter. The primaries should be connected in Δ when the diametrical connection is used for the secondaries because of the suppression of the third-harmonic components in the exciting currents by Y-connected primaries.

The diagram for the diametrical connection of transformers to feed six-phase power is given in Fig. 83. The hexagon at the bottom represents the armature which is to receive six-phase power.

If taps are brought out from the middle points of each of the three secondaries and these taps are interconnected, the diametrical connection becomes the double Y.

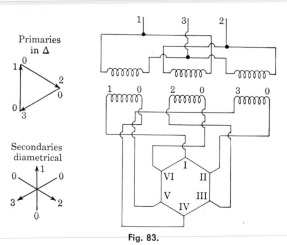

Fig. 83.

Two-phase or Four-phase to Six-phase Transformation. Two-phase or four-phase to six-phase transformation can be accomplished by use of double-T connection on the secondary side of Scott transformers. The connections for this are shown in Fig. 84.

The ratio between the primary and secondary voltages should be the same as for the Scott transformers. If the primaries are also connected

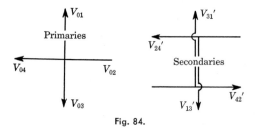

Fig. 84.

in T, the Scott transformers can be used to transform from three phase to six phase.

Interconnected-star or Zigzag Connections. The interconnected-star or zigzag connections can be used to stabilize the neutral when unbalanced loads occur on certain polyphase transformer connections having Y-connected primaries or can be used for deriving multiphase circuits from polyphase circuits of a smaller number of phases.   Six-phase and twelve-phase circuits can be obtained from three-phase circuits by means of interconnected-star or zigzag connections.   Twelve-phase circuits for the operation of large mercury-arc rectifiers are commonly obtained by means of zigzag connections.   Although a six-phase circuit can be obtained from a three-phase circuit by using the zigzag connections, the connection possesses no advantage over the double-Y connection for six phase except to stabilize the neutral when unbalanced line-to-neutral loads occur and there are no windings connected in Δ on the transformers.

Three-phase Zigzag Connection.   The three-phase zigzag connection with primaries in Y is shown in Fig. 85.   For this connection, each trans-

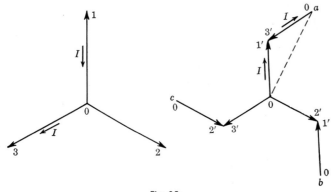

Fig. 85.

former must have two independent secondary windings.   Each phase on the secondary side contains two secondary windings, one from each of two transformers.   For example, phase $0a$ in Fig. 85 consists of a secondary winding from transformer 1 and a secondary winding from transformer 3.

In Fig. 85, 01, 02, 03 represent either the primary windings of the transformers or the voltages impressed on these windings.   Similarly 01′, 02′, 03′ represent either the secondary windings or the voltages across the terminals of these windings.   The effect of the equivalent impedance drops in the transformers is neglected.   The primary voltages are assumed to be balanced.

If the secondary line-to-neutral voltage for the three-phase system is to be V_n, each of the secondary windings must give a voltage equal to V_n divided by $\sqrt{3}$.   Refer to Fig. 85.   The line-to-neutral voltage V_{oa} is

$$V_{oa} = V_{01'} + V_{3'0} \text{ vectorially}$$

Let V_2 be the magnitude of the voltage of each secondary.   The magnitude of V_{oa} is

$$V_{oa} = 2V_2 \cos 30° = \sqrt{3}V_2$$
$$V_2 = \frac{V_{oa}}{\sqrt{3}}$$

If a single-phase load which takes a current I is connected between terminals a and 0, Fig. 85, there is a current I in one secondary winding of transformer 1 and also in one winding of transformer 3.

$$I_{oa} = I_{01'} = I_{3'0}$$

The direction of these currents is shown by arrows (Fig. 85). Neglecting the exciting currents and assuming a ratio of transformation of unity, the primary currents are

$$I_{01} = -I_{01'} \qquad I_{03} = -I_{03'} \qquad I_{02} = 0$$

Since the vector sum of the three primary currents at the common junction 0 is zero, there is no displacement of the neutral point, as is explained on page 114, when the Y-Y connection is used with a line-to-line load. The only displacement of the neutral is that caused by the relatively small equivalent impedance drops in the transformers. There is a triple-frequency voltage in each of the transformer windings owing to the suppression of the third-harmonic components of the exciting currents by the Y-connected primaries. However this third-harmonic voltage does not appear between any line terminal and the neutral on the secondary side since the third-harmonic voltages in the two secondary windings between the neutral and any line terminal are oppositely directed and cancel. The Y-zigzag three-phase connection is objectionable because of the increase in the stress on the transformer insulation caused by the third-harmonic voltage induced in each winding. If Δ connection is used for the primary windings, nothing is gained by using zigzag connection for the secondaries rather than using Y-connected secondaries.

Three-phase to Twelve-phase Transformation. Although the permissible output of a synchronous converter built on a given frame increases with an increase in the number of phases for which it is designed, the gain in going from 6 phases to 12 phases is small and the extra complications and extra cost probably would not be warranted. However, 12 phases are employed generally for the operation of large steel-tank mercury-arc rectifiers in order to decrease the ripple in the voltage on the d-c side. For this reason, three methods of transforming from 3 phase to 12 phase are given. One of these, the double-chord connection, does not give a true 12-phase system and cannot be used for operating a mercury-arc rectifier, though it can supply the equivalent of 12-phase power to any machine which has a closed-circuit armature winding. This method of transformation is given merely as a matter of interest. It is similar to the double-Δ method of transforming from 3 phase to 6 phase. The double-Δ connection is a double-chord connection in which the ends

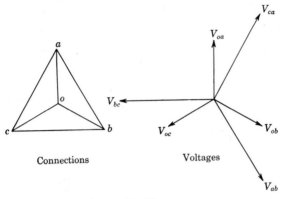

Connections   Voltages

Fig. 86.

of the parallel chords are displaced by 60 deg instead of by 30 deg as in the 12-phase connection.

There is a difference of 30 deg in phase between corresponding Y and Δ voltages of a 3-phase system. Therefore two groups of transformers connected for 3-to-6-phase transformation have their corresponding 6-phase voltages 30 deg apart, provided the primaries of one group are connected in Y and the primaries of the other group are connected in Δ. If the ratios of transformation of Y and Δ groups of transformers are $\sqrt{3}:1$, the 6-phase voltages of both groups are equal in magnitude and can be interconnected to give either a star or mesh 12-phase system. The diagram of connections and the phase relations of the primary voltages are shown in Fig. 86. The secondary connections and the vectors for the 12-phase star connection are given in Fig. 87. To simplify the reference to Fig. 86, the secondary voltages are assumed to be in phase with the primary voltages.

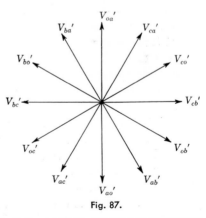

Fig. 87.

The connections in Fig. 87 require six single-phase transformers or two 3-phase transformers with two different ratios of transformation. The complication of such connections would offset as a rule any gain that might be derived from their use.

The equivalent of 12 phases can be obtained for any mesh-connected 12-phase system by the use of a simple double-chord connection which requires only three single-phase transformers or one 3-phase transformer. Each transformer, or each phase of the 3-phase transformer, must have two similar secondary windings. All secondaries must be wound for the same voltage and the same current. The chord connection can be used to supply 12-phase power from a 3-phase system.

Figure 88 shows the connections and the vectors of the voltages for the 12-phase double-chord connection. The chord voltages are approximately 96.5 per cent of the diametrical voltage.

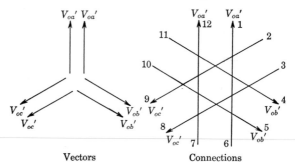

Vectors   Connections

Fig. 88.

The double-chord 12-phase connection of transformers does not give a true 12-phase system as there is no interconnection between phases except through the armature of the machine to which power is supplied. Without the interconnection, the double-chord connection cannot supply power to any loads except those connected across the ends of the chords, *i.e.*, across the points 1 and 6, 2 and 9, 3 and 8, etc. There are no closed paths through the transformers between any other pairs of points.

The interconnected-star or zigzag 12-phase connections of transformers are shown in Fig. 89. The primary windings are connected in Δ. Three identical transformers are required. Each transformer must have five secondary windings, a main winding for 81.7 per cent of the 12-phase line-to-neutral voltage with a tap brought out from its center, and four auxiliary windings, each for 29.9 per cent of the 12-phase voltage. The main windings, connected for 6-phase star transformation, are shown by heavy lines in the right-hand half of the figure. Vectors for the voltages of the three main secondary windings are marked 01, 02, 03. The auxiliary windings used to give the phase displacements necessary for obtaining the 12-phase voltages are shown by light solid lines drawn parallel to the main windings. The voltages of the auxiliary windings are in phase with the voltages of the corresponding main windings. The neutral point on the secondary side is marked n. The left-hand half of the figure is for the primary windings. All solid lines in the figure represent either the actual connections of the windings or the vectors representing the voltages in the windings. The 12-phase voltages are shown by dotted lines and are marked, a, b, c, d, e, f, etc.

Each of the 12-phase voltages is made up of two components which differ in phase by 60 deg; one of the component voltages is from half of

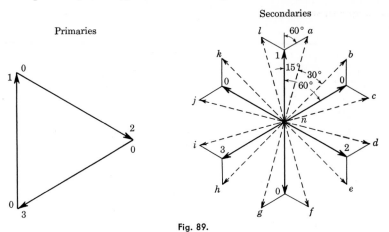

Fig. 89.

the main winding, the other from one of the auxiliary windings. Let V_m represent the magnitude of the voltage of the main winding, V_a the magnitude of the voltage of the auxiliary winding, and V_{12} the magnitude of the 12-phase voltage. Refer to Fig. 89.

$$\tan 15° = \frac{V_a \sin 60°}{V_m + V_a \cos 60°}$$

$$0.2679 = \frac{k0.866}{1 + k0.500}$$

$$k = \frac{V_a}{V_m} = 0.366$$

$$V_{12} \cos 15° = V_m + kV_m \cos 60°$$

$$0.9659V_{12} = V_m(1 + 0.366 \times 0.500)$$

$$V_m = 0.817V_{12} \tag{124}$$

$$V_a = 0.366 \times 0.817V_{12}$$

$$= 0.299V_{12} \tag{125}$$

Polyphase autotransformer connections. Y connection of autotransformers. Δ connection of autotransformers. Extended-Δ connection of autotransformers.

Polyphase Autotransformer Connections. Autotransformers may be used instead of the usual two-winding transformers for transforming polyphase power. Their advantages for this purpose are similar to those discussed in Chap. 8 for single-phase autotransformers. They are frequently used to step up the voltage of a generating station to a higher voltage for transmission purposes, to interconnect two systems of differing voltages, and in the starting compensators used with induction motors. Although there are many ways in which autotransformers may be connected for polyphase transformations, only the more common three-phase connections will be discussed here.

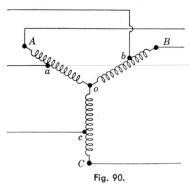

Fig. 90.

Y Connection of Autotransformers. Figure 90 shows the Y connection as applied to autotransformers. This connection may be used either to step up or to step down the voltage. The same difficulties with regard to third harmonics and unbalanced line-to-neutral voltages are experienced with this connection as with the Y-Y connection of two-winding transformers, and these problems may be overcome in the same ways, *i.e.*, either by interconnecting the neutrals of the transformer and source or by using Δ-connected tertiary windings.

The voltages on the primary and secondary sides are in direct proportion to the turns included, that is, $E_{AB}/E_{ab} = N_{Ao}/N_{ao}$. The currents in the series windings are the same as the currents in the high-voltage lines. Since, if the exciting currents are neglected, the total ampere-turns for each transformer must equal zero, the ampere-turns of each common winding must be equal and opposite to the ampere-turns of the series winding of the same transformer. Therefore, each line connected to the low-voltage side carries a current equal to the arithmetical sum of the currents in the common and series windings. If the ratio of transformation were 1.5 and a current of 20 amp was supplied over line A, the current in Aa would be 20 amp, with 10 amp in ao and 30 amp in the line connected to a.

With two-winding transformers the kva capacity is one-half the total kva rating of all windings both when used in single-phase circuits and, under balanced conditions using either Δ or Y connections, in three-phase circuits. For the Y-connected autotransformer, this ratio depends on the ratio of transformation. If a is the ratio of the open-circuit line voltages of the high- and low-tension sides, then

$$\frac{\text{Kva capacity}}{\text{Total kva rating of all windings}} = \frac{a}{2(a-1)}$$

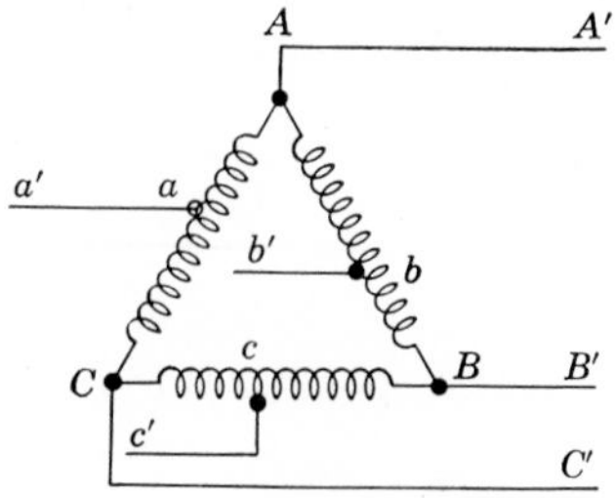

Fig. 91.

Δ Connection of Autotransformers. One form of the Δ connection used with autotransformers is shown in Fig. 91. With this connection the secondary line voltages are not in phase with the primary line voltages and the greatest possible ratio of transformation is 2. Because the volt-

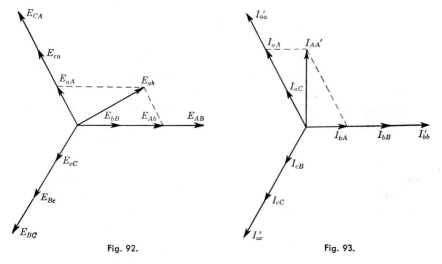

Fig. 92.   Fig. 93.

age and current relationships are more complex than in the Y connection, a special case will be considered where the turns included between A and tap b are two-thirds of the total turns on transformer AB, and similarly for the other two transformers. The voltage relationships are shown in Fig. 92. The line voltage on the low-tension side is the sum of two voltages displaced by an angle of 120 deg. Leakage impedance drops are neglected. For the above case, the low-tension line voltage equals the high-tension line voltage divided by $\sqrt{3}$.

The current relationships are shown in the vector diagram of Fig. 93. In this diagram all exciting currents are neglected. Each low-tension line current divides between the two sections of the winding to which it is connected inversely as the turns and for the above special case is $\sqrt{3}$ greater than the line current on the high-tension side. For this particular ratio of transformation

$$\frac{\text{Kva output}}{\text{Total Kva rating of all windings}} = 0.75$$

While this is better than the 0.5 ratio for two-winding transformers, it is not so good as that for the Y connection.

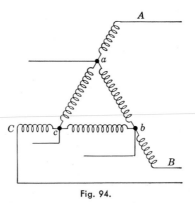

Fig. 94.

Extended-Δ Connection of Autotransformers. The extended-Δ connection is shown in Fig. 94. With this connection ratios of transformation greater than 2 are possible but there is still a phase difference between the primary and secondary voltages. A vector diagram is useful in determining the relationships between the various voltages and currents. If the taps are so located that terminals ab include two-thirds of the turns of transformer aB and similarly for the other transformers, then the ratio of high-tension to low-tension line voltages is approximately 1.8, that is, $E_{AB}/E_{ab} = 1.8$. The current $I_{ab} = I_{Bb}/2$ and the current in any low-tension line is 1.8 times the high-tension line current. For this particular ratio of transformation,

$$\frac{\text{Kva capacity}}{\text{Total kva rating of all windings}} = 1.04$$

This is better than for the other Δ connection, but is not so good as the Y.

Chapter 12

Three-phase transformers. Third harmonics in the exciting currents and in the induced voltages of Y-connected and Δ-connected transformers. Advantages and disadvantages of three-phase transformers. Parallel operation of three-phase transformers or three-phase groups of single-phase transformers. V-connected and Δ-connected transformers in parallel.

Three-phase Transformers. A considerable saving in material and therefore in the cost of transformers required for three-phase circuits can be effected by combining their magnetic circuits.

Core Type. For example, consider three identical single-phase core-type transformers which are to be used on a three-phase circuit. If both

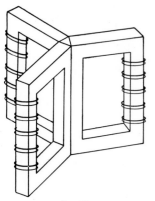

Fig. 95.

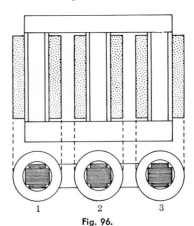

1 2 3

Fig. 96.

the windings on each transformer are placed on only one side of the core and the opposite sides of the iron cores are butted together as in Fig. 95, the component fluxes in the three sides which are placed together are 120 deg apart in time phase and their resultant is zero. The common portion of the iron core can therefore be removed without affecting the operation of the transformers.

The core type of three-phase transformer as actually built has the three parts of the core which carry the windings in one plane as shown in Fig.

145

96.   Any one leg of the iron core carries a flux which is the resultant of the fluxes in the other two legs; consequently the reluctances of the magnetic circuits for the fluxes of phases 1 and 3, Fig. 96, are slightly greater than the reluctance of the magnetic circuit for the flux of phase 2.   The effect of this is a slight unbalancing of the magnetizing currents.   This has little influence upon the operation of the transformer.

The yokes between the portions of the iron core which are surrounded by the windings form a Y coupling for the three magnetic circuits of the

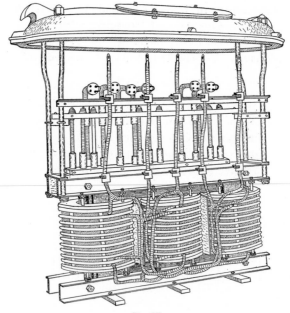

Fig. 97.

three-phase transformer in Fig. 96.   The yokes carry equal fluxes, neglecting the leakage fluxes, and should have the same cross section as the portions of the iron core surrounded by the windings.   The yokes could be arranged in Δ, but this arrangement would be more expensive to construct, would require more space, and would possess no particular advantage.   A three-phase core-type transformer is shown in Fig. 97.

Shell Type.   When the three-phase transformer is of the shell type, the windings are embedded in the iron core instead of surrounding the iron core as in the core type.   The usual arrangement of a shell-type three-phase transformer is shown in Fig. 98, which gives two sectional views. The three groups of windings are 1-1, 2-2, and 3-3.

The resultant mmf producing flux along the whole length of the core, or along the line *abc*, is the vector sum of the three mmfs due to the three

groups of windings. If the three groups are connected in the same direction, the mmfs produced by them are 120 deg apart and their vector sum
taken along abc is zero. The fluxes passing between the two pairs of
adjacent windings, 1 and 2, and 2 and 3, in the spaces d and e and f and
g, are equal to one-half the vector difference of two equal fluxes which
have a phase difference of 120 deg. The fluxes in the spaces d and e and
f and g therefore are equal to $\frac{1}{2} \sqrt{3} = 0.866$ of the flux linking a single
winding.

The mmfs acting to produce fluxes between any pair of coils are in
parallel, instead of in series as in the core-type transformer. The magnetic circuits of the three phases of a shell-type transformer are therefore
much more independent of one another than the magnetic circuits of a

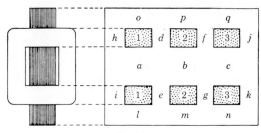

Fig. 98.

core-type transformer. If the flux is prevented from passing through the
windings of any one phase of a shell-type transformer, there are still
magnetic circuits for the fluxes of the other two phases and they can be
operated in open Δ. The two windings of a core-type transformer cannot be operated in open Δ if the flux is prevented from passing through the
core of the third phase, since both of the active windings would have to
carry the same flux instead of carrying two fluxes differing in phase by
120 deg. The performance of a shell-type transformer under these
conditions is of some commercial importance since it permits such a
transformer to be operated temporarily with one winding out of service.
If one winding of a shell-type transformer is injured, the remaining two
windings can be operated in open Δ giving 58 per cent of the normal
capacity of the transformer, provided the injured winding is disconnected
and either its primary or its secondary winding, or preferably both, are
short-circuited. If the injured winding is short-circuited, any current
which flows in it has no electric circuit upon which to react and therefore is
wholly magnetizing current. As a result, any flux tending to pass
through the injured winding is opposed and only a small current flows in
the short-circuited phase. The voltage induced in this phase is equal
merely to the impedance drop due to this small current. If one phase of

a core-type transformer is short-circuited, the remaining two phases cannot operate in open Δ since their magnetic circuits are in series.

Some iron can be saved in the construction of a shell-type transformer by reversing the connections of the middle phase, phase 2, Fig. 98. If the connections of the middle winding are reversed, the fluxes carried by the portions of the core between the windings, the portions $d, e, f, g,$ are as before equal to one-half the vector difference of the fluxes linking two adjacent windings, but in this case the fluxes threading two adjacent windings differ in phase by 60 instead of by 120 deg. Their vector difference is therefore numerically equal to either flux, and the parts $d, e,$

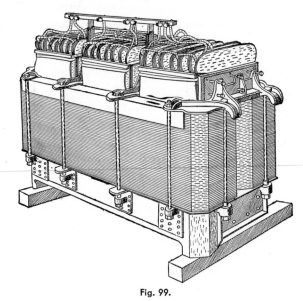

Fig. 99.

f, g of the core carry fluxes which are equal to one-half the flux through any one winding instead of being equal to 0.866 of this flux as when the middle winding is not reversed. If the middle winding is reversed, the cross section of the magnetic circuit throughout the transformer should be the same. It should be remembered that certain portions of the magnetic circuit consist of two parallel paths. With the middle winding reversed, the cross sections of $a, b, c, h + i, d + e, f + g, j + k, o + l,$ $p + m, q + n$ should be equal if they are to operate at equal flux densities. If the phases are all connected alike, the cross sections of $d + e$ and $f + g$ must be $\sqrt{3} = 1.73$ times the cross section of the other parts of the magnetic circuit for equal flux densities.

A three-phase shell-type transformer is shown in Fig. 99.

Third Harmonics in the Exciting Currents and in the Induced Voltages of Y-connected and Δ-connected Transformers. What has been said about third harmonics in single-phase transformers connected for three-phase transformation applies equally well to three-phase shell-type transformers but does not apply to three-phase core-type transformers. The portions of the core about which the windings of a three-phase core-type transformer are placed are joined in Y without a common return corresponding to the neutral conductor of a Y-connected electric circuit. This should be made clear by referring to Fig. 95, page 145, remembering however that the central portion of the core shown in this figure, the portion made by the three sides which are butted together, is left out in a three-phase transformer. The three third-harmonic fluxes which must be present in the three legs of the core to produce third-harmonic voltages in the windings are in phase. There are no return paths for these fluxes except from one yoke to the other through the high-reluctance nonmagnetic paths surrounding the windings. As these return paths are high-reluctance paths, the resulting third-harmonic fluxes are much less than in single-phase transformers or in three-phase shell-type transformers under similar conditions, where the return paths for the third harmonic fluxes are in iron and are consequently low-reluctance paths. The effect of these high-reluctance paths in three-phase core-type transformers is to reduce the third-harmonic voltages to a few per cent of the fundamental voltages instead of being 50 to 90 per cent of the fundamental voltages as in corresponding single-phase transformers. There can therefore be no third-harmonic voltages in the windings of a core-type three-phase transformer under the condition of balanced impressed voltages except those due to the third-harmonic leakage fluxes between the yokes of the core. These fluxes between the yokes must be very small. Neither can there be any third-harmonic components in the magnetizing currents in any of the windings, no matter how connected.

In what follows, balanced impressed voltages are assumed and the effect of the leakage fluxes is neglected. Sinusoidal impressed voltages are assumed. Assume that the primaries are connected in Y with the neutral connected to the neutral of the source of power and that the secondaries are open. Under these conditions the primary windings receive power independently of one another. The neutral conductor carries the combined third-harmonic currents for the three phases, if such currents exist.

Refer to Fig. 95, page 145. Consider the common central leg of the core removed. There is then no common return path for the third-harmonic flux. Let the instantaneous values of the mmfs of the three phases at any instant due to the magnetizing currents be $\mathfrak{F}_1$, $\mathfrak{F}_2$, $\mathfrak{F}_3$ and let $\mathfrak{R}_1$, $\mathfrak{R}_2$, $\mathfrak{R}_3$ be the corresponding instantaneous values of the reluctances of the three magnetic circuits up to their common junction. Let φ_1, φ_2, φ_3 be the instantaneous values of the fluxes.

Then

$$\mathfrak{F}_1 - \mathfrak{R}_1\varphi_1 - \mathfrak{F}_2 + \mathfrak{R}_2\varphi_2 = 0$$
$$\mathfrak{F}_1 - \mathfrak{R}_1\varphi_1 - \mathfrak{F}_3 + \mathfrak{R}_3\varphi_3 = 0$$
$$\varphi_1 + \varphi_2 + \varphi_3 = 0$$

Solving for φ_1,

$$\varphi_1 = \frac{(\mathfrak{F}_1 - \mathfrak{F}_2)\mathfrak{R}_3 + (\mathfrak{F}_1 - \mathfrak{F}_3)\mathfrak{R}_2}{\mathfrak{R}_1\mathfrak{R}_2 + \mathfrak{R}_2\mathfrak{R}_3 + \mathfrak{R}_3\mathfrak{R}_1} \tag{126}$$

The expression for the flux φ_1 is general. Similar expressions hold for φ_2 and φ_3.

If the primaries are in Y with neutral connection, the third-harmonic components in the magnetizing currents which would ordinarily be necessary to produce a sinusoidal flux can come in over the neutral. As a matter of fact, no such third-harmonic components are necessary for a core-type transformer as shown in Eq. (126).

Assume that φ_1 is sinusoidal and that $\mathfrak{F}_1$, $\mathfrak{F}_2$, $\mathfrak{F}_3$ contain the necessary third harmonics. These third harmonics are in phase. They are equal since the reluctances of the three magnetic circuits are equal. They affect equally all three mmfs, and as the mmfs enter in the expression for the flux as differences, the third harmonics can be eliminated without altering the flux [(Eq. (126)]. Hence the third harmonics are not required in the mmfs of a core-type three-phase transformer in order to produce sinusoidal flux variation in each phase. No third-harmonic component exists in the magnetizing currents or in the primary neutral. Opening the neutral does not alter the fluxes or the induced voltages in the transformer. If the secondaries are in Δ with the primaries in Y without neutral, no third-harmonic magnetizing currents exist in the transformer such as existed in the secondaries of a three-phase shell-type transformer or in the secondaries of three single-phase transformers under similar conditions.

The transformer in Fig. 95 has a core which is symmetrical with respect to the three phases. The core of the ordinary three-phase core-type transformer is like that shown in Fig. 96 and is unsymmetrical. The reluctances for the three phases are not equal. As a result the magnetizing currents are somewhat unbalanced and there is a current of fundamental frequency in the neutral conductor if the primaries are Y-connected with the neutral connection. If there is no neutral connection, the Y voltages are slightly unbalanced. There can be small third-harmonic magnetizing currents with the neutral closed. When the neutral is open, there are small corresponding harmonics in the induced voltages. If either the primary or the secondary windings are connected in Δ, a small zero-sequence current can flow in the closed path formed by the Δ. Such a current would largely eliminate the very slight unbalancing of the voltages caused by slight dissimilarity in the magnetic circuits for the three phases.

If the alternator supplying the transformer has a third harmonic in its phase voltage and the primaries of the transformers are in Y with their neutral connected to the neutral of the alternator, there may be third-harmonic currents in the transformers, the lines, and the neutral connection. Since the third-harmonic mmfs for the three phases are in phase, they cannot produce any mutual flux. The conditions, so far as third harmonics are concerned, are the same as those existing in a single-phase core-type transformer with two equal sections of the primary winding bucking and on opposite sides of the core. The third-harmonic voltages impressed on each phase are short-circuited through the resistance and the third-harmonic leakage reactance of each phase. The third-harmonic leakage flux which causes the third-harmonic leakage reactance in this case is not like the ordinary leakage flux for the fundamental which passes between the primary and secondary windings, but is a leakage flux which links both primary and secondary windings, passes through the upright legs of the core, and returns through the high-reluctance air paths between the yokes. It is a leakage flux for the core but not for the windings. The leakage reactance caused by this third-harmonic leakage flux is much higher than the ordinary leakage reactance for the fundamental partially on account of the higher frequency of the third harmonic but chiefly because of the much lower reluctance of the magnetic circuit for the third-harmonic leakage flux. The third-harmonic leakage flux links both primary and secondary windings and induces third-harmonic voltages in each.

Advantages and Disadvantages of Three-phase Transformers. *Advantages.*

Three-phase transformers require less material for a given output than three single-phase transformers. They are lighter, cost less, require less floor space, and have higher efficiencies than three single-phase transformers of equal total capacity.

The windings of a three-phase transformer can be connected for Y or Δ inside the containing tank, thus reducing the number of high-tension leads which have to be brought out through the tank. Only three high-potential leads need to be brought out, but with three single-phase transformers six leads must be brought out for Δ connection and six for Y connection, except in the case of high-potential transformers when one terminal of their high-potential windings can be grounded on the tank. As high-potential transformers are always connected in Y on their high-potential sides and the neutral point is grounded, there is no object in insulating from the tank more than one end of the high-potential winding.

Disadvantages. The three principal disadvantages of three-phase transformers or of polyphase transformers in general are the greater cost of spare units, the greater cost of repairs, and the greater derangement of service in case of breakdown.

In small distributing systems having few transformers of the same rating, the relative cost of spares with single-phase and three-phase transformers is the relative cost of one single-phase transformer as compared with one three-phase transformer. When however a distributing system is large and requires many transformers, the number of spares necessary compared with the total number of transformers in service is much smaller than in a small system, and the increase in the cost of spares is low compared with the saving in cost of transformers required for the whole system. In such a case, the total cost of three-phase transformers with the necessary spares is usually less than the cost of an equivalent capacity in single-phase transformers including spares. The gain in efficiency and the decreased cost of transportation due to the decrease in total weight for a given capacity and the decrease in cost of installation on account of the simplification of wiring are important items favoring three-phase transformers.

With three-phase transformers, there is a possibility of greater damage with a bad short circuit on one phase, but this is not an important item except when transformers are used in exposed places where they are liable to be subjected to severe strains from lightning or other causes.

Although a three-phase transformer weighs less and occupies less space than three single-phase transformers of the same total capacity, its greater weight and size as compared with one of the single-phase transformers may prevent its use in places difficult of access on account of the greater difficulties of transporting it as compared with transporting three single-phase transformers.

Parallel Operation of Three-phase Transformers or Three-phase Groups of Single-phase Transformers. The conditions to be fulfilled for the parallel operation of single-phase transformers must also be fulfilled for the parallel operation of three-phase transformers. These conditions are equal ratios of transformation, phase voltages which give equal line voltages, equal percentage impedance drops, and equal ratios of equivalent resistance to equivalent reactance. The last condition is unimportant when the ratio of equivalent resistance to equivalent reactance for the transformers is small.

Three-phase transformers or groups of three single-phase transformers, supplied from a common source, cannot be paralleled indiscriminately, even when the conditions just stated are fulfilled, since there is a phase difference between corresponding secondary voltages with certain connections. For example, if the primaries of two groups of transformers are connected in Δ and the secondaries of one group are in Y and the secondaries of the other group are in Δ, there is a phase difference of 30 deg between corresponding secondary voltages. This is shown in Fig. 100 which gives vector diagrams of the voltages obtained. The secondary line voltages given by the two connections can be made equal by using proper ratios of transformation, but the line voltages cannot be brought into phase. The smallest difference in phase between the secondary line voltages given by the two connections is 30 deg. A Y-Δ system cannot be paralleled with a Y-Y system, or with a Δ-Δ system, but a Δ-Δ system can be paralleled with a Δ-Δ system or with a Y-Y system. The secondary

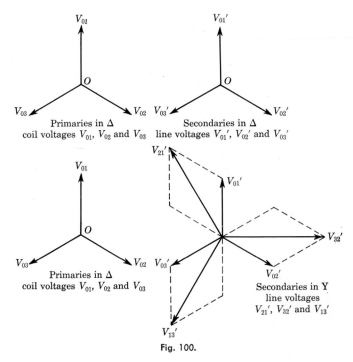

Fig. 100.

voltages of a Y-Δ system and a Δ-Y system are either in phase or are 60 deg out of phase, depending upon the way the connections are made. Two such systems can therefore be paralleled.

Any group of three-phase transformers fed from independent sources can be paralleled on their secondary sides, provided the magnitudes of the secondary line-to-line voltages are the same and these voltages are brought into phase by adjusting the phase relation between the sources of power supplying the primaries. In this case the phase relation of the sources of power can be made to accommodate itself to the conditions imposed by the transformer connections.

In order to show the magnitude of the short-circuit circulatory current produced if two groups, each of three single-phase transformers supplied from the same source of power, are put in parallel when their connections are such that they cannot properly be paralleled, consider a particular case. Let the primaries of both groups of transformers be in Δ. Let the secondaries of one group be in Δ and of the other in Y. Assume that the transformers in the two groups are wound for the same primary voltage and are identical except in their ratios of transformation which are in the ratio of $1 : \sqrt{3}$. Let the impedance voltages of the transformers be 5 per cent of their rated voltages at full-load current. The connections are shown diagrammatically in Fig. 101.

Let the ratio of transformation of the Δ-Δ-connected group be unity and the ratio of transformation of the Δ-Y-connected group be $\sqrt{3}$.

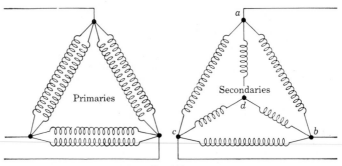

Fig. 101.

Assume that the primary and secondary impedances are equal when referred to the same winding. If z is the leakage impedance of the primaries, the equivalent leakage impedances of the Δ-Δ-connected group and of the Δ-Y-connected group referred to the secondaries are $2z$ and $\frac{1}{3}z + \frac{1}{3}z = \frac{2}{3}z$.

$$E_{bd} + E_{da} + E_{ab} = I_{bd}\tfrac{2}{3}z + I_{da}\tfrac{2}{3}z + I_{ab}2z \qquad (127)$$

For the short-circuit circulatory currents, by Kirchhoff's laws,

$$I_{da} = I_{ab} + I_{ac}$$
$$I_{bd} = I_{ab} + I_{cb}$$

Since the system is symmetrical, the currents in the two groups of transformers are balanced, i.e., the phase currents in each group are equal in magnitude and differ in phase by 120 deg. There is the usual phase displacement between the Δ and Y currents. Therefore

$$\begin{aligned}
I_{da} &= \sqrt{3}\, I_{ab}(\cos 30° - j \sin 30°) \\
&= \sqrt{3}\, I_{ab}(0.866 - j0.500) \\
I_{bd} &= \sqrt{3}\, I_{ab}(\cos 30° + j \sin 30°) \\
&= \sqrt{3}\, I_{ab}(0.866 + j0.500)
\end{aligned}$$

Substituting these values of I_{da} and I_{bd} in Eq. (127),

$$(E_{bd} + E_{da}) + E_{ab} = zI_{ab}[\tfrac{2}{3} \sqrt{3}(0.866 + j0.500)$$
$$+ \tfrac{2}{3} \sqrt{3}(0.866 - j0.500) + 2]$$
$$= 4I_{ab}z$$

The voltage $(E_{bd} + E_{da})$ is equal in magnitude to the voltage $-E_{ab}$ but differs from it in phase by 30, 90, or 150 deg according to the way the Δ-Δ-connected and Δ-Y-connected groups of secondaries are connected together.

Consider the connection which gives the least phase difference and therefore the least short-circuit current. For this phase difference of 30 deg, the voltage $[(E_{bd} + E_{da}) + E_{ab}]$ is equal to

$$2 \cos \left(\frac{180° - 30°}{2} \right) E_{ab} = 0.518 E_{ab}$$

Hence for this particular phase difference

$$0.518 E_{ab} = 4I_{ab}z$$
$$I_{ab} = I_{\Delta cir} = E_{ab} \frac{0.518}{4z} \tag{128}$$

where $I_{\Delta cir}$ is the circulatory current in the Δ-Δ-connected group.

Since a 5 per cent impedance drop at full-load current is assumed, the full-load current $I_{\Delta fl}$ in each of the Δ-Δ-connected transformers can be found from

$$I_{\Delta fl}(2z) = 0.05 E_{ab}$$
$$I_{\Delta fl} = \frac{0.05}{2z} E_{ab} \tag{129}$$

Substituting the value of E_{ab} from Eq. (129) in Eq. (128),

$$I_{\Delta cir} = 5.18 I_{\Delta fl}$$

or the short-circuit circulatory current is 5.18 times the rated full-load current. The current in each winding of each transformer is 5.18 times its rated full-load value. If a Δ-Y-connected group is paralleled with a Y-Δ-connected group, the short-circuit circulatory current is either about twice as great, as in the case discussed, or is zero, depending on the connections between the two groups of transformers. These discussions assume that the primary impressed voltages remain fixed.

V-connected and Δ-connected Transformers in Parallel. If two identical transformers connected in V-V are operated in parallel with three identical Δ-Δ-connected transformers of the same type and rating, and if the transformers of the two groups have identical equivalent impedances, the combined permissible output of the system is only $33\frac{1}{3}$ per cent greater than the rated output of the Δ-Δ-connected group, although the total capacity involved is 66⅔ per cent greater. Let Fig. 102 represent the connections of both primary and secondary sides of the transformers. On two phases, two transformers are in parallel. A single transformer is connected to the third phase.

The limit of output is fixed by the transformer which first reaches full load. This is the transformer across the open side of the V.

The Δ-Δ-connected transformers in parallel with the V-V-connected transformers are equivalent to a single bank of Δ-Δ-connected transformers, two of which have equivalent impedances which are half as large as the equivalent impedance of the third transformer. Refer to Fig. 67, page 116, and to Eqs. (118), (119), (120), page 117, which apply to Δ-Δ-connected transformers having unequal equivalent impedances. Let the branches bc and ca, Fig. 102, correspond to the branches carrying currents I''_a and I'''_a in Fig. 67. Since the five transformers are identical,

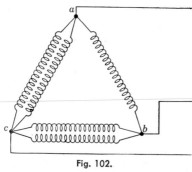

Fig. 102.

$$z''_e = z'''_e = 0.5z'_e$$

Assume the currents I_1, I_2, I_3, Fig. 67, page 116, to be balanced and the current I_1 to be leading the current I_2. Call I the magnitude of the balanced line currents.

$$I_1 = I(1 + j0)$$
$$I_2 = I(-0.5 - j0.866)$$
$$I_3 = I(-0.5 + j0.866)$$

From Eq. (118), page 117,

$$\begin{aligned}
I_{ab} = I'_a &= \frac{I_1 z'''_e - I_2 z''_e}{z'_e + z''_e + z'''_e} \\
&= \frac{I(1 + j0)0.5z'_e - I(-0.5 - j0.866)0.5z'_e}{0.5z'_e + 0.5z'_e + z'_e} \\
&= \frac{0.5}{2}(1.5 + j0.866)I
\end{aligned}$$

$$I = I'_a \frac{4}{\sqrt{(1.5)^2 + (0.866)^2}} = \frac{4}{\sqrt{3}}I'_a \text{ in magnitude}$$

If I'_a is the full-load current of one transformer, the permissible line current from the Δ-Δ-connected transformers alone is $\sqrt{3}\,I'_a$. Therefore the ratio of the permissible output from a Δ-Δ-connected bank of transformers in parallel with two identical transformers connected in V-V is $(4/\sqrt{3})(1/\sqrt{3}) = 1.33$. The gain in output by putting the V-V-connected transformers in parallel with the Δ-Δ-connected transformers is 33⅓ per cent.

PART TWO

SYNCHRONOUS GENERATORS

Chapter 13

Types of synchronous generators. Frequency. Armature cores. Field cores. Armature insulation. Field insulation. Cooling. Cooling air. Hydrogen as a cooling medium. Permissible temperatures for different types of insulation.

Types of Synchronous Generators. Synchronous generators do not differ in principle from generators for direct current. Any d-c generator, with the exception of the unipolar generator, is in fact a synchronous generator in which the alternating voltage set up in the armature inductors is rectified by means of a commutator. Although any d-c generator, with the exception of the unipolar generator, can be used as a synchronous generator by the addition of collector rings electrically connected to suitable points of its armature winding, it is found more satisfactory, both mechanically and electrically, to interchange the moving and fixed parts when only alternating currents are to be generated. It is not only a distinct advantage mechanically to have the more complex part of the machine stationary, but it is easier with this arrangement to protect and insulate the armature leads which usually carry current at high potential. The only moving contacts required are those necessary for the field excitation, and these carry current at low potential.

Synchronous generators may be divided into three classes which differ mainly in the disposition and arrangement of their parts. The three classes are (1) generators with revolving fields, (2) generators with revolving armatures, and (3) inductor-type generators.

All modern synchronous generators, with few exceptions, belong to the first class for reasons which have already been stated. Inductor-type generators differ from the other two types by having the variation in the flux through their armature windings produced by the rotation of iron inductors. The windings of both the armature and the field of this type of synchronous generator may be stationary. A distinguishing feature of an inductor-type generator is that any one set of armature coils is subjected to flux of only one polarity. This fluctuates between the limits of zero and maximum but does not reverse. Figures 103 and 104 illustrate the first two classes of synchronous generators in their simplest forms. Figures 105 and 106 show two views of a form of synchronous generator of the third class. Figure 105 is a portion of a section taken through the axis of the shaft about which the inductor revolves. Figure 106 is a side view. The letters on these figures have the following significance:

F = field coil
A = shaft
CC = armature coils
NIS = inductor

By referring to Figs. 103 and 104 it is evident that both sides of the coils on the armatures of the revolving-field and the revolving-armature

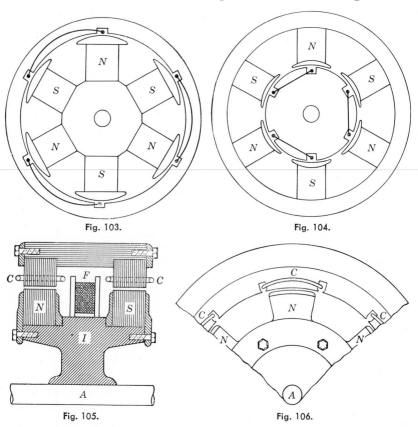

Fig. 103.   Fig. 104.

Fig. 105.   Fig. 106.

types of generators are in active parts of the field at the same time, and since the opposite sides of the coils are under poles of opposite polarity at each instant, the voltages induced in them are in phase with respect to the coil. The conditions are different in the case of the inductor type of generator. In this generator only one side of an armature coil is in an active part of the field at any time. The product of turns and flux must be doubled in order to get the same voltage as would be obtained if the flux through the armature winding reversed as it does in the two other types of synchronous generators. An inductor generator is usually characterized by large armature reaction, relatively high magnetic density, small air gap, and greater weight than synchronous generators of the other types. The difficulties in the design of a satisfactory inductor generator have caused this type of generator to go out of use except for limited applications.

Frequency. The commercial frequencies that are most common in the United States are 60 and 25 cps, although other frequencies are used. In California and in Mexico, a frequency of 50 cps is used for some large power systems. In Europe, 50 and 40 cps are common. A frequency of 25 cps is best for long-distance power transmission, because of the decrease with frequency of the troublesome effects due to the inductance and capacitance of transmission lines, but so low a frequency is not suitable for lighting on account of the noticeable flicker produced by it on arc lights and on all incandescent lamps except those with filaments of large cross section. A frequency of 25 cps or less is best adapted for single-phase motors of the series type such as are used for traction purposes.

The frequency given by any synchronous generator depends upon its speed and the number of poles and is equal to

$$f = \frac{pn}{2(60)} \tag{130}$$

where f, p, and n are the frequency in cycles per second, the number of poles, and the speed in revolutions per minute. The speed and therefore the number of poles for which a synchronous generator of a given frequency is designed depend upon the method of driving it. Engine-driven generators as well as generators driven by water wheels operated from low heads must run at relatively low speeds and consequently must have many poles. On the other hand generators driven by steam turbines operate at high speeds and must have few poles, usually two to six according to their frequency and size. Low-frequency generators, as well as nearly all low-frequency power and conversion apparatus, are always heavier and therefore in general more expensive than high-frequency generators of the same rating and speed, but the advantages of low frequency for certain classes of work, notably power transmission and traction, in some cases more than balance the higher cost of the use of low frequency.

Armature Cores. The armature core of a synchronous generator is built up of thin sheet-steel stampings with slots for the armature coils on one edge. The opposite edge usually has either two or more notches for

keys which are inserted in the frame in which the laminations are built
up, or projections which fit in slots cut in the frame. Notches cut in the
sides of the teeth serve to hold the wedges driven between adjacent teeth
to keep the armature coils in place.

Typical armature stampings are
shown in Figs. 107 and 108, which
illustrate stampings for a slow-speed
or moderate-speed generator and
for a turbine generator. The holes

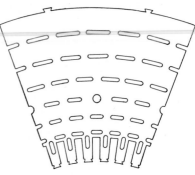

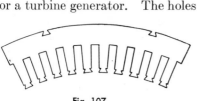

Fig. 107.   Fig. 108.

through the laminations for the turbine generator form passages, when
the laminations are built up, through which air is forced for cooling the
armature.

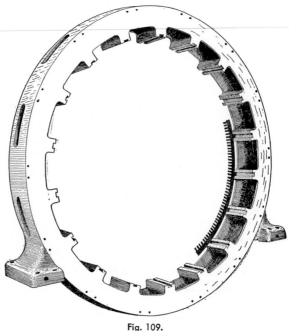

Fig. 109.

The armature stampings are built up with lap joints in a frame or yoke
ring, which formerly was of cast steel but which now is usually built up
of fabricated steel plates electrically welded. The stampings are held

Fig. 110. Welded-steel frame for stationary armature for General Electric alternating-current turbine generator.

Fig. 111. Wound stationary armature for General Electric alternating-current turbine generator.

from slipping either by keys in the frame or by projections on the laminations. The stampings are securely bolted together and to the frame between the end plates. These plates usually have projecting fingers to support the teeth. Figure 109 shows a typical cast-steel frame for an engine-driven generator. Figure 110 shows a fabricated steel frame.

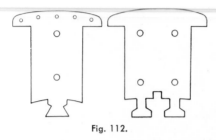

Fig. 112.

The frame or yoke which supports the laminations is hollow and is provided with openings for ventilation. The armature laminations are separated in two or more places by the insertion of spacing pieces in order to provide radial air ducts for cooling the armature. Large slow-speed generators, which necessarily have many poles and frames of large diameter, have their frames or yoke rings made in two or more sections which are bolted together and can be separated for transportation.

A typical frame for a turbine generator with the laminations and armature winding in place is shown in Fig. 111. As turbine generators require forced ventilation, they must be completely enclosed.

Field Cores. All slow-speed synchronous generators of standard design have laminated salient or projecting poles built up of steel stampings. These are bolted together and either keyed or bolted to a steel

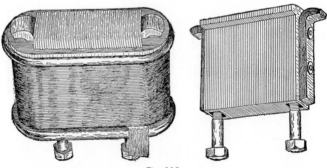

Fig. 113.

spider which is itself keyed to the shaft. Figure 112 shows typical pole stampings. Figure 113 shows the core of a complete pole of the bolted-on type both with and without the winding. A complete field with the poles and winding in place is shown in Fig. 114.

The field structure of a high-speed turbine generator does not have projecting poles. It is cylindrical in form and has slots cut in its surface for the field winding. Such generators have two to six poles, according to their size and the frequency.   Projecting or salient poles would cause excessive windage loss and in addition would make a high-speed generator very noisy.   Moreover it would be difficult, if not impossible, to make a field structure with salient poles sufficiently strong to stand safely the high speeds used for turbine generators.

Fig. 114.   (*Courtesy of Westinghouse Electric Corp.*)

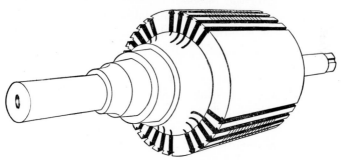

Fig. 115.

The field structure for a turbine generator is often a solid steel forging. For a large machine it is sometimes built up of several forged-steel sections or of thick disks cut from forged-steel plates.   The shaft does not pass through the core, as the hole required in the core for this purpose would remove too much metal back of the slots which receive the field winding and thus weaken the structure.   The shaft is in two pieces fastened to end plates securely bolted to the core.   The distribution of flux over the

pole faces is determined by the distribution of the field coils which are
placed in slots cut in the core.   Figure 115 illustrates the construction of
a two-pole field.

Armature Insulation.   The conductors which form the coils of synchronous generators, as well as the coils themselves, must be insulated in much
the same way as the conductors and coils of d-c generators.   On account
of the high voltages at which synchronous generators usually operate,
they require much more insulation than d-c machines.   The materials
used in the insulation of d-c generators are often not suitable for synchronous generators on account of the higher voltages of the latter and the
higher temperatures often reached by their windings.

Vulcanized fiber, horn fiber, fish paper, varnished cambric and paper,
mica, and other similar materials are used in the insulation of synchronous
generators.   When high temperatures do not have to be resisted by
the armature windings, double-cotton-covered or triple-cotton-covered
wire is used for the coils.   These are thoroughly impregnated with insulating compound after being wound and are then covered with several
layers of varnished cambric or some other similar material.   With such
coils it is necessary to insulate the slots with fiber or mica.   Vulcanized
fiber has a tendency to absorb moisture, which causes it to expand and
also reduces its insulating properties.   For this reason it should not be
relied upon to insulate high-voltage machines.

Mica is the only reliable insulation when high voltage or high temperature is to be encountered.   The objection to mica is its poor mechanical properties, and for this reason it has to be used with other materials.
For insulating slots it is split into thin flakes which are built up with lap
joints into sheets with varnish or bakelite to cement the mica together.
It is then baked under pressure.   When built up in this way, the finished
sheet can be molded hot into U-shaped troughs or into other shapes for
insulating the slots or other parts of the machine.

Mica is now exclusively used for the insulation of the portions of the
armature windings which lie in the slots of high-voltage generators.
With the high speeds necessary for turbine generators, comparatively few
armature turns are required for a given voltage.   The voltage per turn
is therefore increased and more insulation is necessary between turns.
Large machines often have only one or two turns per coil.   Although
fibrous insulation could still be applied which would withstand the higher
voltage between turns, mica is the only substance which can withstand
the high temperatures reached by certain parts of the coil which are
embedded in the armature iron.   The portions of the coils which lie outside of the slots, *i.e.*, the end connections, are insulated with varnished
cambric, mica tape, or some similar material.

The mica insulation is now generally applied and rolled hot on the straight portions of the conductors and coils. In this way the mica is so tightly rolled on the coil that it forms a solid mass of insulating material of minimum thickness free from air spaces and having good heat conductivity. Mica insulation applied in the ordinary way has a heat conductivity of only 60 or 70 per cent of that of varnished cambric and similar materials.

The static discharge which was often encountered between the armature copper and iron in the earlier high-voltage generators is avoided when rolled-on mica insulation is used. The effect of the static discharge was most marked where there were sharp edges, as at the edges of the radial ventilating ducts. Its effect was to pit the outside insulation of the coils, weakening or even destroying the insulation.

Field Insulation. Since the fields of synchronous generators are always wound for low voltage, 125 to 250 volts, the problem is not so much one of providing insulation as of providing a mechanical separation between turns which is mechanically strong and can withstand high temperature. Neither the problem of mechanical strength nor the problem of high temperature is serious in the case of slow-speed generators, since the stresses and temperatures which have to be withstood by the field windings of such machines are not great.

In case a synchronous generator becomes short-circuited, the field winding may be subjected to high voltage during the initial rush of armature current because of the transformer action which takes place between the armature and the field. This action as a rule is not serious. It is least in synchronous generators with non-salient poles and low field reactance. Sufficient insulation must be provided on the field winding to guard against breakdown due to this cause.

Synchronous generators with salient poles usually have their fields wound with double-cotton-covered wire with insulating strips between layers. After being wound, the coils are impregnated with insulating compound and taped. They are then placed on insulating spools of fiber or similar material and slipped over the pole pieces. Fields are often wound with flat strip copper wound on edge. In this case the successive turns are insulated from one another by strips of thin asbestos paper or other material. The copper at the outside surface of edgewise-wound field coils may be left bare to facilitate cooling.

The windings of cylindrical fields, such as are used for turbine generators, are subjected to much greater stresses, on account of the high speed at which they operate, than the windings of fields having salient poles. At times of short circuit the stresses in the field windings of large turbine generators become very great. Ordinary cotton insulation would not have sufficient strength to withstand the severe crushing stresses existing at such times, especially if the insulation had become slightly carbonized by the high temperatures at which the fields of such machines generally operate. The only material which can withstand the high temperature, and which is at the same time sufficiently strong, is mica. The slots of the cylindrical fields of turbine generators are insulated with mica troughs, and the separate turns of the field windings, which consist of flat strips of copper laid in the slots by hand, are separated from one another by thin strips of asbestos or mica paper.

Cooling. All synchronous generators, with the exception of those in which hydrogen is used as a cooling medium, are air-cooled by either natural or forced ventilation. There are four things which must be considered in the cooling or ventilation of any generator: the total losses to be dissipated, the surface exposed for dissipating these losses, the quantity of air required, and the temperature of the cooling air. The rate at which heat is lost from any heated surface depends upon the difference in temperature between the heated surface and the cooling medium, which is air except in the case of a few hydrogen-cooled machines. If the quantity of air supplied is too small, the cooling air reaches a temperature which is nearly the same as the temperature of the surface to be cooled and little heat is carried off. If on the other hand the quantity of air is large, its temperature is only slightly increased. Any increase in the volume of air beyond this point produces little further gain in cooling and is wasteful.

There is little difficulty in cooling slow-speed generators. By providing proper ventilating ducts in the armature laminations and openings in the frame, with fans added to the rotors in some cases, the cooling of such generators can be handled without difficulty. The conditions are different in the case of high-speed turbine generators.

The output of turbine generators is great per unit volume, and the quantity of heat which must be dissipated per unit area of available cooling surface is large. Forced ventilation must be used for such generators, and even with this it is difficult to get sufficient air through the air gap and ventilating ducts for cooling. For this reason large turbine generators must operate at a higher temperature than low-speed generators of smaller output and the insulation used in their construction must be such as to withstand the higher temperature. Mica insulation is universally used for such machines.

One kilowatt acting for one minute raises the temperature of 100 cu ft of air by approximately 18°C. Assuming a 50,000-kva synchronous generator delivering its rated kilovolt-amperes at unity power factor with an efficiency of 97 per cent, 1,500 kw must be taken up by the cooling air. If the increase in the temperature of the air in passing through the machine is not to exceed 20°C, then $1,500 \times 100 \times {}^{18}\!/_{20} = 135,000$ cu ft of air are required per minute. If this air has a velocity of 5,000 fpm, ventilating ducts of 27 sq ft cross section are required. Since in the case of such a machine the air is passed in from both ends, ducts of only half this cross section are required. With the cooling air passed in from both ends and with velocities as high as 1,500 to 2,500 fpm, such as are used in practice, it would be exceedingly difficult to provide ventilating ducts of sufficient cross section. The ventilating duct formed by the air gap between the field and armature alone would not be sufficient. To use forced ventilation, it is obviously necessary to enclose a machine.

Three methods, which are designated according to the way the cooling air is passed through the machine, have been used for artificially ventilating turbine generators. These are radial ventilation, circumferential ventilation, and axial ventilation. Air-gap ventilation is used in conjunction with these.

Radial Ventilation. In the simple radial method of cooling generators, the air is passed in along the air gap from both ends and out through radial ducts made in the armature core by inserting spacing pieces between the armature laminations. As a rule, when radial-slot rotors are used they are provided with radial ducts. Air is passed through the rotor under the slots, out through these radial ducts, and thence through the stator ducts. All of the air passes out through the radial ducts in the stator. The air gap alone, with any reasonable air velocity, is not sufficient in most cases to allow the passage of sufficient air for cooling the stator. Simple radial ventilation has been used with success, but when applied to a large generator it is difficult to pass sufficient air to keep the stator cool.

It is impossible to cool very large turbine generators, which necessarily have long rotors, by passing air in from the two ends. The cross section of the air gap of such a generator is not sufficient to allow the passage of the quantity of air required for cooling. The problem of cooling such a machine has been solved by using multiple inlets in the ventilating system. The cooling air is passed in through the air gaps at the two ends and through certain radial slots in the stator, then along the air gap and out through other radial slots in the stator. This method is illustrated in Fig. 116. The arrows in this figure indicate the direction of air flow.

Circumferential Ventilation. When the circumferential method of ventilation is used, the air for cooling the stator is supplied to one or more openings in the circumference of the stator and passes around through ducts in the stator core in two directions from each opening and out other openings also in the circumference of the stator, without entering the air gap. If air is admitted at only one point on the circumference, it passes out at a point diametrically opposite. In addition to the air for cooling the stator, air must also be supplied to the air gap for cooling the rotor.

Axial Ventilation. A common objection to both the radial and circumferential methods of cooling is that the heat developed in the stator must pass transversely across the laminations to the air ducts in order to be carried off by the cooling air. The rate of heat conduction across a pile of laminations is not over 10 per cent as great as along them. Since in both the radial and circumferential methods of cooling the heat must pass across the laminations to the air ducts, neither of these methods is so efficient as one where the heat passes along the laminations to the air ducts. This is the way the heat passes in the axial method of cooling. For this method numerous holes are punched in the armature stampings. When the stampings are built up, these holes form ducts in the armature core which are parallel to the axis of the machine, and which may extend

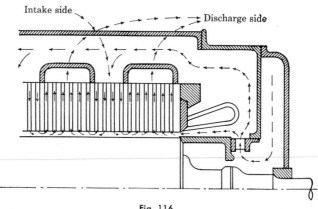

Fig. 116.

either uninterruptedly from one side to the other or from each side to one or more large central radial channels or ducts which form the outlet. The holes in the stator stamping shown in Fig. 108 are for axial ventilation. Air-gap ventilation is used for cooling the rotor.

Cooling Air. In the early turbine generators, the cooling air was generally taken from outside the station and discharged into the station. The quantity of air necessary is so great that, even when it is reasonably clean, enormous quantities of foreign matter are carried by it through the ventilating ducts in the course of a year and the deposit of even a very small percentage of this is serious. If any moisture or oil gets into the passages, the dirt collects quickly. The film of dirt on the surfaces of the ducts materially cuts down the heat transfer to the cooling air and also decreases the effective cross section of the ducts. Such a collection of dirt would necessitate frequent cleaning, and therefore effective steps must be taken to prevent its forming. This is especially true in places where there is considerable dust or dirt, such as near coal mines or smelting plants. It is also true in most large cities.

To avoid this difficulty, it was customary to clean the cooling air by passing it through sprays of water before it entered the generator. This method of cleaning had the advantage of decreasing the temperature of the air 5 or 10 deg, and it appreciably increased the permissible output of the machine with which it was used. Air washing also increased the humidity of the cooling air but this had little effect on the permissible output, since the difference between the specific heat of dry air and that of saturated air is small.

It is now common practice to use totally enclosed ventilating systems for large turbine generators. When totally enclosed systems are used, the cooling air is circulated through the generator and through a surface cooler consisting of tubes through which water is circulated.

Totally enclosed systems of cooling not only eliminate the possibility of accumulating dirt in the ducts, but they also minimize the fire hazard, because the amount of oxygen which might support combustion is limited to that in the cooling system. Means are generally provided for releasing carbon dioxide in the cooling system in case of fire.

Hydrogen as a Cooling Medium. Hydrogen has certain marked advantages over air as a cooling medium. As compared with air, hydrogen has a thermal conductivity which is about 7 times as great and a density which is only about 0.07 as great. Its specific heat at constant pressure is a little over 14 times as great as that of air, and its heat transfer is about 1.5 times as great. The ability of a gas to absorb heat is proportional to the product of its specific heat and density. This product is known as its heat capacity. The heat capacity of hydrogen is almost the same as that of air under similar conditions. Actually it is about 0.5 per cent less, an amount too small to be of importance. One of the chief advantages of hydrogen as a cooling medium is its very low density. This reduces the windage loss to 7 or 8 per cent of that with air and also reduces the windage noise. Not only is the windage loss greatly reduced, but the power required to drive the fans or blowers is reduced by a similar amount. Because of the high thermal conductivity of hydrogen, it transfers about 30 to 50 per cent more heat units than air for a given temperature difference. On account of this, heat passes across small spaces in the insulation and between the laminations more readily with hydrogen than with air.

Because of the danger of explosion if the oxygen content of the hydrogen becomes too high, leaks in the ventilating system must be reduced to a minimum and the pressure in the machine must be kept slightly higher than that outside to prevent leakage of air into the machine. A mixture of hydrogen and air does not explode if the hydrogen content is lower than about 10 per cent or higher than about 70 per cent. In practice it is possible to maintain a hydrogen purity of 98 per cent or even better. Indicators are used to give the oxygen content in the cooling system.

Hydrogen cannot support combustion; therefore, if an internal short circuit should occur in a hydrogen-cooled generator, much less damage would be done than if the generator were air-cooled. Although corona in hydrogen starts at a slightly lower potential gradient than in air, the detrimental effects of corona in a hydrogen-cooled machine are practically absent because of the absence of oxygen and nitrogen. The chief difficulty with the use of hydrogen as a cooling medium for generators has been to get tight joints where the shaft passes through the casing. This difficulty has been overcome by the use of special oil-sealed glands.

At present (1952) there are many hydrogen-cooled synchronous condensers and synchronous generators in operation. By the use of hydrogen cooling, the kva output of a machine can be increased 30 to 40 per cent.

Permissible Temperatures for Different Types of Insulation. All insulating materials are injured or destroyed by high temperature. The continued application of a temperature which would not injure an insulating material if applied for a short time causes it to deteriorate slowly and ultimately to be destroyed. The continued application of even moderate temperatures to cotton, silk, varnishes, and other similar materials commonly used for insulating electrical apparatus causes them to carbonize and to lose their insulating qualities and mechanical strength. Mica, asbestos, and spun glass are three substances used for insulation which can withstand high temperatures. Mica alone can seldom be used without being built up into sheets or strips with some form of binder. The AIEE Standards divide insulating materials into five general classes. Class O includes cotton, silk, paper, and similar organic substances, when neither impregnated nor immersed in oil. Class A includes cotton, silk, paper, and similar organic substances when either impregnated or immersed in a liquid dielectric and includes also enamel applied to conductors. Class B includes inorganic materials such as mica, fiber glass, and asbestos in built-up form combined with organic binding substances. Class C includes inorganic substances such as pure mica, porcelain, and quartz. Class H includes inorganic materials such as mica, asbestos, and fiber glass combined with binding substances composed of silicone compounds which may be in rubbery or resinous forms. The temperature limits recommended in degrees centigrade are: for class O, 90°; for class A, 105°; for class B, 130°; for class H, 180°. No temperature limit has yet been set (1952) for class C.

Chapter 14

Induced voltage. Phase relation between a flux and the voltage it induces. Shape of flux and voltage waves when the coil sides are 180 electrical degrees apart. Calculation of the voltage induced in a coil when the coil sides are not 180 electrical degrees apart.

Induced Voltage. The voltage induced in a d-c generator is equal to the average voltage induced in the coils in series between brushes. This depends upon the speed of the generator, the number of armature inductors connected in series between brushes, and the total flux per pole and is independent of the manner in which the flux is distributed, provided the brushes are in the neutral plane. In a synchronous generator, however, the voltage induced is the effective or rms voltage. This depends not only upon the number of armature inductors between terminals but also upon the way in which the flux is distributed, as well as upon the number of conductors and the pole flux. The same total pole flux can be made to give different values of maximum and of rms voltages by merely changing its distribution. The value of the voltage also depends upon the pitch and coil spread of the winding.

The instantaneous voltage induced in any coil on the armature of an alternator is given by

$$e = -N\frac{d\varphi}{dt}$$

where N, φ, and e are the number of turns in the coil, the flux enclosed by the coil at the instant considered, and the instantaneous voltage. A coil consists of a number of turns which are laid in a single pair of slots. The terms coil and phase must not be confused. A phase usually consists of a number of coils which occupy different pairs of slots and which are generally connected in series. Let the flux from the poles be distributed in such a way that the flux linking any armature coil varies as some function of the maximum flux φ_m through the coil and its angular displacement ρ from the position where it contains maximum flux. The angle ρ is expressed in electrical radians or electrical degrees, where 2π electrical radians or 360 electrical degrees are equivalent to the distance between the centers of consecutive poles of like polarity. If the synchronous

171

generator is bipolar, 360 electrical degrees correspond to 360 space degrees. The flux φ linking the coil in any position is

$$\varphi = f(\varphi_m, \rho)$$

and

$$e = -N \frac{d}{dt} f(\varphi_m, \rho)$$

The rms voltage per coil is

$$E = -N \left\{ \frac{1}{\pi} \int^{\pi} \left[\frac{d}{dt} f(\varphi_m, \rho) \right]^2 d\rho \right\}^{\frac{1}{2}}$$

If the flux in the air gap between the pole faces and the armature is distributed in such a way that the flux through the coil varies as the cosine of the angular displacement ρ of the coil from the position where the flux through it is a maximum,

$$e = -N \frac{d}{dt} \varphi_m \cos \rho$$

$$= -N \frac{d}{dt} \varphi_m \cos \omega t$$

where ω is the angular velocity of the armature in electrical radians per second and t is the time in seconds required for it to move through the angle $\rho = \omega t$.

$$e = \omega N \varphi_m \sin \omega t$$

$$E = \omega N \left(\frac{\omega}{\pi} \int_0^{\pi/\omega} \varphi_m^2 \sin^2 \omega t \, dt \right)^{\frac{1}{2}}$$

$$= \omega N \varphi_m \frac{1}{\sqrt{2}}$$

$$= 2\pi \frac{p}{2} \frac{n}{60} N \varphi_m \frac{1}{\sqrt{2}}$$

where p is the number of poles and n is the speed in revolutions per minute. Therefore

$$E = 4.44 N f \varphi_m \text{ volts} \qquad (131)$$

This equation holds only when the flux distribution in the air gap is such as to produce a sinusoidal voltage wave in the armature coils. When there are several armature coils per pair of poles, they are located in different pairs of slots. Their voltages are therefore not in phase and must be added vectorially when the coils are connected in series. This vector addition is usually taken care of by the use of a breadth factor. Breadth factors are considered in Chap. 15. If however the winding is concentrated in one slot per pole per phase, the voltages in all coils belonging to one phase are in phase and add arithmetically. Therefore, if N is taken as the total turns in series per phase, Eq. (131)

gives the voltage per phase for the complete winding. Moreover, if the sides of each coil are 180 electrical degrees apart, giving what is known as a full-pitch winding, the maximum flux φ_m through the winding is the total mutual flux of one pole in webers. Equation (131) is very useful for calculating the voltage of a concentrated full-pitch winding.

Phase Relation between a Flux and the Voltage It Induces. The difference in phase between a flux and the voltage it induces, when both are considered with respect to a coil and when both are considered with respect to an inductor, should not be overlooked. An inductor is one of the two active sides of each turn of a coil. Its length is equal to the length of that portion of the coil side which actually cuts flux.

Assume that the two active sides of a coil are 180 electrical degrees apart. Under this condition, when the coil is directly over a pole and contains a maximum flux, its two active sides are midway between the poles and are in zero fields. They are cutting no flux, and the voltages induced in them are zero. When the coil has moved forward 90 electrical degrees, the flux through it becomes zero, but the inductors are now directly under the centers of opposite poles and are in the strongest parts of the fields. The voltages induced in the two coil sides have maximum values and are in the same direction around the coil. The effective voltage in the coil is always equal to the vector difference between the effective voltages induced in its two sides.

It follows that, while a voltage in a coil is in time quadrature with respect to the flux through it, the voltages in the inductors and the intensity of the fields or the flux densities at the inductors are in time phase. The inductors move at uniform speed across the field, and the voltage induced in them must at every instant be proportional to the strength of the field in which they are moving at the instant in question. The voltage is equal to the flux density multiplied by the length of the inductor and by the component of its velocity at right angles to the field.

Shape of Flux and Voltage Waves When the Coil Sides Are 180 Electrical Degrees Apart. If the sides of a coil are 180 electrical degrees apart and the distribution of the air-gap flux which they cut is a sine function of the distance measured from a point midway between the poles, the variation of flux through the coil with respect to time is a cosine function and the voltage induced in the coil is sinusoidal. If the flux has any other distribution, the time variation of flux through the coil has a wave form different from that of the space distribution of the flux. The wave form of the voltage induced in the coil however is the same as the wave form of the space distribution of the flux in the air gap, since the voltages in the two coil sides are opposite and at each instant proportional to the strength of the fields in which they move.

Figure 117 shows curves of the space variation of flux density in the air gap, the corresponding time curves of flux through a coil, and the voltage induced in a coil with sides 180 electrical degrees apart, by a sine, a rectangular, and a triangular space distribution of the air-gap flux density. In curves C and E of Fig. 117, the position of the coil is considered to be fixed by the position of its center with respect to the centers of the poles. Note that the voltage curves are displaced by 90 electrical degrees with respect to the flux-density curves and that the voltage curve in each case has the same shape as its corresponding curve of flux den-

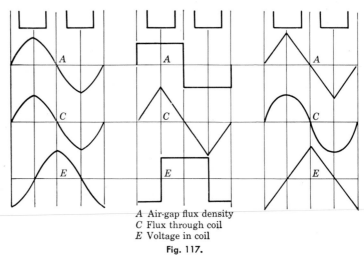

A Air-gap flux density
C Flux through coil
E Voltage in coil

Fig. 117.

sity. This is always the case for a full-pitch coil and for each phase of a concentrated full-pitch winding.

If a nonsinusoidal flux-density curve is split into its fundamental and harmonic components, the effective voltage produced by each component may be calculated by Eq. (131). The frequency of the qth harmonic is, of course, q times the frequency of the fundamental. Let the space distribution of flux, $i.e.$, the flux density, in the air gap of a synchronous generator be

$$\mathfrak{B} = \mathfrak{B}_1 \sin \alpha + \mathfrak{B}_3 \sin 3\alpha + \mathfrak{B}_5 \sin 5\alpha$$

where the $\mathfrak{B}$'s represent the maximum flux densities for the fundamental and harmonic components and α is the angular distance in electrical degrees measured around the air gap from the reference point midway between the poles. The flux per pole equals

$$K \int_0^\pi \mathfrak{B}d\alpha = 2K \left(\mathfrak{B}_1 + \frac{\mathfrak{B}_3}{3} + \frac{\mathfrak{B}_5}{5} \right) = \varphi_{m_1} + \varphi_{m_3} + \varphi_{m_5}$$

K is a constant involving the inductor length and the choice of units. φ_{m_1}, φ_{m_3}, and φ_{m_5} are the maximum values of the flux linking a full-pitch coil on the armature due to the fundamental, third- and fifth-harmonic components, respectively, of the flux-density curve. Then the qth harmonic in the voltage wave has an effective value $E_q = 4.44qfN\varphi_{mq}$.

Calculation of the Voltage Induced in a Coil When the Coil Sides Are Not 180 Electrical Degrees Apart. If the sides of a coil are not 180 electrical degrees apart, the voltages in them are not in phase at every instant when considered around the coil. The effective voltage in the

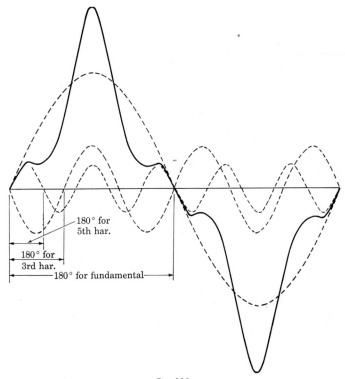

180° for 5th har.

180° for 3rd har.

180° for fundamental

Fig. 118.

coil however is still equal to the vector difference of the effective voltages induced in its active sides.

If the distribution of the flux density is not sinusoidal but its distribution in the air gap is known in terms of a fundamental and a series of harmonics, the fundamental and the harmonic effective voltages in the coil can be found by taking the vector differences of the effective voltages induced in the sides of the coil by the fundamental and each of the harmonics separately. Figure 118 gives a distribution of flux density which contains a fundamental and third and fifth harmonics. Inspection of this curve should make it clear that any change ρ in the angular distance between the two sides of a coil corresponds to a change in phase between the voltages in the two sides of ρ for the fundamental and $q\rho$ for the qth harmonic.

Let the space distribution of flux, *i.e.*, the flux density, in the air gap of a synchronous generator measured from a point midway between the poles be

$$\mathcal{B} = \mathcal{B}_1 \sin \alpha + \mathcal{B}_3 \sin 3\alpha + \mathcal{B}_5 \sin 5\alpha$$

where the $\mathcal{B}$'s represent the maximum flux densities for the fundamental and harmonics and α is the angular distance in electrical radians measured

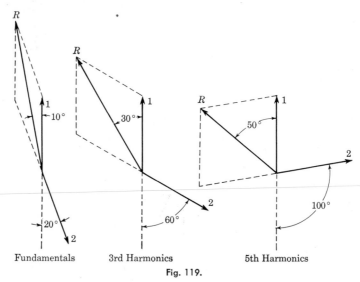

Fig. 119.

around the gap from the reference point midway between the poles.   If the inductors of the coil are 160 electrical degrees apart, the fundamentals of the voltages in the two inductors are 20 deg out of phase opposition, the third harmonics are $3 \times 20 = 60$ deg, and the fifth harmonics are $5 \times 20 = 100$ deg.   The vectors for the fundamentals and the harmonics are shown in Fig. 119.   In this figure the R's are the resultant voltages; 1 and 2 are the voltages in coil sides 1 and 2.

If the coil contains N turns and moves with a velocity of v m per sec and the length of the inductors that cut flux is L meters, the instantaneous voltage in volts induced in the coil referred to the voltage in inductor 1 is

$$e = 2LvN[\mathcal{B}_1 \cos 10° \sin (\alpha + 10°)$$
$$+ \mathcal{B}_3 \cos 30° \sin (3\alpha + 30°) + \mathcal{B}_5 \cos 50° \sin (5\alpha + 50°)]$$

The rms value of this voltage is equal to the square root of one-half the sum of the squares of the maximum values of the fundamental and harmonics.   When rms voltages are desired, it is easier to make use of pitch factors.   Pitch factors are considered in Chap. 15.

Chapter 15

Open-circuit and closed-circuit windings. Bar windings and coil windings. Concentrated and distributed windings. Whole-coiled and half-
coiled windings. Spiral, lap, and wave windings. Single-phase and
polyphase windings. Multicircuit windings. Pole pitch and coil pitch.
Pitch factor. Effect of pitch on harmonics. Equation of voltage generated in a fractional-pitch coil. Effect on wave form of distributing a
winding. Integral-slot windings. Breadth or distribution factor for an
integral-slot winding. Two-layer windings. Fractional-slot windings.
Breadth factor for fractional-slot windings. Harmonics in three-phase
synchronous generators. Harmonics due to slots.

Open-circuit and Closed-circuit Windings. All a-c windings can be
divided into two general groups: (1) open-circuit windings and (2) closed-
circuit windings.

An open-circuit winding, as its name signifies, is not closed on itself.
In an open-circuit winding there is a continuous path through the conductors of each phase on the armature which terminates in two free ends.

A closed-circuit winding has a continuous path through the armature
which re-enters on itself, forming a closed circuit. All closed-circuit
windings have at least two parallel paths between their terminals.

All modern d-c windings are closed-circuit windings. Either open-
circuit or closed-circuit windings can be employed for synchronous
generators, but except in a few special cases, open-circuit windings are
better adapted and are always used. Synchronous-generator armature
windings can have two or more parallel paths through the armatures, but
such windings are not re-entrant, *i.e.*, are not closed-circuit windings.
Windings with two or more parallel paths between terminals are multicircuit windings. A d-c winding can be used for a synchronous generator,
but an a-c winding, since it is not re-entrant, cannot be used for a d-c
generator.

Bar Windings and Coil Windings. Armature windings can be divided
into two general classes according to the way in which the coils are placed
in the slots: bar windings and coil windings. In bar windings, insulated
rectangular copper bars, which are usually laminated, are placed in the
armature slots and are suitably connected by brazing or welding to form

coils.   In coil windings, coils of either rectangular or round insulated wire are wound on forms, insulated, and placed in the slots.   When closed or nearly closed slots are used, it is sometimes necessary to wind the coils by hand directly on the armatures by threading the wire through the slots. Form-wound coils are more reliable and are generally used, except where nearly closed slots are required.   Whether bar windings or coil windings are employed, the slots must be properly insulated by pressboard, mica, or other suitable material.

Figure 120 shows two types of coils for coil-wound armatures.   When the type of coil shown on the left is used, all the armature coils are of the

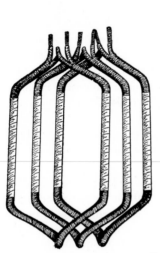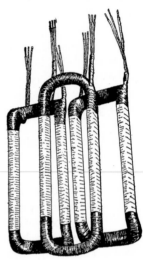

Fig. 120.

same size and shape irrespective of the phases in which they are placed and their positions on the armature.   This type of coil from its shape is known as the diamond-shaped coil.   Two different shapes of coils are required for the type of coil shown on the right, and this type has the additional disadvantage of not permitting as good bracing of the end connections.

Concentrated and Distributed Windings.   Concentrated windings have all the inductors of any one phase, which lie under a single pole, in a single slot.   Better results can usually be obtained by distributing the inductors among several slots.   Such windings are called distributed windings.   They are completely distributed or partially distributed according as they are spread over the entire armature surface or over only a portion of it.   Distributed windings diminish armature reaction and armature reactance and give better wave forms and better distribution of the heating due to the armature copper losses than concentrated windings.

Whole-coiled and Half-coiled Windings. The one common require-
ment for all windings is that all conductors must be connected in such a
way that their voltages assist. Figure 121 shows a six-pole alternator
with two inductors per pole. The short lines over the poles represent
diagrammatically the armature inductors, and the arrows on these lines

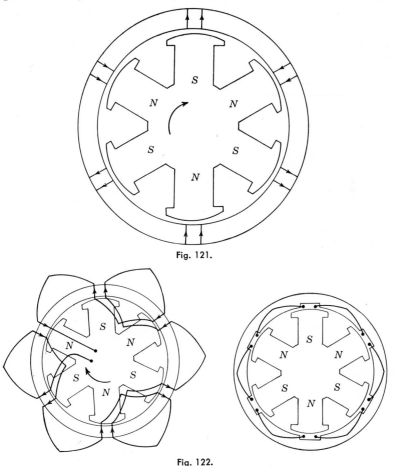

Fig. 121.

Fig. 122.

show the directions of the voltages induced in them for clockwise rotation
of the field. An inductor extending out of the paper is represented by a
line drawn radially outward. Each slot on the armature is assumed to
contain two inductors. These are shown side by side in Fig. 121. They
can be connected in two ways as illustrated by Figs. 122 and 123. Electrically, the connections shown in Figs. 122 and 123 are identical. The
right-hand halves of these figures represent the connections on the fronts
of the armatures as they would actually appear.

Figure 122 represents what is known as a whole-coiled winding. Figure 123 shows a half-coiled winding. Whole-coiled windings have as many coils per phase as there are poles. Half-coiled windings have only one coil per phase per pair of poles. The two turns per pair of poles

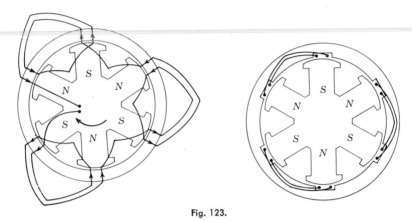

Fig. 123.

shown in Fig. 123 are assumed to be in a single coil. The only real difference between the two types of winding lies in the method of making the end connections between the inductors in the slots.

Spiral, Lap, and Wave Windings. When the windings are distributed, they can be connected in three different ways, giving what are known

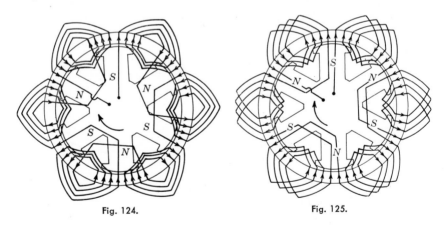

Fig. 124.   Fig. 125.

as (1) spiral windings, (2) lap windings, and (3) wave or progressive windings.

Lap and wave windings can also be used in concentrated windings. The differences in these three types of windings should be made clear by referring to Figs. 124, 125, and 126, which show a spiral winding, a lap winding, and a wave winding. All three figures show distributed single-phase windings with eight slots per pole. The winding shown in Fig. 125 is called a barrel winding.

The lap winding lends itself better than the spiral winding to the use of machine-wound formed coils, as in the lap winding all the coils are the same. If formed coils are used for a spiral winding, there have to be as many different widths of coils, *i.e.*, coil pitches, as there are slots per pole per phase for a half-coil winding, but only one-half as many for a whole-coil winding.

If spiral, lap, or wave windings are arranged so that the same armature inductors are included in each phase belt, the use of the three types of windings on a given synchronous generator produces equal voltages and identical wave forms for a given pole flux and flux distribution. The only electrical difference in the types of windings, so far as the generation of voltage is concerned, lies in the order in which the armature conductors are connected in series between terminals.

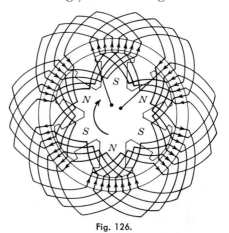

Fig. 126.

Single-phase and Polyphase Windings. A single-phase winding has only one group of inductors per pole. These may be in a single slot or in several slots, according to whether the winding is concentrated or distributed. A polyphase winding may be considered to consist of a number of single-phase windings displaced by suitable angles from one another. The electrical space displacement between the single-phase elements must be the same as the phase differences between the voltages to be induced. For example, the corresponding elements of the winding of a three-phase synchronous generator must be displaced 120 electrical degrees from one another. Although the single-phase windings which make up the polyphase winding are independent of one another, the windings are always interconnected in either star or mesh. The number of leads brought out is equal to the number of phases, except for star connection, where an additional lead may be brought out from the common junction or neutral point of the phases. In the case of three-phase synchronous generators, the star and mesh connections are the Y and Δ connections. Most modern three-phase synchronous generators are connected in Y. Y connection permits the neutral point to be grounded and gives a higher voltage between terminals for the same phase voltage than the Δ connection. It also gives a higher slot factor, *i.e.*, the ratio of the cross section of the copper in a slot to the slot area is greater than for the Δ connection. High-voltage synchronous generators are invariably Y-connected. When there is no consideration such as high voltage to determine whether Y or Δ connection should be used, the method of connecting the phases is sometimes fixed by the number of slots in the standard armature stampings which are available, the frequency, the voltage, and the permissible range of flux density. For the same voltage between terminals and the same line current, Y and Δ connections require the same amount of copper, but the Y connection requires fewer total turns than the Δ connection ($1/\sqrt{3} = 0.58$ as many), and since the thickness of insulation required on the wires depends chiefly upon the voltage and not upon the size of the wires, the ratio of space occupied in a slot by the copper to the space occupied by insulation is greater for the Y connection than for the Δ connection. In other words, the slot factor of a Y-connected synchronous generator is higher than the slot factor of a Δ-connected generator.

Figure 127 shows a simple six-pole two-phase half-coiled winding with two inductors per slot. Figure 128 shows a similar three-phase winding. The phases 1, 2, and 3 of the three-phase winding are indicated by full lines, lines of dots and dashes, and dotted lines.

Multicircuit Windings. The multicircuit windings, which have been commonly used for many years to reduce the currents individual circuits have to carry, are not suitable for independent loading of the circuits because of the magnetic unbalance and bad mechanical stresses produced when the circuits are unequally loaded. To allow for independent loading of the circuits, a type of two-circuit winding has been developed for use on large turbine generators which meets the conditions of independent loading satisfactorily. On these machines the two windings for each phase are connected to adjacent sections of ring bus bars and not only are capable of operating independently without producing unbalanced magnetic conditions or abnormal mechanical stresses but are so interlinked that they can act as a high reactance transformer for transferring power from one bus-bar section to the other. The voltages generated in the two circuits for each phase of these windings not only are equal but are in phase.

There are two types of this kind of winding, the alternate-slot winding and the alternate-belt winding. In the alternate-slot type, the two circuits in each phase lie in alternate slots, with no coil sides of the two circuits for a phase in the same slot. In the alternate-belt type, each phase belt is divided into two equal portions, one of which is assigned to each circuit. Both types of windings are arranged so that the voltages generated in the two circuits for each phase are equal and in phase. The

general arrangement of the two types of winding is best understood from
Figs. 129 and 130, which show one phase for the alternate-slot type and
one for the alternate-belt type.

Pole Pitch and Coil Pitch.   The pole pitch is the distance between the
centers of adjacent north and south poles.   The distance between the two
sides of any armature coil is called the coil pitch.   Coil pitch is usually

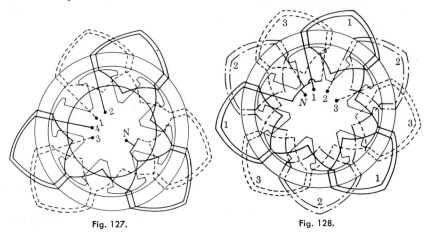

Fig. 127.   Fig. 128.

expressed as a fraction of pole pitch, but it is sometimes convenient to
express it in electrical degrees or in slots.   For example, a coil pitch of
$\frac{2}{3}$ is equivalent to a pitch of 120 electrical degrees, or if there are 12 slots
per pole, to a pitch of 8 slots.   A winding having a coil pitch of less than
180 electrical degrees is called a fractional-pitch or a chorded winding.
Since the two sides of a coil of a fractional-pitch winding do not lie under

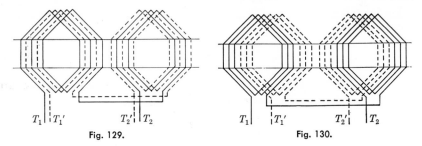

Fig. 129.   Fig. 130.

the centers of adjacent poles at the same instant, the voltages induced in
them are not in phase when considered around the coil.   The voltage
produced in a fractional-pitch winding is therefore less than that produced in a full-pitch winding having the same number of turns.   Fractional-pitch or chorded windings are often used.   They decrease the
length of the end connections and thus the amount of copper required.
They also somewhat reduce slot reactance and give a means of improving

wave forms of synchronous generators.   A fractional-pitch winding can be used to eliminate any one harmonic from the voltage wave, as well as to reduce other harmonics.   A fractional-pitch winding requires a few more turns or a greater flux for the same voltage than a winding having a full pitch.

Pitch Factor.   The voltage generated in any single turn on the armature of a synchronous generator is the vector difference of the voltages generated in the two inductors which form the active sides of the turn. With a full-pitch winding, these two voltages are 180 deg out of phase and their vector difference is equal to twice the voltage of one inductor.

In the case of a fractional-pitch winding, the active sides of the coil are less than 180 electrical degrees apart and the voltages generated in them are therefore out of phase by an angle of less than 180 deg.   If ρ is the pitch expressed in electrical degrees, the difference in phase for the fundamental of the two voltages is ρ.   In general, since the displacement for any harmonic such as the qth must be q times the phase displacement for the fundamental, the difference in phase between the harmonics of any order, such as the qth, generated in the two active sides of any coil of a fractional-pitch winding is $q\rho$.

Since the voltage in a coil is the vector difference of the voltages generated in its active sides, the voltage E_q of the qth harmonic generated in a coil is

$$E_q = 2E'_q \sin \frac{q\rho}{2}$$

where E'_q is the value of the qth harmonic voltage in the coil side.

The pitch factor is the ratio of the voltage E_q, induced in a fractional-pitch winding, to the voltage $2E'_q$, that would be induced if the winding had a full pitch.   The magnitude of the pitch factor for any harmonic, such as the qth, is

$$k_p = \sin \frac{q\rho}{2} \tag{132}$$

For odd harmonics, the sign and magnitude of the pitch factor are given in terms of pitch deficiency $(180 - \rho)$ by

$$k_p = \cos \frac{q(180 - \rho)}{2}$$

Effect of Pitch on Harmonics.   Any harmonic can be eliminated from the voltage generated in an armature coil by choosing a pitch that makes the pitch factor zero for that harmonic.   To eliminate the qth harmonic,

$$k_p = \sin \frac{q\rho}{2} = 0 \quad \text{or} \quad \rho = \frac{2m\pi}{q} \tag{133}$$

where m is any integer.

For example, to eliminate the fifth harmonic, pitches of $\frac{2}{5}$, $\frac{4}{5}$, $\frac{6}{5}$, etc., can be used. Since any departure from full pitch diminishes the fundamental by an amount which increases progressively with the departure of the pitch from unity, in practice the shortened pitch which is nearest to unity, in this case $\frac{4}{5}$, is used. Pitches which are greater than unity require more copper for the coil-end connections than shortened pitches and possess no compensating advantage.

Eliminating any one harmonic from the voltage induced in the armature coils of a synchronous generator not only eliminates that particular harmonic but diminishes other harmonics and also the fundamental, usually by different amounts, and may change the phase relations of the harmonics with respect to the fundamental.

A $\frac{5}{6}$ pitch is particularly satisfactory for eliminating harmonics as it nearly cuts out both the fifth and seventh harmonics. It will be shown later that there can be no third harmonic or multiples of the third harmonic between the terminals of a three-phase Y-connected generator. Therefore by using a $\frac{5}{6}$ pitch and Y connection, there can be no third, ninth, or fifteenth harmonics and only a small fifth and seventh between the line terminals. Even harmonics usually do not occur in the voltage waves of synchronous generators. The first harmonic occurring then is the eleventh, and harmonics of as high an order as this are seldom present in sufficient magnitude to have much effect on the wave form.

The magnitudes of the harmonics in fractional-pitch windings, as compared with their magnitudes in a full-pitch winding having the same number of turns, are given in Table 5.

TABLE 5

Pitch	Harmonic				
	1	3	5	7	11
$\frac{2}{3}$	0.866	0.000	0.866	0.866	0.866
$\frac{4}{5}$	0.951	0.588	0.000	0.588	0.951
$\frac{5}{6}$	0.966	0.707	0.259	0.259	0.966
$\frac{6}{7}$	0.975	0.782	0.434	0.000	0.782

The effect of pitch on the relative phase of the harmonics and the fundamental can be seen best by drawing the vectors which represent the voltages in the coil sides for the fundamental and the harmonics. Let Eq. (134) represent the voltage generated in the leading side of a coil of a fractional-pitch winding.

$$e = E_1 \sin \omega t + E_2 \sin 2\omega t + E_3 \sin 3\omega t + E_4 \sin 4\omega t + E_5 \sin 5\omega t \quad (134)$$

Let the pitch be $\frac{2}{3}$. This pitch cuts out the third harmonic and all multiples of the third harmonic in the coil voltage. Therefore these harmonics need not be considered in the coil voltage in what follows.

The voltage in the trailing coil side is

$$e' = E_1 \sin(\omega t - 120°) + E_2 \sin 2(\omega t - 120°) + E_3 \sin 3(\omega t - 120°)$$
$$+ E_4 \sin 4(\omega t - 120°) + E_5 \sin 5(\omega t - 120°)$$
$$= E_1 \sin(\omega t - 120°) + E_2 \sin(2\omega t - 240°) + E_3 \sin(3\omega t - 0°)$$
$$+ E_4 \sin(4\omega t - 120°) + E_5 \sin(5\omega t - 240°)$$

The vectors for the voltages in the coil sides for the fundamental, the second, the fourth, and the fifth harmonics are given in Fig. 131.

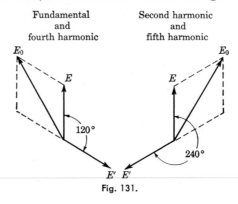

Fig. 131.

The displacements of the voltages in the coil with respect to the voltages in the leading side of the coil are, from Fig. 131:

Fundamental	Second harmonic	Fourth harmonic	Fifth harmonic
+30°	−30°	+30°	−30°

Shift the reference point of the coil voltages to make the phase angle of the fundamental equal to zero by subtracting 30 deg from the phase displacement of the fundamental and $q \times 30$ deg from the phase displacements of the harmonics. This gives

Fundamental	Second harmonic	Fourth harmonic	Fifth harmonic
30° − 30° = 0°	−30° − 2 × 30° = −90°	30° − 4 × 30° = −90°	−30° − 5 × 30° = −180°

The equation of the coil voltage, considering the phase angle of the fundamental as zero, is therefore

$$e = 1.73E_1 \sin \omega t + 1.73E_2 \sin(2\omega t - 90°)$$
$$+ 1.73E_4 \sin(4\omega t - 90°) + 1.73 \sin E_5(5\omega t - 180°)$$

In general, even harmonics do not exist in a-c machines and therefore need not be considered when studying such machines.

The effect of pitch on the phase of any harmonic in a fractional-pitch winding, as compared with its phase in a full-pitch winding, depends on the order of the harmonic. All odd harmonics either are not changed in phase or are reversed. All even harmonics are shifted by either plus or minus 90 deg.

Equation of Voltage Generated in a Fractional-pitch Coil. The effect of pitch on the magnitude and phase of a harmonic can be expressed in a single equation. Let the fundamental instantaneous voltage generated in the leading side of a coil in a fractional-pitch winding be

$$e_1 = E_1 \sin \omega t$$

Let the pitch be ρ. Then the instantaneous fundamental voltage generated in the trailing side of the coil is

$$e_1' = E_1 \sin (\omega t - \rho)$$

The fundamental coil voltage is

$$e_{1R} = e_1 - e_1' = E_1[\sin \omega t - \sin (\omega t - \rho)]$$

Let $\omega t = \left(\omega t - \frac{\rho}{2} \right) + \left(\frac{\rho}{2} \right)$ and $(\omega t - \rho) = \left(\omega t - \frac{\rho}{2} \right) - \left(\frac{\rho}{2} \right)$

Then

$$
\begin{aligned}
e_{1R} &= E_1 \left\{ \sin \left[\left(\omega t - \frac{\rho}{2} \right) + \left(\frac{\rho}{2} \right) \right] - \sin \left[\left(\omega t - \frac{\rho}{2} \right) - \left(\frac{\rho}{2} \right) \right] \right\} \\
&= E_1 \left\{ \left[\sin \left(\omega t - \frac{\rho}{2} \right) \cos \left(\frac{\rho}{2} \right) + \cos \left(\omega t - \frac{\rho}{2} \right) \sin \left(\frac{\rho}{2} \right) \right] \right. \\
&\quad \left. - \left[\sin \left(\omega t - \frac{\rho}{2} \right) \cos \left(\frac{\rho}{2} \right) - \cos \left(\omega t - \frac{\rho}{2} \right) \sin \left(\frac{\rho}{2} \right) \right] \right\} \\
&= E_1 2 \cos \left(\omega t - \frac{\rho}{2} \right) \sin \left(\frac{\rho}{2} \right) \\
&= 2E_1 \sin \left[\omega t + \left(90° - \frac{\rho}{2} \right) \right] \sin \left(\frac{\rho}{2} \right) \quad (135)
\end{aligned}
$$

For the qth harmonic, by a similar process,

$$
\begin{aligned}
e_{qR} &= E_q[\sin q(\omega t) - \sin q(\omega t - \rho)] \\
&= 2E_q \sin \left[q\omega t + \left(90° - \frac{q\rho}{2} \right) \right] \sin \left(\frac{q\rho}{2} \right) \quad (136)
\end{aligned}
$$

Shift the reference point from which the angles are measured so as to make the phase angle of the fundamental coil voltage given by Eq. (135) equal to zero, by subtracting the angle $[90 - (\rho/2)]$ from the phase angle of the fundamental. An angle $q[90 - (\rho/2)]$ must then be subtracted from the qth harmonic. Making this change gives

$$e_{qR} = 2E_q \sin\left[q\omega t + \left(90° - \frac{q\rho}{2}\right) - q\left(90° - \frac{\rho}{2}\right)\right] \sin \frac{q\rho}{2}$$

$$= 2E_q \sin\left[q\omega t + (1 - q)90° \right] \sin \frac{q\rho}{2} \qquad (137)$$

The magnitude of the term $\sin(q\rho/2)$ is the magnitude of the pitch factor for the qth harmonic. For the fundamental, $q = 1$. The shift in phase produced by the pitch is given by $(1 - q)90$ deg multiplied by the algebraic sign of the term $\sin(q\rho/2)$. The sign of this term may be either plus or minus, depending on the order of the harmonic and the pitch.

Effect on Wave Form of Distributing a Winding. When a winding is distributed, *i.e.*, when it occupies more than one slot per pole per phase, the voltages generated in the turns of a single phase which occupy different pairs of slots are not in phase. For the fundamental of the voltage wave, this difference in phase is equal to the angle between two adjacent slots. For the third harmonic it is three times this angle; for the fifth, five times; for the seventh, seven times. All angles are measured in electrical degrees.

The general effect of distributing a winding is to smooth out the wave form by diminishing the amplitude of the harmonics with respect to the fundamental. This can be made clear by considering a specific case. Take a synchronous generator having a distribution of flux in its air gap which gives a voltage containing a third and a fifth harmonic in each turn of the armature winding. Let the equation of this voltage be

$$e = E(\sin \omega t + \tfrac{1}{3} \sin 3\omega t + \tfrac{1}{5} \sin 5\omega t)$$

Let there be four turns per pole per phase. If all four turns are placed in a single pair of slots, the resultant voltage generated in them is

$$e_r = E(4 \sin \omega t + 1.33 \sin 3\omega t + 0.8 \sin 5\omega t)$$

and the harmonics have the following relative magnitudes:

$$1\text{st}:3\text{d}:5\text{th} = 1:0.33:0.2$$

Suppose the four turns are distributed among four pairs of slots which are 15 electrical degrees apart. This corresponds to the distribution of the armature winding of a three-phase synchronous generator having four slots per pole per phase and gives a phase spread of 60 deg.

Let e_1, e_2, e_3, and e_4 be the voltages generated in the four turns referred to the center of the phase belt. Then

$e_1 = E[\sin(\omega t - 22.5°) + \frac{1}{3}\sin 3(\omega t - 22.5°) + \frac{1}{5}\sin 5(\omega t - 22.5°)]$
$e_2 = E[\sin(\omega t - 7.5°) + \frac{1}{3}\sin 3(\omega t - 7.5°) + \frac{1}{5}\sin 5(\omega t - 7.5°)]$
$e_3 = E[\sin(\omega t + 7.5°) + \frac{1}{3}\sin 3(\omega t + 7.5°) + \frac{1}{5}\sin 5(\omega t + 7.5°)]$
$e_4 = E[\sin(\omega t + 22.5°) + \frac{1}{3}\sin 3(\omega t + 22.5°) + \frac{1}{5}\sin 5(\omega t + 22.5°)]$

Adding these vectorially gives the resultant voltage e_r equal to

$$e_r = E(3.84 \sin \omega t + 0.869 \sin 3\omega t + 0.164 \sin 5\omega t)$$

The relative magnitudes of the harmonics in this resultant wave are

$$1st:3d:5th = 1:0.226:0.043$$

With all four turns in the same pair of slots, the rms voltage is

$$E_{rms} = \frac{4E}{\sqrt{2}} \sqrt{(1)^2 + (\frac{1}{3})^2 + (\frac{1}{5})^2}$$

$$= 4.28 \frac{E}{\sqrt{2}}$$

With the turns distributed, this voltage is

$$E'_{rms} = \frac{E}{\sqrt{2}} \sqrt{(3.84)^2 + (0.869)^2 + (0.164)^2}$$

$$= 3.94 \frac{E}{\sqrt{2}}$$

Distributing the winding has diminished the voltage by about 8 per cent. Therefore either 8 per cent more turns or 8 per cent more flux is required in this particular case for the same voltage. The disadvantage of increasing the flux or the turns is usually more than balanced by the smoothing out of the wave form by diminishing the harmonics. The distribution of the armature copper loss is also improved. In the particular example just given, distributing the winding reduced the third harmonic about 31 per cent and the fifth about 79 per cent.

The effect on the wave form of a synchronous generator of distributing its winding can most easily be obtained by comparing the magnitudes of the breadth factors (see page 190) for the harmonics with unity, which is the value these factors would have in a concentrated winding. The signs of the breadth factors indicate whether or not distributing the winding has reversed any of the harmonics.

Integral-slot Windings. If the number of slots per pole divided by the number of phases is an integer, there is a whole number of slots per pole and the winding belts of conductors for all phases and poles are identical. The windings for all phases are symmetrical not only as a whole but with respect to each pole. Such a winding may be either full pitch or fractional pitch. A three-phase six-pole winding with 54 arma-ture slots is such a winding. In the winding, there are nine slots per pole and three slots per pole per phase. This is an integral-slot winding.

Breadth or Distribution Factor for an Integral-slot Winding. The volt-ages induced in the separate coils of a distributed winding are not in exact phase, and their resultant is therefore less than would be produced in a concentrated winding having the same number of turns. The ratio of the voltages produced by distributed and concentrated windings having the same number of turns is called the breadth or distribution factor. The breadth factor for any form of winding can be found by calculating the voltage induced in each turn or group of turns occupying a single pair of slots and then adding vectorially the voltages produced in all pairs of slots over which the phase is distributed. The ratio of this voltage to the voltage which would be produced if all the turns were concentrated in a single pair of slots is the breadth factor. Consider a three-phase syn-chronous generator having six slots per pole, *i.e.*, two slots per pole per phase. Let N be the number of turns per coil. From Eq. (131), page 172, the voltage per coil is

$$E = 4.44 N f \varphi_m$$

The angle between adjacent slots is $180 / 6 = 30$ electrical degrees. This is the phase angle between the voltages produced by the inductors in the two groups of coils. The sum of these two voltages is

$$E' = 2E \cos \frac{30°}{2}$$
$$= 1.932E$$

If all the turns are in the same pair of slots, the voltage is

$$E'' = 2E$$

The breadth factor is therefore

$$\frac{E'}{E''} = \frac{1.932}{2} = 0.966$$

The breadth factor for a winding with n slots per phase belt can be found as follows. Let α be the angle in electrical radians between adja-cent slots of a winding having n slots per phase belt containing series-connected inductors. If all inductors of one phase are in series, then n is the number of slots per pole per phase. Let Z be the number of induc-tors per slot. Assume a sinusoidal voltage. Then if E is the effective voltage per inductor, the voltage per phase belt is

$$E_B = ZE(1 + \epsilon^{j\alpha} + \epsilon^{j2\alpha} + \ldots \epsilon^{j(n-1)\alpha})$$

The terms in parenthesis constitute a geometrical series whose sum is given by the formula

$$S = \frac{ar^n - a}{r - 1}$$

in which a is the first term of the series, r is the ratio of adjacent terms, and n is the number of terms. Applying this relationship

$$E_B = ZE \frac{\epsilon^{jn\alpha} - 1}{\epsilon^{j\alpha} - 1}$$

$$= ZE \frac{\cos n\alpha + j \sin n\alpha - 1}{\cos \alpha + j \sin \alpha - 1}$$

The effective value of the resultant voltage is therefore

$$E_B = ZE \sqrt{\frac{(\cos n\alpha - 1)^2 + \sin^2 n\alpha}{(\cos \alpha - 1)^2 + \sin^2 \alpha}}$$

$$= ZE \sqrt{\frac{2 - 2 \cos n\alpha}{2 - 2 \cos \alpha}}$$

$$= ZE \frac{\sin (n\alpha/2)}{\sin (\alpha/2)}$$

With all the inductors of the phase belt concentrated in a single slot, giving nZ conductors per slot, the voltage per phase belt is nZE. The ratio of the voltages of the distributed and concentrated windings, which is the breadth factor, is therefore

$$k_b = \frac{\sin (n\alpha/2)}{n \sin (\alpha/2)} \tag{138}$$

Equation (138) applies only to sinusoidal voltages.

With a nonsinusoidal flux distribution, the fundamental and harmonic components of the voltage produced in a concentrated full-pitch winding can be determined as indicated in Chap. 14. Multiplying these component voltages by the appropriate pitch factors gives the fundamental and harmonic components of the voltage if a concentrated fractional-pitch winding is used. If these components are then each multiplied by the breadth factor, the component voltages of the more usual fractional-pitch distributed windings are obtained. The breadth factor, like the pitch factor, differs for each harmonic component. Since a displacement α between the slots for the fundamental is $q\alpha$ for the qth harmonic, the breadth factor for the qth harmonic becomes

$$k_b = \frac{\sin (nq\alpha/2)}{n \sin (q\alpha/2)} \tag{139}$$

The breadth factors for the fundamentals for a few uniformly distributed windings, assuming sinusoidal voltages, are given in Table 6.

TABLE 6

Number of slots per belt	Breadth factor				
	Width of phase belt in fractional part of the pole pitch				
	¼	⅓	½	⅔	Whole
2	0.980	0.966	0.924	0.866	0.707
3	0.977	0.960	0.911	0.844	0.666
4	0.976	0.958	0.906	0.836	0.653
Infinite	0.975	0.955	0.901	0.827	0.637

The sign of the breadth factor may become negative for certain harmonics. This occurs when $\sin q(n\alpha/2)$ becomes negative. The angle $q(\alpha/2)$ is always too small for $\sin q(n\alpha/2)$ to become negative for the fundamental or low-order harmonics. When this factor is negative for a harmonic, that harmonic in the voltage is opposite in phase to what it would be with the same flux distribution and a concentrated winding.

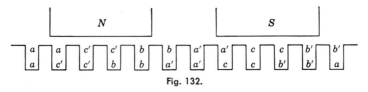

Fig. 132.

Two-layer Windings. In Fig. 132 is shown a two-layer integral slot winding having six slots per pole and two slots per pole per phase. Only two poles are shown in the figure although any even number of poles could be used as the same layout is repeated for each pair of poles. Two-layer windings are common, since the use of fractional-pitch coils is less restricted than with a single layer and the coils may be made all alike and of the diamond shape shown in Fig. 120, page 178. One side of each coil is placed in the top of a slot, while the other side is placed in the bottom of some other slot. For instance, in Fig. 132, a single coil would consist of coil side a in the slot at the left and coil side a' in the sixth slot from the left. The coil pitch would then be five slots or 150 deg. The three phases are indicated by the letters a, b, and c. In passing around the armature, the phase belts occur in the order a, c', b, a', c, b', etc. The significance of the primes on certain of the letters a, b, and c is evident from an inspection of the order in which the coil sides occur for the three phases in Fig. 128, page 183, that is, a and a' are on opposite sides of a coil, and therefore the voltages produced in a and a' substract vectorially.

It should be noted that the winding distribution in the top layer is exactly repeated in the lower layer although one layer may be displaced with respect to the other because of the fractional pitch. The breadth factor for the winding may, therefore, be determined from the distribution in either layer.

Fractional-slot Windings. In many cases the number of slots per pole of modern synchronous generators is not an integer. Although in this case the phase belts are not alike, yet if the winding is to give balanced polyphase voltages it must be capable of being divided into as many identical sections as there are phases which are displaced the equivalent of $360/n$ electrical degrees apart, where n is the number of phases.

If the ratio of the number of slots to the number of poles is reduced to its lowest terms, the denominator gives the number of poles before the winding repeats, $i.e.$, the number of poles in a repeatable section, and the numerator gives the number of slots per section multiplied by the number

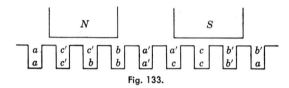

Fig. 133.

of phases. If the repeatable section is not divisible by the number of phases, a balanced polyphase winding cannot be obtained. A section is equal to a repeatable section divided by the number of phases. If a 20-pole synchronous generator has 84 slots, the ratio of the number of slots to the number of poles is $84/20 = 21/5$. Each phase of this winding repeats itself every five poles, and since the numerator 21 is divisible by 3, the winding gives balanced three-phase voltages.

Fractional-slot windings give a better control of wave form than integral-slot windings and have the mechanical advantage of requiring fewer slots and coils than integral-slot windings producing similar electrical and mechanical results.

Breadth Factor for Fractional-slot Windings. To determine the breadth factor of a fractional-slot winding, a repeatable section must be considered. If a two-layer winding is used, the breadth factor may be determined from the distribution in only one layer because as in the integral-slot winding the distribution in each layer is the same although one layer may be displaced with respect to the other because of fractional-pitch coils. In Fig. 133 is shown a repeatable section of a three-phase winding having $9/2$ slots per pole or $3/2$ slots per pole per phase. In each layer there are three slots containing coil sides of any one phase. For phase a in the upper layer, these coil sides are in slots 1, 5, and 6 counting from the left. In connecting coils, the order of connection does not influence the phase voltage, but if for one coil the winding starts with a coil side marked a, then all coils must start with side a and end with side a' in proceeding around the winding. The phase voltage will therefore be the vector sum of the voltages in all coil sides marked a minus the voltages of coil sides marked a' and similarly for the other phases. Since slots 5 and 6 are displaced by 160 and 200 deg, respectively, from slot 1 and contain coil sides a', the reversed vectors for slots 5 and 6 give voltages in the three coil sides which, when added vectorially, are displaced from one another by 20 deg. This is the same condition as found in the phase belt of a three-phase integral-slot winding having three slots per pole per phase. The breadth factor for this fractional-slot winding is therefore the same as that of the integral-slot winding and may be calculated from formula 139, page 191, using $n = 3$ and $\alpha = 20$ deg. The breadth factor for any properly laid out fractional-slot winding may be found in a similar manner. The number of slots per pole per phase for the corresponding integral-slot winding is in each case equal to the numerator of the fraction giving the slots per pole per phase of the fractional-slot winding when the fraction is reduced to its lowest terms.

As another example of the above, consider an 84-slot 20-pole 3-phase winding. The number of slots per pole is $84/20 = 2\frac{1}{5}$. This winding repeats itself every five poles, and since 21 is divisible by the number of phases, a balanced three-phase winding can be arranged. The number of slots per pole per phase equals $\frac{7}{5}$, and the distribution of this winding is equivalent to that of an integral-slot winding having seven slots per pole per phase. With such a winding the angle between slots, α, would be

$$\frac{180°}{7 \times 3} = 8\tfrac{4}{7}°$$

The breadth factor for the fundamental is

$$k_b = \frac{\sin 7\,\dfrac{8\tfrac{4}{7}°}{2}}{7 \sin \dfrac{8\tfrac{4}{7}°}{2}} = \frac{\sin 30°}{7 \sin 4.286°} = 0.956$$

A more complete analysis of this winding is given by referring to Fig. 134. In Fig. 134 the numbers represent the slots for five poles. The vertical lines separate the three sections. Minus signs are used in place of primes.

1	2	3	4	5	6	7	8	9	10	11	12	13	14	15	16	17	18	19	20	21
a	$-c$	$-c$	b	$-a$	$-a$	c	$-b$	a	a	$-c$	b	b	$-a$	c	$-b$	$-b$	a	$-c$	$-c$	b

Figure 134

Starting at slot 1, the successive belts for phase a contain 1, 2, 2, 1, and 1 slots. Starting at slot 8, which is the beginning of the second section, the belts for phase b contain 1, 2, 2, 1, and 1 slots, and starting at slot 15, which is the beginning of the third section, the successive belts for phase c contain 1, 2, 2, 1, and 1 slots. The signs of the successive belts for the three phases a, b, c do not occur in the same order. This makes no difference since when connecting the belts for any phase in series they must be connected to give the maximum voltage. Since starting at slots 1, 8, and 15 the arrangement of the belts for the phases a, b, and c is the same, the phase differences of the voltages in the three groups of belts must be the same as for the voltages in the conductors at the beginning of the sections, $i.e.$, in the conductors 1, 8, and 15.

The angle between adjacent slots is $(20 \times 180)/84$ electrical degrees. The phase angle between the voltages in the conductors in slots 8 and 1 is

$$\frac{20 \times 180}{84} \times 7 = 300 \text{ deg}$$

This is equivalent to a displacement of $300 - 360 = -60$ deg, but $-b$ is in slot 8. Reversing the connections of phase b makes it $180 - 60 = 120$ deg from phase a.

The phase angle between the voltages in the conductors in slots 15 and 1 is

$$\frac{20 \times 180}{84} \times 14 = 600 \text{ deg}$$

This displacement is equivalent to a displacement of $600 - 360 = 240$ deg. Conductor c is in slot 15. This makes phase c 240 deg behind phase a. The phases are therefore in the correct phase relation, and since the grouping of the belts for the three phases is the same, the phases have voltages that are equal in magnitude.

The complete winding for the synchronous generator considered must be a two-layer winding. Figure 134 shows only one layer. The second layer must be similar to the first, but it must be slipped by the first by an amount that gives the desired pitch. If a pitch of three slots, $i.e.$, $(20 \times 180)/84 \times 3 = 128\frac{4}{7}$ electrical degrees, is desired, the second layer must start with $-a$ in slot 4 followed by c, c, $-b$, etc., in succeeding slots. This duplicates the first layer with the exception that it is displaced three slots and opposite coil sides are used as indicated by the change in signs.

As a check on the breadth factor proceed as follows. In the repeatable section shown in Fig. 134, reduce the angles between the first coil side of phase a and the other coil sides of the phase to small equivalent angles by subtracting 360 deg or some multiple of 360 deg from the angles for the positive coil sides and subtracting 180 deg or 180 deg plus some multiple of 360 deg from the angles for the negative coil sides. These subtractions do not change the relative phase angles. This process is shown below for the 84-slot 20-pole winding.

Slot	Sign of coil side	Angle, degrees between the coil side considered and the first coil side	Equivalent phase angle, degrees
1	+	0	
5	−	$4 \times 42\frac{6}{7} = 171\frac{3}{7}$	$171\frac{3}{7} - 180 = -8\frac{4}{7}$
6	−	$5 \times 42\frac{6}{7} = 214\frac{2}{7}$	$214\frac{2}{7} - 180 = +34\frac{2}{7}$
9	+	$8 \times 42\frac{6}{7} = 342\frac{6}{7}$	$342\frac{6}{7} - 360 = -17\frac{1}{7}$
10	+	$9 \times 42\frac{6}{7} = 385\frac{5}{7}$	$385\frac{5}{7} - 360 = +25\frac{5}{7}$
14	−	$13 \times 42\frac{6}{7} = 557\frac{1}{7}$	$557\frac{1}{7} - 360 - 180 = +17\frac{1}{7}$
18	+	$17 \times 42\frac{6}{7} = 728\frac{4}{7}$	$728\frac{4}{7} - 720 = +8\frac{4}{7}$

The coil sides can now be arranged in the following to form a regular fan.

Slot...............	9	5	1	18	14	10	6
Angle...............	$-17\frac{1}{7}$	$-8\frac{4}{7}$	0	$+8\frac{4}{7}$	$+17\frac{1}{7}$	$+25\frac{5}{7}$	$+34\frac{2}{7}$

With seven slots displaced effectively by $8\frac{4}{7}$ deg from each other, the breadth factor should be equal to that of an integral-slot winding as calculated on page 194.

This latter method of analysis, besides serving as a check on the calculations for the particular winding considered, is a good general method for determining the breadth factor of a fractional-slot winding. The method of page 194 is simpler but applies only when the winding is so laid out as to give the maximum breadth factor. This, however, is the most common arrangement of fractional-slot windings and may be accomplished if the coil sides in the top of each slot are assigned to the phases of a three-phase winding in accordance with the following procedure. Starting with a reference point midway between a pair of slots, and proceeding always in the same direction around the armature, determine the angle in electrical degrees between this reference point and the center of each slot. Angles of 360 deg or over should be reduced to equivalent smaller angles by subtracting from them 360 deg or some multiple of 360 deg so that the resulting angle will always be less than 360 deg. The coil sides are then assigned to each phase in accordance with the accompanying table.

Angle, degrees *between coil side* *and reference point*	*Phase to which* *coil side is* *assigned*
0 to less than 60..................	*a*
60 to less than 120.................	*c'*
120 to less than 180................	*b*
180 to less than 240................	*a'*
240 to less than 300................	*c*
300 to less than 360................	*b'*

Harmonics in Three-phase Synchronous Generators. There can be neither a third harmonic nor any multiple of a third harmonic in the voltages between the terminals of a three-phase synchronous generator, but such harmonics can exist between any one of the three terminals and the neutral point if the generator is Y-connected.

Let the phase voltages of a three-phase synchronous generator be given by

$$e_1 = E_1 \sin \omega t + E_3 \sin 3\omega t + E_5 \sin 5\omega t + E_7 \sin 7\omega t + \ldots$$
$$e_2 = E_1 \sin (\omega t - 120°) + E_3 \sin 3(\omega t - 120°) + E_5 \sin 5(\omega t - 120°)$$
$$+ E_7 \sin 7(\omega t - 120°) + \ldots$$
$$e_3 = E_1 \sin (\omega t - 240°) + E_3 \sin 3(\omega t - 240°) + E_5 \sin 5(\omega t - 240°)$$
$$+ E_7 \sin 7(\omega t - 240°) + \ldots$$

The angular displacement between any harmonic of any one phase and the corresponding harmonic of phase 1 is given in Table 7.

TABLE 7

Phase	Displacement, electrical degrees				
	1st	3d	5th	7th	9th
1	0	0	0	0	0
2	120	3(120) = 360 = 0	5(120) 600 = 240	7(120) = 840 = 120	9(120) = 1,080 = 0
3	240	3(240) = 720 = 0	5(240) = 1,200 = 120	7(240) = 1,680 = 240	9(240) = 2,160 = 0

Referring to Table 7, it is evident that all the third harmonics are in phase. The ninth harmonics are also in phase. All multiples of the third harmonic are in phase. The fifth harmonics are 120 deg apart, but they occur in inverted order, *i.e.*, in the order 1, 3, 2. The seventh harmonics are 120 deg apart and in natural order. In general, starting with the fifth harmonic and neglecting those harmonics which are in phase, the sequence in which the harmonics of any order occur in the three phases alternates from the order 1, 3, 2 to the order 1, 2, 3.

Consider a Y-connected synchronous generator. Figure 135 represents a space-phase diagram of the connections of the phases of a Y-connected synchronous generator and a time-phase diagram of the induced voltages.

The voltages across the three pairs of terminals 1-2, 2-3, and 3-1 are

$$e_{12} = e_{10} + e_{02} = e_{10} - e_{20}$$
$$e_{23} = e_{20} + e_{03} = e_{20} - e_{30}$$
$$e_{31} = e_{30} + e_{01} = e_{30} - e_{10}$$

The voltage between any pair of terminals is the vector difference of the phase voltages. Since the third harmonics and all the multiples of the

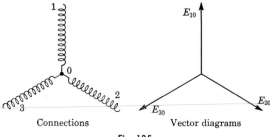

Connections   Vector diagrams

Fig. 135.

third harmonics are in phase, they cancel in the differences. Therefore there can be no third harmonic or any multiple of it in the line or terminal voltage of a three-phase Y-connected synchronous generator. The third harmonics and their multiples existing between the terminals and neutral point, however, are in phase.

A study of the phase differences between harmonics of the same order for the three phases shows that the voltages e_{12}, e_{23}, and e_{31} of a Y-connected synchronous generator, when referred to e_{10}, are

$$e_{12} = \sqrt{3}\, E_1 \sin\left(\omega t + 30°\right) + 0 + \sqrt{3}\, E_5 \sin 5\left(\omega t - \frac{30°}{5}\right)$$
$$+ \sqrt{3}\, E_7 \sin 7\left(\omega t + \frac{30°}{7}\right) + 0 + \ldots$$
$$e_{23} = \sqrt{3}\, E_1 \sin\left(\omega t - 120° + 30°\right) + 0 +$$
$$\sqrt{3}\, E_5 \sin 5\left(\omega t - 120° - \frac{30°}{5}\right)$$
$$+ \sqrt{3}\, E_7 \sin 7\left(\omega t - 120° + \frac{30°}{7}\right) + 0 + \ldots$$

$$e_{31} = \sqrt{3}\, E_1 \sin (\omega t - 240° + 30°) + 0$$
$$+ \sqrt{3}\, E_5 \sin 5 \left(\omega t - 240° - \frac{30°}{5} \right)$$
$$+ \sqrt{3}\, E_7 \sin 7 \left(\omega t - 240° + \frac{30°}{7} \right) + 0 + \ldots$$

Consider the conditions existing in a Δ-connected synchronous generator. The voltage acting around the closed Δ is $e_{10} + e_{20} + e_{30}$. By referring to Table 7, page 197, it is evident that the three components of the third harmonic voltage are in phase. They are therefore short-circuited in the closed Δ and cannot appear between the terminals of the generator. The ninth and all other multiples of the third harmonic are also short-circuited in the Δ. The vector sums of all other harmonics, and also the fundamental, are zero when taken around the closed Δ. The three line or terminal voltages of a Δ-connected synchronous generator are

$$e_{12} = E_1 \sin \omega t + 0 + E_5 \sin 5\omega t + E_7 \sin 7\omega t + 0 + \ldots$$
$$e_{23} = E_1 \sin (\omega t - 120°) + 0 + E_5 \sin 5(\omega t - 120°)$$
$$+ E_7 \sin 7(\omega t - 120°) + 0 + \ldots$$
$$e_{31} = E_1 \sin (\omega t - 240°) + 0 + E_5 \sin 5(\omega t - 240°)$$
$$+ E_7 \sin 7(\omega t - 240°) + 0 + \ldots$$

Although the terminal voltages of a synchronous generator, when connected in Y and in Δ, contain the same harmonics in the same relative magnitudes, the wave forms given by the two connections are different, because of the phase displacement of 30 deg which occurs in the harmonics of a Y-connected synchronous generator.

The rms voltages given by the Y and Δ connections are in the ratio of $\sqrt{3}:1$, but the maximum voltages are not in this ratio, since the phase relations between harmonics are different for the two connections.

The effective value of the circulatory current caused by the third harmonic and its multiples in the armature of a Δ-connected synchronous generator is

$$\frac{1}{\sqrt{2}} \sqrt{\left(\frac{3E_3}{3z_3} \right)^2 + \left(\frac{3E_9}{3z_9} \right)^2 + \ldots}$$

where the z's are the effective impedances of the armature per phase for the different harmonics.

The effective reactance of the armature of a synchronous generator for any harmonic is not the effective or synchronous reactance of the armature for the fundamental multiplied by the order of the harmonic, but it is considerably less than this on account of the difference between the armature reactions produced by the harmonics and the fundamental.

Δ connection is objectionable for synchronous generators unless their wave forms are free from third harmonics and their multiples. If third harmonics are present in any great magnitude, there is a large short-circuit current in the closed Δ formed by the armature winding. This current combined with the load current may cause dangerous heating. Most modern synchronous generators are Y-connected. The effect of the third harmonic in a Δ-connected synchronous generator is only one of several things that make Y connection preferable as a rule.

Harmonics Due to Slots. Figure 206, page 336, shows two positions of an armature core relative to a pole. The reluctance of the magnetic circuit is obviously less for the position shown at the right, where there are four slots over the pole, than for the position shown at the left, where

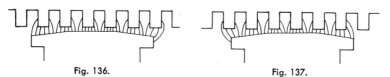

Fig. 136.   Fig. 137.

there are only three slots over the pole. The movement of the slots across the pole therefore causes the pole flux to pulsate with a frequency of $2nf$, where n is the number of slots per pole and f is the normal frequency of the machine. If the effective width of the pole is equal to any integer times the width of a slot plus a tooth, there is no variation in the reluctance of the magnetic circuit as a whole, because of the relative movement of the slots and poles. The effective width of the poles is slightly greater than the actual width on account of the fringing of the flux at the pole edges.

Because of the slots entering and leaving the polar region, there is a periodic variation in the fringing of the flux at the pole tips which causes an oscillation of the field axis with respect to the axis of the poles. Figures 136 and 137 make this clear.

For the relative position of the slots and field pole shown in Fig. 136, the fringing is greater at the right side of the pole than at the left. Figure 137 shows the condition after a relative movement of the pole with respect to the slots equal approximately to the width of a slot. In this case the fringing is greater at the left side than at the right. This change in the fringing produces an oscillation of the axis of the flux with respect to that of the poles at a frequency which equals $2nf$.

The slots also cause ripples in the flux wave (Fig. 171, page 272) which move across the pole faces. The movement of these ripples cannot cause any harmonics in the armature voltage since there is no relative movement between them and the armature inductors. Either the variation in the magnitude of the field or in the position of its axis with respect to the poles may induce tooth harmonics in the armature voltage.   The voltage induced by either the oscillation of the axis or the variation in the field strength is of the form

$$e = k(\sin 2n\omega t) \sin \omega t$$
$$= \tfrac{1}{2}k[\cos (2n - 1)\omega t - \cos (2n + 1)\omega t]$$

Either effect may therefore produce harmonics of two different frequencies in the armature winding of the orders $(2n - 1)$ and $(2n + 1)$.

Chapter 16

Rating. Regulation. Magnetomotive forces and fluxes concerned in the operation of a synchronous generator. Armature reaction. Armature reaction of a synchronous generator. Effect of armature reaction of a non-salient-pole or cylindrical-rotor synchronous generator with balanced load. Effect of armature reaction of a salient-pole synchronous generator. Graphical plot of the armature reaction of a polyphase synchronous generator. Armature leakage reactance. Equivalent armature leakage flux. Effective resistance. Single-phase rating.

Rating. The maximum output of any synchronous generator is limited by its mechanical strength, by the temperature of its parts produced by its losses, and by the permissible increase in field excitation which is necessary to maintain rated voltage at some specified load and power factor. Usually the limit of output is fixed by the temperature rise under load.

The maximum voltage any synchronous generator can give at a definite frequency depends upon the permissible pole flux and field heating. The maximum safe temperature of the hottest spots in the armature winding fixes the safe armature current. The kw output depends not only upon the voltage and the armature current but also upon the power factor. The armature copper loss is independent of the power factor, but the field heating at a given armature current, voltage, and frequency and, to some extent, the core loss are dependent upon the power factor.

Synchronous generators may be rated in kilovolt-amperes at a definite frequency and voltage or they may be rated in kilowatts at a stated power factor. In the latter case, the frequency and the voltage must also be included in the rating. The permissible output of a polyphase synchronous generator depends also upon the balance of the load, *i.e.*, how the load is distributed among the phases. It is a maximum for a balanced load and a minimum for a single-phase load.

The limiting safe temperatures for different kinds of insulating materials are given on page 170. The methods for determining the temperatures of the windings of synchronous generators are given in the Standardization Rules of the AIEE.

Regulation. The regulation of a synchronous generator at any given power factor is the percentage rise in voltage, under the conditions of constant excitation and frequency, when the rated kva load at the given power factor is removed. The change in voltage produced under this condition depends not only upon the magnitude of the load and the constants of the synchronous generator but also upon the power factor of the load. The regulation is positive for both a noninductive and an inductive load, since each causes a rise in voltage when it is removed. A capacitive load, on the other hand, if the angle of lead is sufficiently great, may cause a fall in voltage instead of a rise. Under this condition the regulation is negative. The increase in field excitation that is necessary to maintain rated voltage under load conditions is more important than the voltage regulation of a synchronous generator, but both depend upon the same factors. The inherent regulation is the regulation on full noninductive load.

The regulation of a synchronous generator depends upon four factors: (1) armature reaction, (2) armature reactance, (3) armature effective resistance, and (4) the change in the pole leakage with change in load.

Some of the four factors produce similar effects and are combined in certain approximate methods for determining regulation and the increase in field excitation necessary to maintain rated voltage under load. The relative magnitudes of the effects produced upon the terminal voltage of a synchronous generator by these four factors depend not only upon the magnitudes of the factors but also upon the power factor of the load. At unity power factor with respect to the generated voltage, armature reaction and leakage reactance have their minimum effects. Their maximum effects occur at zero power factor. Just the opposite is true in regard to the effect produced by resistance. The actual magnitudes of reaction, reactance, and resistance are fixed by the design and may be varied over rather wide limits, but considering merely the component change in voltage produced by each when acting separately, the magnitudes of their effects are usually in the order named. In general, the armature resistance drop in a large synchronous generator is small, and at full-load current it is only a minor percentage of the rated phase voltage.

Since it is now the almost universal practice to operate synchronous generators with automatic voltage regulators to hold the voltage constant under varying loads, a knowledge of the voltage regulation of a synchronous generator is of less importance than the knowledge of the change in the field excitation which is required to maintain constant voltage under varying loads. As has been stated, the determination of voltage regulation and of the change in field excitation necessary to maintain constant voltage involves the same factors. In general the same methods are used for both field current and voltage regulations, and the determination of one involves the determination of the other.

Magnetomotive Forces and Fluxes Concerned in the Operation of a Synchronous Generator. There are two distinct mmfs and three component fluxes to be considered in the operation of any synchronous generator. The two mmfs are (1) the mmf of the impressed field, (2) the mmf due to the armature currents, *i.e.*, the armature reaction. Although both of these mmfs may be expressed either in ampere-turns per pole or per pair of poles, it is usually more convenient, especially when dealing with synchronous generators with more than two poles, to express them in ampere-turns per pole.

The three component fluxes are (1) the flux which is common or mutual to the armature and the field, which is called the air-gap flux; (2) that portion of the total armature flux which links only the armature inductors; (3) the field leakage flux. This last is the portion of the field flux which passes between adjacent north and south poles without entering the armature. The ratio of the maximum flux in a pole to the portion of that flux which enters the armature is called the leakage coefficient or the leakage factor of the field. This coefficient varies from about 1.15 to 1.25, according to the design and type of the synchronous generator. If the leakage coefficient were constant and independent of the load, the field leakage would produce no effect on the regulation of a synchronous generator. The field leakage is inversely proportional to the reluctance of the path of the stray field and is directly proportional to the magnetic potential between the poles. This potential is made up of two parts: (1) the drop in the magnetic potential necessary to force the flux through the armature and the air gap, (2) the opposing ampere-turns of armature reaction.

Armature Reaction. When a synchronous generator operates at no load, the only mmf acting is that of the field winding. The flux produced by this winding depends only upon the current it carries, the number of turns and their arrangement, and the total reluctance of the path through which the mmf acts. Neglecting the effect of armature slots, the distribution of the no-load air-gap flux of a salient-pole synchronous generator depends upon the shape of the pole shoes. In the case of a non-salient-pole machine, *i.e.*, one with a cylindrical or drum type of field core, neglecting the effects of the field and armature slots, the distribution of the no-load air-gap flux depends upon the distribution of the field winding.

When load is applied to a synchronous generator, the mmf of the armature current modifies the flux produced by the field winding. The effect of the armature mmf, or armature reaction, depends not only upon the arrangement of the armature winding, the current it carries, and the reluctance of the magnetic circuit, but also upon the power factor of the load.

Consider a salient-pole synchronous generator. Neglecting field distortion, the voltage generated in any coil or turn on the armature of a single-phase synchronous generator has its maximum value when the center of the coil lies midway between two adjacent poles. It is zero when the center of the coil is directly opposite the center of a pole. If the power factor is zero with respect to the voltage produced by the air-gap flux, the maximum current occurs when the voltage is zero or when the coil is directly opposite a pole. Under this condition the axis of the magnetic circuit for the armature reaction coincides with the axis of the magnetic circuit for the field winding, and the resultant mmf acting to produce the field flux is the algebraic sum of the mmfs of armature reaction and field excitation. In this case the armature reaction either strengthens or weakens the field without producing distortion. The armature reaction caused by a lagging current opposes the mmf of the field winding and weakens the field. A leading armature current strengthens the field.

If the coil, instead of lying with its center opposite a pole when the current in it is a maximum, lies with its center midway between two poles, it covers half of two adjacent poles (a full-pitch winding is assumed) and produces a demagnetizing action on half of one pole and a magnetizing action on half of the other pole. It follows that one half of each pole is strengthened and the other half is weakened by the action of the armature current. If the change in the reluctances produced under the foregoing conditions by armature reaction in the two halves of the poles were negligible, there would be no change in the total pole flux. Actually there is a slight decrease in the total pole flux. The principal effect of armature reaction under the foregoing conditions, however, is field distortion. The application of the corkscrew rule to the direction of the current carried by the armature coils shows that the trailing pole tips are strengthened and the leading pole tips are weakened by armature current which is in phase with the excitation voltage. The effect is chiefly a shift in the flux from the leading pole tip to the trailing pole tip. The condition just described, i.e., with the center of the armature coil midway between two poles when the current in it is a maximum, corresponds approximately to unity power factor with respect to the terminal voltage.

The approximate distributions of the flux at the instant when the armature current is a maximum for a reactive load of zero power factor and for a power factor of unity, both with respect to the excitation voltage, are shown in Figs. 138 and 139. Figure 140 shows the distribution at no load.

In the preceding discussion, only the instant when the current is a maximum is considered. While the field is moving through a distance corresponding to 360 electrical degrees, the current in any armature coil,

as *ab*, Fig. 139, goes through a complete cycle, and consequently the value
of the total flux from a pole and its distribution also goes through a complete cycle.   The average distribution of flux, however, is about the same
as when the current passes through its maximum value.   A corresponding

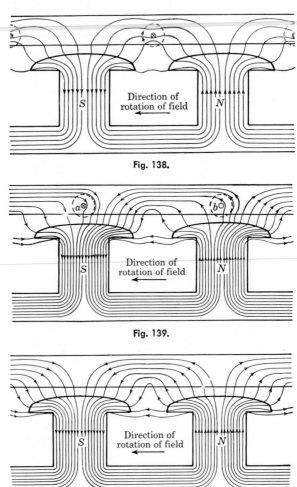

Fig. 138.

Fig. 139.

Fig. 140.

variation of the flux does not occur in a polyphase synchronous generator
which carries a balanced load, since the armature reaction of such a
generator under these conditions is fixed in magnitude and in direction
with respect to the poles.   The effect is the same as that which occurs in
a single-phase synchronous generator at the instant when the current has
its maximum value.

To sum up, the general effect of armature reaction is as follows: with a noninductive load, it distorts the field without appreciably changing the total field flux; with an inductive load of zero power factor, it weakens the field without distorting it; with a load having a power factor between unity and zero, it both distorts the field and changes its strength.

In addition to the general distortion of the field already considered, there is a local distortion in the neighborhood of each inductor. This distortion is limited to the region about the slots and to the air gap and does not extend to any appreciable depth into the pole faces. It is equivalent to a little ripple in the flux about each inductor and may be considered to be due to the superposition upon the main field of local fluxes which surround the armature inductors. These local fluxes are indicated by the dotted lines in Figs. 138 and 139. Although the local fluxes have no real existence except about the end connections of the coils, it is convenient to consider them separately as components of the main flux. They are alternating fluxes and are very nearly in time phase with the currents which cause them. They are the so-called armature leakage fluxes and give rise to a voltage of self-inductance in the inductors which they link. This voltage alternates with the same frequency as the armature current and lags 90 deg behind that current. The reactance corresponding to this voltage of self-inductance plus the reactance due to end connections of the armature winding is the leakage reactance of a synchronous generator. More will be said of this under reactance.

A knowledge of armature reaction is necessary in order to predetermine the regulation of a synchronous generator and also to determine the number of field ampere-turns required at full load to maintain the rated voltage at different power factors. In synchronous generators with salient or projecting poles, such as are illustrated in Figs. 138, 139, 140, armature reaction produces a distortion of the air-gap flux except when the power factor is zero, a condition which is impossible in practice and which is not even approached under ordinary operating conditions.

The distortion of the air-gap flux which takes place in a synchronous generator with salient poles is caused almost entirely by the difference in the reluctance of the magnetic circuits for the armature reaction and the impressed field. Except when the power factor of the load is zero, the mmfs of the field and armature do not act along the same axis. They are not in space phase and the axis of their resultant does not coincide with the axis of either. Since flux always distributes itself so as to follow the path of minimum reluctance, the flux caused by the combined action of the mmfs of the armature and field currents still clings to the poles, but is crowded toward one side instead of being symmetrical about the pole axes. In generators with non-salient poles, the air gap is uniform in all directions about the armature except for the effect of the armature and the field slots.   Under this condition, there is no distortion of the magnetic field under load, provided the field and armature winding both give sinusoidal distributions of magnetic potential in the air gap.   This condition cannot be fulfilled exactly in practice.

Armature Reaction of a Synchronous Generator.   A single-phase two-pole synchronous generator with a concentrated winding will be considered first.   The mmf of the armature winding under this condition,

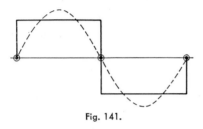

Fig. 141.

i.e., its armature reaction, is constant in its space distribution over the coil area and can be represented by a rectangular wave.   Resolve this rectangular space wave into its Fourier series and neglect all terms of the series except the fundamental, *i.e.*, neglect the harmonics.   The terms that are neglected are considered under armature leakage reactance.   With a distributed winding and especially one with a fractional pitch, such as is ordinarily used in practice, the harmonics are greatly reduced as compared with those for a concentrated winding. The amplitude of the fundamental of the Fourier series which represents a rectangular wave is $4/\pi$ times the amplitude of the rectangular wave. The rectangular wave and the fundamental of the series that represents it are shown in Fig. 141, where the small circles represent the coil sides. Both the rectangular wave and the fundamental vary sinusoidally in magnitude with time, with respect to the frequency of the circuit, but they are stationary in space with respect to the armature coil.   The fundamental of the wave can be represented by an oscillating vector which lies along the axis of the coil and which has a maximum length equal to the maximum value of the fundamental.

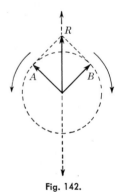

Fig. 142.

Any sinusoidally oscillating vector can be replaced by two equal oppositely rotating component vectors, each with a constant magnitude equal to one-half the maximum value of the vector they replace and rotating with a frequency equal to that of the oscillating vector.   An inspection of Fig. 142 should make this clear.   The vertical dotted line of this figure represents the line along which the original vector oscillates.   *A* and *B* are the two oppositely rotating component vectors.   Their resultant *R* is equal at every instant to the corresponding value of the original vector.

Let the oscillating vector, which represents the fundamental of the Fourier series, be replaced by two equal and oppositely rotating component vectors. These two oppositely rotating vectors represent oppositely rotating, sinusoidally distributed mmf waves, each of which has a constant amplitude equal to one-half the amplitude of the original sinusoidally oscillating mmf wave. Since these two oppositely rotating waves have the same wave length as the original wave and are sinusoidal in their space distribution, their sum is always a sinusoidal wave and is equal to the original sinusoidally oscillating wave. When the original wave has its maximum value, the two component waves coincide. When the original wave has its zero value, the two component waves are 180 deg apart and their sum is zero.

Both of the rotating vectors revolve at synchronous speed with respect to the armature, one clockwise, the other counterclockwise. One of these vectors rotates in the same direction as the field poles of the generator and is stationary with respect to them. The other rotates at double synchronous speed with respect to the field poles.

Let N be the number of armature turns per pole and let I_m be the maximum value of the armature current. The amplitude of the rotating components of the sinusoidally distributed armature reaction is $\frac{1}{2}(4/\pi)NI_m$ ampere-turns per pole. Replacing I_m by its rms or effective value gives

$$A = \sqrt{2}\,\frac{1}{2}\frac{4}{\pi}\,NI = 0.90NI \qquad (140)$$

where I is the effective value of the armature current.

The component of the armature reaction that is stationary with respect to the field poles changes the field strength and in a salient-pole machine, except at zero power factor, distorts the field. The other component, which rotates at twice synchronous speed with respect to the poles, sets up in the poles double-frequency or second-harmonic flux variations. The effect is most marked in a salient-pole machine, because of the large variation in the reluctance of the magnetic circuit around the air gap. This double-frequency component of the field flux, in combination with the rotation of the field poles, induces a third-harmonic voltage in the armature turns, which is present across the terminals of the synchronous generator. This double-frequency component of the flux in the poles and hence the third-harmonic voltage between terminals can be much reduced by the reaction of currents in dampers similar to those used on synchronous motors (see page 332). This third harmonic cannot be eliminated by the use of fractional pitch.

The voltage generated in any armature turn is

$$e = k\varphi \sin \omega t$$

where k is a constant and φ is the pole flux. Ordinarily φ is constant for any given excitation and load. If it varies with time, it must be inserted in the formula for the voltage as a function of time.   Assume that the double-frequency flux variation due to armature reaction is sinusoidal.

Then if $\varphi = \varphi_m \sin 2\omega t$ is the value of this flux at each instant, the voltage induced by it in the armature is

$$e = k\varphi_m \sin 2\omega t \sin \omega t$$
$$= \tfrac{1}{2}k\varphi_m(\cos \omega t - \cos 3\omega t)$$

The double-frequency component of the flux therefore produces voltages of both fundamental and triple frequency in each armature turn. The actual variation in the flux produced by armature reaction in the poles of a single-phase synchronous generator is much reduced by the self-inductance of the field winding, by eddy currents in the poles, and by any short-circuited damper there may be on the field structure.

A single-phase synchronous generator should be provided with a damper in its pole faces.   This damper is like those used on all synchronous motors.   It consists of copper bars inserted in holes punched in the pole faces and short-circuited by bolting or welding copper straps to the ends of the bars.   A damper is illustrated in Fig. 205, page 333. The double-frequency flux generates currents in the damper which, according to the law of Lenz, oppose the change in flux producing them. These currents nearly damp out the flux variation.   Eddy currents induced in the pole faces by this flux also have a damping effect.

If the armature current is in phase with the excitation voltage, i.e., the voltage which would be produced by the field excitation acting alone, the axis of the component of the armature reaction which is fixed with respect to the field lies midway between the two poles.   If the lag of the current behind the voltage is 90 deg, the axis of this component field is along the axis of the poles.   In general the space angle between the axes of the resultant air-gap flux and the armature reaction is equal to 90 deg plus the angle of lag between the current and the excitation voltage.   An angle of lead is equivalent to a negative angle of lag.

If the generator has a distributed fractional-pitch winding, Eq. (140), page 209, for the armature reaction must be multiplied by breadth and pitch factors.   These factors are the same as for voltage.   Some pitch and breadth factors are given in Tables 5 and 6 on pages 185 and 192. The complete formula for the armature reaction of a single-phase synchronous generator in ampere-turns per pole is

$$A = 0.90NIk_bk_p \tag{141}$$

A is the maximum value of the armature-reaction wave and is the length of the vector which represents the armature reaction which is fixed with respect to the poles.

In a polyphase synchronous generator, each phase can be treated like the one phase of a single-phase synchronous generator.   Consequently each produces a fixed reaction on the field poles which is equal to $0.90NIk_bk_p$, where N is the number of armature turns per pole per phase and I is the effective phase current, *i.e.*, the effective current carried by the armature conductors.   Besides this fixed reaction, each phase also produces a double-frequency reaction on the poles.   For equal-phase power factors, the fixed parts of the reactions of all phases on any pole for a balanced load lie along the same axis and can be added directly.   The armature reaction becomes

$$0.90NnIk_bk_p \qquad (142)$$

where n is the number of phases.   The number of turns per pole per phase multiplied by the number of phases is equal to the total number of armature turns per pole.   Therefore, the armature reaction of any synchronous generator carrying a balanced load is given by Eq. (141), provided N is taken as the total number of armature turns per pole in all phases. If the synchronous generator has a multicircuit armature winding, the current I must be the current per turn, *i.e.*, the phase current divided by the number of parallel paths in the armature winding.

The variable or double-frequency reactions of all the phases neutralize and are zero for a balanced load.   For example take a three-phase generator.   The current waves of the three phases differ in phase by 120 deg. When referred to phase 1, their phase differences are 0, 120, and 240 deg. The phase relations between the double-frequency reactions produced by the three phases are 0, 2(120), and 2(240 deg), which are equivalent to 0, 240, and 120 deg.   Since the vector sum of three equal vectors which differ by 120 deg is zero, the variable parts of the reactions of the three phases of a three-phase synchronous generator carrying a balanced load neutralize one another.   In a four-phase generator, the phase relations between the double-frequency reactions are 0, 2(90), 2(180), and 2(270) deg, which are equivalent to 0, 180, 0, and 180 deg.   The vector sum of these is obviously zero.   Since the variable components of the armature reaction neutralize, the armature reaction of a polyphase synchronous generator which carries a balanced load is fixed in direction and in magnitude with respect to the poles.   This assumes that the mmf of each phase is sinusoidal in its distribution; in other words, that all harmonics in the mmf of each coil are neglected.

The effect of the armature reaction of a polyphase synchronous generator may be explained in another way.   Let Fig. 143 represent the developed field and armature of a three-phase generator.   The armature coils for the three phases are 1-1′, 2-2′, and 3-3′.   The poles are shown as if they were salient, merely to indicate their positions.

Take a reference point a at the zero point of the flux which the field current alone would produce. Assume that this flux is sinusoidal in its space distribution with its zero points midway between the poles. The currents carried by the three phases, referred to a as a reference point from which to measure time, are $I_m \sin(\omega t - \theta)$, $I_m \sin(\omega t - 120° - \theta)$, and $I_m \sin(\omega t - 240° - \theta)$, where θ is the angle of lag of the currents behind their corresponding excitation voltages, $i.e.$, the voltages which would be produced in the coils by the flux due to the field current acting alone. As in the previous discussion of armature reaction, neglect all harmonics in the space distribution of the mmf of each coil. Assume concentrated full-pitch armature windings. Let N be the number of armature turns per phase per pole and i the instantaneous phase current.

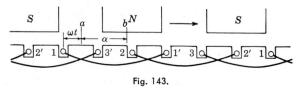

Fig. 143.

Then the maximum value of the magnetomotive force due to each phase when the phase carries a current i is $(4/\pi)Ni$ ampere-turns per pole.

The mmf due to any one phase at a point b in the air gap varies with time, on account of the variation in the current in the phase and also because of the change in the position of the phase with respect to the point a. If for the moment the currents in the phases are assumed constant and equal to I_1, I_2, I_3, the mmf at any point b at any instant of time t, due to all three phases, is

$$\frac{4}{\pi}\{NI_1 \sin(\alpha + \omega t) + NI_2 \sin[\alpha + (\omega t - 120°)]$$
$$+ NI_3 \sin[\alpha + (\omega t - 240°)]\}$$

Putting the actual values of the currents in place of I_1, I_2, I_3 gives for the mmf at the point b

$$\frac{4}{\pi}\{NI_m \sin(\omega t - \theta)\sin(\alpha + \omega t)$$
$$+ NI_m \sin(\omega t - 120° - \theta)\sin[\alpha + (\omega t - 120°)]$$
$$+ NI_m \sin(\omega t - 240° - \theta)\sin[\alpha + (\omega t - 240°)]\} \quad (143)$$

Since

$$\sin x \sin y = \tfrac{1}{2}[\cos(y - x) - \cos(y + x)]$$

Eq. (143) can be reduced to

$$\frac{2}{\pi}NI_m[\cos(\alpha + \theta) - \cos(2\omega t + \alpha - \theta) + \cos(\alpha + \theta)$$
$$- \cos(2\omega t + \alpha - \theta - 240°) + \cos(\alpha + \theta)$$
$$- \cos(2\omega t + \alpha - \theta - 480°)]$$

The three terms involving $2\omega t$ are three second harmonics which differ in phase by 120 deg. Their vector sum is zero and Eq. (143) reduces to

$$\frac{6}{\pi} N I_m \cos (\alpha + \theta) \qquad (144)$$

This is independent of time and varies only with the position of the point b in the air gap. Its maximum value occurs when b is at such a distance from a that $\alpha = -\theta$. The maximum value of the reaction is $(6/\pi)N I_m$. Replacing I_m by its rms value and letting N be the total number of turns on the armature per pole in all phases instead of per phase as in the preceding equations, this reduces to $(2/\pi)N \sqrt{2}I = 0.90NI$, which is the same as the expression previously found.

From Eq. (144) it is obvious that the maximum value of the mmf of armature reaction lies at a distance from the reference point a which is equal to the angle of lag θ of the current behind the excitation voltage, i.e., the voltage which would be produced in a phase by the flux due to the field current acting alone. With an angle of lag of 90 deg, the vector representing the mmf of armature reaction lies along the field axis, and since the expression for the armature reaction is negative, it opposes the mmf of the impressed field. Therefore, as has been shown already, a lagging armature current weakens the field.

Although armature reaction is fixed in direction and in magnitude with respect to the field poles and thus is a space vector with respect to the field poles, it revolves at synchronous speed with respect to the armature and is a time vector when considered with respect to the armature. The maximum excitation voltage occurs in any coil on the armature when its center is displaced 90 deg from the maximum point of the flux due to the field current acting alone. The maximum current in any armature coil occurs after the coil has been still further displaced by an angle θ. The maximum current in the coil and the maximum mmf through it due to armature reaction therefore occur at the same instant. Considering armature reaction as a time vector, armature reaction and armature current are in phase. If the armature winding is distributed and has a fractional pitch, N in the above formulas must be multiplied by the breadth and pitch factors.

Effect of Armature Reaction of a Non-salient-pole or Cylindrical-rotor Synchronous Generator with Balanced Load. The field winding of a cylindrical-rotor synchronous generator is always distributed and produces an mmf which is roughly sinusoidal in its space distribution. If the harmonics in the mmf of the field winding, as well as those in the armature reaction caused by a balanced load, are neglected, these two mmfs, which acting together produce the resultant or air-gap flux, always add to give a resultant flux which has a sinusoidal space distribution.

This follows from the well-known fact that any two space sinusoidal waves of the same wave length always add to give a resultant wave which is a sinusoid and is of the same wave length as the components. This is independent of the phase displacement of the component waves. The phase displacement of the mmfs of the impressed field and the armature reaction depend upon the power factor of the load. These mmfs are in phase only for a power factor of zero with respect to the excitation voltage.

Since the air gap of a cylindrical-rotor synchronous generator is uniform, except for the effect of the slots on the armature and field structure, the shape of the flux wave, which is produced by the resultant sinusoidal mmf of the impressed field and armature reaction, is independent of the direction of the armature reaction with respect to the field axis. In a cylindrical-rotor synchronous generator, therefore, armature reaction produces little field distortion. It affects only the resultant field strength for any given field current and displaces the axis of the resultant field from the axis of the field poles.

The armature reaction of a cylindrical-rotor or non-salient-pole generator which carries a balanced load may be treated as a space vector with respect to the field structure and may be combined vectorially with the mmf of the impressed field, which also may be represented by a space vector. Both may be considered as time vectors with reference to the armature. The vectors in both cases are equal to the maximum values of the mmf waves. Armature reaction cannot be combined vectorially with the impressed field in the case of a salient-pole generator, as will be understood from what follows.

Effect of Armature Reaction of a Salient-pole Synchronous Generator. What has been said in regard to the armature reaction of a generator with non-salient poles or a cylindrical rotor does not apply to a salient-pole generator, since the air gap of such a generator is not uniform and therefore its magnetic circuit is not the same in all directions with respect to the field structure. A given number of ampere-turns of armature reaction produce different effects on the flux according to the direction in which they act with respect to the field poles. The armature reaction is due to a distributed winding, but the impressed field mmf is due to a concentrated winding. The latter cannot be treated as a vector and combined vectorially with the vector which represents the ampere-turns of armature reaction without producing serious error.

Figures 144, 145, and 146 show the distribution of the component fluxes caused by the impressed field, the armature reaction, and the resultant flux in a small three-phase Y-connected salient-pole synchronous generator with a distributed armature winding. Figure 144 is for unity power factor, Fig. 145 for zero power factor, and Fig. 146 for 0.7 power factor. Compare Figs. 144 and 145 with Figs. 139 and 138, page 206.

In order to represent more nearly the actual conditions existing in synchronous generators with salient poles when calculating their regulation or field current for a given load, armature reaction is sometimes

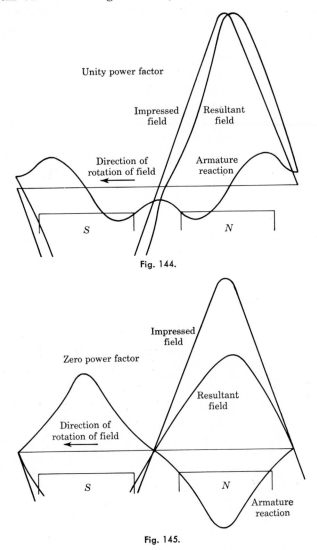

Fig. 144.

Fig. 145.

divided into two quadrature components, one along the field axis and the other at right angles to it or midway between the poles. By taking into consideration the difference between the reluctances of the magnetic circuits in the two directions, fairly satisfactory results can be obtained. This method of splitting armature reaction into two components at right angles to each other was suggested by André Blondel and is known as the Blondel or double-reaction method for calculating the performance of salient-pole machines.

The magnitude of the armature reaction of a synchronous generator expressed in ampere-turns depends only on the number of armature conductors, their distribution on the armature, and the currents they carry, but the effect on the field of a given number of ampere-turns of armature reaction for any fixed load and power factor depends upon the ratio of the ampere-turns of armature reaction to the field ampere-turns. To

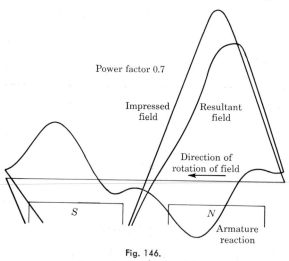

Fig. 146.

reduce the effect of armature reaction, this ratio must be decreased. This may be accomplished by increasing the radial length of the air gap or by increasing the saturation of the magnetic circuit. Neither of the changes affects the armature ampere-turns for a given load, but both increase the reluctance of the magnetic circuit and make an increase in the field ampere-turns necessary in order to maintain the same flux. The higher the degree of saturation, the less is the effect of a given number of ampere-turns of armature reaction, but high saturation in the field circuit increases the field-pole leakage. Increasing the length of the air gap has a similar effect so far as armature reaction is concerned, but it does not increase the field leakage to so great an extent as does increasing the degree of saturation of the magnetic circuit.

Graphical Plot of the Armature Reaction of a Polyphase Synchronous Generator. A three-phase synchronous generator with two slots per pole per phase and a ⅚-pitch armature winding will be used. Let the load be balanced with I effective amperes per phase. Assume that the currents are sinusoidal. The armature must have a two-layer winding with two coil sides in each slot, in order to have ⅚ pitch. The winding is developed at the bottom of Fig. 147, where the Arabic numbers represnt the slots and the Roman numbers I, II, III indicate the coil sides for the phases 1, 2, 3. Circles around the Roman numbers represent trail-

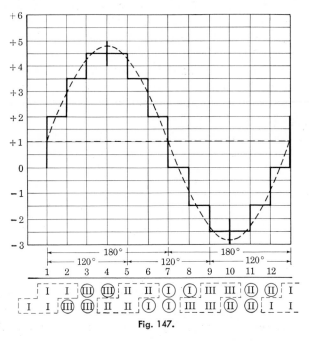

Fig. 147.

ing coil sides. Angular distances are shown on the diagram. The coil sides marked I, which lie in the top of slot 1 and in the bottom of slot 6, are in the same coil. They must be ⅚ of 180 deg or 150 deg apart, since the pitch is ⅚. The corresponding coil belts of conductors for the three phases are displaced 120 deg from one another, since the winding is three-phase. The belts of conductors which con- tain leading coil sides are enclosed by dotted lines.

Plot the mmf wave of armature reaction for the instant of time when the current in phase 1 has a maximum positive value. Let the current in phase 2 lag that in phase 1. The vectors which represent the currents are shown in Fig. 148 at the instant when current 1 has its maxi-

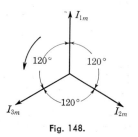

Fig. 148.

mum positive value. The currents in phases 2 and 3 are negative at this instant and have one-half their maximum values.

For convenience in plotting, let the mmf of a single coil, when it carries maximum current, be represented by one unit on the scale of ordinates in Fig. 147. This mmf is equal to NI_m, where N is the number of turns in a coil and I_m is the maximum current it carries. Assume that positive currents flow out of the figure in leading coil sides. Take the position of slot 1 as an arbitrary zero of the armature reaction. The real zero is the axis of symmetry of the armature-reaction curve which is to be plotted. Assume that the currents in the conductors in any coil side are concentrated at the center of the coil side. According to this assumption, whenever a coil side is passed in going around the armature, the mmf of armature reaction immediately changes by an amount corresponding to that produced by the current in the coil side passed.

Start at slot 1. In passing this slot, going from left to right, the mmf changes positively by two units on the arbitrary scale chosen, since phase 1 is assumed to carry its maximum positive current. The mmf acts upward, since the currents in both coil sides in this slot are in phase 1 and both carry maximum positive currents. This mmf is plotted in Fig. 147. In passing slot 2, the mmf increases positively by one unit because of the current in coil side I. It also increases positively by one-half unit because of coil side III which carries one-half its maximum current. Coil side III in slot 2 is a trailing coil side and carries a negative current. The fact that the current is negative as well as in a trailing coil side makes the change in mmf caused by it positive.

This process is continued until 360 electrical degrees of the armature are covered. This angle is equivalent to twice the pole pitch. The complete mmf wave is shown in Fig. 147 and is a step curve. A smooth curve drawn through the step curve is not far different from a sinusoid. The greater the number of slots and phases, the more nearly does the resultant mmf wave approach a sinusoid. The phase spread and the pitch chosen for this example have largely eliminated the fifth and the seventh harmonics which were present in the component rectangular waves. The balanced three-phase load has eliminated the third harmonics and all harmonics which are multiples of the third. The sinusoidal mmf wave, which is obtained when all harmonics in the rectangular component waves are neglected, is shown dotted on the plot. The maximum value of the sinusoidal wave is given by $0.90NIk_bk_p$, in which the letters have the significance stated when this equation for armature reaction was developed.

If successively later times are used in plotting, the armature-reaction wave is found to advance in position with time and to move a distance equal to two pole pitches in the time of a complete cycle. The shape of the wave changes as it advances, because of the effect of the harmonics which have not been wholly eliminated, in the component waves of the phases, by phase spread and coil pitch. With a three-phase generator, the shape of the wave of armature mmf repeats every 60 deg, *i.e.*, every sixth of a cycle.

A graphical study of armature reaction is of importance because it gives a visual conception of the notion of rotating magnetic fields. Rotating magnetic fields occur not only in polyphase synchronous machines, because of their armature reaction, but also in polyphase induction motors. The operation of the polyphase induction motor depends upon the rotating field set up by its armature windings when they carry polyphase currents.

If it were not for the change in reluctance of the magnetic circuit, due to a change in flux density, the flux which the armature reaction acting alone would produce in a polyphase cylindrical-rotor generator, which carries a balanced load, could be added vectorially to the flux which would be produced by the field current, also acting alone, to give the resultant flux. The two fields would be stationary with respect to each other. The above statement is not true for an unbalanced load, since the armature reaction of an unbalanced load is fixed neither in magnitude nor in direction with respect to the poles. The statement also neglects the effect of the slots in producing ripples in the flux waves. What is actually done in the case of a polyphase cylindrical-rotor generator with a balanced load is to combine the mmfs of the armature reaction and the impressed field and then to find what flux the resultant mmf produces.

Armature Leakage Reactance. When a synchronous generator is loaded, mmfs are produced by the armature currents which modify the flux that the field current alone would produce. Part of this effect has been discussed under armature reaction, which considers the effect of the fundamental component of the resultant mmf set up by the armature currents. The direction of this component, for any given power factor, is fixed with respect to the poles, as has been shown. The armature-reaction mmf, combined with that due to the field current, produces a flux which links both the armature and the field windings and is known as the air-gap flux, although other fluxes exist in the air gap. These other component fluxes along with the slot and end connection fluxes caused by the armature currents are called armature leakage fluxes. The voltages induced in the armature windings by the air-gap flux, as defined above, are called the air-gap voltages. The leakage fluxes also induce voltages of fundamental frequency in the armature windings. These are taken into account by the introduction of leakage-reactance drops.

Most of the reluctances of the magnetic circuits for the armature leakage fluxes are due to air paths. The fluxes are therefore nearly proportional to the armature currents producing them and are in phase with these currents. For this reason, the voltages they induce in the armature

windings can be taken into account by the use of constant leakage reactances for the phases, which, multiplied by the phase currents, give the
component voltages induced in the phases by the leakage flux. The
voltages are the leakage-reactance drops and are 90 deg ahead of the
currents producing them.

For the purpose of computing the leakage reactance of a synchronous
generator, the leakage flux is divided into the following components: (1)
slot leakage flux, (2) end-connection leakage flux, and (3) air-gap leakage
flux.

The slot leakage flux includes all the flux which links or surrounds the
portions of the armature conductors which are embedded in the iron and
which flux does not enter the air gap. This
portion of the leakage flux is assumed to pass
directly across the slots and to complete its
path through the teeth and the iron below
the teeth. The path of this portion of the
leakage flux is indicated in Fig. 149.

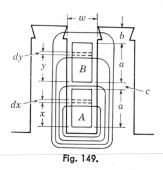

Fig. 149.

The slot leakage reactance may be calculated as follows. Assume that the leakage
flux passes directly across the slot and that
all the reluctance of the leakage-flux paths
is in the slot. Assume that there are two
equal conductors, A and B, in each slot. The dimensions of the slot and of
the conductors and their positions in the slot are shown in Fig. 149. The
effect of the notches for the wedge is neglected. In the following calculation of slot leakage reactance, the current density in the conductors in
the slots is assumed to be constant. This is not actually the case, but
with the types of laminated conductors now used for all large synchronous generators the error is small. In these conductors the laminations are insulated from one another, are connected in parallel at their
ends, and are arranged in a slot in such a manner that each lamination in
passing through the slot occupies successively the positions of the other
laminations. In this way the inequality in reactance drops in different
parts of the conductors, to which the lack of constant density is due, is
largely eliminated.

Consider the conductor A in the lower part of the slot. Let the axial
length of the slot be l. At a distance x from the bottom of the conductor
A, the flux density per ampere in A is

$$\beta_x = \frac{\mu_0(x/a)}{w}$$

where μ_0 is the permeability of free space in rationalized units.

The flux within the elementary depth dx is

$$l\beta_x dx = \frac{\mu_0 lx}{aw} dx$$

This flux links the portion x/a of the current in A. The total self-flux linkage with A, due to the flux within it, is

$$\int_0^a \frac{\mu_0 lx}{aw} \frac{x}{a} dx = \frac{\mu_0 la}{3w} \text{ henrys} \qquad (145)$$

The flux density above A, due to 1 amp in it, is constant and is equal to μ_0/w. Hence the self-flux linkage with A, due to the flux it produces above itself, is

$$\frac{\mu_0}{w} l(c + a + b) \qquad (146)$$

The total self-flux linkage for A is equal to the sum of the linkages given by Eqs. (145) and (146). This is its self-inductance L_A.

$$L_A = \frac{\mu_0 l}{w} \left(\frac{4a}{3} + c + b \right) \text{ henrys} \qquad (147)$$

Consider the conductor B in the upper part of the slot. The flux density at a point y above the bottom of B, per ampere in B, is

$$\beta_y = \frac{\mu_0(y/a)}{w}$$

The flux within the elementary depth dy is

$$l\beta_y dy = \frac{\mu_0 ly}{aw} dy$$

This flux links the portion y/a of the current in B. The total self-flux linkage with B, due to the flux within it, is

$$\int_0^a \frac{\mu_0 ly}{aw} \frac{y}{a} dy = \frac{\mu_0 la}{3w} \text{ henrys} \qquad (148)$$

The flux above B, due to 1 amp in it, is μ_0/w. Hence the self-flux linkage with B, due to the flux it produces above itself, is

$$\frac{\mu_0 l}{w} b \qquad (149)$$

The total self-flux linkage for conductor B is equal to the sum of the linkages given by Eqs. (148) and (149). This is the self-inductance L_B.

$$L_B = \frac{\mu_0 l}{w} \left(\frac{a}{3} + b \right) \qquad (150)$$

The mutual inductance between A and B can be found in a similar manner. The flux within the elementary depth dy, due to 1 amp in A, is

$$\frac{\mu_0 l}{w} \, dy$$

This links y/a of the current in B. Hence the mutual-flux linkage with B, due to the flux produced within it by 1 amp in conductor A, is

$$\int_0^a \frac{\mu_0 l}{w} \frac{y}{a} \, dy = \frac{\mu_0 l a}{2w} \qquad (151)$$

The flux above B, due to 1 amp in A, is $(\mu_0 l/w)b$. This flux links all the current in B. Therefore, the mutual-flux linkage with B, due to this flux, is

$$\frac{\mu_0 l}{w} b \qquad (152)$$

The total mutual inductance M of A on B is therefore the sum of Eqs. (151) and (152).

$$M = \frac{\mu_0 l}{w}\left(\frac{a}{2} + b\right) \qquad (153)$$

In a similar manner it can be shown that the mutual inductance of B on A is also

$$M = \frac{\mu_0 l}{w}\left(\frac{a}{2} + b\right)$$

Therefore for a winding in which the conductors in the slots are in the same phase and are identical, the total slot inductance is

$$L_A + L_B + 2M = \frac{\mu_0 l}{w}\left[\left(\frac{4a}{3} + c + b\right) + \left(\frac{a}{3} + b\right) + 2\left(\frac{a}{2} + b\right)\right]$$
$$= \frac{\mu_0 l}{w}\left(\frac{8a}{3} + c + 4b\right) \text{ henrys} \qquad (154)$$

If the conductors A and B in Fig. 149 represent coil sides, each with $Z/2$ series connected conductors, the inductances given by the above equations must be multiplied by $(Z/2)^2$ in order to get the inductances for the coil sides. This neglects the space occupied by the insulation between turns, but the error introduced is negligible.

The slot reactance in ohms is found by multiplying the above slot inductance in henrys by $2\pi f$ and by the number of slots in series per phase, s. The slot reactance therefore equals

$$2\pi f s \frac{Z^2}{4} \frac{\mu_0 l}{w}\left(\frac{8a}{3} + c + 4b\right) \qquad (155)$$

If the dimensions are given in inches, then μ_0 equals 3.19 and the factor 10^{-8} must also be included. The reactance in ohms due to the slot leakage flux then equals

$$2\pi fs \frac{Z^2}{4} \frac{3.19l}{10^8 w} \left(\frac{8a}{3} + c + 4b\right) = \frac{20}{10^8} fs Z^2 l \left(\frac{2a}{3w} + \frac{c}{4w} + \frac{b}{w}\right) \quad (156)$$

In a fractional-pitch winding with an integral number of slots per pole per phase, there are slots that have coil sides in different phases. In this case the mutual inductances cannot be added directly to the self-inductances to get the slot inductances.

Consider a three-phase synchronous generator with a $\frac{5}{6}$-pitch winding and six slots per pole. Let the phases be I, II, III. Such a winding is shown at the bottom of Fig. 147. In any belt of conductors, such as the first belt from the left for phase III, the middle slot contains coil sides in the same phase and the outer slots in the belt contain coil sides in different phases. The left-hand slot in the belt has for a bottom coil side a trailing side of phase III and the top side is a leading side of phase I. The right-hand slot in the belt has for a top coil side a trailing side of phase III and the bottom coil side is a leading side of phase II. The leading coil sides for phases I and II carry currents which are 120 deg out of phase (a balanced load is assumed), one leading the current in the trailing side of III by 60 deg, the other lagging it by 60 deg. The two mutually induced voltages produced in this case are 120 deg apart in phase, but their resultant, which is equal in magnitude to one of the voltages, is in phase with the self-induced voltages for the trailing sides of phase III. Therefore for the four coil sides of phase III the slot-belt leakage inductance is

$$2L_A + 2L_B + 2M + 2M \cos \frac{120°}{2} = 2(L_A + L_B + \tfrac{3}{2}M) \text{ henrys}$$

The computation of the component parts of leakage reactance due to the end-connection and air-gap leakage fluxes is not a simple matter, and for this reason only a general discussion of the problem is given and the reader is referred for such computation to articles on the subject in technical literature.[1]

The end-connection leakage flux, as its name implies, is the flux which links only the end connections of the armature winding. Since the end connections are not imbedded in the iron, the path of this flux is largely through air. The component leakage reactance due to this flux is depend-

[1] P. L. Alger, "The Calculation of Armature Reactance of Synchronous Machines," *AIEE Transactions*, Vol. 47, pp. 493–513, April, 1928.

F. A. Kilgore, "Calculation of Synchronous Machine Constants," *AIEE Transactions*, Vol. 50, pp. 1201–1214, December, 1931.

ent on the length of the end connections. Its computation is complicated because adjacent conductors may belong to different phases. The coil form also influences the closeness of the conductors to adjacent magnetic material.

The air-gap leakage flux includes all the space harmonics of the flux in the air gap, due to the armature current, which produce voltages of fundamental frequency in the armature. This includes the flux which crosses from tooth to tooth in the air gap and in the interpolar spaces and takes account of the flux not included in the slot leakage flux.

The harmonics of like frequency in the space mmfs due to the armature currents of a polyphase synchronous generator combine to produce component rotating harmonic fluxes in the same way that the fundamentals combine to produce a fundamental rotating mmf. Although this mmf, which is the armature reaction, rotates at synchronous speed with respect to the armature and is stationary with respect to the field poles, the rotating component harmonic fluxes do not rotate at synchronous speed with respect to the armature and are not stationary with respect to the field poles.

Since the time variation of all the space harmonics, as well as the fundamentals, in the mmf waves produced in all phases is the same as that of the currents and is therefore of fundamental frequency, and since the number of waves for any harmonic such as the qth in the space distributions of the mmfs of the phases is q times that for the fundamentals, the speed of rotation with respect to the armature of the component rotating field caused by any harmonic such as the qth harmonic is $1/q$ times that due to the fundamentals. The phase orders of the harmonics in the mmfs of the phases are not all the same as that of the fundamentals. In a three-phase synchronous generator with a balanced load, the third harmonics in the mmfs cancel. All multiples of the third harmonic also cancel. The phase order of the fifth, eleventh, etc., harmonics is opposite to that of the fundamentals. The phase order of the seventh, thirteenth, etc., is the same as that of the fundamentals. Therefore the direction of rotation of the component rotating fields caused by the seventh, thirteenth, etc., harmonics is the same as that produced by the fundamentals. The speed of rotation of these component fields with respect to the field poles is therefore $[1 - (1/q)]$ times synchronous speed. The speed of rotation, with respect to the field poles, of the component rotating fields, due to the fifth, eleventh, etc., components, is given by $[1 + (1/q)]$. The frequencies of the flux density variations produced at any point on the poles by the two groups of components are $(1 - q)f$ and $(1 + q)f$.

Because of the large number of waves of the harmonic fluxes per pole pitch—there are q for the qth harmonic at any instant in a distance of two pole pitches—and also because of the relatively high frequency of the flux variations produced in the pole faces by the harmonic fluxes, the harmonic-flux variations are confined chiefly to the air gap. The presence of a damper in the pole faces tends to damp out the harmonic-flux variations in them. Eddy currents induced by the harmonic-flux variations in the pole faces, especially when unlaminated poles are used, also tend to damp out the flux variations.

Although the speed of rotation of the component harmonic fields with respect to the armature is $1/q$ times synchronous speed, the fact that the numbers of poles for the harmonic fields are q times that for the fundamental makes the time variations of the component harmonic fluxes equal to fundamental frequency, when considered with respect to any armature conductor. They therefore produce voltages of fundamental frequency in the armature conductors. The component harmonic fluxes, when considered with respect to the armature conductors, constitute the armature air-gap leakage flux.

Leakage reactance depends upon the size and shape of the slots, the number of inductors per slot, the number of slots per pole, the armature winding pitch, and the number of phases. Other things being equal, the slot leakage reactance of a winding in wide shallow slots is less than that of a winding in narrow deep slots. Leakage reactance is also influenced by the type of winding used, its spread and its pitch, and whether it is concentrated or distributed and has an integral or a fractional number of slots per pole. In general, increasing the number of slots and the number of phases decreases the harmonics in the space distributions of the armature mmf and therefore reduces the air-gap leakage flux. The use of a fractional-pitch winding also reduces leakage reactance, since when a fractional-pitch winding is used, there is an overlapping of the phase belts and as a result some slots contain coil sides which are not in the same phase. The resultant mmf of these slots is produced by two currents which in the case of a three-phase synchronous generator differ in phase by 60 deg. The slot reactance of these slots is less than it would be if the coil sides contained in them were in the same phase (see page 223). A fractional-pitch winding also tends to smooth out the resultant mmf wave of the armature due to the armature currents and hence to reduce the harmonics in it. This reduces the air-gap leakage reactance. The shortened end connections required for a fractional-pitch winding further reduce the leakage reactance by decreasing the end-connection reactance.

The leakage reactance of a synchronous generator with salient poles is influenced by the power factor of the load, since the positions, with respect to the poles, of the phase belts of armature conductors when they carry maximum current depend upon the power factor. At unity power factor, the phase belts are opposite the poles when they carry maximum current.

At zero power factor, they are between the poles when they carry maximum current. The air-gap leakage flux and therefore the air-gap leakage reactance are greatest at unity power factor. The magnitude of the change in leakage reactance with a change in power factor depends upon the axial length of the air gap and upon the relative magnitudes of the parts of the leakage reactance which are due to the air-gap leakage flux and the slot leakage flux. The change in leakage reactance of a salient-pole generator for ordinary changes in load power factor is not great as a rule.

Equivalent Armature Leakage Flux. For a given type of armature winding in slots of a definite size and shape and with an air gap of a given length, the combined slot and air-gap leakage reactance varies as the length of the embedded armature conductors and nearly as the square of the number of armature conductors per slot. Under similar conditions, the end-connection leakage reactance varies approximately as the square of the number of conductors per armature coil and the length of the end connections. Both the combined slot and air-gap leakage reactances and the end-connection leakage reactance vary as the number of armature coils in series per phase and inversely as the number of parallel paths per phase. When dealing with a given type of armature winding in slots of a definite size and shape, it is permissible and often convenient to make use of equivalent leakage fluxes. These may be defined as follows. The equivalent slot and air-gap leakage flux is that flux per ampere per unit length of embedded armature conductor which, if linked with all the conductors in a slot, would produce a reactance which would be equal to the actual combined slot and air-gap leakage reactance. The equivalent end-connection leakage flux is, similarly, the leakage flux per ampere per unit length of conductor in the end connections which, if it linked all the turns in the end connections of a coil, would produce a reactance equal to the actual end-connection reactance. The end-connection leakage reactance can sometimes be taken into account by suitably increasing the equivalent slot leakage flux. The value of the equivalent armature leakage flux varies from 2.5 to 10 lines per ampere per inch of embedded conductor. The equivalent end-connection leakage flux is much smaller and is of the order of magnitude of 2 lines per ampere per inch of conductor.

If φ_e is the total equivalent leakage flux per ampere per unit length of embedded inductor, $i.e.$, the armature leakage flux increased to take care of the end-connection leakage flux, the equivalent leakage flux per slot is

$$l\varphi_e Z$$

where l and Z are the length of the embedded inductors and the number of inductors in series per slot. The slot linkages due to this flux are

$$l\varphi_e Z^2$$

and if the winding is a single-circuit winding and there are s slots in series per phase, the total phase reactance in ohms is

$$x_a = 2\pi f l \varphi_e Z^2 s 10^{-8} \tag{157}$$

Equation (157) assumes that the pitch of the winding is unity. If the pitch is less than unity, some slots in each phase contain coil sides which are not in the same phase. In this case, the fact that the resultant equivalent leakage flux in these slots is the resultant of two component equivalent leakage fluxes which are not in phase must be taken into consideration. Although the reactive drops in these slots are not in quadrature with the current in the winding considered—they are more than 90 deg ahead of the current in some slots and less than 90 deg ahead of the current in others—the shift in phase due to the fractional pitch cancels out between phase terminals for a balanced load and leaves the phase reactive drop in quadrature with the current.

Effective Resistance. The apparent or effective resistance of a circuit to an alternating current is greater than its resistance to a steady current, and it may be many times greater. When an electric circuit carries an alternating current, hysteresis losses are produced in any adjacent magnetic material, and eddy-current losses occur in neighboring conducting mediums and in the conductor itself.

The losses increase the total power which it is necessary to supply to the circuit and produce an increase in the energy component of the voltage drop through the circuit. This increase in the voltage drop is equivalent to an apparent increase in the resistance of the circuit.

When a conductor in air carries an alternating current, the distribution of the current over the cross section of the conductor is not uniform. Less current is carried by the central portions of the conductor than by the outer portions. Since the loss in any element of a conductor is proportional to the square of the current it carries, any lack of uniformity in the distribution over its cross section of the current in a conductor increases its apparent resistance by increasing the ratio of the power lost in the conductor to the square of the current it carries. This is the ordinary skin effect. It is much exaggerated when the conductor is surrounded or partially surrounded by iron, as in the case of the armature conductor of a synchronous generator. The difference between the current density at the top and at the bottom of a conductor in the armature slot of a generator may be great, unless something is done to prevent it, especially when the cross section of the conductor is large and it is in a narrow, deep slot.

Refer to Fig. 150, which shows a single conductor in a slot. A few slot leakage-flux lines are shown as full lines. Consider the conductor to be divided into elements parallel to the bottom of the slot as indicated by the dotted lines. The flux produced by the element in the bottom of the slot surrounds this element. It passes across the slot above the element and returns in the iron below it. Only a negligible amount of this return flux passes through the slot below the element, because of the high reluctance of the air path as compared with that of the iron. The flux produced by the next element passes across the slot above the element and returns in the iron below the slot. All the flux produced by both elements surrounds or links the lower element. In general, all the flux produced by any element links all the elements below it. As a result, the number of flux linkages with elements increases in passing from the

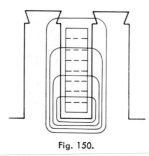

top to the bottom of the slot. For this reason, the reactance of the elements also increases in going from the top to the bottom of the slot. These statements neglect flux caused by any element which lies within the element itself, but this flux becomes negligible as the thickness of the elements is decreased. Although the elements do not actually exist as such, the effect is the same as if they did and were joined at their ends. The current would

Fig. 150.

divide between them inversely as their impedances, and as the impedances of the elements would increase in passing from the top to the bottom of the slot, more current would be carried by the upper elements than by the lower. In other words, the current would be forced toward the top of the slot. This effect may be considered equally well as due to eddy currents in the conductor which are set up by the slot leakage flux, which has its greatest density at the top and its least density at the bottom of the slot. A study of the conditions shows that the eddy currents would be in such a direction as to add to the load current in the top of the slot and to subtract from it in the bottom of the slot.

If nothing is done to even up the current density over the cross section of large armature conductors, their effective resistance due to skin effect alone is many times their d-c resistance. Merely laminating an armature conductor, without the use of some device for equalizing the reactances of the laminations, has little effect. If each conductor in an armature coil is laminated horizontally, with the laminations insulated from one another, and the end connections are twisted so as to make a lamination, which lies at the top of the coil side in one slot, lie at the bottom of the coil side in the other slot, the skin effect is much reduced. But with the ordinary two-layer windings used for most synchronous generators, the conditions are not the same for both coil sides, as one is in the top of a slot and the other is in the bottom of a slot. The twist in the end connections of the ordinary barrel-type armature coils (Fig. 120, page 178) accomplishes the required transposition of laminations.

The most effective method of reducing the ratio of the a-c to the d-c resistance of the armature conductors of synchronous generators, a method which is frequently used for generators having conductors of large cross section, is to laminate the armature conductors both horizontally and vertically and then to transpose the laminations in such a manner that, in passing through a slot, each lamination occupies on the average the same position as every other lamination. Each lamination must occupy successively and for equal distances every possible position between the top and the bottom of the slot. The two types of conductors which accomplish the required transposition are the Punga and the Roebel conductors. The Roebel conductor has two vertical rows of laminations, each row consisting of five conductors. By a simple and ingenious scheme of crossing the conductors from the right-hand to the left-hand row, and vice versa, in passing from one layer to the next, each conductor is made to occupy every one of 10 positions successively in passing from one end to the other end of the slot. The Punga conductor accomplishes the same thing. Laminating a conductor in this way sacrifices a small amount of slot space, because of the small amount of insulation required on the laminations and also because of the slot space lost where the laminations cross. The reduction of the skin effect accomplished by the transposition of the laminations more than compensates for the loss of slot space. Conductors of this type reduce the increase in resistance caused by skin effect to 10 per cent or less of the ohmic or d-c resistance.

Single-phase Rating. The ratio of the single-phase and three-phase ratings of a three-phase generator for fixed inductor copper loss is 16 per cent greater for Y connection than for Δ connection. Let I, V, and $\cos \theta$ be the phase current, the phase voltage, and the load power factor. For Y connection

$$\frac{\text{Single-phase output}}{\text{Three-phase output}} = \frac{(\sqrt{3}V)I \cos \theta}{3VI \cos \theta} = 0.58$$

For Δ connection

$$\frac{\text{Single-phase output}}{\text{Three-phase output}} = \frac{V(\tfrac{3}{2}I) \cos \theta}{3VI \cos \theta} = 0.50$$

Chapter 17

Vector diagram of a synchronous generator with non-salient poles. Vector diagram applied as an approximation to a synchronous generator with salient poles. Calculation of regulation and field current from the vector diagram. Synchronous impedance and synchronous-impedance vector diagram. Unsaturated synchronous reactance. Use of saturated synchronous reactance. Open-circuit and short-circuit characteristics. Potier triangle for determining leakage reactance and armature reaction. Example of calculation of armature reaction and leakage reactance from Potier triangle. Saturated synchronous reactance. Saturated synchronous reactance from open-circuit saturation curve and zero-power-factor curve. Example of calculation of field current and regulation using saturated synchronous reactance. General statements in regard to the methods outlined for determining field current, regulation, and power angle. Zero-power-factor curves for different armature currents. The mmf method. Determination of field current and open-circuit voltage by method of ASA. Application of the ASA method. Two-reactance method of determining field current, regulation, and power angle of a salient-pole generator. Determination of direct-axis and quadrature-axis synchronous reactances.

Vector Diagram of a Synchronous Generator with Non-salient Poles. Consider a polyphase synchronous generator having a distributed armature winding and non-salient poles. Assume that the field winding is distributed in such a way as to produce a sinusoidal distribution of mmf in the air gap. Then if the load is balanced and the armature current is sinusoidal, the armature reaction and the impressed field can be treated as vectors and combined as such, provided proper consideration is given to their phase relation.

A field structure with non-salient poles and a properly distributed winding can be made to give approximately a sinusoidal flux distribution. A spiral winding such as is shown in the upper portion of Fig. 151 can be used for this purpose. Any form of distributed winding with the inductors properly placed answers equally well. The distribution of mmf produced by this winding is shown in the lower part of the figure. In Fig. 151, the change in mmf due to passing a coil side is assumed to occur at a uniform rate over the slot instead of suddenly at the center of the coil, as was assumed in the graph of the armature mmf wave of a three-phase synchronous generator with a balanced load, shown in Fig. 147, page 217. The flux distribution corresponding to the mmf wave would have the sharp corners rounded.

The sinusoidal curve which best represents the stepped curve of Fig. 151 is the sinusoidal wave which would be obtained by adding vectorially the fundamentals of the rectangular waves which represent the space distributions of the mmf produced by the individual

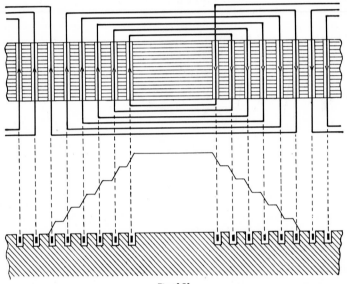

Fig. 151.

field coils. The amplitude of the resultant fundamental mmf wave is found by adding vectorially the vectors which represent the amplitudes of the fundamental components of the mmf waves of the individual field coils. This addition is best performed by making use of the breadth and pitch factors for the field winding. The harmonic components in the mmf waves of the field coils, when added vectorially, largely cancel. The effect of their resultants is to produce small harmonic voltages in the armature coils which in general are much reduced between armature terminals by the spread and the pitch of the armature winding. In a three-phase synchronous generator, all third-harmonic voltages and all voltages having frequencies that are multiples of triple frequency are zero between terminals. Since the effect of the harmonics in the mmfs of the field coils on the terminal voltages is small, the neglect of these harmonics is justifiable.

The amplitude of the resultant fundamental sinusoidal component of the mmf produced by a distributed field winding, expressed in ampere-turns per pole, is

$$\frac{4}{\pi} N_f I_f k_b k_p \tag{158}$$

where N_f is the total number of field turns per pole and I_f is the field current per turn. The other two factors are the breadth and pitch factors for the field winding. These two factors have the same significance as when used in connection with an armature winding.

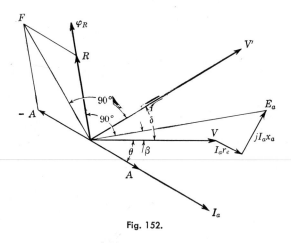

Fig. 152.

The expression $(4/\pi)N_f I_f k_b k_p$ is the length of the vector which may be used to represent the field ampere-turns on the vector diagram of a synchronous generator which has a non-salient-pole field structure.

Figure 152 is the vector diagram of a synchronous generator with non-salient poles. All currents and all voltages on the vector diagrams of generators must be per phase. The mmf of armature reaction, on the other hand, must always be for all phases. Since the reactions of all phases combine to modify the resultant field, they are directly additive (balanced load assumed) and affect the voltages of all phases alike. All mmfs are expressed in ampere-turns per pole.

Referring to Fig. 152, V is the terminal voltage per phase, i.e., the voltage between terminals if the generator is Δ-connected, and the voltage between terminals divided by $\sqrt{3}$ if the generator is connected in Y. I_a is the phase current. This is the same as the line current or the current per terminal for Y connection, or the line current divided by $\sqrt{3}$ for Δ connection.

The angle θ is the angle of lag between the phase current and the phase terminal voltage. $I_a r_e$ is the effective resistance drop and $I_a x_a$ is the leakage reactance drop. Adding these drops vectorially to V gives E_a, which is the voltage rise generated by the air-gap flux φ_R. This is the flux which is produced by the combined action of the impressed field and the armature reaction and is called the resultant field. It must lead the voltage rise E_a in the armature by 90 deg in time. Let R be the resultant mmf required to produce the flux φ_R. All vectors so far mentioned are time vectors. R is also a space vector when considered with respect to the field structure. Since the armature reaction is constant and fixed in direction with respect to the field, it is in this sense a space vector. It is also a time vector when considered with respect to the armature coils. Both R and the armature reaction A must be considered in the same sense, but it is immaterial whether both are considered as time vectors or as space vectors. The phase relation between them is the same in each case.

If it were not for armature reaction, R would be the mmf of the impressed field. The armature reaction, as has been explained, is in phase with the current and is shown by A on the diagram. On account of the armature reaction, the impressed field must have a component $-A$ to balance it. Adding R and $-A$ vectorially gives the field mmf F which is required to produce the terminal voltage V. This assumes that the coefficient of field leakage is unaffected by a change in load. It also assumes that the air gap is uniform. This latter assumption is very nearly correct in this case, since the generator is assumed to have non-salient poles. The effect of the change in the leakage coefficient can be taken into account by making use of an open-circuit saturation curve which has been corrected for field leakage. The open-circuit voltage, when the load is removed, is the voltage V' corresponding to the excitation F on the open-circuit characteristic (Fig. 154, facing page 236). It lags F by 90 deg.

Vector Diagram Applied as an Approximation to a Synchronous Generator with Salient Poles. The vector diagram given in Fig. 152 and the method of calculating the regulation of a synchronous generator from this diagram are correct only when applied to generators with non-salient poles. For reasons which are given under Armature Reaction, page 214, it must be considered as an approximation when applied to other generators, but in spite of this the regulation and the field current for a given load of a generator with salient poles calculated from the vector diagram shown in Fig. 152 are often satisfactory. When the vector diagram is applied as an approximation to a salient-pole generator, the constant 0.75 should be used instead of 0.90 when finding the armature reaction. This partially takes account of the shape of the pole shoes. For a salient-pole generator, the actual number of field turns per pole should be used when finding the number of field ampere-turns per pole instead of Eq. (158), page 232.

Calculation of Regulation and Field Current from the Vector Diagram.
The following example illustrates the method of calculating from the vector diagram the regulation and field current of a generator. For want of a better name, this method of calculating the regulation and field current will be referred to as the general method.

A three-phase cylindrical-rotor Y-connected 60-cycle 13,200-volt four-pole 1,800-rpm synchronous generator, rated at 93,750 kva at 0.8 power factor, is used.

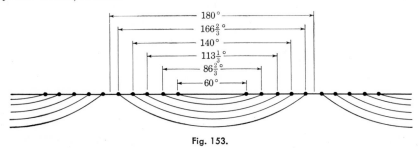

Fig. 153.

The armature has a two-circuit winding with 72 slots and 2 inductors per slot. The winding pitch is ⅔.

The field has a spiral winding with two circuits in parallel. There are 10 slots per pole and 25 conductors per slot. The distribution of the field coils is given below.

Coil 1 (inner coil) has a spread of 60 electrical degrees
Coil 2 has a spread of 86⅔ electrical degrees
Coil 3 has a spread of 113⅓ electrical degrees
Coil 4 has a spread of 140 electrical degrees
Coil 5 (outer coil) has a spread of 166⅔ electrical degrees

The developed field winding is shown in Fig. 153.

The armature resistance at 25°C is 0.00200 ohm and the armature leakage reactance is 0.248 ohm, both per phase. The ratio of the a-c to the d-c armature resistance at 25°C and 60 cycles is 1.82.

The resistance of the field winding between terminals is 0.156 at 25°C.

The complete characteristic curves of the generator at rated frequency are given in Fig. 154, facing page 236. For this problem only the open-circuit saturation curve is required.

The full-load phase current is $\dfrac{93,750 \times 1,000}{\sqrt{3} \times 13,200} = 4,100$ amp and the phase voltage is $13,200/\sqrt{3} = 7,620$ volts. The armature reaction for a balanced load at rated current is found from Eq. (141), page 210.

The current in this equation is the current per turn which in this case is one-half the phase current, since the armature has a two-circuit winding.

$$A = 0.90NIk_bk_p$$

$$N = \frac{72 \times 2}{2 \times 4} = 18 \text{ turns per pole (in all phases)}$$

$$k_p = \sin\frac{\rho}{2} = \sin\frac{120°}{2} = 0.866$$

$$k_b = \frac{\sin(n\alpha/2)}{n\sin(\alpha/2)}$$

$$n = \frac{72}{4 \times 3} = 6 \text{ slots per phase per pole}$$

$$\alpha = \frac{360° \times 2}{72} = 10 \text{ deg between slots}$$

$$k_b = \frac{\sin\dfrac{6 \times 10°}{2}}{6\sin\dfrac{10°}{2}} = 0.956$$

$$A = 0.90 \times 18 \times 0.956 \times 0.866 \times \frac{4,100}{2}$$

$$= 27,500 \text{ amp-turns per pole} \tag{159}$$

A, as found, is the magnitude of the vector which represents the fundamental of the armature reaction. It is A on the vector diagram given in Fig. 152, page 232. All harmonics are neglected.

The amplitude of the vector which represents the fundamental of the field ampere-turns per ampere of field current is found from Eq. (158), page 232, by putting I_f equal to unity.

Since the field has a spiral winding (page 230), the axes of all field coils for any given pole coincide. In this case the breadth factors are unity. Since all coils carry equal currents, the average pitch factor for the coils may be used for k_p in Eq. (158). The same result is obtained if the actual field winding is assumed to be replaced by a full-pitch distributed lap winding with 10 slots in a belt of conductors.

Coil spreads	Pitch factors	
60°.	$\sin \frac{1}{2}(60°)$	= 0.5000
86⅔°.	$\sin \frac{1}{2}(86\frac{2}{3}°)$	= 0.6862
113⅓°.	$\sin \frac{1}{2}(113\frac{1}{3}°)$	= 0.8354
140°.	$\sin \frac{1}{2}(140°)$	= 0.9397
166⅔°.	$\sin \frac{1}{2}(166\frac{2}{3}°)$	= 0.9933
		3.9546

$$\text{Average } k_p = \frac{3.9546}{5} = 0.791$$

If the field winding is replaced by a full-pitch distributed winding, $k_p = 1$ and

$$k_b = \frac{\sin \dfrac{n\alpha}{2}}{n \sin \dfrac{\alpha}{2}} = \frac{\sin 10 \dfrac{13\tfrac{1}{3}°}{2}}{10 \sin \dfrac{13\tfrac{1}{3}°}{2}} = \frac{0.9182}{10 \times 0.1161} = 0.791$$

The number of field turns per pole $= \dfrac{10 \times 25}{2} = 125$

$$\frac{4}{\pi} Nk_p I_f = \frac{4}{\pi} Nk_b I_f = \frac{4}{\pi} 125 \times 0.791 I_f = 125.8 I_f \qquad (160)$$

The foregoing expression gives the amplitude of the vector which can be used on the vector diagram to represent the fundamental of the impressed field ampere-turns. I_f in the equation is the current per turn and for the generator considered is one-half the field current per terminal, since the field winding has two circuits in parallel.

When the effective resistance of the armature of a generator is given at one temperature and is used at another temperature, it is usually assumed that the part of the effective resistance which is due to local losses, skin effect, etc., is not changed by a change in temperature. This is not strictly true but is nearly enough true so far as the resistance drop in the armature is concerned, since the per cent resistance drop in the armature of a large generator is small. Assume an operating temperature of 75°C. The armature effective resistance at 75°C is

$$r_{oh}(\text{at } 25°\text{C}) \frac{1 + 0.0042 \times 75}{1 + 0.0042 \times 25} + r_{oh}(\text{at } 25°\text{C})(1.82 - 1)$$
$$= 0.0020 \times 1.19 + 0.0020 \times 0.82 = 0.00402 \text{ ohm per phase}$$

Refer to Fig. 152, page 232, and take V on this figure as the reference axis.

$$\begin{aligned}
E_a &= V(1 + j0) + I(\cos\theta - j\sin\theta)(r_e + jx_a) \\
&= 7{,}620(1 + j0) + 4{,}100(0.8 - j0.6)(0.00402 + j0.248)* \\
&= (7{,}620 + j0) + (623 + j804) \\
&= 8{,}243 + j804 \\
|E_a| &= \sqrt{(8{,}243)^2 + (804)^2} = 8{,}280 \text{ volts per phase} \\
\sqrt{3}\,|E_a| &= \sqrt{3} \times 8{,}280 = 14{,}340 \text{ volts between terminals}
\end{aligned}$$

R is the air-gap field expressed in terms of the equivalent field current and is the current correponding to the voltage $\sqrt{3}\, E_a = 14{,}340$ on the open-circuit saturation curve (see Fig. 154).

* See page 247 for calculation for $x_a = 0.248$.

$$R \text{ for } 14{,}340 \text{ volts} = 535 \text{ amp}$$
$$F = A(-\cos\theta + j\sin\theta) + R(-\sin\beta + j\cos\beta)$$
$$\sin\beta = \frac{804}{8{,}280} = 0.0972 \qquad \cos\beta = \frac{8{,}243}{8{,}280} = 0.996$$

Since R is expressed in terms of the field current, A must be in terms of the equivalent field current. This equivalent current is twice the equivalent current found by dividing A in ampere-turns per pole by the field ampere-turns per pole per ampere of field current. These field ampere-turns are 125.8 [Eq. (160)].

$$A = 2\frac{27{,}500}{125.8} = 437 \text{ equivalent field amperes}$$
$$\begin{aligned}
F &= 437(-0.8 + j0.6) + 535(-0.0972 + j0.996) \\
&= (-349.6 + j262.2) + (-52.0 + j532) \\
&= -402 + j794
\end{aligned}$$
$$|F| = \sqrt{(402)^2 + (794)^2} = 890 \text{ amp}$$

A field current of 890 amp is required to maintain rated voltage at rated frequency when the generator carries an inductive full kva load at 0.8 power factor.

The voltage found from the open-circuit saturation curve (Fig. 154) corresponding to a field current of 890 amp is the terminal voltage the generator would have at no load if the field current and frequency were kept constant at their full-load values when the load is removed. This voltage is the voltage V' on the vector diagram given in Fig. 152.

The regulation of the generator for the given load is therefore

$$\text{Regulation} = \frac{17{,}500 - 13{,}200}{13{,}200} \times 100 = 32.6 \text{ per cent}$$

Synchronous Impedance and Synchronous-impedance Vector Diagram. It is seldom that the regulation of a large synchronous generator can be determined by actual measurement on account of the expense of making such a test as well as the difficulty of obtaining the necessary load and the impossibility of having sufficient power in most shops to operate large generators under full-load conditions. To avoid the necessity for loading a generator in order to determine its regulation and its field current under any given load condition, there are approximate methods for determining regulation and field current which require only such measurements as can be made with the generator operating on open circuit, on short circuit, or on a load of zero power factor. Such tests require comparatively little power.

For the present, the effect of saturation is neglected, and a non-salient pole or cylindrical-rotor generator is assumed. When the effect of saturation is neglected, the saturation curve (Fig. 154) becomes a straight line, and the change in flux and the corresponding change in voltage at rated frequency produced by any change in mmf are both proportional to the change in mmf. Also, since a cylindrical rotor is assumed, the magnetic circuit, neglecting the effect of rotor and stator slots, is the same in all directions about the air gap, and consequently the component change in flux produced by a given number of ampere-turns of armature reaction is independent of the direction of the armature reaction with respect to the field axis.

According to these assumptions, the effect of the mmfs of the impressed field and the armature reaction are the same as if they acted alone and

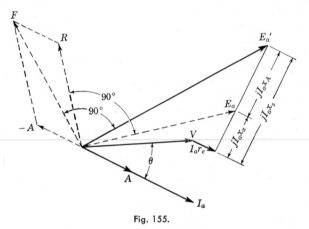

Fig. 155.

may be replaced by the voltages they would cause if acting separately. If this substitution is made, there is nothing left on the vector diagram but voltages. The mmfs of the impressed field and the armature reaction are replaced by equivalent voltages in Fig. 155.

The two mmfs R and $-A$, which have been replaced by voltages, are shown dotted in the diagram. The voltage drop $I_a x_A$, which replaces $-A$, is 90 deg behind $-A$ and 90 deg ahead of the current. It is therefore in phase with the voltage drop $I_a x_a$, which is the leakage reactance drop in the armature. The voltage drop $I_a x_A$ may be considered as due to a fictitious reactance x_A and may be combined with $I_a x_a$ to form a reactance drop $I_a x_s$. The reactance x_s is known as the synchronous reactance. It includes both the leakage reactance and a fictitious reactance x_A which replaces the effect of armature reaction at a definite frequency. E'_a is the open-circuit voltage and F is the impressed field, i.e., the field required for the given load. The field F corresponds to the voltage E'_a on the open-circuit saturation curve which has been assumed

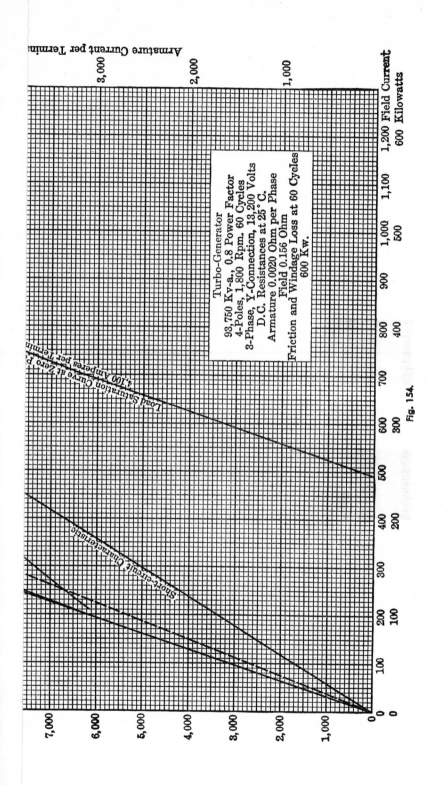

Turbo-Generator
93,750 Kv-a., 0.8 Power Factor
4-Poles, 1,800 Rpm. 60 Cycles
3-Phase, Y-Connection, 13,200 Volts
D.C. Resistances at 25° C.
Armature 0.0020 Ohm per Phase
Field 0.156 Ohm
Friction and Windage Loss at 60 Cycles
600 Kw.

Load Saturation Curve at Zero P. F. 4,100 Amperes per Terminal

Short-circuit Characteristic

Armature Current per Terminal

Field Current
Kilowatts

Fig. 154.

to be a straight line.   The fictitious reactance x_A is not equivalent to a reactance except under steady operating conditions (Chap. 19, page 286).

The saturation of a synchronous generator cannot be neglected in practice.   For this reason, synchronous reactance cannot be considered constant.   It varies with the degree of saturation of the magnetic circuit, decreasing with an increase in saturation.   The part x_a of the synchronous reactance is nearly constant for ordinary generators with open slots and with the usual air gaps.   The part x_A is far from constant.   It varies with the degree of saturation of the magnetic circuit, since x_A is proportional to the component change in flux caused by armature reaction.   The mmf of armature reaction produces different amounts of flux at different degrees of saturation of the magnetic circuit.

If the synchronous-impedance method is used for determining the field current or the regulation of a salient-pole generator, an error of considerable magnitude is introduced because of the nonuniformity of the air gap of a non-salient-pole generator.   Because of the differences in the reluctance of the magnetic circuit in the various directions about the air gap, a fixed number of ampere-turns of armature reaction does not produce the same component flux at different power factors.   The ampere-turns produce a maximum effect at zero power factor when the axis of the magnetic circuit for the armature reaction coincides with the axis of the field poles.   Their minimum effect is at unity power factor when the axis of the magnetic circuit for armature reaction lies midway between two adjacent poles.

The change in the magnitude of synchronous reactance with power factor can be taken into consideration by dividing the phase current into an active component and a reactive component, with respect to excitation voltage, and then using different synchronous reactances for these two components.   These reactances are the direct-axis synchronous reactance and the quadrature-axis synchronous reactance. They are used with the reactive and the active components of current as defined.   The use of the two reactances is really a modification of Blondel's two-reaction method for determining the regulation of a salient-pole generator.   The direct-axis and the quadrature-axis synchronous reactances are considered later.

Unsaturated Synchronous Reactance.   The unsaturated synchronous reactance of a synchronous generator is its synchronous reactance when operating on the straight part of its saturation or magnetization curve. It is measured with the generator operating on short circuit with any suitable current.   Under this condition, the terminal voltage is zero, and the vector diagram of Fig. 155, page 238, collapses into that of Fig. 156.   Since the terminal voltage is zero and unsaturated conditions are assumed, the open-circuit voltage per phase is equal to the vector sum of the phase effective resistance drop and the phase unsaturated synchronous-reactance drop.

The unsaturated synchronous impedance z_s is the ratio of the phase voltage E'_a to the short-circuit phase current I_a, where E'_a is the phase open-circuit voltage at normal frequency corresponding to the field excitation required to produce the current I_a on short circuit, assuming the open-circuit saturation curve to be the straight line drawn through the lower straight part of the actual curve.

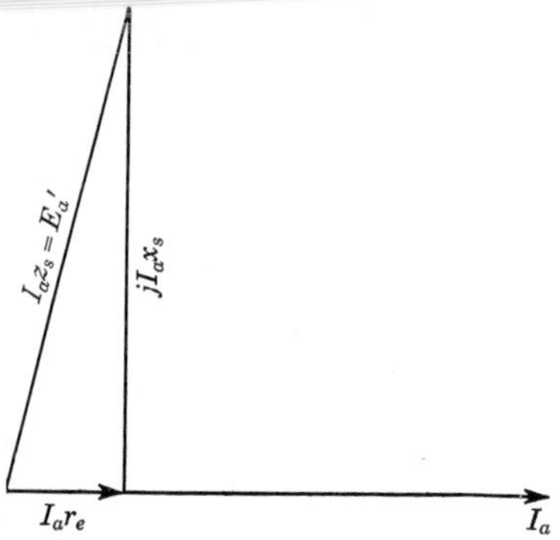

$$z_s = \frac{E'_a}{I_a}$$

and $\qquad x_s = \sqrt{z_s^2 - r_e^2}$

Referring to Fig. 154, facing page 236, the unsaturated synchronous reactance of the 93,750-kva synchronous generator used in the problem on page 234 is the ratio of the voltage divided by $\sqrt{3}$, found from the air-gap line for a given field current, to the current found from the short-circuit characteristic for the same field current.

Fig. 156.

$$x_s \text{ (unsaturated)} = \frac{15,500}{\sqrt{3} \times 4,200} = 2.13 \text{ ohms per phase}$$

The effective resistance r_e is seldom as large as one-tenth of the synchronous reactance and may be neglected when finding x_s. The synchronous reactance x_s and the synchronous impedance z_s are approximately equal in magnitude.

Since the armature reactions of all phases of a synchronous generator affect the flux, all phases must be short-circuited when determining the unsaturated synchronous reactance.

The unsaturated synchronous reactance of a synchronous generator is much larger than the synchronous reactance under the condition of normal load saturation. For this reason, it is of little use for determining the operating characteristics of a generator under normal load conditions.

Use of Saturated Synchronous Reactance. The saturated synchronous reactance is the synchronous reactance under the conditions of load saturation of the magnetic circuit. The load synchronous-reactance drop is equal to the leakage-reactance drop plus the component change in voltage which the armature reaction acting alone would produce if the saturation of the magnetic circuit were constant and equal to that under load conditions. Refer to Fig. 157, which is the general diagram of a non-salient pole or cylindrical-rotor generator on which the saturated synchronous reactance drop has been superposed. If the synchronous reactance used on the vector diagram is that for load saturation conditions, E'_a is not the no-load terminal voltage. It is the no-load voltage which would be produced by the impressed mmf F if the saturation of the magnetic circuit at no load were the same as under load conditions. R is the net mmf under load and is effective in producing the air-gap flux. This is the flux which exists in the armature back of the teeth and is the flux which would exist in the poles and the yoke if it were not for

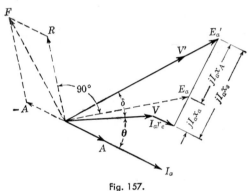

Fig. 157.

the field-pole leakage. For all essential purposes, the air-gap flux may be assumed to fix the saturation of the magnetic circuit. At no load the armature reaction is zero and the impressed mmf corresponding to F fixes the air-gap flux. To get the actual magnitude of the open-circuit voltage, the voltage E'_a shown on the vector diagram must be corrected for change produced in it by the change in the saturation of the magnetic circuit from the load to the no-load condition. V' on the diagram is the actual no-load voltage. It is less than E'_a for the inductive load shown.

The method of determining the correct value of the saturated synchronous reactance and the method of correcting E'_a for saturation is taken up after the Potier triangle has been discussed.

Open-circuit and Short-circuit Characteristics. The two typical characteristic curves of a synchronous generator which have been mentioned so far are the open-circuit characteristic or no-load saturation curve and the short-circuit characteristic or short-circuit saturation curve.

Open-circuit Characteristic. The open-circuit characteristic or open-circuit or no-load saturation curve is a curve plotted for rated frequency with open-circuit voltages as ordinates and the corresponding field excitations as abscissas. Although either terminal voltage or phase voltage may be plotted, it is customary to plot terminal voltage, *i.e.,* voltage between line terminals. The excitation may be expressed in amperes or in ampere-turns. Since with any fixed excitation the open-circuit voltage of a generator varies directly as the speed, it is possible to apply a correction to the measured voltages in case the frequency cannot be maintained constant. The open-circuit characteristic should always extend from zero excitation up to the maximum excitation for which the generator is designed.

Short-circuit Characteristic. The short-circuit characteristic shows the relation between the short-circuit armature current and the field excitation. Although this curve may be plotted with either phase current or current per terminal as ordinates, it is customary to use current per terminal. This curve should always extend to at least one and one-half times the full-load current and as much further as is possible without overheating the generator.

The magnitude of the steady short-circuit current of a synchronous generator at normal excitation depends upon its design and its size. This current lies between one and one-half and four times the rated full-load current. It is limited by the unsaturated synchronous impedance, since on short circuit the saturation of the magnetic circuit is low. The instantaneous rush of current, which takes place at the instant of short circuit, is limited by the resistance and by a reactance which is similar to the equivalent leakage reactance of a transformer considering the armature winding of the generator as the primary and its field circuit short-circuited as the secondary. It is somewhat larger than the armature leakage reactance of the generator. This current rush may be ten to thirty times as large as the normal full-load current, depending on the design and type of the generator (Chap. 19).

Measurements for a short-circuit characteristic should be made at rated frequency, but a considerable variation in the frequency produces a relatively small effect on the armature current. Both the voltage induced in the armature and the synchronous reactance at fixed saturation vary directly as the frequency. Therefore if it were not for the armature resistance, which is always small compared with the synchronous reactance, a change in the frequency would have little or no effect on the short-circuit current.

Short-circuit characteristics are straight lines over the range of saturation through which it is possible to carry them. Although the impressed field may be large, the resultant field, which determines the degree of saturation, is small on account of the large armature reaction caused by the relatively large short-circuit armature current. The effect of the armature reaction is moreover a maximum on account of the large angle of lag between the current and the generated voltage, and as a result the armature reaction almost directly opposes the impressed field.

An open-circuit saturation curve and a short-circuit characteristic of a 93,750-kva synchronous generator are shown in Fig. 154, facing page 236.

Potier Triangle for Determining Leakage Reactance and Armature Reaction. The terminal voltage of a synchronous generator under load differs from its open-circuit voltage at the same field excitation on account of the change in the field which is caused by the armature reaction and also on account of the drop in voltage through the armature produced by the leakage reactance and the armature effective resistance.

The relative influence of the three factors depends upon the power factor of the load. With a reactive load at zero power factor, the decrease in the terminal voltage is due almost entirely to the armature reaction and the armature leakage reactance. Under this condition, the effective resistance drop is in quadrature with the terminal voltage, and since it is small in magnitude, it has little influence on the change in the terminal voltage caused by a change in load. This is made clear by the vector diagram given in Fig. 158, which is for a reactive load of zero power factor. The resistance drop on this diagram is much exaggerated in order to increase the angles between some of the vectors which would almost overlap if the resistance drop were drawn to scale.

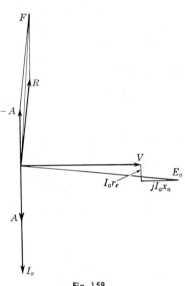

Fig. 158.

The resultant field R is almost exactly equal to the algebraic difference between F and A, and the terminal voltage V is nearly equal to the algebraic difference between E_a and $I_a x_a$. Under these conditions, the armature reaction subtracts almost directly from the impressed field and the armature leakage-reactance drop subtracts almost directly from the generated voltage. The armature resistance drop has no appreciable effect on the terminal voltage, as has just been stated. It follows from this that if an open-circuit characteristic OB and a curve CD, showing the variation in the terminal voltage with excitation for the condition of constant armature current at a reactive power factor of zero, are plotted as in Fig. 159, the two curves are so related that any two points, as E and F, which correspond to the same degree of saturation and consequently to the same generated voltage, are displaced from each other horizontally by an amount equal to the armature reaction and vertically by an amount equal to the leakage-reactance drop.

GF represents the armature reaction in equivalent field amperes, provided the excitation is plotted in amperes, and GE represents the leakage-reactance drop in volts.

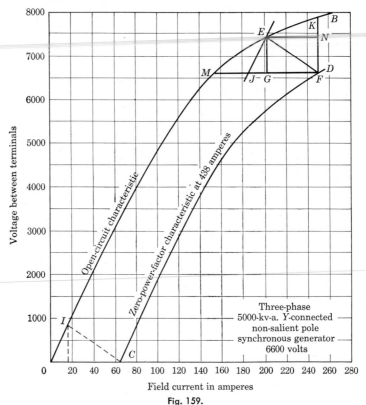

Fig. 159.

Let the curve CD be for an armature current I'. Then the armature reaction for any current I_a is

$$A = I_a \frac{GF}{I'} = I_a K$$

and since the plot is for a Y-connected generator, the armature leakage-reactance voltage per phase for the current I_a is

$$I_a x_a = I_a \frac{EG}{\sqrt{3}\, I'}$$

and

$$x_a = \frac{EG}{\sqrt{3}\, I'}$$

is the armature leakage reactance per phase.

If the leakage reactance is constant and the increase in field current necessary to balance a given number of ampere-turns of armature reaction is independent of the saturation of the magnetic circuit, Potier triangles, drawn between the open-circuit saturation curve and the zero-power-factor curve at points corresponding to different degrees of saturation, would be identical. Under these conditions, the zero-power-factor curve, i.e., the load-saturation curve at zero power factor, would have the same shape as the open-circuit saturation curve but would be displaced from the open-circuit curve by a distance equal to the length of the hypotenuse of the Potier triangle and in the direction of it. It has been shown experimentally that the two curves have approximately the same shape. This would be expected, since the leakage reactance of generators with open slots and relatively large air gaps, such as are commonly used, is nearly independent of the degree of saturation of the armature teeth, and also because, since the axis along which the armature reaction acts at zero power factor coincides with the field axis, the effect on the field mmf F of a fixed number of ampere-turns of armature reaction should be independent of the degree of saturation of the armature and field, except for the change in the field-pole leakage caused by a change in the saturation. Actually, the field-pole leakage increases somewhat with an increase in saturation, and for this reason the number of field ampere-turns which are necessary to balance a fixed number of ampere-turns of armature reaction is not quite constant. In spite of this change in field-pole leakage, the curves have nearly enough the same shape for practical purposes and are assumed to have the same shape when determining the Potier triangle.

In order to make use of the Potier method, it is necessary to find some means of locating two points, one on each curve, corresponding to the same generated voltage and the same resultant field. There are two ways by which this can be done.

First Method. Make a tracing of the zero-power-factor curve CD and the coordinate axes, and mark the point F, Fig. 159, which corresponds to rated voltage, on both the zero-power-factor curve and its tracing. Lay the tracing on the plot. Keeping the axes on the tracing parallel with the axes on the plot, slide the tracing about until the traced curve coincides with the open-circuit characteristic OB. Then prick the point F through onto the open-circuit characteristic, fixing the point E. By drawing the right-angle triangle EGF with its base parallel to the axis of abscissas, the armature reaction and the leakage-reactance drop can be determined.

The complete load characteristic is not necessary. Two points on this curve are sufficient, provided one of them, F, is well up on the bend of the curve. The other point is preferably the point C. This point corresponds to the condition of short circuit. A tracing is made as before but with the points C and F on it instead of the load characteristic. The tracing is now moved parallel to itself until the two points C and F lie on the open-circuit curve. Transfer the point F to the open-circuit characteristic. This locates the point E. This method is not very satisfactory as the position of the tracing which makes the zero-power-factor curve coincide with the open-circuit characteristic is not always definite. The second method which follows is better.

Second Method. Since the two curves, Fig. 159, are parallel, the small right-angle triangle EGF fits anywhere between them. Let it be moved down until its base lies on the line OC. It is shown dotted in this position. A new triangle OIC is formed with the lower part of the open-circuit characteristic. This new triangle has a definite base OC. From the point F draw a line FJ parallel and equal to OC. Through J draw another line parallel to the lower part of the open-circuit characteristic. The intersection of this latter line with the open-circuit curve locates the point E of the desired triangle. It is seen that, unless the point F is taken well up on the bending part of the curve, the line JE is nearly parallel to the open-circuit characteristic and the intersection of JE and the open-circuit characteristic is not definite.

The Potier method for determining experimentally the armature reaction and armature leakage reactance of a synchronous generator determines these quantities under approximately normal saturation but at a power factor which is very much below that met in practice. For this reason the value of the armature reaction obtained is too large for generators with salient poles.

In practice a power factor which is sufficiently low for determining the zero-power-factor curve of a generator can be obtained by using an underexcited synchronous motor which operates at no load as a load for the generator.

The Potier triangle for a large turbine generator is shown on Fig. 154, facing page 236.

The values of the leakage reactance and the armature reaction of a generator found from its Potier triangle can be used on the complete vector diagram of the non-salient-pole generator for determining by the general method the field current and regulation for any given load condition. The equivalent leakage flux per ampere per unit length of embedded conductor can be found from the leakage reactance determined by the Potier triangle and the winding data of the armature.

Example of Calculation of Armature Reaction and Leakage Reactance from Potier Triangle. Refer to Fig. 154, facing page 236, which gives the characteristic curves of a 93,750-kva three-phase Y-connected synchronous generator. The distances corresponding to EG and GF on Fig. 159 are 1,750 volts and 426 field amperes. Since the generator is

Y-connected and the zero-power-factor curve is for an armature current of 4,100 amp per terminal,

$$x_a = \frac{1,750}{\sqrt{3} \times 4,100} = 0.248 \text{ ohm per phase}$$

$A = 426/4,100 = 0.104$ equivalent field ampere per ampere of balanced armature current per terminal.

Saturated Synchronous Reactance.[1] As is stated on page 238, the synchronous reactance of a synchronous generator is equal to its leakage reactance plus a fictitious reactance which replaces the effect of the armature reaction on the voltage. This fictitious reactance is equal to the change in voltage produced by armature reaction divided by the armature current, all quantities considered per phase. The magnitude of the fictitious reactance x_A, which replaces the effect of armature reaction, depends on the degree of saturation of the magnetic circuit at which it is measured. For ordinary machines the leakage reactance x_a of a synchronous generator is nearly independent of the saturation of the magnetic circuit and is so assumed.

When the synchronous reactance of a generator is obtained from short-circuit conditions, the generator is operating on the straight part or unsaturated part of its saturation curve. The synchronous reactance obtained from short-circuit conditions is the unsaturated synchronous reactance. The same number of ampere-turns applied at a higher degree of saturation of the magnetic circuit produces a smaller change in flux than at a lower degree of saturation and therefore results in a smaller fictitious reactance x_A.

The synchronous reactance decreases with an increase in saturation. If the synchronous reactance of a cylindrical-rotor generator can be found under the degree of saturation at which it operates, the field current under load and the regulation calculated from the synchronous-impedance vector diagram shown in Fig. 155 should check the same quantities found from the complete vector diagram used for calculating field current and regulation by the general method.

The synchronous reactance under load-saturation conditions is called the saturated synchronous reactance. Refer to Fig. 160. The line *OFD* is the open-circuit saturation curve of a generator. *OB* is the air-gap line, *i.e.*, a straight line drawn through *O* and coinciding with the lower straight part of the open-circuit saturation curve. This air-gap line shows the relation between the voltage which would be produced by any field current and that current if the magnetic circuit remained unsaturated. *OFC* is a straight line drawn through the point *F* on the

[1] Charles Kingsley, Jr., "Saturated Synchronous Reactance," *AIEE Transactions*, Vol. 54, pp. 300–305, 1935.

open-circuit saturation curve and the point O and shows the relation that would exist between the voltage produced by any field current and that current if the saturation of the magnetic circuit remained constant at the value corresponding to the point F on the open-circuit saturation curve. The ratio of the voltage E_B to the voltage E_F is the ratio of the voltages produced by the excitation OG under the unsaturated condition and under the condition of saturation corresponding to the point F. This ratio is also equal to the ratio of the fluxes produced by OG under the two conditions and is the saturation factor for the saturation corresponding to the point F on the open-circuit saturation curve. A curve of saturation factors can be plotted against per cent rated air-gap voltages. Such a curve is shown in Fig. 161.

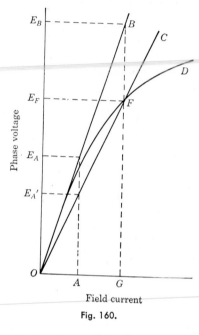

Fig. 160.

Let x_a be the leakage reactance of a synchronous generator per phase and A its armature reaction for any phase armature current such as I. Let the armature reaction be expressed in terms of the equivalent field current, *i.e.*, as the armature reaction in ampere-turns per pole divided by the number of effective field turns per pole. For a non-salient-pole

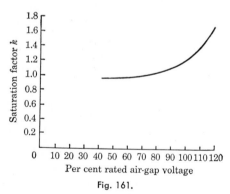

Fig. 161.

or cylindrical-rotor generator, the number of effective field turns is given by $(4/\pi)N_f k_p k_b$ where N_f is the actual number of field turns per pole [see Eq. (158), page 232]. When the synchronous-reactance method for calculating field current or regulation is applied to a salient-pole generator as an approximation, the factor 0.75 should be used instead of 0.90 when finding the armature reaction, and the actual number of field turns per pole should be used when finding the armature reaction in terms of the equivalent field current (see page 233).

Refer to Fig. 160, page 248. E_A is the phase voltage which armature reaction A would produce under unsaturated conditions. A in the figure is the equivalent field current. The unsaturated synchronous reactance is therefore

$$x_s \text{ (unsaturated)} = x_a + \frac{E_A}{I} = x_a + x_A$$

Under the condition of saturation represented by the point F on the open-circuit saturation curve, A would produce a voltage E'_A. The saturated synchronous reactance corresponding to the saturation represented by the point F is

$$x_s \text{ (saturated)} = x_a + \frac{E'_A}{I} = x_a + x'_A = x_a + \frac{1}{k} x_A$$

where k is the saturation factor.

As is stated on page 241, the air-gap voltage of a synchronous generator may be considered as a measure of the degree of saturation at which a generator is operating. The air-gap voltage is the voltage which is obtained by adding the resistance and the leakage reactance drops vectorially to the phase terminal voltage. This air-gap voltage should be used to locate the point F in order to find the saturated synchronous reactance for the given load.

The voltage E'_a found from the synchronous-impedance diagram (Fig. 157, page 241), when the saturated synchronous reactance is used, is the open-circuit voltage that would be obtained when the load is removed if the saturation remained constant at the value fixed by the point F. This voltage entered on the line OFC gives the field current required for the given load. The actual open-circuit voltage is the voltage found on the open-circuit saturation curve corresponding to the field current found. The angle between the voltages E'_a and V on the synchronous-impedance diagram (Fig. 157, page 241), when the saturated synchronous reactance is used, is the true power angle. More is said later of the power angle of a synchronous machine.

Saturated Synchronous Reactance from Open-circuit Saturation Curve and Zero-power-factor Curve. If the Potier reactance is assumed to be equal to the leakage reactance, the saturated synchronous reactance of a synchronous generator can be found from its open-circuit saturation curve and its zero-power-factor curve taken with lagging current. The Potier reactance and the armature effective resistance are used to determine the air-gap voltage under the load condition. This is the voltage that is considered as a measure of the saturation at which the generator operates. The reactance obtained from the Potier triangle is usually slightly larger than the leakage reactance, chiefly because of the change in the field leakage from no load to load condition. This change in the field leakage is neglected in determining the Potier triangle. With a cylindrical-rotor generator, the change of the field leakage with load is not

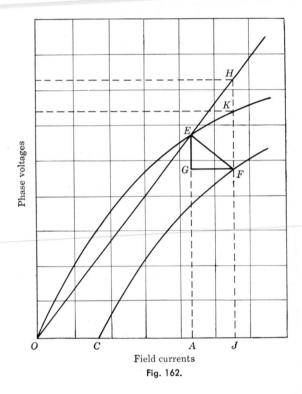

Fig. 162.

large and the Potier reactance is approximately equal to the leakage reactance.

Refer to Fig. 162, which shows the open-circuit saturation characteristic OK of a synchronous generator and also the zero-power-factor curve CF for lagging current. For convenience in the following discussion, assume that phase voltages and phase currents are plotted in Fig. 162 instead of voltages between terminals and current per terminal, as is the usual practice. When terminal values are plotted, the proper constants must be applied to the currents and voltages as read from the graph, in order to reduce them to phase values.

Construct a Potier triangle FGE in such a position that the voltage represented by the line AE is the air-gap voltage E_a under the required load conditions as fixed by adding the Potier reactance and the effective resistance drops for the required load current to the rated phase terminal voltage.

Under the condition of zero power factor, which is the condition for which the Potier triangle is drawn, the resistance drop has negligible effect on the terminal voltage, as is seen by referring to Fig. 158, page 243. The resistance drop in Fig. 158 is much exaggerated in order to separate the vectors V and E_a, which would nearly overlap if the resistance drop were drawn to scale. Under the conditions of zero power factor, the terminal voltage may be considered to be equal to the air-gap voltage E_a minus the Potier reactance drop where the subtraction is made in an arithmetical sense.

$$V = E_a - Ix_a$$

Referring to Fig. 162, OJ is the actual excitation required for the zero-power-factor load and OA is the net excitation, i.e., it is the actual excitation minus the armature reaction FG and is the excitation which would be required to produce the air-gap voltage E_a if the armature reaction were zero. The subtraction of the armature reaction FG from the impressed field OJ may be made arithmetically, as is seen by referring to Fig. 158.

If the zero-power-factor load is removed without altering the excitation OJ, the phase terminal voltage JF, Fig. 162, increases to JK, but in changing from JF to JK the saturation of the magnetic circuit has increased. The distance FK, which is the change in voltage when the zero-power-factor load is removed, is equal to the leakage reactance drop plus a voltage drop which corresponds to the change in flux caused by the change in armature reaction. It is the synchronous-reactance drop for a saturation which is not constant and has no real significance when used on a vector diagram. Refer to the vector diagram of a non-salient-pole generator which is shown in Fig. 157. If FK, Fig. 162, has the significance of a reactance drop, it must be perpendicular to the current vector on the vector diagram, but if drawn perpendicular to the current vector, its extremity does not lie at the end of the vector V', which represents the actual open-circuit voltage in both magnitude and phase.

The saturated synchronous-reactance drop at any given saturation is equal to the drop in voltage $I_a x_a$, caused by the armature leakage reactance, plus a drop in voltage which corresponds to the change in flux which would have been produced by the change in armature reaction if the saturation of the magnetic circuit had remained the same as under the load condition. The saturated synchronous-reactance drop is FH, Fig. 162. This is $I_a x_s$ on the vector diagram of Fig. 157.

If the saturation of the magnetic circuit had remained constant when the zero-power-factor load was removed, the magnetization curve would have been the straight line drawn through O and E. Under this condition, the open-circuit voltage would have been JH.

$$x_s = \frac{FH}{I_a}$$

therefore is the saturated synchronous reactance corresponding to the saturation of the magnetic circuit when the air-gap voltage is $AE = E_a$. As has been stated, the saturated synchronous reactance of a synchronous generator is not constant. It is different for different degrees of saturation fixed by the air-gap voltage E_a. Data for a graph of saturated synchronous reactance against excitation voltage can be obtained by placing the Potier triangle in different positions on Fig. 162.

Example of Calculation of Field Current and Regulation Using Saturated Synchronous Reactance. The 93,750-kva three-phase synchronous generator, for which the field current and regulation were calculated by the general method on page 232, is used in this illustration. The constants and the characteristic curves of this generator are given on Fig. 154, facing page 236. An inductive, 93,750-kva load at 0.8 power factor at rated frequency and voltage is used.

The Potier reactance of this generator is 0.248 ohm per phase (see page 247). From page 236, the effective resistance at 75°C is 0.00402 ohm per phase. The air-gap voltage for the given load is

$$
\begin{aligned}
E_a &= V + I(r_e + jx_a) \\
&= \frac{13,200}{\sqrt{3}}(1 + j0) + 4,100(0.8 - j0.6)(0.00402 + j0.248) \\
&= 7,620(1 + j0) + (623 + j804) \\
&= 8,243 + j804 \\
|E_a| &= \sqrt{(8,243)^2 + (804)^2} = 8,280 \text{ volts per phase} \\
\sqrt{3}\,|E_a| &= \sqrt{3} \times 8,280 = 14,340 \text{ volts between terminals}
\end{aligned}
$$

In Fig. 154 draw a straight line through O and the point corresponding to 14,340 volts on the open-circuit saturation curve. This line represents the saturation at which the generator operates for the given load. Construct a new Potier triangle, shown dotted on Fig. 154, with its upper angle at the voltage 14,340 volts. The saturated synchronous reactance at which the generator operates is

$$x_s \text{ (saturated)} = \frac{13,270}{\sqrt{3} \times 4,000} = 1.867 \text{ ohms per phase}$$

The excitation voltage is

$$E'_a = 7{,}620(1 + j0) + 4{,}100(0.8 - j0.6)(0.00402 + j1.867)$$
$$= 7{,}620(1 + j0) + (4{,}605 + j6{,}110)$$
$$= 12{,}225 + j6{,}110$$
$$|E'_a| = \sqrt{(12{,}225)^2 + (6{,}110)^2}$$
$$= 13{,}650 \text{ volts per phase}$$
$$\sqrt{3}\,|E'_a| = \sqrt{3} \times 13{,}650 = 23{,}620 \text{ volts between terminals}$$

The load field current is the field current corresponding to 23,620 volts on the saturation line OH, Fig. 154. This current is 890 amp, which checks, as it should, the field current found by the general method on page 237, since the magnitude of the leakage used is the same in both cases.

The open-circuit voltage for this field current, from the open-circuit saturation curve, is 17,500 volts between terminals. The regulation is

$$\frac{17{,}500 - 13{,}200}{13{,}200} \times 100 = 32.6 \text{ per cent}$$

General Statements in Regard to the Methods Outlined for Determining Field Current, Regulation, and Power Angle. The general method for determining the field current, the regulation, and the power angle, page 234, of a non-salient-pole synchronous generator is based on the correct vector diagram of such a machine and should give accurate results except as the results are influenced by the values used for leakage reactance and armature reaction. These may be calculated from the dimensions of the generator and its winding data or may be determined experimentally from a Potier triangle. The Potier triangle does not give the exact leakage reactance and armature reaction of a non-salient-pole generator, chiefly because of the neglect of the change in field-pole leakage with change in load when determining the triangle. The error in general is small in the case of a non-salient-pole generator. The reactance of a salient-pole generator, as obtained from its Potier triangle, may be appreciably larger than the true leakage reactance, not only because of the neglect of the change in the field-pole leakage with load when determining the Potier triangle but also because the triangle gives both the reactance and the armature reaction for a very low power factor. The part of the leakage reactance which is caused by the harmonics in the air-gap flux is larger than under normal operating power factors because of the positions of the armature conductors with respect to the poles when the conductors carry maximum current. The effect of the change in the field-pole leakage with load on the air-gap field R and on the impressed field F used in the general method can be taken into consideration by correcting the open-circuit saturation curve for field-pole leakage.

When the general method, as outlined for a non-salient-pole generator, is applied to a salient-pole machine, it is not correct because the armature reaction under these conditions of a salient-pole generator cannot correctly be considered as a vector (see page 214). At the best, the results obtained are uncertain and may be subject to large errors.

The use of unsaturated synchronous reactance, obtained from open-circuit and short-circuit saturation curves, in the synchronous-impedance vector diagram of a synchronous generator, gives results which are subject to large errors. This is because the synchronous reactance so obtained is for much too low a saturation of the magnetic circuit and is therefore much too large, especially when found for the field current on short circuit which produces the desired load current. Better results are obtained by using the synchronous reactance corresponding to the largest field current that it is safe to use with the machine short-circuited. The synchronous-impedance method using the unsaturated synchronous reactance cannot be expected to give satisfactory results even for a non-salient-pole generator, as should be understood from the explanation of the saturated synchronous reactance of a generator (page 240).

The synchronous-reactance method for finding the field current of a generator under load and its voltage regulation when the saturated-synchronous reactance is used should give the same results as the general method when applied under similar conditions. Whether the general method or the saturated synchronous-reactance method is the better in any case is largely a matter of convenience or ease of application in the particular problem. The results given by both methods should be the same.

Zero-power-factor Curves for Different Armature Currents. The zero-power-factor curve of a synchronous generator, *i.e.*, its load saturation curve at zero power factor, can be obtained experimentally from actual measurements made with the generator operating at rated frequency on inductive loads of zero power factor, with fixed balanced armature currents and with different field excitations. Although there is difficulty in obtaining a zero-power-factor inductive load, a load with a power factor sufficiently close to zero can be obtained by using as a load a synchronous motor which delivers no mechanical power, provided a large enough motor is available and it is operated with sufficiently low field excitation, *i.e.*, is underexcited. Only two points on the zero-power-factor curve are needed to determine the complete curve, provided one of these is well up on the saturation curve and the other is for short circuit. From these two points and the open-circuit characteristic of the generator, the Potier triangle can be determined, as explained on page 245.

Refer to Fig. 163. If the hypotenuse of the Potier triangle is moved parallel to itself, keeping the point E on the open-circuit curve, the point F traces the zero-power-factor curve. This is illustrated in the figure where the hypotenuse of the triangle is shown in several different positions.

The Potier triangle can be obtained from the leakage reactance and the armature reaction of the generator calculated from the dimensions of the machine and its winding data. This triangle and the calculated or

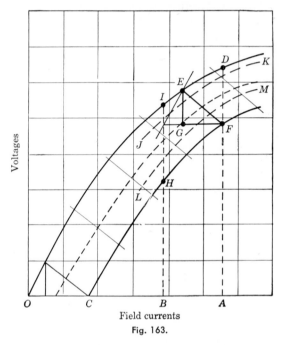

Fig. 163.

measured open-circuit curve make it possible to determine the zero-power-factor curve for any desired current. The determination of the zero-power-factor curve in the manner outlined neglects the change in the field-pole leakage with a change in the saturation of the magnetic circuit of the generator.

Since the leakage reactance drop and the armature reaction of a synchronous generator are proportional to the armature current, the size of the Potier triangle is also proportional to the armature current. This statement neglects any slight change there may be, with a change in the saturation of the magnetic circuit, in the leakage reactance, and also in the impressed field ampere-turns which are necessary to balance the armature reaction.

Assume that the zero-power-factor curve of Fig. 163 is for full-load current. Then for any other load current, such as half-load current, the Potier triangle is half the size shown. The zero-power-factor curve for half-load current is the curve drawn through the middle points of the diagonal lines drawn on the figure and is shown dotted.

The Mmf Method. The method for determining the voltage regulation and field current of a synchronous generator which was first approved by the American Standards Association in 1936 and which is included in the revised standards of 1943 is an mmf method. Figure 164 is an mmf vector diagram and is derived from the general vector diagram of Fig. 152, page 232, by replacing the reactance voltage $jI_a x_a$ by an mmf which, acting alone, would be just sufficient to produce this voltage.

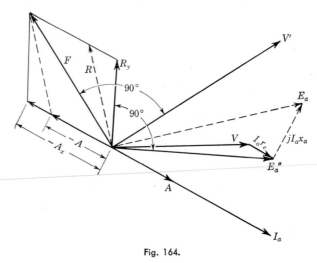

Fig. 164.

This mmf leads $jI_a x_a$ by 90 deg and consequently is in phase with $-A$. The sum of $-A$ and this fictitious mmf is shown as $-A_x$. In the general vector diagram the vector sum of F and A gives the resultant mmf R which produces the generated voltage E_a. In Fig. 164, the vector sum of F and A_x gives an mmf R_y which, since it differs from R by the fictitious mmf referred to above, produces a voltage E_a'' which differs from E_a by the leakage-reactance voltage $jI_a x_a$, and which therefore is the sum of the voltages V and $I_a r_e$. In Fig. 164 the replaced values are shown dotted. Solving this vector diagram for F gives the field current required for any given load and voltage. The voltage corresponding to this field current on the open-circuit characteristic is the no-load voltage from which the regulation can be found.

Determination of Field Current and Open-circuit Voltage by Method of ASA. If the vector diagram of Fig. 164 is rotated clockwise until R_y is horizontal and R_y, $-A_x$, and F are replaced by I_y, and I_x, and I_f,

these mmfs will be related as indicated in Fig. 166.   In this method the
mmfs are determined under unsaturated conditions from the field air-gap
line and the short-circuit characteristic and so are usually in terms of
equivalent field amperes, which is the reason for the change in nomenclature.   A correction is made to take saturation into account by adding
to the field current as found the difference between the field currents from
the air-gap line and the open-circuit saturation curve corresponding to

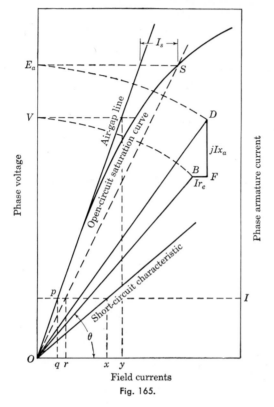

Fig. 165.

the air-gap voltage for the given load.   This air-gap voltage may be
considered a measure of the saturation at which the generator operates.

Referring to Fig. 165, OB is the rated phase terminal voltage drawn
at an angle θ, the power-factor angle, with the base line of the figure taken
as the direction of the current vector. BF is the phase resistance drop
and FD is the phase Potier reactance drop drawn in the proper phase
relations with respect to the current reference line. $OD = E_a$ is then
the air-gap voltage at which the synchronous generator is operating.
This voltage is considered a measure of the degree of saturation of the
magnetic circuit. $OB = V$ is the terminal voltage.

Now refer to Fig. 166.   The line I_y is the field current Oy from the air-gap line on Fig. 165 corresponding to the terminal voltage V.   Add vectorially to I_y, Fig. 166, the field current I_x from Fig. 165 corresponding to the rated armature current from the short-circuit characteristic.   The current I_x is added at an angle θ, the load power-factor angle, with a perpendicular drawn to I_y.   This makes the angle between I_x and I_y the same as that which exists between the terminal voltage and the synchronous-reactance drop on the regular synchronous-impedance vector diagram.   The effect of the armature resistance is neglected on this diagram.   In a large synchronous generator it is in general too small to be of importance.   Its effect would be to increase the angle θ by a very small angle, the tangent of which is equal to the ratio of the voltage drops

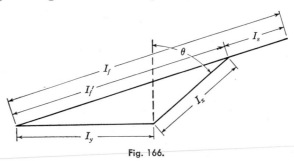

Fig. 166.

in the armature caused by the armature resistance and the synchronous reactance.   This angle is too small to be of importance.   Neglecting saturation, I_f' is the field current for the given load and power factor. To take saturation into account, I_s is added directly to the current I_f' to give the actual field current I_f. I_s is the increase in the field current (Fig. 165) corresponding to the air-gap voltage E_a which is necessary to take saturation into account.   It is the difference between the field currents required for the air-gap voltage E_a on the open-circuit saturation curve and on the air-gap line which represents the relation which would exist between the open-circuit voltage and the field current under unsaturated conditions.

Since I_x is the mmf $-A_x$ of Fig. 164, page 256, in equivalent field amperes, it consists of the armature reaction plus the fictitious mmf replacing the leakage-reactance voltage.   In the synchronous-impedance method an mmf (armature reaction) is replaced by a voltage whose magnitude is dependent on saturation.   The reverse, replacing a voltage by an mmf, likewise gives a value of $-A_x$ dependent on saturation.   To obviate this difficulty in this approved method, both I_y and I_x are determined under unsaturated conditions giving the value of field current I_f' with saturation effects neglected.   The added field current required because of saturation, I_s, is determined at a voltage E_a which accurately measures the saturation of the armature iron. The calculation of E_a, however, involves determining the leakage reactance from a Potier triangle. Because the saturation curves for many modern machines show much less curvature than indicated by some of the plots in this text (for instance Fig. 159, page 244), the reactance determined from the construction of the triangle is often not very reliable. In the methods previously discussed involving this Potier reactance, this makes considerable difference. In this method, the reduced curvature of the saturation curve results in a reduced value of I_s. Since I_s then becomes a very small part of I_f, the resulting percentage error in I_f is small even when the error in x_a is large.

The component field current I_s in Fig. 166 which is added to I_f' to take care of the load saturation should not be added exactly in phase with I_f' as shown. If the field currents I_y and I_x in Fig. 166 were for load-saturation conditions, their vector sum would be the correct field current for the given load, and it would not be necessary to add any extra component current such as I_s to take care of saturation. All of the current I_y is affected by the saturation of the magnetic circuit, but only that part of I_x which takes care of the leakage-reactance drop is similarly affected. The rest of I_x balances the ampere-turns of armature reaction and, neglecting the change in field-pole leakage with change in saturation, is uninfluenced by the saturation of the magnetic circuit. To take account of saturation, I_y should be taken from the load-saturation line OS, and I_x should be increased by qr (Fig. 165). In Fig. 165, Ox is the field current corresponding to the synchronous-reactance drop for the armature current I under unsaturated conditions. The portion qx of this current balances armature reaction. The portion Oq, under the unsaturated conditions represented by the air-gap line, takes care of the leakage-reactance drop $pq = Ix_a$. To take care of this drop under the conditions represented by the load-saturation line SO, a field current Or would be required. Therefore to care for load saturation, the field current Ox should be increased by qr.

The vector diagram of the ASA and a similar vector diagram using field currents corresponding to load saturation are superimposed in Fig. 167. In this diagram the saturated values are shown dotted and double primes are used on the field currents representing saturated conditions. The current Oe is the correct field current for the given load. The component current I_s which is added in the ASA diagram to take care of saturation should be added in the direction de instead of in phase with I_f. However, the error of adding it in phase with I_f is negligible.

In practice, the values of field current, open-circuit voltage, and regulation determined by the ASA Method appear to be reliable for both salient and non-salient pole machines. The power angle, discussed in Chap. 19, may also be determined by this method. This angle is shown as the angle δ in Fig. 152, page 232. In Fig. 166, it is the angle between I_y and I_f. The power angle determined from Fig. 166 gives good results when used for nonsalient pole machines, but is considerably in error when applied to salient-pole machines.

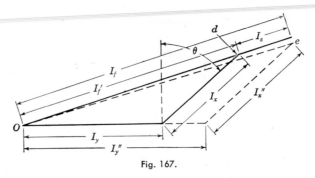

Fig. 167.

Application of the ASA Method. The 93,750-kva synchronous generator is again used. The characteristic curves for this generator are given in Fig. 154, facing page 236. Since field currents are used, these currents are the same whether taken for phase voltages or terminal voltages. An inductive 0.8 power-factor load with rated armature current and rated terminal voltage is used.

The field currents for rated voltage (13,200 volts) and rated short-circuit armature current (4,100 amp) from the air-gap line and the short-circuit characteristic are 427 and 490 amp. Refer to Fig. 166.

$$I'_f = I_y(1 + j0) + I_x(\sin \theta + j \cos \theta)$$
$$= 427(1 + j0) + 490(0.6 + j0.8)$$
$$= 427 + 294 + j392 = 721 + j392$$
$$I'_f = \sqrt{(721)^2 + (392)^2} = 821 \text{ amp}$$

I_s from Fig. 154 for an air-gap voltage $E_a = 14,340$ volts between terminals, which is the air-gap voltage found for the given load on page 236, is

$$I_s = 535 - 465 = 70 \text{ amp}$$
$$I_f = I'_f + I_s = 821 + 70 = 891 \text{ amp}$$

The open-circuit voltage corresponding to this current on the open-circuit saturation curve, Fig. 154, facing page 236, is 17,530 volts between terminals.

The field current found by the general method and also by the saturated synchronous-reactance method, pages 237 and 253, is 890 amp.

Two-reactance Method of Determining Field Current, Regulation, and Power Angle[1] of a Salient-pole Generator. The armature reaction of a synchronous generator with nonsalient poles shifts the axis of the field flux and modifies the strength of the field without producing any great amount of field distortion. With a salient-pole synchronous generator, however, the armature reaction not only modifies the field strength, but except in the case of zero power factor with respect to the excitation voltage, it also distorts the field by crowding the flux toward one pole tip. The effects of the armature reaction of synchronous generators with salient poles have already been discussed and are shown in Figs. 138, 139, 144, 145, and 146, Chap. 16.

If the armature current of a synchronous generator is in phase with its excitation voltage, i.e., with the voltage which would be produced on open circuit by the impressed field, the armature reaction caused by that current distorts the field without appreciably modifying its strength. On the other hand, if the power factor is zero with respect to the excitation voltage, the armature reaction modifies the strength of the field without producing distortion. It is therefore convenient to resolve the armature reaction of a generator with salient poles into two quadrature components, one producing only distortion, the other producing only a change in the field strength. To take account of the two effects of the armature reaction of generators with salient poles, Blondel suggested the two-reaction theory.[2] Since mmf of armature reaction is in phase with the armature current, resolving it into the two components just mentioned is equivalent to considering the armature current to be resolved into two quadrature components. If I_a is the armature current, the two components into which it should be resolved are

$$I_a \sin (\delta + \theta) \quad \text{and} \quad I_a \cos (\delta + \theta)$$

where $(\delta + \theta)$ is the phase angle between the current I_a and the voltage corresponding to the field excitation, i.e., the excitation voltage. The angle θ is the load power-factor angle, i.e., the angle between the phase current and the phase terminal voltage. The angle δ is the power angle which is considered in Chap. 19. The mmf of the first component $I_a \sin (\delta + \theta)$ acts on the same magnetic circuit as the coils on the field poles, and its effect is the same as adding an equivalent number of ampere-turns to the field winding. It merely strengthens or weakens the field according as the current leads or lags. It must be balanced by an equal but opposite number of ampere-turns in the field circuit. The mmf caused by the second component $I_a \cos (\delta + \theta)$ of the armature current

---

[1] See Chap. 19.

[2] *Transactions of the International Congress at St. Louis*, 1904, p. 635.

is in quadrature with the field axis and therefore cannot produce any net reaction with the impressed field. It adds ampere-turns to one half of each pole and subtracts an equal number of ampere-turns from the other half. It produces field distortion by strengthening one half of each pole and weakening the other half. It may be considered as producing component fluxes in the poles which have the general paths indicated by the dotted lines in Fig. 168. The full lines on this figure show the general paths of the component flux which the armature mmf caused by the component $I_a \sin (\delta + \theta)$ of the armature current may be considered to produce. To simplify the figure, only two of these lines for

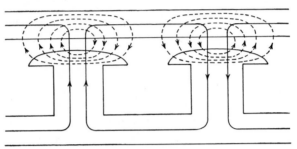

Fig. 168.

each pole are shown. The armature mmf caused by the component $I_a \cos (\delta + \theta)$ of the armature current produces some flux in the interpolar space, but this is relatively small because of the high reluctance of its path. Each of the paths for the two component fluxes, which may be considered to be produced by the component currents into which the armature current is divided, contains two air gaps, but the path for the flux caused by $I_a \cos (\delta + \theta)$ contains less iron than the path for the flux caused by $I_a \sin (\delta + \theta)$ and is therefore less affected by the saturation of the magnetic circuit of the generator although not independent of it.

The fundamental of the component flux which is caused by $I_a \cos (\delta + \theta)$ is stationary with respect to the poles and is fixed in magnitude for any given load in the case of a polyphase machine carrying a balanced load. It therefore rotates at synchronous speed with respect to the armature and induces a fundamental voltage in each phase of the armature winding. Since this component flux does not link the main field circuit, it may be considered as a component leakage flux. The leakage reactance corresponding to this flux may then be combined with the ordinary leakage armature reactance and used with the component $I_a \cos (\delta + \theta)$ of the armature current.

In the Blondel method,[1] the armature effective resistance and the

[1] V. Karapetoff, "Magnetic Circuit," and R. R. Lawrence, "Principles of Alternating Current Machinery," 2d ed.

ordinary leakage-reactance drops are added to the terminal voltage in the usual way. The armature-reaction ampere-turns A_d due to the component $I_a \sin (\delta + \theta)$ of the armature current which is in quadrature with the excitation voltage are combined directly with the ampere-turns of the impressed field. The ampere-turns A_q due to the component $I_a \cos (\delta + \theta)$ of the armature current, which is in phase with the excitation voltage, are taken care of by replacing these ampere-turns by a component voltage drop on the vector diagram. This component voltage drop is the voltage which would be caused in each phase of the armature by the flux that would be produced by the armature-reaction ampere-turns A_q acting alone. This voltage drop is found by making use of the air-gap line of the open-circuit characteristic, *i.e.*, a line drawn through the straight part of the open-circuit characteristic. It is 90 deg ahead of $I_a \cos (\delta + \theta)$.

The double-reactance method of Doherty and Nickle for calculating the performance of a synchronous generator is a modification of the Blondel double-reaction method. It makes use of two synchronous reactances, a direct-axis synchronous reactance and a quadrature-axis synchronous reactance. The direct-axis synchronous reactance corresponds to the synchronous reactance for an armature current that is in quadrature with the excitation voltage and produces an armature reaction which acts directly along the pole axis or along the path of least magnetic reluctance. This reactance is used with the component $I_a \sin (\delta + \theta)$ of the armature current. It is the synchronous reactance which is found from an open-circuit characteristic and a zero-power-factor curve for lagging current. This last statement assumes that the ordinary zero-power-factor curve is for zero power factor with respect to the excitation voltage. The error of making this assumption is negligible. The quadrature-axis synchronous reactance is the synchronous reactance for an armature current that produces an armature reaction which acts midway between the poles, *i.e.*, the armature reaction that is produced by the component $I_a \cos (\delta + \theta)$ of the armature current. It is used with this component of the armature current. The voltage drop corresponding to it is the voltage drop caused by the component current $I_a \cos (\delta + \theta)$ with the regular leakage reactance plus the voltage drop induced in the armature winding by the component flux which is shown dotted in Fig. 168.

The direct-axis and the quadrature-axis synchronous reactances are not constant. They both depend upon the saturation of the magnetic circuit, but because of the fact that more iron is included in the magnetic circuit for the component flux which enters into the direct-axis synchronous reactance than for the component flux which enters into the quadrature-axis synchronous reactance (see Fig. 168), the variation of the direct-axis synchronous reactance with the saturation of the magnetic circuit is somewhat greater than the variation of the quadrature-axis synchronous reactance.

As has been stated, *direct-axis* and *quadrature-axis* synchronous reactances are used with the *reactive* and *active* components of the armature current considered with respect to the field *excitation* voltage, i.e., with $I_a \sin (\delta + \theta)$ and $I_a \cos (\delta + \theta)$. To be in accordance with the notation used in the AIEE definitions of the two reactances, the *active* component of the current with respect to the excitation voltage will be called the *quadrature* component because it produces a component armature reaction which is in *quadrature* with the field axis. This component acts along the path of maximum magnetic reluctance. The *reactive* component of the current with respect to the *excitation* voltage will be called the *direct* component because it produces a component armature reaction which lies *along* the field axis. This component acts along the path of minimum magnetic reluctance.[1]

The direct-axis and quadrature-axis synchronous reactances are always used with the direct component and the quadrature component of the armature current as defined above. The two reactances can be determined experimentally.

The quadrature-axis synchronous reactance for a salient-pole machine is smaller than the direct-axis synchronous reactance. For a non-salient-pole machine the two reactances are nearly equal, and nothing is gained by their use as compared with a single synchronous reactance. The two reactances would be exactly equal for a non-salient-pole machine if it were not for the slots on the rotor which carry the field winding. These make the reluctance of the magnetic circuit slightly different along the field axis and at right angles to it and give the effect of slight saliency.

The power angle δ, between the terminal and excitation voltages, plays an important part in the operation of a synchronous machine. No change can occur in the load carried by such a machine which operates with fixed terminal voltage, fixed frequency, and fixed excitation, whether it operates as a motor or a generator, without a change in the power angle. No change can occur in the excitation of a synchronous machine which operates with fixed terminal voltage, fixed frequency, and fixed load without a corresponding change in the power angle. For a non-salient-pole machine, which operates with fixed terminal voltage, fixed frequency, and fixed excitation, the internal power developed is proportional to the sine

---

[1] Care must be taken not to confuse the direct component and the quadrature component of current as used with the direct-axis and the quadrature-axis synchronous reactances with the direct (active) and the quadrature (reactive) components of current with respect to voltage which are used when active and reactive powers are considered.

of the power angle, if the resistance drop is negligible.   This assumes constant synchronous reactance.   For a salient-pole machine operating under similar conditions, the curve of internal power versus power angle contains a second harmonic (see Chap. 19).

The vector diagram for a salient-pole synchronous generator is shown in Fig. 169.   On this figure x_d, x_q, I_d, and I_q are the direct-axis synchronous reactance, the quadrature-axis synchronous reactance, the direct component $I_a \sin (\delta + \theta)$, and the quadrature component $I_a \cos (\delta + \theta)$ of the armature current as already defined.

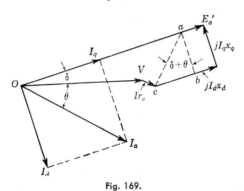

Fig. 169.

Draw the lines ac and ab perpendicular to the lines OI_a and OE_a'.   The triangles abc and OI_qI_a are similar and the angles I_qOI_a and bac are equal. The angle bac is therefore equal to $(\delta + \theta)$.

$$ca = \frac{ab}{\cos (\delta + \theta)} = \frac{I_q x_q}{\cos (\delta + \theta)} = \frac{I_a \cos (\delta + \theta)x_q}{\cos (\delta + \theta)} = I_a x_q$$

The tangent of the power angle δ can now be determined by finding the vector Oa referred to V as a reference axis.   If V is the reference axis,

$$Oa(\cos \delta + j \sin \delta) = V(1 + j0) + I_a(\cos \theta - j \sin \theta)(r_e + jx_q)$$
$$= A + jB$$
$$\tan \delta = \frac{B}{A} = \frac{I_a(x_q \cos \theta - r_e \sin \theta)}{V + I_a(r_e \cos \theta + x_q \sin \theta)} \tag{161}$$

E_a' may now be calculated for any given load condition, provided the two reactances and the effective resistance are known.   Using E_a' as the reference axis,

$$E_a'(1 + j0) = V(\cos \delta - j \sin \delta) + I_a[\cos (\delta + \theta) - j \sin (\delta + \theta)]r_e$$
$$+ [I_a \cos (\delta + \theta)](jx_q) + [-jI_a \sin (\delta + \theta)](jx_d)$$
$$= \{V \cos \delta + [I_a \cos (\delta + \theta)]r_e + [I_a \sin (\delta + \theta)]x_d\}$$
$$+ j\{-V \sin \delta - [I_a \sin (\delta + \theta)]r_e + [I_a \cos (\delta + \theta)]x_q\}$$

The angle θ in the above equations is considered positive for a lagging current since the operator $(\cos \theta - j \sin \theta)$ is used with the current. For a leading current θ is negative.

The imaginary or j part of the above equation must be zero, since E'_a was taken as the reference axis. The complex expression for E'_a therefore cannot have a j component. Hence if the excitation voltage is used as a reference axis,

$$E'_a = V \cos \delta + [I_a \cos (\delta + \theta)]r_e + [I_a \sin (\delta + \theta)]x_d \qquad (162)$$

Only the quadrature reactance x_q appears in the determination of the power angle δ.

Determination of Direct-axis and Quadrature-axis Synchronous Reactances. The unsaturated values of the two synchronous reactances of a polyphase machine can be determined by applying subnormal balanced voltages to its armature and then driving its rotor mechanically, with its field circuit open, at a speed which differs slightly from the speed of rotation of the fundamental of the armature-reaction mmf wave. Under these conditions, the field poles gradually slip through the rotating mmf wave of armature reaction. The permeance of the path for this mmf changes at slip frequency from a maximum, when the axis of the poles and the axis of the armature-reaction wave coincide, to a minimum, when the axes are at right angles. If the voltages impressed on the armature are constant, the currents taken by the armature fluctuate from a minimum when the two axes coincide to a maximum when they are at right angles. Usually the voltages impressed on the armature vary slightly in the opposite sense to the armature currents because of the voltage drops in the line supplying the machine and in the apparatus giving the subnormal voltages. If the slip of the rotor with respect to the armature-reaction field is low enough to avoid appreciable transient effects and inertia effects in the measuring instruments, the maximum ratio of the phase impressed voltage to the phase armature current as read on indicating instruments is the unsaturated direct-axis synchronous reactance. The minimum value of this ratio is the unsaturated quadrature-axis synchronous reactance. The voltage across the open-circuited field winding is a measure of the rate of change of flux through the field circuit. This is zero at the instant when the axes of the field poles and armature-reaction wave coincide and is a maximum when the two axes are at right angles. Oscillograph waves of the field voltage, the voltage impressed on the armature, and the armature current are similar to the curves shown on Fig. 170. On this figure the vertical lines marked b and d indicate instants when the axes of the armature-reaction mmf and the field poles are at right angles. The lines a, c, and e indicate instants when the axes coincide.

The direct-axis synchronous reactance obtained by the above method is the same as the synchronous reactance which would be obtained from open-circuits and short-circuit tests.

Saturated values of the direct-axis synchronous reactance can be obtained from an open-circuit characteristic and a zero-power-factor curve by the same method as is used for determining the synchronous reactance of a cylindrical-rotor generator.

A second method, called the pull-out method, for determining the quadrature-axis synchronous reactance requires a source of power which is large enough compared with the size of the machine to be tested to

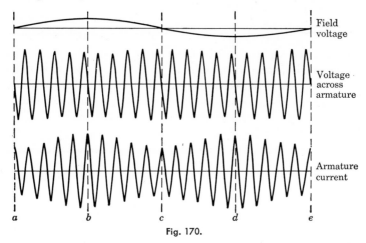

Fig. 170.

approximate the condition of an infinite bus. It must also be capable of having its voltage varied over a wide range. In the pull-out method of obtaining x_q, the machine to be tested is operated as a synchronous motor. The relation between power and power angle for a salient-pole machine is given by Eq. (174), page 284. This equation neglects the effect of the armature resistance. The equation is

$$P = \frac{VE'_a}{x_d} \sin \delta + \frac{V^2(x_d - x_q)}{2x_d x_q} \sin 2\delta$$

To determine x_q by the pull-out method, operate the machine as a synchronous motor at rated frequency at no load. Gradually reduce the field excitation to zero and at the same time decrease the impressed voltage in order to keep the armature current within safe limits. When the field excitation is zero, the first term on the right-hand side of the equation for power versus power angle is zero and the motor is operating as a reluctance motor. The first term on the right-hand side of the equation represents synchronous motor power, the second term represents reluctance motor power.   Now slowly reduce the impressed voltage until the motor drops out of synchronism as indicated by a stroboscope. At this instant, the reluctance power is a maximum and sin 2δ is unity. Read the input P and the impressed voltage at the instant the motor pulls out of synchronism.   The quadrature-axis reactance can then be calculated from the second term of the equation for power versus power angle if the unsaturated direct-axis synchronous reactance is known. The unsaturated direct-axis synchronous reactance can be found in the usual way from a short-circuit curve and an air-gap line.   This method of determining x_q neglects the effect of armature resistance.   This can be taken into account when necessary.

An approximate correction can be made for the effect of saturation on x_q by assuming that the part of x_q which is due to the quadrature-axis armature reaction varies in the same way as the part of x_d which is due to the direct-axis armature reaction.   If this assumption is made, the same saturation factor is used for correcting the magnetizing reactances of both the quadrature-axis and the direct-axis synchronous reactances.

Chapter 18

Efficiency. Statements in regard to certain losses in a synchronous generator. Measurement of the losses by the use of a motor. Measurement of effective resistance. Correction of armature effective resistance for temperature. Retardation method of determining the losses. Calculated efficiency of a synchronous generator. An example of the calculation of the conventional efficiency of a synchronous generator.

Efficiency. The efficiency of a synchronous generator is the ratio of its useful electrical output to the total input. The standard conditions under which efficiency is determined are rated voltage, frequency, and power factor. Unless a specific load is stated, rated load is understood. The standard reference temperature in the standards mentioned below is 75°C.

The efficiency of a synchronous generator can be determined from simultaneous measurements of the mechanical power input and the useful electrical power output, or it can be calculated from an assumed output and the corresponding losses. The losses can be found from the design data of the machine, or they can be measured by the methods recommended by the ASA in American Standard Rotating Electrical Machinery approved in 1943. When the efficiency is determined in the latter manner, it is called the conventional efficiency since it is calculated from the losses measured by specific conventional methods.

Except in a small synchronous generator, it is usually impractical and often impossible to determine the efficiency from direct measurements of input and output on account of the amount of power involved and the impossibility in most cases of obtaining the required load, and also because of the difficulty of measuring the input accurately. Any error in the measurement of either input or output enters directly into the efficiency as found by this method. When the efficiency is calculated from an assumed output and the corresponding losses, an error in the losses does not affect the efficiency proportionately. A given error in the determination of the losses produces a much smaller error in the efficiency, especially when the losses are small, as in a large synchronous generator. If the full-load losses of a synchronous generator are 5 per cent of its output, an error of 5 per cent, *i.e.*, 1 part in 20, in the determination of the losses causes only a $(5 \times 0.05)/100 \times 100 = 0.25$ per cent error in the efficiency.

The losses in a synchronous generator are the copper losses in the armature and the field windings, the core losses, and the mechanical losses   In American Standards for Rotating Electrical Machinery the losses are divided into the following groups for convenience in determining them:

1. I^2r loss in the field circuit
2. Loss in the field rheostat
3. Brush-contact loss
4. Exciter loss
5. Friction and windage loss
6. Brush friction loss
7. Ventilation loss
8. Core loss
9. Armature I^2r loss
10. Stray-load loss

1. The I^2r loss in the field is the copper loss in the field winding at the standard temperature of 75°C. It is equal to the square of the field current for the given load multiplied by the field resistance corrected to 75°C.

2. The loss in the field rheostat is the I^2r loss in the rheostat which controls the field current. This loss is charged against the plant of which the generator is a part and not against the generator.

3. The brush-contact loss is the loss in the collector-ring brushes and contacts. It is small and is usually neglected.

4. The exciter loss is the sum of the electrical and mechanical losses in the exciter which is used for the field excitation. As with the loss in the field rheostat, it is charged against the plant.

5. The friction and windage loss is the mechanical loss at rated speed due to windage and friction of bearings and brushes. The windage loss is caused by the fanning action of the rotating parts in the cooling medium.

6. The brush friction loss is the mechanical loss caused by the brush friction. It is usually included as a part of 5.

7. The ventilation loss is the power required to circulate the cooling air through the generator in the case of a machine with a closed ventilating system. It is in addition to the windage loss. This loss is charged against the generator except that part of it which occurs in air ducts that are external to the machine. This latter part is charged against the plant.

8. The core loss is the sum of the eddy-current and hysteresis losses at rated frequency which are caused by the main flux. It does not include any local eddy-current and hysteresis losses caused by the variation of

the armature currents, and it does not include the effect on the core loss of the distortion of the field flux produced by the load. The core loss is taken as the loss caused by the magnetic field at no load which produces a no-load voltage that is equal to the rated phase no-load terminal voltage corrected for the phase armature resistance drop under the load conditions. The correction is made by adding the phase resistance drop vectorially to the rated phase terminal voltage.

9. The armature I^2r loss is the copper loss in the armature caused by the d-c resistance of the armature windings corrected to 75°C.

10. The stray-load loss includes all the eddy-current losses in the copper and all additional core loss in the iron caused by the distortion of the magnetic field produced by the load currents.

Losses 9 and 10 together constitute the loss caused by the effective resistance.

Statements in Regard to Certain Losses in a Synchronous Generator. The bearing friction loss is proportional to the length and the diameter of the bearings and to the $\frac{3}{2}$ power of the linear velocity of the shaft. It depends upon many factors such as the condition of the bearings and the lubrication, and it varies with the load. The loss caused by the bearing friction is small and for this reason it is assumed constant when finding the conventional efficiency.

The brush friction loss in a rotating-field type of generator is caused by the brushes for the field excitation. On account of the few brushes required and the low rubbing velocity of the slip rings against the brushes, this loss is small. In the case of a rotating-armature type of generator, this loss is due to the brushes for the collection of the armature current. It is larger than for a rotating-field type of machine. Very few generators of this type are built except in small sizes.

The windage loss is not great except in turbine generators. It cannot be calculated. All of the friction and windage losses are generally grouped together and are determined experimentally or are estimated from experimental data obtained from measurements made on similar machines. The windage loss in hydrogen-cooled generators is small.

The core loss includes all the eddy-current and hysteresis losses caused by the relative movement of the field flux and armature, the eddy-current and hysteresis losses produced by any part of the field flux that cuts the frame and any other part of the machine, the pole-face losses caused by the movement of the armature slots across the pole faces, and the eddy-current losses in the armature conductors caused by the field flux.

The eddy-current and hysteresis losses in the pole faces, caused by the movement across the pole faces of the ripples in the flux wave due to the armature slots, increase rapidly as the ratio of the width of the slot opening to the length of the air gap increases. Figure 171 shows the distribution of the flux across the pole face at no load at one particular instant in a generator with a $\frac{9}{16}$-in. air gap and armature slots 1 in. wide.

The flux entering a slot is not constant but varies with the position of the slot with respect to a pole. It is a maximum when the slot is

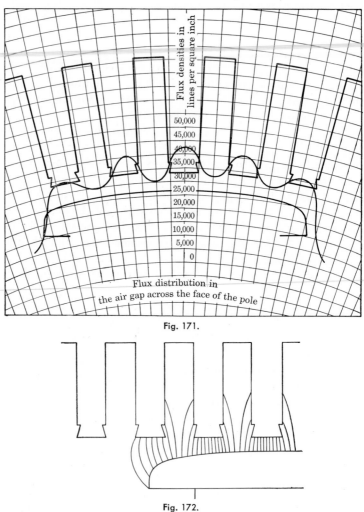

Fig. 171.

Fig. 172.

opposite the center of a pole, and a minimum when the slot is midway between the poles. Figure 172 shows the approximate direction of the flux lines in the slots and in the air gap of a generator at no load. The number of lines per inch represents in a crude way the intensity of the field. The variation in the flux entering a slot sets up eddy currents in the armature conductors. The voltages producing these eddy currents are different on the two sides of the slots and are greater at the top of the slots than at the bottom. To prevent eddy-current losses due to these differences in voltage, it is necessary to laminate the armature conductors both horizontally and vertically. This type of lamination is found in the special laminated conductors now used in large generators. The eddy-current losses are small in these special conductors.

The brush-contact losses are negligible in revolving-field machines. They are caused by the voltage drop at contact of the brushes with the slip rings. When this loss is not negligible, the standard voltage drop used in determining it is 1 volt per collector ring for carbon or graphite brushes with pigtails attached, $1\frac{1}{2}$ volts per collector ring for carbon or graphite brushes with pigtails not attached, and $\frac{1}{4}$ volt per collector ring with metal-graphite brushes.

Many of the losses of a synchronous generator cannot be accurately determined. Among these are the eddy-current losses in the armature conductors due to the local fluxes caused by the load currents and to the distortion of the main flux under load, the core loss caused by the field flux under load, and the tooth frequency losses under load. For this reason more or less arbitrary methods are adopted for determining the losses of a generator when calculating its conventional efficiency.

Measurement of the Losses by the Use of a Motor. The open-circuit losses of a synchronous generator, 5, 6, 8, page 270, can be determined by driving the generator on open circuit by a shunt motor. The open-circuit losses corresponding to any excitation are equal to the input to the armature of the motor minus the brush contact loss, the armature copper loss, and the stray power of the motor. The input to the generator when its field circuit is open is its friction and windage loss. The difference between the open-circuit loss and the friction and windage loss is known as the open-circuit core loss. It is customary to plot this loss against armature terminal voltage. The open-circuit core loss corresponding to the rated terminal voltage corrected for the armature resistance drop is the core loss 8 used in finding the conventional efficiency.

The load loss can be obtained by finding the power required to drive the generator on short circuit. All phases should be short-circuited. This power minus the friction and windage loss is the load loss. The difference between the load loss and the short-circuit copper loss caused by the d-c armature resistance at the temperature of the test is the stray-load loss. This loss depends upon the armature current and should therefore be plotted against that current. This is the loss 10 used in the conventional efficiency. The stray-load loss determined in this way includes a small core loss caused by the resultant air-gap flux. This loss is really not a part of the stray-load loss, but since the resultant air-gap flux on short circuit is merely that necessary to induce sufficient voltage in the armature to overcome the leakage impedance drop, (see short-circuit vector diagram Fig. 173) the flux is small and the core loss caused by it may be neglected. The stray-load loss under normal operating conditions is usually less than the stray-load loss determined on short circuit for the same armature current. The difference between these losses under the two conditions depends upon many factors; it is greatest in high-speed turbine generators with solid cylindrical field structures. Although the stray-load loss measured on short circuit is greater than under operating conditions, the stray-load loss measured as described above is used in calculating the conventional efficiency of synchronous generators and motors.

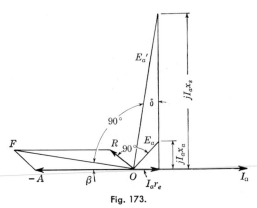

Fig. 173.

Measurement of Effective Resistance. If the local loss produced by a fixed armature current is assumed to be the same on short circuit as under normal conditions, the effective resistance of a generator can be found by dividing the total loss produced by the armature current, when the generator is short-circuited, by the number of phases and the square of the armature phase current, *i.e.*, by dividing the load loss per phase by the square of the phase armature current. This method of determining the effective resistance neglects the core loss mentioned in the preceding paragraph which is caused by the air-gap flux required to induce a voltage in the armature sufficient to overcome the armature leakage impedance drop. The effective resistance determined in this way is for low saturation and is subject to errors from this cause.

Correction of Armature Effective Resistance for Temperature. (See page 236, Chap. 17). The armature effective resistance of a synchronous generator increases with temperature, but not so rapidly as its ohmic or d-c resistance, since the portion of the effective resistance which is caused by the local losses due to the armature leakage fluxes does not vary in the same way with temperature as the part due to the copper resistance.

When it is possible to separate the effective resistance into two parts, one due to the resistance of the armature conductors and the other due to the local eddy-current and hysteresis losses, an approximate correction for temperature is possible. The correction can then be made if the local losses caused by the leakage fluxes are assumed to be independent of temperature. This assumption is not quite justified. The eddy-current losses vary with temperature, but these losses are in general small compared with the hysteresis losses. The hysteresis losses may be assumed constant without appreciable error.

If r_e is the armature effective resistance at a temperature t and r_{oh} is its ohmic resistance at the same temperature, its effective resistance r_e' at a temperature t' is approximately

$$r_e' = r_{oh} \frac{234.5 + t'}{234.5 + t} + (r_e - r_{oh}) \tag{163}$$

This method of correcting the effective resistance of the armature of a generator for temperature not only neglects the change in the eddy-current losses with temperature but also assumes that the part of the effective resistance r_{oh}, which is due to armature copper, is the resistance measured with direct current. In other words, it neglects the skin effect in the armature conductors. The error due to this neglect is large in generators with unlaminated conductors of large cross section, but in generators which have armature conductors of small cross section or having large laminated conductors of the Roebel or Pungar types (see page 229), the error probably is not serious, since the skin effect in such conductors is small.

Retardation Method of Determining the Losses. It is often impossible or impracticable, when dealing with large machines, to drive them by motors to determine their losses. In the case of turbine generators, there is often no projecting shaft to which a motor can be attached. Under this condition the retardation method of determining the losses is a convenient method to use.

The kinetic energy of any rotating body is

$$W = \tfrac{1}{2}\omega^2 \mathfrak{J} \tag{164}$$

where W, ω, and $\mathfrak{J}$ are the kinetic energy, the angular velocity, and the moment of inertia of the rotating part.

Differentiating Eq. (164) with respect to t,

$$\frac{dW}{dt} = \omega \mathfrak{J} \frac{dw}{dt}$$

The derivative of energy with respect to time is power, and the rate of change of angular velocity $d\omega/dt$ is angular acceleration. Replacing dW/dt by power P and $d\omega/dt$ by angular acceleration α gives Eq. (165).

$$P = \mathfrak{I}\omega\alpha \tag{165}$$

Therefore the power causing any change in the angular velocity of a rotating body is equal to the moment of inertia of the body multiplied by its angular velocity and by its angular acceleration at the instant considered. If the rotating body is coming to rest, the acceleration is negative and is called retardation.

The equation $P = \mathfrak{I}\omega\alpha$ can be applied to a motor or a generator to determine the losses, provided the moment of inertia of the rotating part can be found. There are several methods by which the moment of inertia can be determined. It can be calculated from the dimensions of the rotating member of the machine, if they are known, or it can be determined experimentally from the retardation produced by known losses or by a known amount of power applied to the shaft.

If a generator is brought above its synchronous speed with its armature circuit open and its field circuit closed and is then allowed to come to rest, the retarding power causing it to slow down is its friction and windage loss and open-circuit core loss. If the angular retardation α is measured at the instant the generator passes through synchronous speed, the friction and windage loss plus the open-circuit core loss corresponding to the excitation used can be calculated from Eq. (165), provided the moment of inertia is known. If the generator comes to rest without field excitation, the formula gives the friction and windage loss only.

The chief source of error in the application of the retardation method lies in the determination of the retardation α. In order to find α, it is necessary to take readings for a speed-time curve as the generator slows down. Some form of direct-reading tachometer is necessary for this. The interval required between the successive readings for the speed-time curve depends upon the size and speed of the generator under test and varies from 5 sec for very small machines to as many minutes in the case of the largest turbine generators. A speed-time curve is plotted in Fig. 174.

The simplest and most satisfactory method of finding the moment of inertia is as follows. First measure the open-circuit loss at rated frequency and with a definite field excitation. This can be done by operating the machine as a synchronous motor and adjusting the excitation for unity power factor (see Synchronous Motors, page 298). The power input to the armature under this condition is equal to the friction and windage loss, the core loss corresponding to the excitation used, and a very small armature copper loss, which can usually be neglected if the

power factor is properly adjusted. Having determined the loss, the
speed of the generator is increased 10 or 15 per cent by increasing the
frequency of the circuit from which it is operated or by any other convenient means. The generator is then allowed to come to rest with its
field circuit still closed and its excitation unaltered, and readings are
taken for a speed-time curve as the generator slows down. By substituting in Eq. (165) the friction and windage and core loss as measured at
synchronous speed and the values of ω and α also at synchronous speed,
the moment of inertia can be found.

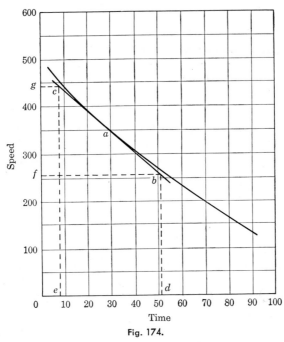

Fig. 174.

Having determined the moment of inertia, the friction and windage
loss can be found by taking measurements for a speed-time curve while
the generator is coming to rest without field excitation. The friction
and windage and core loss corresponding to different field excitations can
be found by allowing the generator to come to rest with different field
excitations. Knowing the friction and windage loss, the open-circuit
core loss corresponding to these field excitations and different open-
circuit voltages can be found.

It is also possible to get the short-circuit loss by letting the generator
come to rest with its armature short-circuited and with a field excitation
which produces the desired short-circuit armature current at synchronous
speed. The power found under this condition, minus the friction and

windage loss and the I^2r loss in the armature due to its ohmic resistance, is the stray-load loss corresponding to the current in the armature when the generator passes through synchronous speed.   The armature current remains nearly constant over a wide range of speed.   The reason for this is given in the discussion of the short-circuit characteristic.

For a small machine which comes to rest quickly, the retardation α can be found by drawing a tangent to the retardation curve (Fig. 174) at the point a which is at rated speed.   From Fig. 174

$$\alpha = \frac{fg}{ed}$$

When a large high-speed machine such as a turbine generator is allowed to come to rest, its angular velocity changes very slowly.   With such a machine the following is the most satisfactory way of applying the retardation method for determining the losses.   From plotted speed-time curves taken with the machine coming to rest under different conditions of retardation, read off the instants of time t_1 and t_2 when the machine passes through angular velocities ω_1, which is a few per cent above synchronous speed, and ω_2, which is an equal number of per cent below synchronous speed.   Let W_1 and W_2 be the kinetic energy of the revolving part at the instants of time t_1 and t_2.   Then

$$\frac{W_1 - W_2}{t_1 - t_2} = \frac{1}{2}\, \mathfrak{J}\, \frac{\omega_1{}^2 - \omega_2{}^2}{t_1 - t_2}$$
$$= \mathfrak{J}\left(\frac{\omega_1 + \omega_2}{2}\right)\left(\frac{\omega_1 - \omega_2}{t_1 - t_2}\right)$$

In this equation $(W_1 - W_2)/(t_1 - t_2)$, $(\omega_1 + \omega_2)/2$, and $(\omega_1 - \omega_2)/(t_1 - t_2)$ are approximately equal to $P = dW/dt$, the angular velocity ω, and the angular retardation $\alpha = d\omega/dt$ in Eq. (165), page 276, at synchronous speed.

When the moment of inertia is determined by the test method just outlined, the units in which P, $\mathfrak{J}$, ω, and α are expressed are of no consequence since the method is purely a substitution method.   If the moment of inertia is determined from dimensions, however, a more convenient form of Eq. (165) using practical units is as follows:

$$P = \frac{746}{550}\left(\frac{2\pi}{60}\right)^2 \frac{Wk^2}{g}\, n\, \frac{dn}{dt} = 462.2 \times 10^{-6}(Wk^2)n\, \frac{dn}{dt} \qquad (166)$$

where P = power, watts
$\quad\ W$ = weight of rotating parts, lb
$\quad\ k$ = radius of gyration of rotating parts, ft
$\quad\ n$ = speed, rpm
$\quad\ g$ = acceleration due to gravity, fps
$\quad\dfrac{dn}{dt}$ = slope of speed-time curve at n, rpm per sec

Calculated Efficiency of a Synchronous Generator. The efficiency of any piece of apparatus is equal to the ratio of its output to its output plus its losses.

$$\text{Efficiency} = \frac{\text{output}}{\text{output} + \text{losses}} \qquad (167)$$

If the losses corresponding to any given output are known, the conventional efficiency corresponding to that output can be calculated by means of Eq. (167). For a three-phase synchronous generator operating under a balanced load, the conventional efficiency is given by

$$\text{Efficiency} = \frac{\sqrt{3}\,VI(\text{pf})}{\sqrt{3}\,VI(\text{pf}) + P_c + 3I_a{}^2 r_{dc} + P_s + I_f{}^2 r_f + P_{f+w} + P_v} \qquad (168)$$

This may be written

$$\text{Efficiency} = \frac{\sqrt{3}\,VI(\text{pf})}{\sqrt{3}\,VI(\text{pf}) + P_c + 3I_a{}^2 r_e + P_{f+w} + I_f{}^2 r_f + P_v} \qquad (169)$$

where the letters in both Eqs. (168) and (169) have the following significance:

V = terminal voltage

I = line current

I_a = phase current

P_c = open-circuit core loss for the terminal voltage corrected for the armature resistance drop

P_s = stray-load loss

P_{f+w} = friction and windage loss

P_v = ventilation loss

r_{dc} = armature d-c resistance per phase

r_e = effective resistance of the armature per phase

r_f = resistance of the field winding

I_f = field current

pf = power factor

The American Standards for Rotating Electrical Machinery specify that the core loss used in the determination of the conventional efficiency shall be that found on open circuit for the rated terminal voltage corrected for the armature resistance drop. Probably a somewhat more correct value is that found on open circuit corresponding to the rated terminal voltage corrected for the leakage impedance drop, *i.e.*, corresponding to the air-gap voltage for the given load, as this voltage is a fairly satisfactory measure of the saturation of the magnetic circuit at any given frequency. In many cases the leakage impedance is not known.

Using a core loss corresponding to a definite rms or effective voltage neglects the effect on core loss of the distortion of the field flux by armature reaction under load. This is negligible in the case of a cylindrical-rotor generator, but it may be appreciable when there are salient poles. At a given frequency, core loss is fixed by the maximum flux density in the core. For a fixed frequency and rms voltage, the maximum flux density is not the same with field distortion as without it.

An Example of the Calculation of the Conventional Efficiency of a Synchronous Generator. The 93,750-kva 13,200-volt 60-cycle synchronous generator for which the characteristic curves and other data are given on Fig. 154, facing page 236, is used for this calculation. The ventilation loss is 350 kw at rated load. The friction and windage loss is 600 kw. From page 236, the effective armature resistance at 75°C is 0.00402 ohm per phase. The armature d-c resistance at 75°C is

$$r_{dc} \text{ (at 75°C)} = 0.0020 \times \frac{1 + 0.0042 \times 75}{1 + 0.0042 \times 25} = 0.00238 \text{ ohm per phase}$$

The rated full-load armature current is

$$I = \frac{93,750 \times 1,000}{\sqrt{3} \times 13,200} = 4,100 \text{ amp per phase}$$

From Fig. 154, the load loss at rated frequency at full-load current is 183 kw.

The stray-load loss at rated armature current is

$$P_s = 183 - 3 \times (4,100)^2 \times 0.0020 \times \frac{1}{1,000} = 82 \text{ kw}$$

In the foregoing calculation of the stray-load loss, the armature resistance at 25°C is used instead of the resistance at 75°C as it is probable that the temperature at which the curves on Fig. 154, facing page 236, were obtained was more nearly 25° than 75°.

The field current found by the general method for the given load is 890 amp (see page 237).

$$E = V + Ir_e = \frac{13,200}{\sqrt{3}} (1 + j0) + 4,100(0.8 - j0.6)(0.00402)$$
$$|E| = 7,633 \text{ volts per phase}$$
$$\sqrt{3} |E| = 13,220 \text{ volts between terminals}$$

From Fig. 154 the core loss corresponding to the above voltage is 357 kw.

Output $= \sqrt{3} \times 13{,}200 \times 4{,}100 \times 0.8\ldots = 75{,}000{,}000$ watts

Losses

Armature copper loss $= 3(4{,}100)^2 \times 0.00238 =$	120,000 watts
Core loss............................... $=$	357,000
Stray-load loss.......................... $=$	82,000
Field loss $= (890)^2 \times 0.186\ldots\ldots\ldots =$	147,300
Friction and windage loss................ $=$	600,000
Ventilation loss......................... $=$	350,000
	1,656,300 watts

$$\text{Efficiency} = \frac{75{,}000}{75{,}000 + 1{,}656} = 0.978 \text{ or } 9.78 \text{ per cent}$$

Chapter 19

Power angle. Transient reactance and subtransient reactance. Determination of the subtransient and transient reactances from oscillograph records of the armature currents and the voltage when a synchronous generator which is operating at no load is short-circuited. Determination of the subtransient reactance from a blocked test. Maximum amplitude of the short-circuit current. Unbalanced conditions. Measurement of negative-sequence impedance. Measurement of zero-sequence impedance. Per-unit values. An example of the use of per-unit values.

Power Angle. Consider first a non-salient-pole machine. Figure 175 shows the synchronous-impedance vector diagram of a non-salient-pole synchronous generator for an inductive load. The angle δ between the excitation voltage E_a' and the terminal voltage V is known as the power

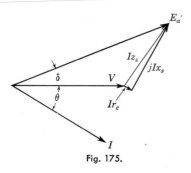

Fig. 175.

angle and plays an important part in the operation of a synchronous machine whether it acts as generator or motor. The excitation voltage E_a' is fixed by the field current and the frequency. The electrical power delivered per phase is $VI \cos \theta$ where V and I are the phase voltage and the phase current. The angle θ is the phase power-factor angle. For power to be delivered, *i.e.*, for generator action, $VI \cos \theta$ must be positive. It must be negative for motor action. Note that the voltages V and E_a' are voltage rises when considered through the machine. Voltage rises are positive in the above equation for power. For generator action the power-factor angle θ must be less than 90 deg. For motor action it must be greater than 90 deg. Motor action is considered in a later section, but the equations for power in terms of the power angle which are to be developed apply to either generator or motor action depending upon the sign of the power angle. For any given values of the excitation voltage E_a', the terminal voltage V, and the frequency, the electrical power $VI \cos \theta$ cannot be changed without altering the power angle δ.

282

Take the terminal voltage rise as the reference axis.   Then

$$V = |V|(1 + j0)$$

$$I = \frac{E_a' - V}{z_s}$$

$$= \frac{|E_a'|(\cos \delta + j \sin \delta) - |V|(1 + j0)}{r_e + jx_s}$$

$$= \frac{(|E_a'| \cos \delta - |V|) + j(|E_a'| \sin \delta)}{r_e + jx_s} \qquad (170)$$

Rationalizing Eq. (170), the expression for the current becomes

$$I = \frac{(r_e|E_a'| \cos \delta + x_s|E_a'| \sin \delta - r_e|V|)}{r_s{}^2 + x_s{}^2}$$
$$+ j \frac{(-x_s|E_a'| \cos \delta + r_e|E_a'| \sin \delta + x_s|V|)}{r_e{}^2 + x_s{}^2} \qquad (171)$$

Since power is equal to the algebraic product of the real parts of the current and the voltage plus the algebraic product of the j parts of the current and the voltage,

$$P = \frac{|V|(r_e|E_a'| \cos \delta + x_s|E_a'| \sin \delta - |V|r_e)}{r_e{}^2 + x_s{}^2}$$

By considering the two terms $r_e|V|E_a'| \cos \delta$ and $x_s|V|E_a'| \sin \delta$ in the equation for P as two stationary right-angle vectors, they can be combined into a single term $|V|E_a'|z_s \sin \left(\delta + \tan^{-1} \dfrac{r_e}{x_s} \right)$ and the equation can then be written

$$P = \frac{|V|E_a'|}{|z_s|} \sin \left(\delta + \tan^{-1} \frac{r_e}{x_s} \right) - \frac{|V^2|r_e}{|z_s|^2} \qquad (172)$$

In general r_e is small compared with x_s.   Therefore the error introduced by neglecting the terms involving r_e in Eq. (172) is small.   If these terms are neglected, Eq. (172) becomes

$$P = \frac{|V|E_a'| \sin \delta}{x_s} \qquad (173)$$

If the armature resistance r_e is negligible, the electrical power is zero when the power angle is zero.   When this angle is positive, P is positive and electrical power is delivered, *i.e.*, there is generator action.   When the power angle is negative, the electrical power is negative and power is absorbed, *i.e.*, there is motor action.

It is evident from Eqs (172) and (173) that with fixed values of terminal voltage, excitation voltage, and frequency there can be no change in the electrical power without a change in the power angle.

Equations (172) and (173) apply only to a non-salient-pole machine. If x_s is assumed constant, the curve showing the relation between power and power angle is a sinusoid if r_e is negligible compared with x_s, Eq. (173). A power-angle curve for a non-salient-pole machine for constant x_s is shown in Fig. 176.

Neglecting r_e, maximum power occurs for a power angle of 90 deg. The power is zero when the power angle is zero.

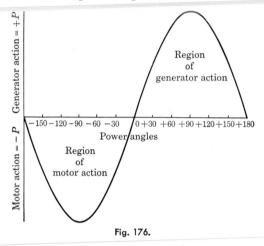

Fig. 176.

Consider a salient-pole machine using the direct and the quadrature synchronous reactances. Figure 177 shows the vector diagram of such a machine. The resistance drop in the armature is neglected in the figure and in what follows.

$$P = |I|\,|V|\,\cos\theta$$

Since

$$|I|\,\cos\theta = |I_q|\,\cos\delta + |I_d|\,\sin\delta$$
$$P = |V|(|I_q|\,\cos\delta + |I_d|\,\sin\delta)$$

$$|V|\cos\delta = |E_a'| - |I_d|x_d \qquad |I_d| = \frac{|E_a'| - |V|\cos\delta}{x_d}$$

$$|V|\sin\delta = [I_q]|x|_q \qquad |I_q| = \frac{|V|\sin\delta}{x_q}$$

$$P = |V|\left(\frac{|V|\sin\delta}{x_q}\cos\delta + \frac{|E_a'| - |V|\cos\delta}{x_d}\sin\delta\right)$$

$$= \frac{|V||E_a'|}{x_d}\sin\delta + |V|^2\left(\frac{\sin\delta\cos\delta}{x_q} - \frac{\sin\delta\cos\delta}{x_d}\right)$$

$$= \frac{|V||E_a'|}{x_d}\sin\delta + \frac{|V|^2}{x_d x_q}(x_d - x_q)\sin\delta\cos\delta$$

$$= \frac{|V||E_a'|}{x_d}\sin\delta + \frac{|V|^2(x_d - x_q)}{2x_d x_q}\sin 2\delta \qquad (174)$$

As is shown by Eq. (174), the power-angle curve of a salient-pole machine contains a second harmonic as well as the fundamental. This curve is shown in Fig. 178. The resultant power-angle curve is shown by a full line. The fundamental and second harmonic are shown dotted.

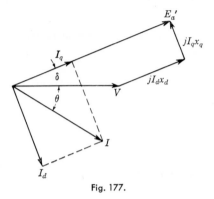

Fig. 177.

Since in general the quantities $|V|E_a'|$ and $|V|^2$ of the two numerators in Eq. (174) are not very different in magnitude and since the second term has the product of the two reactances for its denominator instead of having a single reactance as in the first term, the second harmonic in the power-angle curve of a salient-pole machine is much smaller than the fundamental.

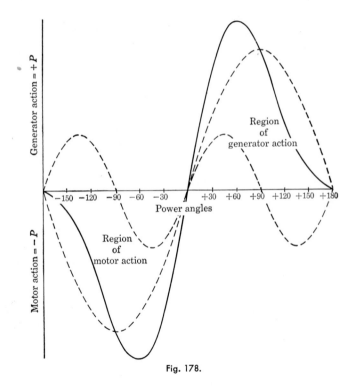

Fig. 178.

As is evident from Eq. (174), the power of a salient-pole machine, neglecting armature resistance, is zero when the power angle is zero but is a maximum before the power angle reaches 90 deg. With a non-salient-pole machine, the power does not reach a maximum until the power angle reaches 90 deg. When δ is positive, electrical power is delivered, and when δ is negative, electrical power is absorbed. As is seen from Eq. (174) and Fig. 178, page 284, with fixed excitation voltage E_a', fixed terminal voltage V, and fixed frequency, as in a non-salient-pole machine, there cannot be any change in the electrical power developed or absorbed without a change in the power angle. Use is made of the power angle in the discussions of the parallel operation of synchronous generators and the operation of synchronous motors.

Transient Reactance and Subtransient Reactance. Under the conditions of a sustained short circuit, the steady-state armature current of a polyphase synchronous generator is equal to the excitation voltage divided by the synchronous impedance. In a short-circuited synchronous generator, the current lags the voltage by a large angle and as a result the axis of the armature-reaction ampere-turns nearly coincides with the field axis. The synchronous reactance, which limits the short-circuit current, is therefore the direct-axis synchronous reactance (page 263). For a synchronous generator with a cylindrical rotor, i.e., with non-salient poles, there is no difference between the direct-axis and the quadrature-axis synchronous reactances except that caused by the slight saliency due to the absence of rotor slots over the middle of the pole faces. The effect of armature reaction under steady conditions is to change the net number of ampere-turns acting on the field circuit, and under this condition its effect may be replaced by an equivalent reactance which added to the leakage reactance constitutes the synchronous reactance. For any given armature current and saturation, this equivalent reactance produces a voltage drop in the armature which is equal to that caused by the change in the air-gap flux produced by armature reaction. The synchronous reactance, whether direct-axis or quadrature-axis is considered, exists only under steady-state conditions. Under steady-state conditions, the armature reaction is fixed both in magnitude and direction with respect to the poles. Although it modifies the field strength and may be replaced by a fictitious reactance in so far as its effect on the voltages induced in the armature windings is concerned, it cannot produce any induced voltages and therefore cannot produce any currents in any part of the field structure such as the field winding or in the damper if one exists, since its rate of change with respect to the field structure is zero.

When a sudden change occurs in the armature currents, such as is caused by a short circuit, the armature reaction is no longer constant in magnitude with respect to the poles and also it probably changes in direction with respect to them. Under this condition voltages are induced by the changing armature reaction in the field winding and in any other closed windings on the field structure. These voltages cause transient currents in these parts.

Any change in a magnetic circuit is always accompanied by a corresponding reaction which tends to prevent the change. If the current increases in a winding on a magnetic circuit which has other closed windings associated with it, the currents in the other windings must change in such a way as to tend to maintain constant the total number of ampere-turns acting on the magnetic circuit and to prevent a change in the flux which links the circuit.

The resistance and inductance of the circuit supplying the current to the field winding of a synchronous generator are small compared with the resistance and self-inductance of the field winding, and therefore for transient conditions the field circuit may be assumed to be virtually short-circuited. Dampers are always short-circuited. During transient conditions, the armature and the field windings and other windings on the field structure act like the windings of a transformer where the armature winding is the primary and the other windings are the secondaries. Each winding possesses self-inductance and there is mutual induction between windings. As in the iron-core transformer, the conditions are simplified by treating the problem from the standpoint of the leakage reactances and the resistances of the windings. If no damper exists, then during the transient period the armature winding and the field winding act like the windings of a short-circuited transformer. The sudden increase in the short-circuit armature-reaction ampere-turns is resisted by an increase in the field ampere-turns which reduces the net flux per ampere of armature current and decreases the apparent reactance which limits the short-circuit current. The condition is analogous to that existing in a short-circuited transformer. The short-circuit current is limited by an impedance which corresponds to the equivalent impedance of a short-circuited transformer (see page 79). For the synchronous generator, however, the reactance, instead of being constant as in the transformer, increases from its initial transient value to the synchronous reactance when steady conditions are reached.

If a synchronous generator is short-circuited, the current lags the voltage producing it by nearly 90 deg because of the large ratio of armature apparent inductive reactance to armature resistance. As a consequence the axis of the armature reaction lies nearly along the field axis. The mutual induction between the armature winding and the field winding is a maximum. Under this condition the reactance which limits the transient short-circuit current is a direct-axis reactance. If there is considerable resistance in the path of short circuit, because the short circuit occurs at an appreciable distance from the generator, the alternating components of the transient armature currents do not lag the voltages causing them by 90 deg. Under these conditions the transient alternating currents must be divided into direct-axis and quadrature-axis components, and direct-axis and quadrature-axis reactances must be considered for the transient currents.

When an alternating voltage is short-circuited through an inductance and a resistance in series, the short-circuit current always consists of two components, a unidirectional or d-c component which decreases logarithmically to zero and an alternating component which is fixed by the voltage, the reactance, and the resistance of the circuit and is constant when the resistance and the reactance are constant. In the short-circuited synchronous generator the reactance is not constant. It increases from its initial transient value to a value corresponding to synchronous reactance when steady-state conditions are reached. Under this condition of short circuit, the alternating components of the armature currents of a short-circuited generator decrease from their initial transient values to their steady-state values which are fixed by synchronous reactance.

In a polyphase synchronous generator the alternating components of the short-circuit armature currents produce armature reaction which is fixed in direction with respect to the poles but decreases from an initial value, determined by the initial values of the transient alternating components of the armature currents, to a final value, fixed by the steady-state short-circuit currents. To balance the increase in armature reaction due to the alternating components in the armature current when a short circuit occurs, there must be an increase in the field current. This increase in field current is in the same direction as the initial field current since the armature reaction caused by lagging currents is demagnetizing. The change in field current decreases and becomes zero when steady-state conditions are reached.

The unidirectional components in the armature currents produce a resultant mmf which is fixed with respect to the armature but has fundamental frequency with respect to the field circuit. To balance this, the field current must contain an alternating component of fundamental frequency. This alternating component in the field current produces an mmf of fundamental frequency in the air gap which is fixed in direction with respect to the field poles. This mmf can be resolved into two oppositely rotating components, each of which rotates at synchronous speed with respect to the poles (see Armature Reaction, page 208). The component whose rotation is opposite to that of the poles balances the armature mmf caused by the unidirectional components in the armature currents. The other rotating component rotates at double synchronous speed with respect to the armature winding and must be balanced by second harmonic components in the transient armature currents.

All the components in the armature currents are maximum at the instant of short circuit, and all except the fundamental alternating components decrease to zero when steady-state conditions have been reached. The fundamental alternating components become the steady-state short-circuit currents.

If there is a damper on the synchronous generator, transient currents are induced in it similar to the transient currents in the field winding. Transient eddy currents in such parts as the pole faces may also be appreciable. The transient currents in these parts have the same effects as the transient currents in the field circuit and diminish the apparent reactance during the initial or early part of a short circuit. The ratios of resistance to reactance of the damper and of the eddy-current paths are much higher than for the field circuit, and their time constants are therefore smaller. As a result, the effects of the transient currents in these parts die out much more quickly than the effects of the transient currents in the field circuit and in general disappear after a few cycles. They influence the apparent reactance during only the early part of a short circuit.

If a synchronous generator operating at no load is short-circuited, the ratio of the open-circuit effective phase voltage to the initial effective value of the fundamental of the alternating component of the phase transient armature current is the subtransient reactance. This includes the effects of reaction on the armature of the transient currents in the field circuit, and in all parts in which transient currents may be induced, as in dampers and in the pole faces. The transient reactance is the ratio of the open-circuit phase voltage to the initial fundamental of the alternating component of the transient armature current, which would have existed if only the field circuit and the armature had carried transient currents.

Both the transient and the subtransient reactances determined as just defined are direct-axis reactances since the alternating components of the short-circuit armature currents lag the voltages producing them by nearly 90 deg, with the result that the armature reaction caused by them lies nearly along the pole axis.

Quadrature transient and subtransient reactances are important under conditions where the alternating components of the transient short-circuit currents do not lag the voltages causing them by nearly 90 deg. An example of this occurs in the study of the effect of a short circuit in a distribution system where there is appreciable resistance between the short circuit and the generators. In such a case both direct-axis and quadrature-axis reactances must be considered. In general, the direct-axis transient and subtransient reactances determine the short-circuit currents.

Subtransient reactance in general is used to determine the initial amplitude of short-circuit currents. Transient reactance is used in studying questions of stability. In many cases, as in questions of the heating caused by a short circuit, the rate of decay of the transient currents must be considered. This rate is determined by time constants. Both the subtransient reactance and the transient reactance can be obtained from an oscillograph record of the short-circuit currents and the voltage of a synchronous generator. The subtransient reactance can also be obtained from measurements made with the generator at rest.

Determination of the Subtransient and Transient Reactances from Oscillograph Records of the Armature Currents and the Voltage When a Synchronous Generator Which Is Operating at No Load Is Short-circuited. Figure 179a is an oscillograph record of a sudden three-phase short circuit at 50 per cent rated voltage of a 1,250-kva two-pole 60-cycle

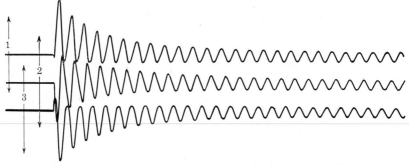

Fig. 179a. General Electric turbine generator, 450 volts, 2 poles, 60 cks, 1250 kva, 0.8 power factor. Sudden three-phase short circuit at 50 per cent rated voltage. Calibration for curves 1, 2, and 3 = 2 X 15,960 amps.

450-volt synchronous generator. Figure 179b is an oscillograph record of a line-to-neutral short circuit at 10 per cent rated voltage of an 1,875-kva three-phase two-pole 2,400-volt synchronous generator. Figure 180 shows the short-circuit current in one phase of a three-phase synchronous generator which has all three phases short-circuited at no load.

Draw the envelope of the wave (see Fig. 180) and also draw a line *ef* midway between the two sides of the envelope. This line is the d-c component of the current for phase 1. The d-c component has different magnitudes in the three phases. It is zero in one phase, if the short circuit for that phase occurs at the point on the voltage wave which corresponds to the zero of the resultant of the alternating components of the transient current. The alternating components are the same in all phases. The envelope of the alternating component is redrawn in Fig. 181 with its axis horizontal, *i.e.*, with the d-c component eliminated.

The subtransient reactance is given by the effective phase voltage of the generator, at the instant of short circuit, divided by the initial effective value of the alternating component of the phase current. This current is equal to $ac/2\sqrt{2}$, Fig. 180, where ac is found by extrapolating the sides of the envelope to an ordinate drawn through zero time.

The lines gb and hd, Fig. 181, represent the envelope of the alternating component, ignoring the first few cycles. The difference between the

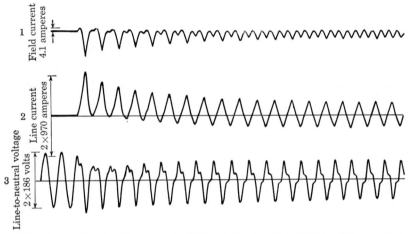

Fig. 179b. General Electric turbine generator, 2400 volts, 2 poles, 60 cks, 1875 kva, 0.8 power factor. Line-to-neutral short circuit at 0.1 rated voltage. No external resistance. Curve 1, field current, calibration $= 4.1$ amps. Curve 2, line current, calibration $= 2 \times 970$ amps. Curve 3, line-to-neutral voltage, calibration $= 2 \times 186$ volts.

two envelopes is caused by the currents in the damper or in other short-circuited paths exclusive of the field circuit. The transient currents in these paths die out much more quickly than the transient current in the field circuit. These currents are the cause of the difference between the subtransient and the transient reactances. The transient reactance is equal to the initial value of the effective phase voltage divided by $gh/2\sqrt{2}$.

The alternating components of the transient currents may be considered to consist of two superimposed transient alternating currents, one caused by the reaction of the currents in the field circuit, the other caused by the reactions of the currents in the damper and other short-circuited paths exclusive of the field circuit. The two alternating components of the transient current decay logarithmically but at different rates, that due to the currents in the damper and other short-circuited paths exclusive of the field circuit practically disappearing in comparatively few cycles. If the logarithms of each of the components of the transient alternating currents are plotted against time, straight lines result. The intercepts with the current axis of these lines extended give the initial values, *i.e.*, the values at zero time, of the two component currents.

The logarithms of the amplitudes of the alternating components of the transient currents are plotted to an enlarged scale against time in Fig. 182.

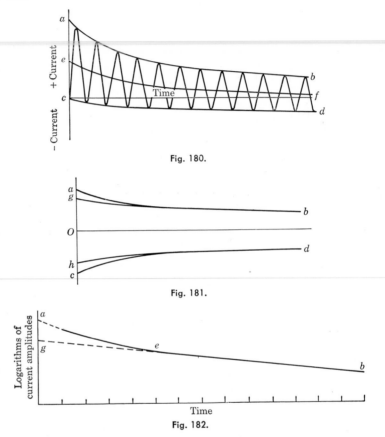

Fig. 180.

Fig. 181.

Fig. 182.

The straight part *eb* of the curve in Fig. 182 is caused by the alternating currents due to the transient reactance. The intercept of the straight part of this line extended with the axis of currents gives the initial value of the alternating component of the transient current due to the transient reactance. It locates the points *g* and *h* in Fig. 181.

The logarithm of the differences between the ordinates in Fig. 182 of the curve *ab* and the straight line *gb* are plotted to arbitrary scale against time in Fig. 183.

This plot is a straight line. The intercept of this line extended with the vertical axis gives the initial difference between the two currents and

locates the points a and c in Fig. 181. Referring to Fig. 181 the transient
and the subtransient reactances are

$$x' = \frac{E}{og/\sqrt{2}} \tag{175}$$

and

$$x'' = \frac{E}{oa/\sqrt{2}} \tag{176}$$

where E is the initial effective value of the phase voltage of the generator.
The reactances given by Eqs. (175) and (176) are direct-axis reactances
since they are obtained from short-circuit
conditions where the alternating components of the transient armature currents
are lagging the voltages which cause them
by nearly 90 deg. In practice, semi-
logarithm paper can be used to advantage
in making Figs. 182 and 183.

**Determination of the Subtransient
Reactance from a Blocked Test.** The subtransient reactance of a synchronous machine can be obtained by short-circuiting
the field circuit and applying voltage at

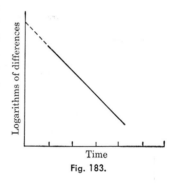

Fig. 183.

rated frequency between any two armature terminals (a three-phase
Y-connected machine assumed) and measuring the voltage and current.
The ratio of one-half the effective voltage to the effective current is the
direct-axis subtransient reactance when the armature is so placed that
the resultant mmf of the two excited phases lies along the pole axis.
For the three-phase machine this occurs when the axis of the unexcited
phase coincides with the pole axis. When the resultant mmf of the
two windings is midway between the poles, the ratio of one-half the
effective impressed voltage to the effective current is the subtransient
quadrature-axis reactance. There is no resultant torque due to the currents acting to turn the armature with respect to the field if the axis of
the armature mmf is correctly lined up with the poles. If it is not correctly lined up, there is a strong torque to rotate the armature with
respect to the field to the position where the axis of the resultant armature mmf lines up with the field axis. The reactances obtained as outlined are unsaturated reactances.

Maximum Amplitude of the Short-circuit Current. When a sinusoidal
alternating potential is impressed on a circuit containing constant resistance and constant reactance, the greatest possible amplitude of the current that can occur is twice the amplitude of the steady-state current.
This maximum would occur if the circuit were closed at the point on the

voltage wave which corresponded to the maximum value of the steady-state component of the current and if the d-c transient did not decrease with time (see Fig. 184).

The maximum current would occur in the second loop of the steady-state current. Actually the maximum value is always somewhat less than twice the maximum value of the steady-state current because of the decrease of the d-c transient with time. When a synchronous generator is short-circuited, the maximum current occurs in a phase when the short circuit takes place at the point on the voltage wave of that phase which corresponds to the maximum value of the alternating component of the transient current. It is fixed by the subtransient reactance, but since the subtransient reactance increases with time, the maximum amplitude of the short-circuit current in any phase is somewhat less than

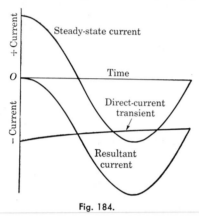

Fig. 184.

$$I_m = \frac{2\sqrt{2}\,E}{x''}$$

where E is the effective value of the air-gap voltage when the short circuit occurs. It is less than the value given by the equation because the d-c transient is decreasing with time and also because the subtransient reactance is increasing and is approaching the transient reactance. The maximum current may easily be ten or fifteen times the rated full-load current.

Unbalanced Conditions. When a synchronous generator is carrying an unbalanced load, its operation may be analyzed by using symmetrical components. The unbalanced currents are resolved into their positive-, negative-, and zero-sequence components. The impedances of most generators are alike for all phases and therefore constitute zero-sequence components. In a rotating machine, such as a generator, however, the positive-, negative-, and zero-sequence components of current each encounter balanced impedances of differing magnitudes. This is largely because of armature reaction. In a synchronous generator, the positive-sequence currents, as in the balanced case, produce an armature reaction which is constant with respect to the field poles. The component currents, therefore, encounter the impedances previously discussed for balanced loads. The negative-sequence currents produce an armature reaction which rotates around the armature in the opposite direction to the direction of rotation of the field poles and hence rotates past the field at twice synchronous speed inducing currents in the field and damper windings and eddy currents in the iron. The balanced impedance encountered by these negative-sequence currents is called the negative-sequence impedance of the generator.

The zero-sequence currents produce an armature reaction in each of the three phases of a three-phase machine, but the combined armature reaction is zero. The impedance encountered by these component currents is therefore different from that encountered by positive- and negative-sequence components and is called the zero-sequence impedance of the machine. Since most power companies attempt to keep the system loads fairly well balanced, an unbalanced condition is most likely to be due to a short circuit at some point in the system. This results in unbalancing the system as long as the short circuit is maintained. It should be kept in mind that the negative and zero-sequence impedances are the impedances encountered by negative-sequence and zero-sequence components of the steady state, and not the transient currents.

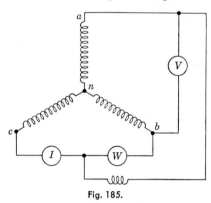

Fig. 185.

Measurement of Negative-sequence Impedance. The negative-sequence impedance may be measured by driving the generator at synchronous speed with its field shorted and a negative-sequence voltage applied to the armature terminals. The ratio of applied voltage and armature current, both per phase, then gives the negative-sequence impedance. If a wattmeter is used to measure the electrical power supplied, this power per phase divided by the phase current squared give the negative-sequence resistance, and the negative-sequence reactance may be found from the impedance and resistance in the usual manner.

Another method of measuring the negative-sequence parameters is to connect the armature terminals as shown in Fig. 185. The machine is then driven at synchronous speed and the field current adjusted until rated current flows in the shorted phases. It can be shown that the negative-sequence component of the voltage V_{an} is one-third of the voltage $V_{ab} = V$, the voltage read on the voltmeter, or $V_{an}^- = V/3$. If I_{an}^- and I are, respectively, the negative-sequence component of the current in phase a and the ammeter reading, then $I_{an}^- = -j\,(I/\sqrt{3})$. The negative-sequence impedance therefore equals

$$\frac{V_{an}^-}{I_{an}^-} = j\left[\frac{V}{\sqrt{3}\,I}\right]$$

Therefore

$$X^- = \frac{W}{\sqrt{3}\, I^2}$$

and

$$R^- = \frac{1}{\sqrt{3}} \sqrt{\left(\frac{V}{I}\right)^2 - \frac{W^2}{I^4}}$$

where W, X^-, and R^- are, respectively, the wattmeter reading, the negative-sequence reactance, and the negative-sequence resistance.[*]

Measurement of Zero-sequence Impedance. The zero-sequence impedance may be measured by connecting the armature windings of the three phases in series and then connecting them to a single-phase source of power. If the machine is then driven at synchronous speed and the field terminals are shorted, the ratio of the applied voltage to the product of the number of phases and the current flowing is the zero-sequence impedance. The electrical power supplied, which may be measured by a wattmeter, divided by 3 and by the square of the current gives the zero-sequence resistance.

Per-unit Values. Instead of expressing the voltage and current of a machine in volts and amperes, it is often convenient to express them as fractions of the rated voltage and current. For example, a machine which is operating at rated voltage and one-half rated current has a 100 per cent voltage and a 50 per cent current. The per-unit values of these are 1 for the voltage and 0.5 for the current. A machine which has an armature resistance of such a magnitude that the resistance drop per phase at rated phase current is 3 per cent of the rated phase voltage is said to have a 3 per cent resistance or a 0.03-per-unit resistance. Reactances are similarly expressed. If the reactance of a synchronous generator is given in ohms, this alone conveys little idea of whether the reactance is high or low. To determine whether it is high or low, the reactance drop per phase at full-load current must be found. This drop expressed as a fraction of the rated phase voltage of the generator, *i.e.*, in its per-unit form, shows whether the generator is a high-reactance or a low-reactance machine. One generator may have a reactance in ohms which is twice as large as the reactance of another generator, and yet the former may be the low-reactance machine if its reactance drop per phase at full-load current expressed as a fraction of the rated phase voltage is smaller than the reactance drop of the other machine similarly expressed, *i.e.*, if it has the smaller per-unit reactance.

A synchronous generator is a high-reaction machine if its armature reaction in ampere-turns per pole at full-load current is a large fraction

[*] See W. V. Lyon, "Applications of the Method of Symmetrical Components," Chap. 9.

of its normal field ampere-turns, or in other words, if its per-unit arma-ture reaction is large. The normal 100 per cent or 1-per-unit field ampere-turns are the field ampere-turns which are necessary to produce rated voltage at no load and rated frequency, *i.e.*, 1-per-unit voltage, at no load and rated frequency.

It is difficult to remember the order of magnitude of the actual values of the constants or parameters of machines of different power ratings and voltages, but it is easy to remember the order of magnitude of the con-stants or parameters expressed in per-unit values. When the actual con-stants are desired in ohms, it is easy to find them from the per-unit values as soon as the current and voltage ratings of the machine are known.

The use of per-unit values is often convenient when making plots of the characteristics of a machine. When per-unit values are plotted, rated phase voltage and rated phase current are plotted as unity. The react-ance and the resistance drops per phase are plotted as fractions of rated. phase voltage. The use of per-unit values on a plot always makes possible the use of convenient scales.

An advantage of the per-unit system is that, when a problem involving a particular machine has been solved using per-unit values, the results may be applied to any other similar machine regardless of its size, pro-vided it has the same per-unit parameters.

An Example of the Use of Per-unit Values. A three-phase synchro-nous generator has the following per-unit constants:

$$\text{Synchronous reactance } x_s = 0.4$$
$$\text{Effective armature resistance } r_e = 0.02$$

Neglecting the effect of saturation on the synchronous reactance, the per-unit open-circuit phase voltage which corresponds to the field exci-tation that produces rated terminal voltage when the generator operates with a full kva inductive load of 0.8 power factor is

$$E'_a = 1.00(0.8 + j0.6) + 1.00(1 + j0)(0.02 + j0.4)$$
$$= 0.8 + j0.6 + 0.02 + j0.4$$
$$= 0.82 + j1.00$$
$$E'_a = \sqrt{(0.82)^2 + (1.00)^2} = 1.29 \text{ per unit}$$

The per cent values are merely the per-unit values multiplied by 100. In the above example the no-load voltage is 129 per cent.

PART THREE

SYNCHRONOUS MOTORS

Chapter 20

Construction. General characteristics. Power factor. V curves. Methods of starting. Explanation of the operation of a synchronous motor.

Construction. Synchronous motors except those for very high speed are always built with salient poles. In other respects there is no essential difference between their construction and the construction of synchronous generators. The differences which do exist do not involve principles of design but are merely changes to adapt the machines more effectively to the particular purposes for which they are to be used. The chief differences are in the relative amounts of armature reaction and in the damping devices. Any synchronous generator can operate as a synchronous motor and any synchronous motor can operate as a synchronous generator. As a rule a synchronous motor has a more effective damping device to prevent hunting than is necessary for a synchronous generator, and its armature reaction is larger than is desirable for a generator.

General Characteristics. A synchronous motor operates at only one speed, the synchronous speed. This speed depends solely upon the number of poles for which the motor is built and the frequency of the circuit in which it is operated. The speed is independent of the load. A change in load is accompanied by a change in phase and in instantaneous speed, but not by a change in average speed. If, because of excessive load or any other cause, the average speed differs from synchronous speed, the average torque developed is zero and the motor comes to rest. A synchronous motor as such has no starting torque.

Power Factor. The power factor of a synchronous motor operating from constant-potential mains is fixed by its field excitation and by its load. At any given load the power factor can be varied over wide limits by altering the field excitation. A motor is said to be overexcited or underexcited according as its excitation is greater or less than normal excitation. Normal excitation is that which produces unity power factor. Overexcitation produces capacitive action and causes a motor to take a leading current. An underexcited synchronous motor takes a lagging current. The field current which produces normal excitation depends upon the load and in general it increases with the load.

V Curves. Since it is possible to operate a synchronous motor at different power factors, curves can be plotted showing the relations between the current per terminal and the excitation for different constant loads. Such curves are called V curves on account of their shape. Lines drawn through points of equal power factor on the V curves are called compounding curves. Figure 186 shows three V curves and three compounding curves.

Curves *I, II, III* are the V curves for three different loads and *A, B, C* are compounding curves. Curve *I* is for the largest load and curve *III*

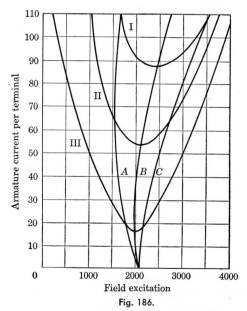

Fig. 186.

is for the smallest load. *B* is the compounding curve for unity power factor and gives the normal excitation for different loads.

Methods of Starting. Since synchronous motors have no starting torque, some auxiliary device must be used to bring them up to speed. Polyphase synchronous motors can be brought up to speed by the induction-motor action produced in their dampers and by the hysteresis and eddy currents in the pole faces. The field winding is usually open while the motor is being started in this way, but in some cases the field winding is short-circuited to diminish the high voltage induced in the field circuit at the instant of starting by the rotating magnetic field produced by the armature currents. When the field is short-circuited, the motor acts like a polyphase induction motor with a single-phase rotor. Such a motor may not accelerate beyond about half synchronous speed. For this reason when the field of a synchronous motor is short-circuited while the motor is being started as an induction motor, the short circuit should be removed before half synchronous speed is reached. The usual practice is either to leave the field open and to insulate it to withstand the high voltage induced in it while the motor is accelerating or to short it through a high resistance. The damper usually consists of copper bars which pass through the pole faces near their surface. The ends of these bars are connected by copper or brass straps. If the synchronous motor is provided with an exciter mounted on its shaft, this exciter can be used as a d-c motor to bring the synchronous motor up to speed. A small

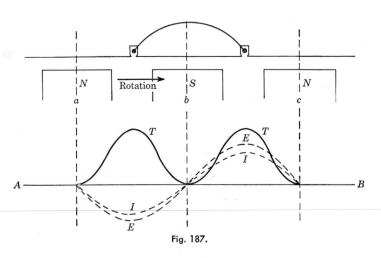

Fig. 187.

induction motor mounted directly on the shaft of the synchronous motor has been used for starting, but this method of starting is now seldom employed. When an induction motor is used for starting, it must have fewer poles—usually two less—than the synchronous motor in order that it may bring the synchronous motor up to synchronous speed.

Explanation of the Operation of a Synchronous Motor. In a generator with a unity power-factor load, the current passes through the armature winding in the direction of the voltage rise. At other power factors, the current leads or lags the voltage rise by the power-factor angle. In a motor operating at unity power factor, in order that it may receive power from the line to which it is connected, the armature current must pass through the winding in the direction of the voltage drop at all times. At other power factors the current leads or lags the voltage drop by the power-factor angle.

In Fig. 187 is shown a developed view of the armature and field of a synchronous machine, and for simplicity, only a single armature coil is shown. Let the direction of rotation be such that the fields are moving at uniform speed from left to right. Consider the voltage and current in the coil to be positive when their direction is out of the paper in the left-hand conductor and into the paper in the right-hand conductor. As the north pole at a moves from left to right, the voltage E generated in the coil is plotted on reference line AB, Fig. 187, against the position of this pole. Let the armature circuit be closed through a load of such constants that the current I in the coil is in phase with the generated voltage. While the north pole moves from a to b, the section of the armature within the coil is magnetized as a north pole. There is a force of repulsion between this section of the armature and the north pole which is approaching it and a force of attraction between it and the south pole which is moving away. As the north pole moves from a to b, there is a torque opposing the motion. The power developed at any instant is equal to the product of the instantaneous values of current and voltage, and since the speed is constant, the torque is also proportional to the product of current and voltage. While the north pole moves from b to c, the current and the induced voltage are both reversed in direction. Their product is still positive and the sign of the torque remains unchanged. The torque curve is marked T in the figure. The torque is intermittent but its average value is positive, and since it opposes the motion of the field poles, it corresponds to generator action, i.e., it represents torque which must be supplied by the prime mover which is driving the synchronous machine. The torque of a polyphase synchronous machine is the sum of the torques developed by all the phases and is constant if the currents and voltages are sine waves and the impressed voltages and currents are balanced.

If the load on a synchronous generator is such that the current is not in phase with the generated voltage, the torque curve has positive and negative loops. The average torque is proportional to the difference between the areas enclosed by these loops. It is positive for any angle of lag or lead less than 90 deg. A study of Fig. 187 shows this. It also shows that in a synchronous generator a lagging current produces a demagnetizing action on the poles and that a leading current produces a magnetizing action on the poles.

Suppose that, while the synchronous generator is running with current and voltage in phase, the current is reversed. The reversed currents could be obtained by connecting the armature terminals to a voltage source. This condition is represented in Fig. 188.

The current and voltage now differ in phase by 180 deg, and their product, which is proportional to the torque, is negative corresponding to motor action.

The current in the coil while the north pole passes from a to b is in a positive direction and the section of the armature included within the coil is magnetized as a south pole. There is a force of attraction between this section of the armature and the north field pole and a force of repulsion between it and the south field pole. These two forces assist the motion of the field and produce motor action.

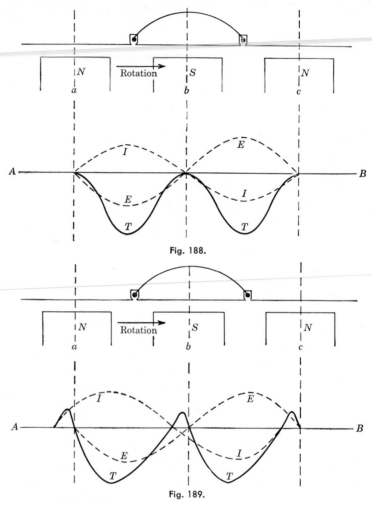

Fig. 188.

Fig. 189.

The conditions existing with a leading current are shown in Fig. 189. The torque has positive and negative loops. For angles of lead between zero and 90 deg, the negative loops are larger than the positive loops and there is a resultant motor torque. The conditions for lagging current are shown in Fig. 190.

The effect of armature reaction on a synchronous motor due to lagging or leading current is exactly opposite to that produced on a synchronous

generator.   The effect of armature reaction depends upon the phase
relation between the current and the generated voltage.   Therefore since
the current of a synchronous motor and the current of a synchronous
generator are nearly opposite in phase, the effect produced on the field by
a leading or lagging current in a synchronous motor is exactly opposite to
the effect produced in a synchronous generator.   A leading current in a
motor demagnetizes and a lagging current magnetizes the field.   This
can be seen by referring to Figs. 189 and 190.   Consider the lagging current in Fig. 190.   When the north field pole is opposite the armature

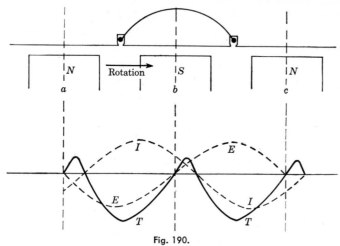

Fig. 190.

coil, the coil has a positive current.   This produces an upward mmf
within the coil.   The mmf of the coil therefore is in the same direction as
the mmf of the field excitation, *i.e.*, a lagging current in a synchronous
motor gives a magnetizing effect.   At constant output, the effect of a
change of field excitation is to alter the armature current and thus to
change the power factor.

A synchronous motor, unlike a d-c motor, can operate with a voltage
generated by the air-gap flux either greater or less than the impressed
voltage.   If it were possible to build a synchronous motor without reactance, it would not operate except with a generated voltage less than the
impressed voltage, and even so it would be unstable.

Chapter 21

Motor and generator action of a synchronous machine and the vector diagrams for the two conditions. Equations for motor and generator action. The effect of the armature reaction of a synchronous motor. Other vector diagrams of a synchronous motor. Change in normal excitation with change in load. Effect of change in load and field excitation. The double-reactance vector diagram of a synchronous motor. An example of the calculation of the field current and the efficiency of a salient-pole synchronous motor using the double-reactance method for finding the field current. The relation between the power developed by a synchronous motor and its power angle. The expression for the electromagnetic power of a cylindrical-rotor synchronous machine. The expression for the electromagnetic power of a salient-pole synchronous machine. Maximum power versus power angle. Effect of a change in the excitation of a synchronous motor on the power angle at which it operates.

Motor and Generator Action of a Synchronous Machine and the Vector Diagrams for the Two Conditions. For generator action, there must be a component of the armature current in phase with the generated voltage. For motor action, there must be a component of the armature current in phase opposition to the generated voltage. The vector diagrams for motor and generator action are similar. They differ in the relative positions of the current and voltage vectors.

Assume as an approximation that the vector diagram for a cylindrical-rotor synchronous generator can be applied to a salient-pole machine. Although the cylindrical-rotor theory and vector diagram when applied to a salient-pole machine do not give correct magnitudes of the changes occurring when a salient-pole machine is operated under different load conditions, the general trend of the changes is correct. As only qualitative results are considered in the following discussion, the cylindrical-rotor vector diagram is assumed to apply.

All the vectors in an a-c circuit are revolving vectors and revolve counterclockwise at speeds which correspond to their frequencies. Since in the following discussion terminal voltage and frequency are assumed constant, the vector which represents the terminal voltage revolves at constant speed. All the other vectors revolve at constant speed under steady operating conditions, but when a change of any sort occurs in the system, the vectors may change their speed long enough to alter their relative phase positions.

Figure 191 is the vector diagram of a cylindrical-rotor synchronous generator. Both V and E_a are voltage rises. Let the generator be connected to a system with constant terminal voltage and frequency. Nothing can be done to the generator that changes the magnitude or the frequency of its terminal voltage. All the vectors in Fig. 191 and in the vector diagrams which follow revolve in a counterclockwise direction. The vector V, which represents the constant terminal voltage rise,

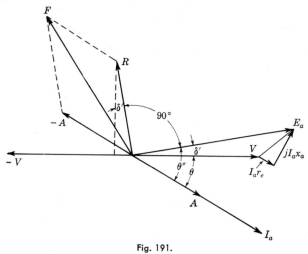

Fig. 191.

revolves at constant speed, but the other vectors can momentarily change their speed long enough to change their relative phase relations to accommodate themselves to new conditions in the system. The letters on the vector diagrams have the same significance as when used previously.

V = terminal voltage per phase
I_a = phase current
E_a = voltage induced per phase by the air-gap flux
r_e = armature effective resistance per phase
x_a = armature leakage reactance per phase
n = number of phases

The input to the generator is $nE_aI_a \cos \theta''$ plus the rotational losses, friction and windage loss, and core loss. $nE_aI_a \cos \theta''$ is the electromagnetic or internal power developed by the machine. If the amount of mechanical power to drive the generator is decreased, its output plus its losses is momentarily greater than its mechanical input and there is no longer equilibrium between power input and power output plus losses. The generator therefore momentarily slows down. The frequency of the vector E_a is momentarily decreased and the vector swings toward the vector V. This decreases the angle δ' and decreases the electrical output (see page 321). E_a continues to swing toward V until equilibrium is again established between mechanical input and electrical output plus losses. Let the driving power be decreased until the terminal output is zero. The vector diagram for this condition is given in Fig. 192. The

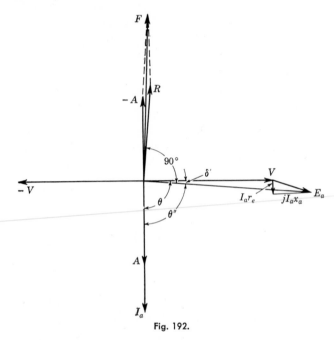

Fig. 192.

current, given by the vector relation $I_a = (E_a - V)/(r_e + jx_a)$, has changed both in magnitude and phase, and since the terminal output is now zero, the angle θ between the phase current I_a and the phase terminal voltage V must be 90 deg.

The differences between the vector diagrams in Figs. 191 and 192 are due solely to the change in the angle δ'. The machine is now running neither as generator nor as motor. It is merely floating on the line, neither taking electrical power from the line nor delivering electrical power to the line. The mechanical input is just sufficient to supply the friction and windage loss, the core loss, and the armature copper loss.

Now let the power which supplies the losses in Fig. 192 be removed and at the same time let a motor load be applied to the shaft. Since the power the machine was receiving before the motor load was applied was just sufficient to take care of the losses in the machine, the power for the motor load and the losses must come momentarily from the kinetic energy of the revolving parts of the machine. The machine therefore momentarily slows down sufficiently for the vector E_a to swing toward lag until equilibrium has again been established between the electrical input and the mechanical output plus the losses. The angle δ' has again changed, has become strongly negative, and motor power is developed. The new condition is represented in Fig. 193. Again the only change in the diagram is due to the change in the angle δ'.

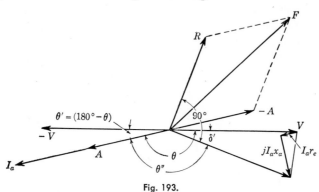

Fig. 193.

Under the new condition, the current I_a has a component in phase with the voltage drop $-V$ across the terminals of the machine. The power input $nVI_a \cos \theta$ is negative as is also the electromagnetic power $nE_aI_a \cos \theta''$, and the machine is operating as a synchronous motor and is delivering mechanical power. The power factor of the motor is the cosine of the phase angle between its phase current and phase voltage. Since this angle is greater than 90 deg for motor action, its cosine is negative. To avoid a negative power factor for the motor, it is customary to take the power-factor angle for motor action as the angle between the phase current and the phase terminal voltage drop $-V$ instead of between the phase current and the phase terminal voltage rise V, which is done in the synchronous machine operating as a generator (see Fig. 193).

It must be clearly understood that a synchronous machine operating with constant impressed voltage and frequency can operate at constant speed only under constant load conditions. This speed is fixed by the frequency f and the number of poles on the machine and is given in revolutions per minute by

$$\text{Speed} = \frac{2f}{p} 60 \tag{177}$$

The change in speed that occurs when the operating conditions are changed is small, not over a small percentage of synchronous speed, and is maintained only long enough for the angle δ' to take up its new value corresponding to the changed operating conditions. When steady conditions have been restored, the speed is the same as before the change in the operating conditions occurred. Unless the average speed of a synchronous motor is constant, the average torque developed is zero and the motor comes to rest.

The vector diagram in Fig. 193 for motor action shows the conditions with a leading current, the current leading the voltage drop across the motor terminals. Lead or lag of the current taken by a synchronous motor is always considered with respect to the voltage drop across its terminals instead of with respect to the voltage rise across the terminals as in the case of a synchronous generator.

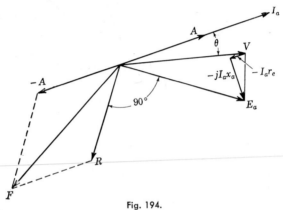

Fig. 194.

The vector diagram for synchronous motor action is frequently drawn with the vector representing the voltage drop across its terminals as positive. When this is done the vector diagram looks very different from the one in Fig. 193. Less confusion and less chance of error are likely to occur if the vector diagrams for the synchronous motor and the synchronous generator are both drawn using the same convention, taking the terminal voltage rise and the air-gap voltage rise as positive. When this convention is used, the resistance and reactance drops are added to the terminal voltage rise to get the air-gap voltage rise, as in Figs. 191, 192, 193. When the voltage drop across the terminals is taken positive, as in Fig. 194, the resistance and reactance drops must be subtracted from the voltage drop across the terminals to get the air-gap voltage drop. Voltage rise across the terminals of a synchronous machine is considered positive unless otherwise stated.

Equations for Motor and Generator Action. In this chapter, the rotating operator $(\cos \beta + j \sin \beta)$ is used to produce both positive (counterclockwise) and negative (clockwise) rotation in order that the equations developed may be used either for motor or for generator action by merely changing the sign of an angle. Whether the operator produces positive or negative rotation depends on the sign of the angle in the operator. Refer to Fig. 191 and take the terminal voltage rise as the axis of reference.

$$
\begin{aligned}
V &= |V|(1 + j0) \\
I_a &= |I_a|(\cos \theta + j \sin \theta) \\
E_a &= V + I_a z \\
&= |V|(1 + j0) + |I_a|(\cos \theta + j \sin \theta)(r_e + jx_a) \qquad (178) \\
&= C + jD
\end{aligned}
$$

where C and D are the real and the j parts of Eq. (178). The angle θ is measured with respect to the terminal voltage rise V and can be either positive or negative according as the current leads or lags the voltage V. The current in Fig. 191 is shown lagging the voltage V. For this condition θ is negative. The algebraic sign of θ must be considered when substituting the values of $\sin \theta$ and $\cos \theta$ in Eq. (178).

$$
\begin{aligned}
\sin \delta' &= \frac{D}{\sqrt{(C)^2 + (D)^2}} \\
\cos \delta' &= \frac{C}{\sqrt{(C)^2 + (D)^2}} \\
R &= |R|(- \sin \delta' + j \cos \delta') \\
A &= |A|(\cos \theta + j \sin \theta) \\
F &= R + (-A) \\
&= |R|(- \sin \delta' + j \cos \delta') - |A|(\cos \theta + j \sin \theta) \qquad (179)
\end{aligned}
$$

The angle δ' can be either positive or negative. The signs of $\sin \delta'$ and $\cos \delta'$ can be determined from the sign of δ' or better from the signs of C and D as determined from Eq. (178).

Equations (178) and (179) apply to either motor or generator action. If the angle θ is less than 90 deg generator action results, and if the angle θ is greater than 90 deg motor action results. The algebraic signs of the sine and cosine depend on the magnitude and sign of the angle θ.

The Effect of the Armature Reaction of a Synchronous Motor. Figures 195 and 196 are the general vector diagrams of a synchronous motor for lagging and leading currents. With a lagging current, the armature reaction has a component which assists the impressed field. It therefore has a magnetizing effect tending to increase the field strength. With a leading current, the armature reaction has a component which opposes the impressed field. It therefore has a demagnetizing effect tending to decrease the field strength. In both cases, the armature reaction has also a component which is in quadrature with the impressed field. This component distorts the field. The effect of the component of the armature reaction of a motor which lies along the field axis, and therefore along the vector which represents the impressed field, is opposite to the effect of the corresponding component in a generator. For a generator, the armature reaction caused by a lagging current is demagnetizing in its effect on the field and is magnetizing in its effect when the current leads. The cause of the difference between the effects of the armature reaction of a synchronous generator and that of a synchronous motor is the difference between the relative positions of current and voltage for motor and generator action. Lead and lag of the current of a synchronous motor are considered with respect to the voltage drop across the motor terminals.

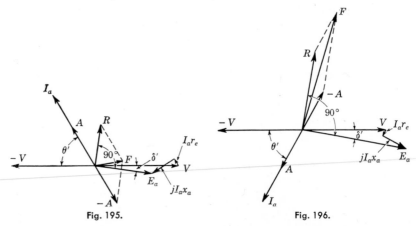

Fig. 195.   Fig. 196.

For a generator, lead and lag of the current are considered with respect to the voltage rise across the generator terminals. For motor action, the current must have a phase angle with respect to the voltage rise across the motor terminals which is greater than 90 deg. For generator action, this angle must be less than 90 deg.

Other Vector Diagrams of a Synchronous Motor. Any of the vector diagrams which are applicable to a synchronous generator can be used for a synchronous motor. They are subject to the same limitations as when used for a generator. Because of its simplicity, the synchronous impedance vector diagram is convenient when explaining the performance of a synchronous motor under different conditions of load and excitation and can be applied except when great accuracy is required in the results obtained from the diagram. The saturated value of the synchronous reactance to be used in the vector diagram can be obtained by the same method as for a synchronous generator from an open-circuit saturation curve and a zero-power-factor curve, both obtained by operating the synchronous motor as a generator (see page 249). The synchronous reactance obtained in this way is the direct-axis synchronous reactance. Because synchronous motors, with the exception of a few high-speed motors for special purposes, have salient poles, the synchronous-impedance diagram can be applied to the synchronous motor as an approximation only. However when the proper value of the synchronous reactance is used, the magnitudes of the excitation voltage and the field current obtained from the diagram are not greatly in error. Although the magnitude of the power angle obtained from the diagram is not correct, the sign of the change in this angle produced by any change in the load conditions is correct.

Figures 197 and 198 are the synchronous-impedance vector diagrams for a synchronous motor, with voltage rise across the terminals taken

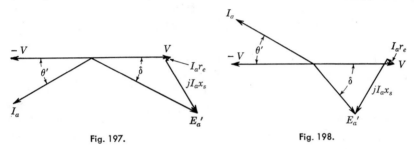

Fig. 197.   Fig. 198.

positive. Figures 197 and 198 are for a leading current and for a lagging current. In both cases the input $VI_a \cos \theta'$, the current I_a, and the terminal voltage V are equal. The power-factor angle θ', which is taken as usual with respect to the voltage drop $-V$ across the motor terminals, has the same magnitude but opposite signs for leading and lagging currents. The synchronous reactance is assumed to be equal in the two cases. In reality the synchronous reactance with the leading current should be somewhat smaller than the synchronous reactance with the lagging current, as should be understood by referring to Figs. 195 and 196, the general vector diagrams for a motor. The air-gap voltage can be considered a measure of the saturation at which the motor operates, as in the generator. With fixed terminal voltage V, the air-gap voltage E_a is greater with leading current (Fig. 196) than with lagging current (Fig. 195). The saturation of the magnetic circuit is therefore greater with leading current than with lagging current, and since that part of the synchronous reactance which replaces the effect of armature reaction decreases with an increase in the saturation of the magnetic circuit, the synchronous reactance for lagging current is the greater.

Change in Normal Excitation with Change in Load. A synchronous motor, unlike a d-c motor, can operate with an internal voltage which is greater or less than its terminal voltage (see Figs. 195 and 196, page 310, and Figs. 197 and 198, page 311). For each particular load, terminal voltage, and frequency, there is one field excitation that makes the phase angle θ' equal to zero and thus causes the motor to operate at unity power factor. This, as is stated in Chap. 20, is called the normal excitation. Any excitation greater than normal excitation is overexcitation and causes the motor to take a leading current, and any excitation less than normal excitation is underexcitation and causes the motor to take a lagging current. With any given impressed voltage and frequency, the field current for normal excitation changes with the load. Under the heading Power Factor, page 298, the statement is made that the field current which produces normal excitation increases with the load except for

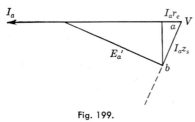

Fig. 199.

small loads. The reason it should decrease with small loads can be seen most readily from Fig. 199, which is the synchronous-reactance vector diagram of a synchronous motor for unity power factor.

Assuming that the synchronous reactance and effective resistance remain constant, the vector $I_a z_s = I_a(r_e + jx_s)$ on the diagram makes a constant angle with the vector V which represents the rise in voltage across the motor terminals. At light loads, the excitation voltage E_a' and the terminal voltage V nearly coincide. As the load is increased, if the power factor remains unity, the terminal end of the vector E_a' travels outward along the line ab and decreases in length until it reaches the position where it is perpendicular to ab. Beyond this position it increases in length. Since E_a' is the excitation voltage, the voltage which the impressed field would produce on open circuit, the impressed field varies in a similar manner. The bottom part of a compounding curve of a synchronous motor is therefore inclined slightly toward low excitation. The point of excitation at which it begins to incline toward higher excitation is where the vector E_a' becomes perpendicular to ab or to $I_a z_s$. This excitation depends upon the ratio of r_e to x_s. It is usually well down on the compounding curve, and with the usual ratios of effective resistance to synchronous reactance found in synchronous motors it is too far down to show at all. If the motor had no reactance, the field excitation for unity power factor would decrease continuously with increasing load. On the other hand, if the motor had no resistance, the field excitation would increase continuously with increasing load.

The explanation just given is not strictly correct since the synchronous-impedance diagram is assumed to apply and since it is also assumed that the synchronous reactance is constant. The latter assumption is nearly correct because with unity power factor the excitation voltage E_a' does not change much with load. It has been shown that the magnitude of the excitation voltage obtained from the synchronous-impedance diagram for a salient-pole machine is not greatly in error.

Effect of Change in Load and Field Excitation. The current taken by a d-c motor is equal to $(V - E_a)/r_a = I_a$. If load is applied, the motor slows down and decreases E_a. It continues to decrease its speed until the current has increased sufficiently to carry the load. The theoretical limit of load can be shown to be reached when $E_a = I_a r_a = \frac{1}{2}V$. Above this limit the decrease in E_a more than balances the increase in I_a. If the field excitation is increased, E_a increases, decreasing $(V - E_a)$ and consequently the current. The power developed by the motor is now too small to carry the load and it starts to slow down. It continues to slow down until the effect on E_a of the decrease in speed balances the effect of the increase in the excitation. The current then has increased to nearly its original value. A d-c motor adjusts itself to change in load or in excitation by changing its speed.

A synchronous motor must run at synchronous speed. It cannot change its average speed to accommodate itself to a change in load or a change in excitation. The current taken by a synchronous motor is

$$I_a = \frac{I_a z_s}{z_s} = \frac{E'_a - V}{z_s}$$

(see Figs. 197 and 198, page 311).

For any given excitation, E'_a is fixed in magnitude, but its phase relation with respect to V can change and thus alter the current. As has been stated, a synchronous motor adjusts itself to change in load by changing the phase of its generated voltage with respect to the voltage impressed across its terminals. Its average speed does not change, but its instantaneous speed changes long enough to permit the required change in phase to take place. If the load is applied, the motor begins to slow down and continues to slow down until sufficient change in phase has been produced. If the motor is not properly damped, it may overrun and develop too much power. It then speeds up and may again overrun. It is now developing too little power and the action is repeated. This is called hunting. Hunting is discussed later in some detail. If the field excitation is changed, the excitation voltage E'_a is also changed. The power developed by the motor consequently must change. Refer to Fig. 200. The heavy lines in this figure form the synchronous-impedance vector diagram of a synchronous motor under steady operating conditions with a leading current. Let the field excitation be increased, increasing E'_a to (E'_a) as indicated by the light lines on the figure. There is nothing to cause E'_a to shift its phase with respect to V except an unbalanced condition between the electrical input and the mechanical output plus

E_a' is the excitation voltage at load saturation as fixed by the air-gap voltage for the given load conditions. The actual field current for the given load is the field current corresponding to a voltage E_a' on the load saturation line, *i.e.*, on a straight line drawn through the zero point of

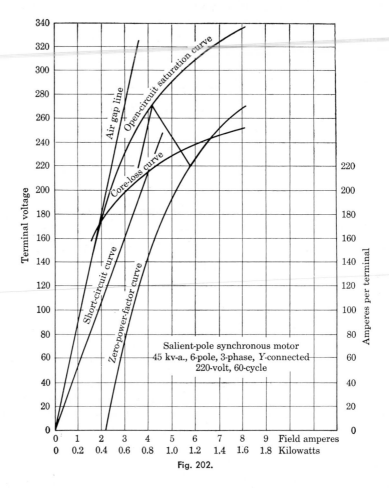

Fig. 202.

the open-circuit saturation curve and the point on this curve which corresponds to the air-gap voltage for the given load.

An Example of the Calculation of the Field Current and the Efficiency of a Salient-pole Synchronous Motor Using the Double-reactance Method for Finding the Field Current. The characteristic curves of a 45-kva six-pole three-phase Y-connected 220-volt 60-cycle salient-pole synchronous motor are given in Fig. 202. The zero-power-factor curve in the figure is for 118 amp per terminal. Other data for the motor are:

Field resistance = 30 ohms at 75°C
Friction and windage loss = 810 watts at 60 cycles
Stray-load loss = 540 watts at 60 cycles and rated armature current
Direct-current armature resistance = 0.050 ohm per phase at 75°C
Unsaturated quadrature synchronous reactance = 0.53 ohm per phase

The quadrature synchronous reactance was obtained by driving the synchronous motor by a d-c motor at a speed which differed slightly from synchronous speed and with sufficient three-phase balanced voltages impressed on the armature terminals to give about full-load current. The minimum ratio of the voltage per phase to the armature current per phase, as determined from oscillograph records of voltage and current, is the unsaturated quadrature synchronous reactance. If the difference between the actual speed and synchronous speed is small enough, the readings of indicating instruments can be used in place of oscillograph records.

$$\text{Rated current} = I_a = \frac{45,000}{\sqrt{3} \times 220} = 118 \text{ amp per phase}$$

$$r_e = \text{effective resistance} = 0.050 + \frac{540}{3(118)^2} = 0.063 \text{ ohm per phase at}$$
$$75°C \text{ and rated current}$$

As an approximation, use the Potier reactance in place of the leakage reactance for finding the air-gap voltage.

$$\text{Potier reactance} = x_p, \text{ assumed equal to } x_a,$$
$$= \frac{271 - 220}{\sqrt{3} \times 118} = 0.250 \text{ ohm per phase}$$

Assume that the motor is operating at rated kva load, with rated voltage and frequency impressed on its terminals, with leading current and power factor of 0.8.

$$E_a = \text{air-gap voltage} = V + I_a(r_e + jx_a)$$
$$= \frac{220}{\sqrt{3}} (1 + j0) + 118(-0.8 - j0.6)(0.063 + j0.250)$$
$$= (127.1 + j0) + (11.8 - j28.1)$$
$$= 138.9 - j28.1$$
$$|E_a| = \sqrt{(138.9)^2 + (28.1)^2} = 141.7 \text{ volts per phase}$$
$$\sqrt{3}\,|E_a| = \sqrt{3} \times 141.7 = 245.5 \text{ volts between terminals}$$

From the open-circuit saturation curve, the field current corresponding to a voltage of 245.5 volts between terminals is $I_f = 3.40$ amp. From the air-gap line, the voltage corresponding to $I_f = 3.40$ amp is 305 volts between terminals.

$$\text{Saturation factor, } k = \frac{305}{246} = 1.24$$

The unsaturated direct-axis synchronous reactance from the short-circuit curve and the air-gap line is

$$x_d' \text{ (unsaturated)} = \frac{198.0}{\sqrt{3} \times 118.0} = 0.969 \text{ ohm per phase}$$

$$x_d \text{ (saturated)} = (x_d' - x_a)\frac{1}{k} + x_a$$

$$= (0.969 - 0.25)\frac{1}{1.24} + 0.25$$

$$= 0.83 \text{ ohm per phase}$$

Lacking better information, assume as an approximation that the saturation factor for the quadrature-axis synchronous reactance is the same as the saturation factor for the direct-axis synchronous reactance.

$$x_q \text{ (saturated)} = (0.53 - 0.25)\frac{1}{1.24} + 0.25$$

$$= 0.476 \text{ ohm per phase}$$

Refer to Eq. (180), page 315.

$$\tan \delta = \frac{-I_a(r_e \sin \theta' + x_q \cos \theta')}{V + I_a(x_q \sin \theta' - r_e \cos \theta')}$$

$$= \frac{-118(0.063 \times 0.6 + 0.476 \times 0.8)}{127.1 + 118(0.476 \times 0.6 - 0.063 \times 0.8)}$$

$$= \frac{-118 \times 0.4186}{127.1 + 118 \times 0.2360} = -0.3191$$

$$\delta = -17.7 \text{ deg}$$

Refer to Fig. 201. The angle between I_a and I_q is $(\theta' + \delta)$.

$$\theta' = \tan^{-1} 0.8 = 36.87 \text{ deg}$$
$$(\theta' + \delta) = 36.87 + 17.7 = 54.57 \text{ deg}$$
$$\sin \delta = 0.3040 \qquad \cos \delta = 0.9527$$
$$\sin (\theta' + \delta) = 0.8148 \qquad \cos (\theta' + \delta) = 0.5797$$

Refer to Eq. (181), page 315.

$$E_a'(1 + j0) = 127.1(0.9527 + j0.3040)$$
$$+ 118.0(-0.5797 - j0.8148)0.063$$
$$- 118.0(0.5797)(j0.476) - j118.0(0.8148)(j0.830)$$
$$= (121.0 + j38.6) + (-4.31 - j6.06) - (j32.5) + (79.8)$$
$$= 196.5 + j0$$
$$|E_a'| = 196.5 \text{ volts per phase}$$
$$\sqrt{3}\, E_a' = \sqrt{3} \times 196.5 = 340.3 \text{ volts between terminals}$$

The voltage 340.3 is the excitation voltage at load saturation. The required field current for the assumed load is the field current found from the saturation line corresponding to 340.3 volts. By extrapolation this field current is

$$I_f = \frac{340.3}{245.5} \times 3.40 = 4.71 \text{ amp}$$

According to the Standardization Rules of the AIEE and also the ASA Standards for Rotating Electrical Machinery, the proper core loss to use when finding the efficiency is that corresponding to a voltage equal to the terminal voltage corrected for the armature resistance drop. This voltage is

$$V + I_a r_e = 127.1 + 118.0(-0.8 - j0.6)(0.063)$$
$$= 121.1 - j4.46 \text{ volts per phase}$$
$$|V + I_a r_e| = \sqrt{(121.1)^2 + (4.46)^2} = 121.2 \text{ volts per phase}$$
$$\sqrt{3}\, |V + I_a r_e| = \sqrt{3} \times 121.2 = 210.0 \text{ volts between terminals}$$

From the core-loss curve, the core loss corresponding to 210.0 volts is 740 watts.

Losses	Watts
Armature copper $3(118)^2 0.050$	2,089
Stray-load.	540
Core.	740
Field copper $(4.71)^2 30$	666
Friction and windage	810
Total.	4,845

$$\text{Efficiency} = \frac{45,000 \times 0.8 - 4,845}{45,000 \times 0.8}$$
$$= 1 - \frac{4,845}{36,000} = 0.866 \text{ or } 86.6 \text{ per cent}$$

The Relation between the Power Developed by a Synchronous Motor and Its Power Angle.

The relation between power angle and power is developed in Chap. 19 for a synchronous generator, but the equations found apply equally well to a synchronous motor. Whether motor or generator action is developed depends only on the sign of the power angle. When the terminal voltage rise is considered positive, as it is in the vector diagrams, the power angle is always negative for motor action and if the resistance drop is neglected it is always positive for generator action. If the resistance drop is not neglected, the power angle can be slightly negative and still have generator action if the power factor is low. This can be seen from a vector diagram of a synchronous generator drawn for a power-factor angle of nearly 90 deg and a lagging current. Under the usual operating conditions, the power angle is positive for generator action.

The power-angle-power relations developed in Chap. 19 are for terminal power. The relations for a cylindrical-rotor machine are given by Eqs. (172) and (173), page 283, and for a salient-pole machine by Eq. (174), page 284. The power relations for a cylindrical-rotor machine neglecting the resistance are plotted in Fig. 176, page 284, and for a salient-pole machine in Fig. 178, page 285. The power-angle relations can be developed equally well with respect to internal or electromagnetic power. The electromagnetic power differs from the terminal power by the armature resistance loss. In motor action, mechanical output at the shaft is equal to the armature terminal input minus the armature effective resistance loss and all of the rotational losses. These latter include the friction and windage loss and the core loss. The only difference between terminal electrical input and electromagnetic power is in the armature effective resistance loss.

$$\text{Terminal electrical power } P = VI_a \cos \theta_{I_a}^{V} \text{ per phase}$$
$$\text{Electromagnetic power } P' = E_a' I_a \cos \theta_{I_a}^{E_a'} \text{ per phase}$$

where the terminal voltage V, the excitation voltage E_a', and the current I_a are per phase. The expressions for P and P' differ by merely $I_a^2 r_e$ where r_e is the effective armature resistance per phase. When the resistance is neglected, the expressions for the terminal electrical power and the electromagnetic power of a synchronous machine are identical.

The Expression for the Electromagnetic Power of a Cylindrical-rotor Synchronous Machine. Refer to Figs. 197 and 198, page 311. In Figs. 197 and 198, which represent motor action, the angle δ is negative with respect to V, the reference axis, and its sign must be considered when substituting its value in the operator $(\cos \delta + j \sin \delta)$. For generator action the power angle δ is positive, for motor action the power angle δ is negative. Take the terminal voltage rise V as the reference axis. Then

$$V = |V|(1 + j0)$$
$$E_a' = |E_a'|(\cos \delta + j \sin \delta)$$
$$I_a = \frac{E_a' - V}{r_e + jx_s}$$
$$= \frac{|E_a'|(\cos \delta + j \sin \delta) - |V|(1 + j0)}{r_e + jx_s}$$
$$= \frac{(|E_a'|r_e \cos \delta + |E_a'|x_s \sin \delta - |V|r_e)}{r_e^2 + x_s^2}$$
$$+ j \frac{(|E_a'|r_e \sin \delta - |E_a'|x_s \cos \delta + |V|x_s)}{r_s^2 + x_s^2}$$

Power is equal to the product of the real parts of the current and voltage plus the product of the j parts of the current and voltage. Therefore the internal or electromagnetic power is

$$P' = |E'_a||I_a| \cos \theta^{E'_a}_{I_a} = \frac{|E'_a|^2 r_e \cos^2 \delta + |E'_a|^2 x_s \sin \delta \cos \delta - |E'_a||V|r_e \cos \delta}{r_e^2 + x_s^2}$$

$$+ \frac{|E'_a|^2 r_e \sin^2 \delta - |E'_a|^2 x_s \sin \delta \cos \delta + |V||E'_a|x_s \sin \delta}{r_e^2 + x_s^2}$$

$$= \frac{|E'_a|^2 r_e - |E'_a||V|r_e \cos \delta + |E'_a||V|x_s \sin \delta}{r_e^2 + x_s^2} \quad (182)$$

Equation (182) can be found also by adding the armature resistance loss to the expression for terminal power given by Eq. (172), page 283.

If the resistance r_e is negligible, Eq. (182) reduces to

$$P' = \frac{|E'_a||V|}{x_s} \sin \delta \quad (183)$$

When the power angle δ is positive, P' is positive, there is generator action, and electrical power is delivered. When the power angle δ is negative, P' is negative, there is motor action, and mechanical power is delivered.

The Expression for the Electromagnetic Power of a Salient-pole Synchronous Machine. The electromagnetic power in a synchronous motor is the electrical power input minus the armature copper loss.

$$P' = |V||I_a| \cos \theta' - |I_a^2|r_e = |V||I_q| \cos \delta + |V||I_d| \sin \delta$$
$$- (|I_q^2| + |I_d^2|)r_e \quad (184)$$

Refer to Fig. 201, page 315.

$$|I_d| = |I_a| \sin (\delta + \theta') = \frac{|E'_a| - |V| \cos \delta - |I_a r_e| \cos (\theta' + \delta)}{x_d}$$

$$|I_q| = |I_a| \cos (\delta + \theta') = \frac{|V| \sin \delta - |I_a r_e| \sin (\theta' + \delta)}{x_q} \quad (185)$$

The substitution of these values of I_d and I_q in Eq. (184) gives an expression not suited for ordinary use. If the armature resistance loss is neglected, the expression is much simplified and reduces to that for the electrical power input. Dropping the terms involving resistance in the expression for the internal power introduces only small error over the ordinary working range of a synchronous machine of usual design. If the armature resistance loss is neglected, the expression for the internal power is the same as that given in Eq. (174), page 284, for the terminal power of a synchronous generator.

$$P = \frac{VE'_a}{x_d} \sin \delta + \frac{V^2(x_d - x_q)}{2x_d x_q} \sin 2\delta \quad (186)$$

For motor action the power angle δ is negative with respect to the terminal voltage rise V. The angle δ is positive for generator action. Power is negative for motor action and positive for generator action. With fixed terminal voltage, fixed excitation voltage, and fixed frequency, any retardation in phase of the excitation voltage E_a' with respect to the terminal voltage V causes a negative change in the power angle and therefore a negative change in the power P. It causes an increase in the motor action if the machine is operating as a motor or a decrease in the generator action if the machine is operating as a generator. Any advance in phase of the excitation voltage E_a' with respect to V causes a positive change in the power angle and therefore a positive change in the power P. It causes an increase in the generator action if the machine is operating as a generator or a decrease in the motor action if the machine is operating as a motor. The change in phase of the excitation voltage caused by a change in operating conditions can be observed readily by means of a stroboscope which illuminates the poles of the machine with a series of flashes of light occurring $2f/p$ times a second, where f is the frequency of the applied voltage and p is the number of poles. If the machine is operating under steady conditions, the field structure as seen by the flashes of light appears to be stationary. If there is a change in the power angle, the field structure can be seen to swing and take up its new position.

Maximum Power versus Power Angle. The maximum power developed by a cylindrical-rotor synchronous motor operating with fixed impressed voltage and fixed frequency occurs when the sine of the power angle has its maximum value. This statement assumes that the resistance loss in the armature is negligible. The power angle must be negative for motor action [see Eq. (183), page 321]. Since the maximum value of the sine of an angle is unity, maximum power is developed by a cylindrical-rotor synchronous motor when the power angle is 90 deg. In practice a synchronous motor cannot operate with as great a power angle as this, since any further increase in the load applied to the motor causes its power angle to increase, and the motor then develops less power to carry the increase in load. This produces a further increase in the power angle and causes the motor to develop still less power to carry the greater load. The effect is cumulative and results in the breakdown of the motor, which comes to rest. A curve of power versus power angle is plotted for a cylindrical-rotor synchronous machine, Fig. 176, page 284. The negative loops of the figure correspond to motor action.

The maximum power developed by a salient-pole synchronous motor operating with constant impressed voltage and constant frequency occurs at a power angle somewhat less than 90 deg because of the effect of the second term on the right-hand side in the equation for motor power [see

Eq. (186), page 321]. The power versus the power angle for a salient-pole synchronous machine is plotted in Fig. 178, page 285. The negative loops of this figure correspond to motor action. The figure neglects armature resistance loss. The effect of an increase in load beyond that which corresponds to maximum power is the same as for a cylindrical-rotor machine.

Effect of a Change in the Excitation of a Synchronous Motor on the Power Angle at Which It Operates. Since the power demanded by the load is a function of the motor speed, the power taken by a given load must remain constant as long as the speed is constant. Since a synchronous motor operating with fixed frequency must operate at constant speed up to its point of maximum power, its mechanical output cannot be influenced by a change in its excitation. With any fixed load, constant impressed voltage, and constant frequency, any change in the excitation of a synchronous motor must cause a change in its power angle to compensate for the change produced in its excitation voltage [see Eqs. (182), (183), (185)]. An increase in the excitation must cause the excitation voltage to advance its phase with respect to the terminal voltage V. This also causes the current to swing toward lead. A decrease in the excitation causes the current to swing toward lag. There is one particular excitation which makes a synchronous motor, operating with constant impressed voltage, constant frequency, and constant load, take a current which is in phase with its impressed voltage. Any greater excitation makes the motor take a leading current, a current that leads the voltage drop across the motor terminals. Any less excitation makes the motor take a lagging current.

Chapter 22

Limitations of the Relations Developed in This Chapter. The relations developed in this chapter are true for only a cylindrical-rotor synchronous motor since they are based on the simple synchronous-reactance vector diagram making use of a single synchronous reactance. Any attempt to develop similar relations for a salient-pole synchronous motor making use of direct-axis and quadrature-axis synchronous reactances leads to long and complicated equations not suited to practical use. The following relations assume that the synchronous reactance and the effective resistance are constant. The assumption that the synchronous reactance is constant is not strictly true even when the terminal voltage is constant, since the magnitude of the synchronous reactance of a synchronous machine depends upon the saturation of its magnetic circuit. The saturation is fixed for all practical purposes by the air-gap voltage, which varies slightly with current and power factor even if the terminal voltage is constant. The influence of the effective resistance on the results is relatively small. Assuming that the effective resistance is constant, even if it varies somewhat with the current, introduces little error. Although the relations given are not strictly correct under operating conditions, they are nevertheless of value.

In all equations in this chapter, E'_a, V, and z_s are used to represent the magnitudes of these quantities rather than their complex expression.

Maximum Motor Power with Fixed E'_a, V, r_e, and x_s. Equation (182), page 321, gives the electromagnetic power developed in a non-salient pole motor. It should be remembered that for a motor the angle δ is negative and that motor power in this equation is negative power. For fixed values of field current and terminal voltage, the maximum power will occur when $dP'/d\delta = 0$. Thus

$$\frac{dP'}{d\delta} = \frac{E'_a V}{z_s{}^2} \left(r_e \sin \delta - x_s \cos \delta \right) = 0$$

Therefore

$$\tan \delta = \frac{\sin \delta}{\cos \delta} = -\frac{x_s}{r_e} \tag{187}$$

This is the angle δ at which maximum internal power is developed. It is somewhat less than minus 90 deg and would equal minus 90 deg if r_e were negligible.

If $\tan \delta = -x_s/r_e$, then $\sin \delta = -x_s/z_s$ and $\cos \delta = r_e/z_s$. Substituting these values in Eq. (182) gives the maximum electromagnetic power.

$$\begin{aligned} \text{Maximum } P' &= \frac{E_a'^2 r_e}{z_s^2} - \frac{E_a' V}{z_s^2} \left(r_e \frac{r_e}{z_s} + x_s \frac{x_s}{z_s} \right) \\ &= \frac{E_a'^2 r_e}{z_s^2} - \frac{E_a' V}{z_s} \end{aligned} \tag{188}$$

A synchronous motor cannot operate with an excitation voltage greater than the impressed voltage unless the motor has reactance. This follows from Eq. (188). If the reactance is zero,

$$\text{Maximum } P' = \frac{E_a'^2}{r_e} - \frac{V E_a'}{r_e}$$

and for any value of E_a' greater than V, P' is positive and represents generator action.

Maximum and Minimum Motor Excitation for Fixed Motor Power and Fixed Impressed Voltage. The synchronous reactance and effective resistance are assumed constant. The limiting values of the excitation voltage, E_a', for any given load as measured by the electromagnetic power developed in the motor are those values of E_a' which make the maximum internal power P' as given in Eq. (188) equal to that required because of the load and losses. Therefore, if P' is this required power and E_a' represents the limiting value of excitation voltage at this load condition, Eq. (188) becomes

$$P' = \frac{E_a'^2 r_e}{z_s^2} - \frac{E_a V}{z_s}$$

and solving for E_a'

$$\begin{aligned} E_a' &= \frac{V z_s}{2 r_e} \pm \frac{z_s}{2 r_e} \sqrt{V^2 + 4 P' r_e} \\ &= \frac{z_s}{2 r_e} (V \pm \sqrt{V^2 + 4 P' r_e}) \end{aligned} \tag{189}$$

This equation gives the maximum and minimum values of E_a' with which the motor will run, for if the excitation is made greater than the larger of these two values or less than the smaller, the maximum power the motor can develop is less than that required.

Maximum Possible Motor Excitation with Fixed Impressed Voltage.
Since P' in Eq. (189) must be negative for motor action, the equation indicates that the range of excitation voltages at which the motor will run, *i.e.*, the difference between the maximum and minimum values, becomes greater as the internal power required becomes less. If P' is reduced to zero, the extreme values of E'_a are then given. The minimum possible value is then indicated to be equal to zero and the maximum possible value is

$$E'_a = \frac{z_s}{2r_e} 2V = \frac{Vz_s}{r_e} \tag{190}$$

This same result could have been obtained directly from Eq. (188) by setting the maximum P' equal to zero and solving for E'_a.

Maximum Possible Motor Activity with Fixed Impressed Voltage.
The maximum electromagnetic power with fixed excitation voltage is given by Eq. (188). The equation shows that this maximum power, however, varies with the excitation. To find the condition when the maximum power has its greatest value, proceed as follows:

$$\frac{d \text{ (maximum } P')}{dE'_a} = 0 = \frac{d}{dE'_a}\left(\frac{E'^2_a r_e}{z_s^2} - \frac{E'_a V}{z_s}\right)$$

$$\frac{2E'_a r_e}{z_s^2} - \frac{V}{z_s} = 0$$

$$E'_a = \frac{Vz_s}{2r_e} \tag{191}$$

This is the excitation voltage at which the motor develops its greatest electromagnetic power. It is one-half of its maximum possible value. Since z_s is usually several times r_e, this gives a large value of E'_a. This value is seldom obtained in practice because of saturation and also because the maximum field current is limited by the resistance of the field winding. This result does indicate however that, within practical limits, increasing the field current increases the load a motor will carry without "falling out of step" or stalling.

Substituting the value of E'_a from Eq. (191) in Eq. (188) gives

$$\text{Maximum possible } P' = -\frac{V^2}{4r_e} \tag{192}$$

This is the greatest possible motor power.

Analysis of Salient-pole Machines. Equation (186), page 321, gives the electromagnetic power of a salient-pole machine if the effective resistance is neglected. This equation does not lend itself to the strict analysis applied to the non-salient-pole machine. However certain comparisons are useful. With nonsalient poles, x_d and x_q are equal and the second term of Eq. (186) disappears, leaving only the first term which is the same as Eq. (183), page 321, for a non-salient-pole machine. The second term $[V^2(x_d - x_q)/2x_d x_q]$ sin 2δ may therefore be considered as added power due to the saliency of the poles. It shows that the salient-pole motor develops power even when the excitation voltage E'_a is reduced to zero. Moreover, because of this additional power in a salient-pole machine, the angle δ at the condition of maximum power is much smaller than with non-salient-poles as is shown clearly in Fig. 178, page 285.

Chapter 23

Hunting. Damping. Methods of starting synchronous motors. The super-synchronous motor.

Hunting. Synchronous machines in which a change in load is accompanied by a change in phase are subject to hunting. Consider a synchronous motor operating under constant excitation and constant load. Under these conditions there is a definite phase angle between the impressed and excitation voltages. Suppose the load is increased. The motor is now developing less power than is demanded by the load, and as a result it slows down. It continues to slow down until the phase displacement between its impressed and excitation voltages corresponds to that required for the new load condition. This decreasing speed can last for several cycles. The phase displacement between the excitation and terminal voltages which corresponds to the maximum electromagnetic output of a cylindrical-rotor synchronous motor is approximately 90 deg. It is less than 90 deg for a salient-pole motor (see page 285). If the change in speed produces a phase displacement between terminal and excitation voltages greater than that at which the motor develops sufficient power to carry the load, the motor breaks down, *i.e.*, it falls out of step and comes to rest. The motor may swing beyond the phase displacement corresponding to maximum output, provided the motor power remains greater than that required by the load. While the motor is slowing down, the increase in load is supplied by the change in kinetic energy of the moving parts of the motor. At the instant the motor passes through the phase displacement corresponding to the new load, the electromagnetic power developed is equal to the load plus the rotational losses of the motor. Because of the inertia of the rotor it cannot instantly take up synchronous speed and it therefore overruns. It is now developing more power than is required for the load plus the rotational losses and it starts to speed up. If it again overruns in the reverse direction, it develops too little power and immediately starts to slow down. This action is called hunting. It is equivalent to an oscillation in speed superposed on a uniform speed of rotation. A slight amount of hunting always takes place when the load on a synchronous motor is changed, but with a properly designed motor operating under good conditions the amount of hunting should be small and not noticeable. In a poorly designed motor or a motor operating under bad conditions, hunting may become excessive. The stability of operation of a synchronous motor depends on the fact that, up to the maximum power the motor can develop, the power angle, between excitation voltage and terminal voltage, is a direct function of the load, according to Eqs. (183) and (186), page 321. If a motor overruns the power angle corresponding to the load, there is no longer equilibrium between power corresponding to load plus motor losses and the electromagnetic power developed by the motor. The difference between them is always in such a direction as to cause the motor to change its speed momentarily and thereby change its power angle in such a way as to restore equilibrium.

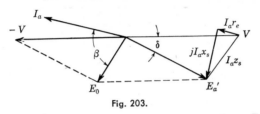

Fig. 203.

The effect of hunting is made clear by a vector diagram. For simplicity, the synchronous-impedance diagram of a cylindrical-rotor motor is used. The vector diagram of a salient-pole motor with direct and quadrature reactances shows the same thing as the synchronous-impedance diagram but is much more complicated, and when only relative changes of power are of interest, nothing is gained by its use. Figure 203 is the synchronous-impedance vector diagram of a synchronous motor to which has been added the drop in voltage $-V$ across the motor terminals.

The resultant voltage $E_o = I_a z_s$ which causes the current I_a in the circuit is equal to the vector sum of $-V$ and E'_a. The current I_a is equal to E_o divided by the synchronous impedance of the motor and lags by an angle $\beta = \tan^{-1}(x_s/r_e)$ behind the voltage E_o. This angle β would be constant if x_s and r_e were constant.

If hunting takes place, the end of the vector E'_a oscillates on the arc of a circle about its mean position, and at the same time E_o and I_a change in magnitude and in phase.

The effect of hunting is shown in Fig. 204. The full lines in this figure represent the stable condition and the dotted and dot-and-dash lines represent the two extreme displacements due to hunting. The positions of the vectors $-V$ and V do not change.

The vector E'_a oscillates from a to b. The resultant voltage E_o oscillates from c to d and at the same time changes its magnitude. The current I_a is proportional to E_o at every instant and swings through an angle equal to the angle through which E_o swings. The minimum power is developed at a when E_a' is ahead of its mean position and has its greatest displacement in the direction of lead. This minimum power is equal to the projection of the motor excitation voltage Oa on the current Og multiplied by that current. The maximum power is developed when E_a' has its extreme displacement in the direction of lag. This maximum power is equal to the product of the current Of and the projection of the motor excitation voltage Ob on that current. It is evident that there can be a large variation in the power developed if hunting occurs.

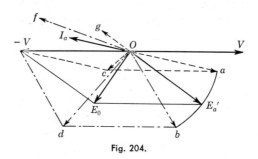

Fig. 204.

The rotating part of the motor acts like a torsional pendulum where the change in the couple producing rotation in the motor corresponds to the torsional couple in the fiber or supporting wire of the pendulum. In the motor, the change in the couple is caused by the displacement of the rotor from its mean position. The moment of inertia of the rotor corresponds to the moment of inertia of the mass of the pendulum.

$$t = 2\pi \sqrt{\frac{\Sigma md^2}{M}} \tag{193}$$

where t, Σmd^2, and M are the time of a complete vibration, the moment of inertia of the rotor, and the restoring couple per unit of angular displacement from the mean position.

From Eq. (186), page 321, the electromagnetic power developed by a salient-pole synchronous motor, neglecting the resistance loss, is

$$P = \frac{VE_a'}{x_d} \sin \delta + \frac{V^2(x_d - x_q)}{2x_d x_q} \sin 2\delta \tag{194}$$

If Eq. (194) is to apply to a polyphase motor, it must be multiplied by the number of phases n.

If p and f are the number of poles on the motor and the frequency, the electromagnetic torque developed by a polyphase motor in pound-feet is

$$T = n\frac{Pp}{4\pi f}\frac{550}{746}$$

$$= n\frac{p}{4\pi f}\frac{550}{746}\left(\frac{VE'_a}{x_d}\sin\delta + \frac{V^2(x_d - x_q)}{2x_d x_q}\sin 2\delta\right) \tag{195}$$

M in Eq. (193) is equal to the derivative of T with respect to δ' where δ' is the displacement in mechanical radians of E'_a from V.

The angle δ in Eq. (195) is in electrical radians.

Since $\delta = \dfrac{p}{2}\delta'$,

$$M = \frac{dt}{d\delta'} = n\frac{p^2}{8\pi f x_d}\frac{550}{746}\left(VE'_a\cos\frac{p}{2}\delta' + V^2\frac{x_d - x_q}{x_q}\cos p\delta'\right) \tag{196}$$

$\Sigma md^2 = \dfrac{wk^2}{g}$ where w = weight of all revolving parts in pounds, k = radius of gyration in feet, and g = acceleration in feet per second per second due to gravity = 32.16 ft per sec per sec.

Substituting these values in Eq. (193) gives for the period of hunting, in seconds, of a polyphase synchronous motor

$$t = 2\pi\sqrt{\frac{8\pi f x_d wk^2}{np^2 g\left(VE'_a\cos\dfrac{p}{2}\delta' + V^2\dfrac{x_d - x_q}{x_q}\cos p\delta'\right)}\frac{746}{550}}$$

$$= 6.45\sqrt{\frac{8\pi f x_d wk^2}{np^2\left(VE'_a\cos\dfrac{p}{2}\delta' + V^2\dfrac{x_d - x_q}{x_q}\cos p\delta'\right)}} \tag{197}$$

Equation (197) is approximate, as it neglects the effect of damping due to currents induced by hunting in the field winding, in the pole faces, and in the damper with which a synchronous motor is provided. Equation (197) also neglects the small effect of armature resistance. This equation is true therefore only when the damping torque produced by the eddy currents in the pole faces and by the currents in the damper, induced by hunting, is negligible when compared with the electromagnetic restoring torque caused by the change in the power angle produced by hunting.

If the period of hunting coincides or nearly coincides with any periodic variation in the load, in the frequency, or in the voltage of the circuit from which the motor is operating, the effect is cumulative, and serious hunting may result which may cause the motor to swing beyond the phase displacement at which it can carry the load, to drop out of synchronism and to come to rest.

From Eq. (197), the period of oscillation of a synchronous motor about its mean angular position depends upon the excitation voltage E_a' and the phase displacement δ of this voltage from the voltage impressed on the motor. Consequently the period depends upon the excitation of the motor and upon the load. Therefore if there is any periodic variation in the load, in the impressed voltage, in the frequency, or in the excitation, hunting may occur at some loads or at some excitations and not at others. Hunting is not likely to occur when well-designed synchronous motors are operated from large power systems with properly controlled frequency and voltage.

Damping. There are two ways by which hunting can be diminished. One way to increase the moment of inertia of the rotor is by adding a flywheel. This method is applicable to either single or polyphase motors. The other way consists in using a short-circuited low-resistance winding, placed in the pole faces, called an amortisseur or damping winding or simply a damper. When a damper is used in a single-phase motor, double-frequency currents are induced in it by the double-frequency flux variation produced in the poles by armature reaction (see page 209, under Synchronous Generators). This double-frequency current increases the copper loss in the damper and tends to damp out the flux variation which produces the current. Its existence is not dependent upon hunting.

Adding a flywheel may decrease the tendency to hunt by making the free period of oscillation of the motor longer than any period which is likely to start hunting. This is an effective way to diminish the tendency to hunt, but it does not accomplish the result by damping action. Except in special cases, flywheels are not used to decrease hunting, mainly on account of their weight. Dampers exert a real damping action and are universally used on polyphase synchronous motors. Besides effectively diminishing hunting, they greatly increase the starting torque of synchronous motors when started as induction motors.

A damper usually takes the form of copper grids placed in the pole faces and copper bridges between the poles. The grids are usually made by placing copper bars in slots or holes in the pole faces near their surfaces and then short-circuiting these bars by brazing or welding them to end straps of brass or copper. A damper is shown in Fig. 205.

For most effective damping, the damper should have as low a resistance as possible, but if this winding is to be used for starting the motor, the resistance which gives the best damping action may be too low to give the best starting torque.

The armature reaction of a polyphase synchronous motor operating under steady conditions is fixed in space phase with respect to the poles. Under this condition, the resultant flux is also fixed with respect to the poles, and the damper is inactive and produces no effect upon the operation of the motor.   However, when hunting starts, the armature reaction is neither constant nor fixed in space phase with respect to the poles, but sweeps back and forth across them with a period equal to the period of oscillation of the rotor.   This causes the resultant flux to cut the damper and to induce currents in it.   Eddy-current and hysteresis losses are produced in the pole faces which assist to some extent in damping out the hunting.   The main damping action is due to the currents induced in the damper which are in such a direction as to oppose the change in the angular velocity of the rotor which produces them.   The reactance of the damper for the period of the current induced in it by

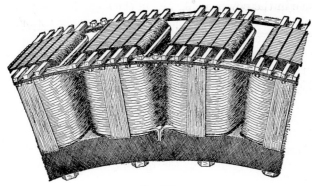

Fig. 205.

hunting is very small.   Assuming this reactance zero, the damping action is a maximum when the rotor swings through its mean position and zero when the rotor has its extreme displacement.   The effect of damping produced in this way is much the same as the damping produced by a viscous fluid on a torsional pendulum.   The braking action produced by a damper is only in part due to the energy dissipated in copper loss in the damper.   On account of the reaction between the currents in the damper and in the armature winding, there is some induction-generator and induction-motor action and energy is returned to the line while the rotor is accelerating and taken from the line while the rotor is retarding.

If a damper is used on a single-phase synchronous motor, currents are induced in it even when the motor is entirely free from hunting.   These currents are caused by the armature-reaction flux which in a single-phase synchronous motor is neither fixed in space phase nor constant in magnitude.   The main effect of a short-circuited winding on a single-phase synchronous machine is to damp out any harmonics in its wave form. The copper loss of such a winding substantially lowers the efficiency.

The armature reaction of a single-phase synchronous motor can be resolved into two revolving vectors, rotating in opposite directions.   One of these is stationary with respect to the rotor.   The other revolves at twice synchronous speed with respect to the rotor. The component which is fixed with respect to the rotor produces no effect on the damper; the other component produces double-frequency currents in the damper. These currents are not large on account of the relatively high reactance of the damper to the double-frequency currents.

The width of the air gap and the magnitude of the leakage reactance have a large influence on the stiffness of coupling of a motor. By stiffness of coupling is meant the tendency of a motor to follow every irregularity in the frequency of the system from which it is operated. The degree of stiffness of coupling depends upon the change in power produced by a given change in the power angle, i.e., in the angle between the impressed voltage and the voltage corresponding to the field excitation. The stiffness of coupling may be defined as the rate of change of power with respect to the power angle and can be found for a salient-pole synchronous motor, neglecting the effect of the armature resistance, by differentiating Eq. (194), page 330, with respect to the angle δ.

$$\frac{dP}{d\delta} = \frac{VE_a'}{x_d} \cos \delta + \frac{V^2(x_d - x_q)}{x_d q_q} \cos 2\delta \qquad (198)$$

The stiffness of coupling, Eq. (198), depends upon x_d, the direct-axis synchronous reactance, and x_q, the quadrature-axis synchronous reactance, but since the second term on the right-hand side in Eq. (198) is relatively small as compared with the first term, the stiffness of coupling of a salient-pole synchronous motor depends chiefly on the direct-axis synchronous reactance. Each of the two reactances is made up of two parts, the armature leakage reactance and an apparent reactance which replaces the effect of armature reaction. The parts of the two reactances which replace the effects of armature reaction in modifying the field strength depend upon the reluctance of the magnetic circuit along the field axis for x_d and at right angles to the field axis for x_q. Both of these reluctances are largely influenced by the radial length of the air gap. Increasing the length of the air gap increases the two reluctances, decreases the effects of the two components of armature reaction along the two directions, and therefore diminishes both x_d and x_q. A large air gap makes both x_d and x_q small and gives a stiff coupling. A small air gap has the opposite effect and gives a soft coupling. With too stiff a coupling, a motor tends to follow too closely the irregularities in the circuit from which it is operated and may be subjected to shocks and strains of considerable magnitude when such irregularities occur. Also, any variation in the power angle of the motor caused by fluctuations in the load is transmitted as surges of current and power to the generating system. With too soft a coupling, there is not sufficient stability and there is danger of a motor dropping out of step when any sudden change occurs in the system. Another objection to a soft coupling is that under the condition of constant excitation there is a large variation in the power factor from no load to full load. A compromise between the two extreme conditions must be made.

What has just been said in regard to stiffness of coupling neglects the effect of the damper and of any damping action that may be produced by eddy-current and hysteresis losses in the pole faces. The damping action of pole-face losses and of a damper increases as the width of the air gap is decreased, since the smaller the air gap the larger is the effect of armature reaction and the greater is the magnitude of the current induced by it in the damper and pole faces when there is a change in phase between the impressed and excitation voltages.

Methods of Starting Synchronous Motors. A synchronous motor is not inherently self-starting. The average synchronous-motor torque is zero at rest and until synchronous speed is reached. Some auxiliary device is necessary to bring a synchronous motor up to speed. Some of the early motors were equipped with auxiliary motors which were designed to be used only for starting and which developed sufficient torque to bring the synchronous motors up to speed. When d-c motors are used for this purpose, they can by proper design be used as generators for field excitation after synchronous speed has been attained. When auxiliary motors are used for starting, the starting torque is small. Synchronous motors which form one unit of a motor generator, where the other unit is a d-c generator, can be started by using the d-c generator as a shunt motor. An objection to starting with an auxiliary motor is that some sort of synchronizing device is necessary to synchronize the motor with the line from which it is to be operated. Polyphase synchronous motors are now started almost universally as induction motors by making use of their dampers as squirrel-cage windings of induction motors. When this method of starting is employed, no auxiliary apparatus or source of power is required, as is the case when d-c motors are used for starting.

The action of the damper during starting conditions will be understood after reading the section on polyphase induction motors. The eddy-current and hysteresis losses in the pole faces, produced by the revolving armature-reaction field of a polyphase motor, produce a starting torque too small to be of use. Single-phase synchronous motors have no revolving field due to armature reaction and cannot be started by means of dampers. Single-phase synchronous motors, except in small sizes, are of little practical importance. In very small sizes they are used in certain spinning operations and for driving electric clocks and other devices which must revolve at fixed speeds.

If a synchronous motor is to be brought up to speed as an induction

motor, care must be taken to design it in such a way that the reluctance
of the air gap under the poles is the same for all positions of the poles
with respect to the armature.   If the reluctance of the air gap under the
poles varies with position, the motor tends to lock in the position of
minimum reluctance when the stator is excited and a large torque is
required to move it from this locked position.

Whether or not the reluctance of the air gap over the poles varies with
the position of the rotor depends upon the spacing of the armature slots.
Figure 206 shows a spacing for which the air-gap reluctance is not constant.   The left-hand half of the figure shows a field pole in the position

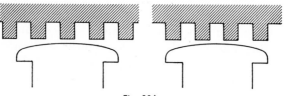

Fig. 206.

which makes the reluctance a maximum.   The position for minimum
reluctance is shown in the right-hand half of the figure.

The armature reaction of a polyphase synchronous motor operating
at synchronous speed is constant and fixed in space phase with respect
to the poles and except when there is hunting produces no effect on the
damper.   When the rotor is at rest or is revolving at any speed other
than synchronism, there is relative motion between the poles and the
field produced by armature reaction, which causes currents to be induced
in the damper.   These currents produce the same effect as the currents
induced in the squirrel-cage winding of an induction motor and cause the
motor to speed up.

A synchronous motor can never reach synchronous speed under the
action of the currents induced in its damper alone, but if the damper is
properly designed, the motor may reach a speed which is near enough
to synchronous speed to pull into step before the field is excited, provided the motor has salient poles.   The flux produced in the field poles
by the armature-reaction field may cause the motor to pull into step
before the field is excited.   Space harmonics in the armature reaction
may produce cusps in the starting-torque curve which tend to make the
motor lock in at some subsynchronous speed, but this difficulty can be
avoided by proper design.

When the motor has reached synchronous speed, the excitation is due
entirely to the armature reaction.   If when the field is closed it happens
to oppose the polarity produced by armature reaction, the motor slips
180 deg and is pulled into step only at the expense of a large rush of

current. To avoid this current rush, it is best to excite the field through a large resistance just before synchronous speed is reached. This causes the motor to pull into step with the correct polarity.

The starting torque of an induction motor depends upon the resistances and the leakage reactances of its rotor and stator windings. For maximum torque at starting, from Eq. 244, page 408,

$$r_2 = \sqrt{r_1{}^2 + (x_1 + x_2)^2}$$

where r_1 and r_2 are the stator and rotor effective resistances at stator frequency, and x_1 and x_2 are the corresponding leakage reactances, also at stator frequency. Since at stator frequency the leakage reactances of the stator and the damper of a synchronous motor are several times as large as the corresponding effective resistances, the resistance of the damper should be large for large starting torque. It should be approximately equal to the sum of the two reactances, both referred to the same winding, as was done in the transformer. The difference between the actual speed of an induction motor and its synchronous speed, i.e., the slip, is directly proportional to the resistance of its rotor winding. If it were possible to make the resistance of the rotor winding zero, an induction motor would operate at synchronous speed at all loads. The requirements for small slip and large starting torque are opposite so far as the resistance of the rotor is concerned. Since a synchronous motor starts as an induction motor, it should have, in order to pull into step easily, a damper of very low resistance, but in order to start readily, especially under load, the resistance of its damper should be high. The conditions for good starting torque are contradictory to the conditions for pulling into step readily, and a compromise is therefore necessary.

The frequency of the current in the damper is very low when synchronous speed has nearly been reached, and the local core loss produced by the current in the winding and also the skin effect are small. The ohmic and effective resistances under these conditions are nearly equal. At the instant of starting, however, the current in the damper is of the same frequency as the voltage impressed on the motor. The local losses produced by this current as well as the skin effect may make the apparent resistance of the damper considerably greater than its ohmic resistance. By making use of this difference between the ohmic and effective resistances, it is possible to design a damper to start a synchronous motor under load. The chief objections to starting a synchronous motor by the use of a damper are the large motor current required and the high voltage induced in the field winding. The large current, which is a lagging current, may seriously disturb the voltage regulation of the system.

The high voltage induced in the field winding during starting is caused by the armature reaction flux sweeping across the pole faces. This voltage is a maximum at the instant of starting and is zero when synchronous speed is reached. To keep this voltage as low as possible, the voltage for the field excitation of a self-starting synchronous motor should be low, corresponding to a small number of field turns. A switch can be provided to sectionalize the field winding during starting, and the voltage strain on the field insulation is then limited to that generated in a single section of the field winding instead of that generated in the entire field winding. This sectionalizing is seldom done because of constructional difficulties. Extra insulation must be provided on the fields of self-starting synchronous motors. The presence of the damper reduces considerably the voltage which would otherwise be induced in the field winding by the reaction of the currents induced in the damper. It might seem probable that short-circuiting the field winding during starting would allow currents to be induced in it which would add to the starting torque. The short-circuited field winding, however, acts like a single-phase winding on the rotor of a polyphase induction motor and tends to hold the motor at about half speed. Somewhat above half speed the starting torque may be slightly increased in this way.

When a synchronous motor is started as an induction motor, the voltage impressed on it should be reduced in starting and while it is coming up to speed. This reduced voltage can be obtained by using a starting compensator or can be obtained from taps on the secondary windings of the transformers supplying the motor, if transformers are used. Transformers are seldom used with a synchronous motor unless the voltage of the line from which it is operated exceeds 13,500 volts. Above that voltage, it is usually more economical to use transformers than to insulate a motor for full-line voltage.

To bring a synchronous motor with a damper up to speed, its field is opened and then about one-half normal full voltage is applied to its armature terminals. The motor should start slowly and if the damper has been properly designed it should speed up with increasing acceleration. The time required to come up to synchronous speed depends upon the fraction of full voltage applied, the size of the motor, and its design. It should not exceed a minute or a minute and a half, for motors of moderate size. When synchronous speed has been attained, which can be determined by the sound, the field circuit should be closed through a moderate amount of resistance and full voltage applied to the motor. The field should then be adjusted to make the motor operate at the desired power factor. Slight overexcitation is more desirable than underexcitation, since it makes the motor take a leading current and in a measure compensate for the reactive components of the currents taken by other loads on the line. Moreover, a slightly overexcited synchronous motor is more stable than one which is underexcited. If the motor is to operate with fixed excitation, the field should be adjusted initially to make the motor operate at approximately unity power factor at its average load unless the conditions under which it is to operate make some other power factor more desirable.

When a synchronous motor is to be started by the induction-motor action of its damper, a short air gap is desirable in order to keep the starting current small. A short air gap gives rise to a soft coupling which may be undesirable. The selection of the best length of air gap for any motor is a compromise and depends upon the particular service demanded.

Although the regular short-circuited type of damper gives sufficient starting torque for most purposes, it does not provide sufficient torque where motors have to be started with considerable load as in certain industrial processes. In order to get sufficient starting torque under such conditions, a torque which may be 200 per cent of full-load torque, polyphase (usually three-phase) coil-wound dampers are used and are connected in Y with the free terminals brought out to slip rings. During the starting period, the slip rings are connected together through adjustable resistances. As the motor speeds up, the resistances are cut out in successive steps by centrifugally controlled contactors, until the slip rings are short-circuited, when the windings are equivalent to an ordinary damper. By the use of the proper resistances the maximum induction-motor torque can be developed at the instant of starting and without excessive currents.

The Supersynchronous Motor. As applied to the synchronous motor this term is a misnomer. The only thing "super" about the motor is that it is capable of developing a very large starting torque with moderate current. The supersynchronous motor is built like an ordinary synchronous motor with the usual damper in the pole faces, but instead of having its armature structure rigidly bolted to the bedplate it is mounted on bearings in such a way that it is capable of revolving. A brake is provided by means of which the rotation of the armature can be prevented. To start the motor, the brake is released and power is applied to the armature terminals. The inertia of the rotating structure and its connected load prevents the field structure from rotating because the torque required to rotate the field structure and its connected load is greater than that required to speed up the armature. The result is that the armature speeds up under the induction action of the damper and pulls into synchronism when the field is excited. If the brake is now applied to the rotating part to slow it down, the field structure and the load gradually speed up and reach synchronous speed when the armature has been brought to rest. In this way a starting torque is available which is equal to the breakdown torque of the synchronous motor.

Chapter 24

Circle diagram of the cylindrical-rotor synchronous motor.   Proof of the
diagram.   Construction of the diagram.   Limiting operating conditions.
Some uses of the circle diagram.

Circle Diagram of the Cylindrical-rotor Synchronous Motor.   Circle
diagrams were first applied to synchronous machines by André Blondel.[1]
Although such diagrams assist in determining the general operating
characteristics of a synchronous motor, they cannot be used for predeter-
mining these characteristics with accuracy, since all circle diagrams are
based upon the assumption of constant resistance and constant synchronous reactance.   Blondel's original circle diagram of the synchronous
motor and synchronous generator was a diagram of voltages.   The circle
diagram of currents, which is merely a modification of the voltage diagram, is in some respects more convenient than the diagram of voltages,
and for this reason only the diagram of currents is considered.

Proof of the Diagram.   Let Fig. 207 be the vector diagram of a synchronous motor in which V represents the drop in voltage through the
motor.   This diagram is similar to the one given in Fig. 203, page 329,
but is rotated through 90 deg.   Let P, P_m, and $E_o = I_a z_s$ represent the
power input, the internal power developed, and the resultant voltage
forcing the current through the circuit.   Everything on the diagram is
per phase.   Take V as a reference axis, drawn vertical for convenience.

$$I_a = \frac{E_o}{z_s}$$

$$\tan \beta = \frac{x_s}{r_e}$$

According to the assumptions made in regard to x_s and r_e, $\tan \beta$ is constant.

If the motor excitation is constant, $ab = E_a'$ is constant and the extremity of the vector E_o swings on the arc of a circle bce as the current varies.
Since the current is proportional to E_o and makes a constant angle

[1] André Blondel, "Moteurs synchrones à courants alternatifs," also *L'industrie
électrique*, February, 1895.

$\beta = \tan^{-1}(x_s/r_e)$ with it, the extremity of the current vector OI_a also swings on the arc of a circle HI_a.

If the motor excitation is decreased, E'_a decreases and the point E_o approaches the point V. It coincides with V, which is the center of the voltage circle bce, when the excitation voltage E'_a is zero. At the same time I_a approaches the center C of its circle and coincides with the

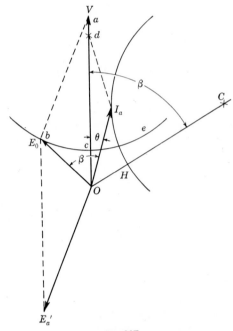

Fig. 207.

center when E_o coincides with V. When E_o and V coincide, OI_a lies along a diameter of the circle HI_a. OC therefore makes an angle

$$\beta = \tan^{-1}\left(\frac{x_s}{x_e}\right)$$

with OV and is equal to the voltage OV impressed on the motor divided by the synchronous impedance.

$$OC = \frac{OV}{z_s} \tag{199}$$

C is the center of a system of concentric circles corresponding to different motor excitations. These circles are the motor-excitation circles of the circle diagram.

If the excitation is constant, I_a travels along the arc of the circle HI_a, and when E_o coincides with c, I_a coincides with H and is then equal to

$$OH = \frac{Oc}{z_s} = \frac{E_o}{z_s}$$

$$HC = OC - OH = \frac{OV}{z_s} - \frac{E_o}{z_s} = \frac{E_a'}{z_s}$$

That is, the radius of any motor-excitation circle is equal to the corresponding excitation voltage divided by the synchronous impedance.

Take any point d on the line OV representing the impressed voltage.

$$(dI_a)^2 = (Od)^2 + (OI_a)^2 - 2(Od)(OI_a) \cos \theta$$

and

$$(Od)^2 - (dI_a)^2 = 2(Od)(OI_a) \cos \theta - (OI_a)^2$$

If Od is made equal to $OV/2r_e$ where

$$\frac{OV}{2r_e} = \frac{\text{impressed voltage}}{\text{twice the effective resistance}}$$

$$(Od)^2 - (dI_a)^2 = \frac{P}{r_e} - I_a^2$$

and

$$(Od)^2 - (dI_a)^2 = \frac{P_m}{r_e} \tag{200}$$

where P is the input and P_m is the internal electromagnetic output of the motor.

If P_m is fixed, Eq. (200) is the equation of a circle having a radius equal to dI_a and a center at a distance $Od = V/2r_e$ from the point O. Hence for any fixed motor power the end of the current vector OI_a must lie on a circle drawn about the point d as a center. The point d is the center of a system of power circles corresponding to different electromagnetic motor powers.

Substituting $Od = V/2r_e$ in Eq. (200) gives

$$\left(\frac{V}{2r_e}\right)^2 - (dI_a)^2 = \frac{P_m}{r_e}$$

$$dI_a = \left[\left(\frac{V}{2r_e}\right)^2 - \frac{P_m}{r_e}\right]^{1/2} \tag{201}$$

Equation (201) gives the radius of any power circle in terms of the electromagnetic motor power P_m, the impressed voltage V, and the effective resistance of the motor r_e.

Od and Cd are equal. This can be proved by showing that the apex of the isosceles triangle, having OC for a base and OV for the direction of one side, coincides with the point d.

From the construction of Fig. 207, the angle COV is equal to the angle

β. The length of the side of the isosceles triangle is therefore

$$\frac{OC}{2}\frac{1}{\cos\beta} = \frac{OC}{2}\frac{z_s}{r_e}$$

and as $OC = OV/z_s$, Eq. (199), the length of the side of the isosceles triangle is $OV/2r_e$. This is equal to the distance Od on the diagram.

The radius of the circle of zero power can be found by putting $P_m = 0$ in Eq. (201). Making this substitution gives $dI_a = V/2r_e$ as the radius of the circle of zero power. Since dO is equal to $V/2r_e$ and dO and dC are

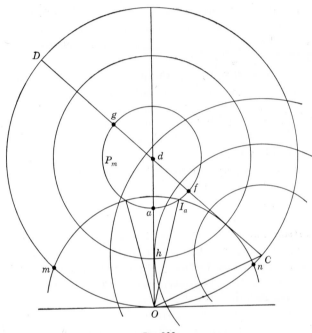

Fig. 208.

equal, this circle passes through the two points O and C. The circle of zero power therefore passes through the center C of the system of excitation circles.

Construction of the Diagram. Choose a suitable current scale. This scale is used for all lines on the diagram. Everything is per phase. Refer to Fig. 208. Lay off the line OC making an angle $\beta = \tan^{-1}(x_s/r_e)$ with the line Od.

Power Circles. The circle OCD of zero power is fixed by the points O and C and the direction of its radius Od. A more accurate method for determining this circle is to find the position of its center d.

$$Od = \frac{V}{2r_e}$$

The diameter of the circle of maximum power is zero. The radii of other power circles are found from Eq. (201) which gives for the radius of any power circle

$$\left[\left(\frac{V}{2r_e}\right)^2 - \left(\frac{P_m}{r_e}\right)\right]^{\frac{1}{2}} = \frac{1}{2r_e}(V^2 - 4r_eP_m)^{\frac{1}{2}}$$

It is usually convenient to have the power circles represent definite electromagnetic outputs, as for example, 100, 200, 300, 400, etc., kilowatts. Three power circles are shown in Fig. 208.

Excitation Circles. A series of concentric circles representing different motor-excitation voltages can be drawn with C as a center. It is sometimes convenient to draw these circles to represent different percentages of the excitation voltage which makes E_a' equal to V. One hundred per cent excitation is represented on the diagram by OC, which is equal to V/z_s. Three excitation circles are shown in the figure. A line drawn from C to I_a represents the motor excitation, corresponding to the current I_a and the power P_m, both in magnitude and in phase, and the angle I_aCO which the line CI_a makes with the line CO is the phase angle between E_a' and $-V$ on Fig. 207.

Current Circles. A series of current circles can be drawn with O as a center. Only one of these, mI_an, is shown. For any fixed electromagnetic motor power, such as is represented by the power circle marked P_m, there can be two motor excitations corresponding to the current I_a. The two corresponding excitation circles are fixed by the intersection of the current circle mI_an with the power circle P_m.

Limiting Operating Conditions. *Maximum and Minimum Excitation for Fixed Motor Power.* The maximum and minimum excitations at which the motor can develop the power P_m are Cg and Cf. The excitation circles corresponding to these are not shown. The points g and f are points of tangency between the motor-power circle and the two excitation circles. The currents corresponding to these excitations are Og and Of. The former leads, the latter lags.

Minimum Power Factor. The power factor for any condition is equal to the cosine of the angle θ, Fig. 207, made by the current line with the line Od. All currents to the right of Od are lagging. All currents to the left of Od are leading. The minimum power factor for any load occurs when the current line is tangent to the power circle for the given load. For fixed excitation and mechanical output, the region above the line Cd is a region of unstable operation and the region below the line Cd is a region of stable operation.

The Maximum Motor Excitation. The maximum motor excitation is CD, where D is the point of tangency of the circle of zero power with a motor-excitation circle. CD is the diameter of the circle of zero power.

This diameter is equal to V/r_e laid off to the scale of currents. The maximum excitation in volts is $(CD)z_s$.

The Maximum Motor Power. The diameter of the circle of maximum motor power is zero. The radius of any power circle according to Eq. (201) is

$$\frac{1}{2r_e} \sqrt{V^2 - 4r_e P_m}$$

If the radius is to be equal to zero, V^2 must be equal to $4r_e P_m$ and

$$P_m = \frac{V^2}{4r_e}$$

The excitation corresponding to this is $Cd = V/2r_e$, which is equal to one-half the maximum excitation.

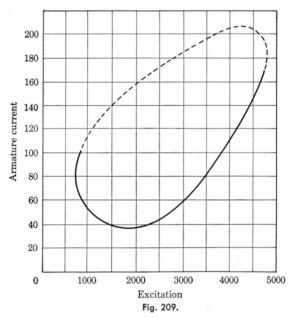

Fig. 209.

Stability. All currents lying above the line CD represent unstable conditions of operation. Any increase in load on a motor causes it to start to slow down, *i.e.*, causes the lag of the motor voltage to increase. If the excitation is fixed, this increase in lag produces an increase in the current (Fig. 208). With fixed excitation, any increase in the current beyond the line CD, Fig. 208, causes the extremity of the current line to move to a power circle of larger radius and consequently of smaller power.

Some Uses of the Circle Diagram. Besides being useful for determining the approximate operating characteristics of a cylindrical-rotor synchronous motor, the circle diagram is useful to show in a simple manner certain peculiarities in the characteristics of such a motor. For example, the possible range of underexcitation with fixed motor power is much less than the range of overexcitation for the same power. Referring to Fig. 208, let the constant motor power be P_m. The range of underexcitation is confined to excitation circles which cut the power circle P_m between a and f, but the range of possible overexcitation includes excitation circles which cut the power circle between a and g. This shows why V curves always extend further on the side corresponding to overexcitation than on the side corresponding to underexcitation.

The complete V curve of a synchronous motor calculated from the circle diagram is plotted in Fig. 209. The dotted part of the curve corresponds to the part of the circle diagram beyond CD and represents unstable conditions.

That the compounding curve for unity power factor should at first bend toward lower excitations as the output is increased can be seen readily from the circle diagram. Why the compounding curve for unity power factor should bend in this way was explained in Chap. 21, page 312. The middle power circle on Fig. 208 corresponds to the power for which the excitation is a minimum for unity power factor. For this power, the excitation circle for unity power factor is tangent to the line Od. For powers either greater or less than this, the excitation for unity power factor is greater. The power for which the excitation is a minimum depends upon the angle COd. This is $\tan^{-1} x_s/r_e$. For most motors the ratio of x_s/r_e is so large that the output at which the bend occurs in the compounding curve for unity power factor is not much above the electromagnetic output on the diagram which corresponds to no load on the motor or may even be less than this output.

Chapter 25

Determination of the V curves of a synchronous motor from tests. Effects on V curve of harmonics in the current wave. Losses and efficiency. Advantages and disadvantages. Uses. Synchronous condensers. Balancing effect of a synchronous motor on the line voltages of a system.

Determination of the V Curves of a Synchronous Motor from Tests. As indicated on page 299, the V curves of a synchronous motor are plots of armature current versus field excitation under conditions of constant terminal voltage and frequency and constant power. Either constant power input or constant power output may be used or, as in the case of the V curve determined from the circle diagram on page 346, constant electromagnetic power. If a sufficient power supply and a suitable method of loading are available, V curves may be determined from a load test. Readings are taken of armature voltage, current, and power, and field current as the field current is varied from a low to a high value. The range of field current should be limited to those values which keep the armature current from becoming excessive. The armature current is usually limited to 1.5 times its normal full-load value. During this test either the power input or the power output is maintained constant. The results, when plotted, give the V curve for the particular load used.

Effects on V Curve of Harmonics in the Current Wave. The lowest point on the V curve should represent the condition of unity power factor. Higher values of field current give a leading armature current and lower values a lagging armature current. When V curves are determined from a load test, it is frequently found that instrument readings taken at the lowest point do not indicate unity power factor. If the measurements are carefully made, the reason is likely to be the presence of harmonics in the current wave. These harmonics may be the result of either a nonsinusoidal impressed voltage or a nonsinusoidal excitation voltage. The magnitude of these harmonics may be determined by the following method. If I is the phase current determined from the ammeter readings, then

$$I^2 = I_f^2 + I_h^2 \tag{202}$$

where I_f is the effective value of the fundamental component of the armature current and I_h is the effective value of the harmonic components. Also

$$I_f{}^2 = (I_f \cos \theta')^2 - (I_f \sin \theta')^2$$

θ' being the power-factor angle as indicated on the vector diagrams such as that shown in Fig. 201, on page 315. Then

$$I^2 = (I_f \cos \theta')^2 + (I_f \sin \theta')^2 + I_h{}^2$$

If W and V are the power per phase and the terminal voltage per phase as determined from the wattmeter and voltmeter readings, then

$$I^2 = \left(\frac{W}{V}\right)^2 + (I_f \sin \theta')^2 + I_h{}^2$$

For the lowest point on the V curve, θ' equals zero, and hence as applied to this lowest point

$$I^2 = \left(\frac{W}{V}\right)^2 + 0 + I_h{}^2 \tag{203}$$

From this equation I_h may be evaluated.

If I_h is assumed to remain constant for all points on the V curve, the fundamental component of the current may be calculated from Eq. (202) and a corrected V curve may be obtained with harmonic components of the current omitted. Since, as the armature current increases, I_h becomes of relatively less importance, the assumption that it is constant is probably justified.

Losses and Efficiency. A synchronous motor does not differ essentially from a synchronous generator. The losses in the two machines are the same and can be determined in the same manner. The same methods can be employed for calculating the efficiency of a synchronous motor and a synchronous generator. The losses and the method of calculating the efficiency of a synchronous generator are discussed in Chap. 18, page 269, under Synchronous Generators. The efficiency of a synchronous generator is calculated in Chap. 18, page 280.

Advantages and Disadvantages. The chief advantage of the synchronous motor is its ability to operate at different power factors and the ease with which its power factor can be adjusted. Its simplicity and the possibility of winding it for high voltages, thus doing away with transformers, are advantages, but the same advantages are possessed by the polyphase induction motor. Under certain conditions, the constant speed of the synchronous motor under varying load is an advantage. Its main disadvantages are its tendency to hunt and its lack of starting torque. In polyphase synchronous motors, neither of these objections is serious and the objections are of little consequence when a motor is provided with a properly designed damper and operates under reasonably good conditions.

A synchronous motor is less sensitive than an induction motor to variations in the impressed voltage. This is an advantage in some cases as it enables a motor to carry its load without falling out of synchronism during periods of reduced voltage arising from some temporary trouble on the line or in the powerhouse. A synchronous motor which has the same breakdown torque as an induction motor continues to carry its full load at a voltage which is considerably lower than that which would cause an induction motor to break down. The maximum output of an induction motor varies as the square of the impressed voltage, but the maximum output of a synchronous motor of the usual design and operating with constant excitation varies nearly as the first power of the impressed voltage, Eqs. (183) and (186), page 321. If a synchronous motor and an induction motor, each having a maximum output equal to twice its rated output, were put on half voltage, the synchronous motor could still develop full load without falling out of synchronism. The induction motor, however, would break down at one-half its rated output.

Uses. A synchronous motor is frequently used for one unit of a motor generator when the other unit is a d-c generator or a synchronous generator. When the other unit is a synchronous generator, the motor generator acts as a frequency changer or frequency converter and converts power at one frequency into power at a different frequency. If the ratio of the number of poles on the two machines is 25:60, the frequency changer converts from 25 to 60 cycles or vice versa according to which machine acts as motor. Frequency changers are often used to interconnect two power systems which have different frequencies. When used in this way, the machine which acts as motor is determined by the direction of flow of energy between the interconnected systems. Synchronous motors are used in industrial work, for the propulsion of electrically driven ships, etc. When operated at unity power factor, a synchronous motor weighs somewhat less and is less expensive than an induction motor of the same speed and output. High-speed synchronous motors are used for driving centrifugal blowers. Such motors are built with ratings of several thousand horsepower, operate on 60 cycles, have two poles, and run at 3,600 rpm. Such high-speed motors usually have rotors which are solid steel forgings with slots milled in their surfaces for the d-c field windings. The starting torque and the damping are obtained from the reaction of the eddy currents induced in the solid pole faces during starting and when there is hunting. There is some hysteresis starting torque, but this is very small. The eddy currents induced in the pole faces have the same effect as the currents induced under similar conditions in dampers.

One important reason for the use of synchronous motors is the possibility of varying their power factors to control the amount of reactive current on power systems. By operating overexcited the synchronous motors connected to a power system it is possible either wholly or partly to neutralize the reactive lagging currents taken by inductive loads connected to the system, hence to maintain high power factor.

Synchronous Condensers. Synchronous motors are frequently operated without load on a transmission system to control the power factor and to improve the voltage regulation. When used in this way, they are called synchronous condensers. Although synchronous motors, if overexcited, take leading currents like static condensers, they do not behave in other respects like static condensers. The reactive current taken by a condenser is directly proportional to the voltage. The reactive current taken by an overexcited synchronous motor operating with fixed excitation decreases with an increase in impressed voltage and becomes zero at a certain voltage. A further increase in the impressed voltage causes the reactive component of the current to reverse and to become a lagging component.

Many large synchronous condensers are hydrogen-cooled. A synchronous condenser is well adapted to hydrogen cooling since it requires no extension of its shaft through the enclosing casing. The advantages of hydrogen as a cooling medium are considered in connection with synchronous generators on page 169.

If a synchronous motor is used as a synchronous condenser to control the voltage of a line, the line must contain reactance. This reactance may be the natural reactance of the line if the line is of sufficient length, or it may be reactance inserted in the line. The control of the voltage is due to the voltage drop in the reactance. The motor serves merely as a means for making the current through the reactance lead or lag. A leading current through a reactance causes a rise in voltage but a lagging current produces a drop in voltage.

A synchronous motor operating with constant excitation tends to maintain constant voltage across its terminals if there is reactance in the power mains supplying it. Suppose a synchronous motor is operating with normal excitation, which produces unity power factor, and the line voltage drops. The excitation of the motor is now higher than normal for the reduced impressed voltage, and the motor takes a leading current as it is now overexcited. This leading current causes a rise in voltage through the line reactance which tends to restore the impressed voltage to its initial value. If the impressed voltage rises, the motor takes a lagging current as it is now underexcited. This results in a drop in voltage in the reactance of the line which tends to restore the voltage to its initial value. The tendency of a synchronous motor to maintain constant voltage at its terminals does not depend upon its initial excitation.   Normal excitation was chosen merely as an example.

Balancing Effect of a Synchronous Motor on the Line Voltages of a System.   A polyphase synchronous motor floated on a circuit carrying an unbalanced load tends to restore balanced conditions in regard to both current and voltage.   If the system is badly out of balance, the synchronous motor may take power from the phases with high voltage and deliver power to the phase or phases with low voltage.   This action of a synchronous motor is due to its short-circuited damper and is best explained by means of the symmetrical-phase components of the unbalanced voltages.   It is discussed after considering the effect of unbalanced voltages on a polyphase induction motor.

PART FOUR

PARALLEL OPERATION
OF SYNCHRONOUS GENERATORS

Chapter 26

General statements. Synchronous generators in parallel. The effect of
a change in the excitation of a synchronous generator which is in parallel
with others. The effect of giving more power to a synchronous generator
which is operating in parallel with others. The parallel operation of
synchronous generators is a stable condition. Inductance is necessary in
the armatures of synchronous generators. An example of synchronous
generators in parallel. The synchronizing power of a synchronous generator. The constants of synchronous generators which are to operate in
parallel need not be inversely proportional to their ratings.

General Statements. Since the terminals of synchronous generators
operating in parallel are connected to common bus bars, the terminal
voltages of the generators must be equal. The load carried by an individual generator and the phase relation between its armature current
and generated voltage must adjust themselves so as to maintain equal
terminal potentials. If the impedance in the cables between the generators and the bus bars or between the generators and the point at which
they are paralleled is zero, the actual potentials at the generator terminals
are equal. Unless otherwise stated, terminal voltage when applied to a
synchronous generator signifies the voltage of the generator measured
at the point of paralleling and the constants of the generators include
the constants of the line or leads up to this point. The terminal voltage
at the point of paralleling is equal to the actual terminal voltage of the
generator minus the drop in the cables between it and the point of
paralleling.

With synchronous generators operating in parallel, not only must the
terminal voltages be equal as measured by a voltmeter, but they must
be equal at every instant, which means that the generators must be in
synchronism and their terminal voltages must be in phase. Fortunately
the natural reactions resulting from a departure from synchronism tend
to reestablish synchronism.

Unless mechanically coupled, synchronous generators cannot ordinarily operate in series. They are stable only when in parallel. If one
of two generators in parallel leads its proper phase with respect to the
other, more load is automatically thrown upon it. At the same time

the other generator, lagging its proper phase, is relieved of some of its load.   The result is that the generator which is leading slows down and the generator which is lagging speeds up until the proper phase relation is restored.   This shift of load between two generators in parallel is equivalent to a transfer of energy from one to the other.   Although it is sometimes convenient to consider the shift of load as an interchange of energy, no actual transfer of energy takes place in reality except when the load on the system is zero or when the load on a generator is less than the change in its load required to restore synchronism.

The ideal operating conditions for synchronous generators in parallel are to have their kva outputs proportional to their kva ratings and to have their current outputs in phase with one another and therefore in phase with the load current.   Under these conditions, the resultant current delivered to the load by the generators in parallel is equal to the arithmetical sum of the current outputs of the individual generators.   The maximum output can be obtained under these conditions without overloading any generator.

The conditions under which synchronous generators can operate satisfactorily in parallel are not so limited as for transformers in parallel. When transformers operate in parallel, they are connected in parallel on both their primary and secondary sides.   All primary voltages must be equal in magnitude and must be in phase.   All secondary voltages must be equal in magnitude and must be in phase.   Under these conditions the relative outputs of the transformers and the power factors at which they operate are fixed by the constants of the transformers.   When synchronous generators are in parallel, their terminal voltages are equal and are in phase.   These voltages correspond to the secondary voltages of the transformers.   Although the excitation voltages of the generators correspond to the primary voltages of the transformers, the excitation voltages need not be equal in magnitude and need not be in phase.   They can be adjusted both in magnitude and in phase by changing the relative amounts of power given to the generators and at the same time changing their relative excitations.   In this way the generators can be made to divide the kilowatt load in any desired way and to operate at any desired power factors.

There are two methods of treating the parallel operation of synchronous generators.   In the first method, the current carried by each generator is divided into two components: one, the current the generator should carry under ideal operating conditions; the other, a circulatory current between the generators present only when the excitation voltages of the generators are not identical.   When these voltages are not identical, there is a resultant voltage acting in the series circuit formed by the machines causing a circulating current to flow among them.   This circulatory current is merely a component current and has no real existence except when the load on the generator is zero. This method of treating synchronous generators in parallel has little significance when there are more than two generators in parallel. The second method is much more satisfactory than the first and is applicable to any number of generators in parallel. It makes use of the relation between the power developed by a generator and its power angle. This method is the only one considered in what follows.

Synchronous Generators in Parallel. The following treatment of parallel operation of synchronous generators is based on the relation between the power developed by a synchronous generator and its power angle, the angle between its terminal voltage and its excitation voltage. This relation between the power developed per phase by a synchronous generator and the power angle is given by Eq. (173), page 283, and Eq. (174), page 284. These equations neglect the effect of armature resistance. Equation (173) applies to cylindrical-rotor machines and Eq. (174) to salient-pole machines. These equations are repeated here as Eqs. (204) and (205).

$$P = \frac{VE'_a}{x_s} \sin \delta \tag{204}$$

$$P = \frac{VE'_a}{x_d} \sin \delta + \frac{V^2(x_d - x_q)}{2x_d x_q} \sin 2\delta \tag{205}$$

where V and E'_a are the magnitudes of the phase terminal voltage and the phase excitation voltage and δ is the power angle. The reactances x_s, x_d, x_q are per phase and are the synchronous reactance, the direct-axis reactance, and the quadrature-axis reactance.

The relation between power output and power angle, Eqs. (204) and (205), is plotted in Figs. 176 and 178, pages 284 and 285. Whether a cylindrical-rotor generator is considered or one with salient poles, the power developed when the terminal voltage V, the excitation voltage E'_a, and the frequency are fixed is a function of the power angle, and the only way that the power can be changed is by changing the power angle. If the power is fixed, any change in either V or E'_a or in both must be accompanied by a corresponding change in the power angle. Any general discussion based on the power angle applies to either a cylindrical-rotor or a salient-pole machine.

Up to the point of maximum power developed by a synchronous generator, the power angle varies in the same way both in sign and in magnitude as the generator power. This makes the parallel operation of synchronous generators stable.

Synchronous generators which operate in parallel must have identical terminal voltages.   These voltages must be equal in magnitude, must be in phase, and must have the same frequency.

All types of prime movers, steam engines, steam turbines, water wheels, etc., which operate with fixed governor settings, for example with fixed tension of governor springs, must decrease in speed slightly in order to carry an increase in load since the governors of the prime movers depend for their action upon a change in speed.   When an increase in load is applied to a prime mover, it slows down.   This causes the governor to act and to admit more steam or water as the case may be.   The prime mover continues to slow down until equilibrium is established between the power it receives and its output plus its losses.   It follows that, if a prime mover must operate at a fixed speed, the only way that it can be made to deliver more or less power is to change the governor setting, usually by changing the tension of the governor spring, in such a way as to increase or decrease the power given to the prime mover at the fixed speed.

Direct-current generators in parallel do not have to operate at any fixed relative speeds.   When a d-c generator in parallel with others is to carry more load, its field excitation is increased.   This increases its generated voltage and causes it to deliver more current.

$$I = \frac{E - V}{r}$$

where E is its generated voltage, V is the terminal voltage, assumed constant, and r is the armature resistance.   The generator is now delivering more power than it receives from its prime mover.   It therefore immediately starts to slow down.   This causes the governor on the prime mover to act and allows the prime mover to deliver more power to the generator. The prime mover continues to slow down until equilibrium is reestablished between the output of the prime mover and the output of the generator plus generator losses.   When this condition is reached, the prime mover and its generator are operating at a slightly lower speed than before the increase in load and continue to operate at this lower speed as long as the load remains constant.

Synchronous generators in parallel must operate at the same relative speeds since they must have the same frequency.   One cannot slow down relatively to the others in order to receive more power from its prime mover.   The only way a synchronous generator can be made to deliver more power at the same speed is to increase the input from its prime mover by changing the governor setting.   This causes the generator to speed up momentarily and thus to increase its power angle.   The fact that d-c generators in parallel need not operate at fixed relative speeds, whereas synchronous generators in parallel must operate at fixed relative speeds, accounts for most of the differences between the parallel operation of the two types of machines.

Direct-current generators in parallel must have equal terminal voltages and should share the power load on the system in direct proportion to their individual ratings. They do not have to operate at any fixed relative speeds.

Synchronous generators which are in parallel must have terminal voltages that are not only equal in magnitude but are also in phase. They must operate at the same frequency and therefore must operate at fixed relative speeds. Like d-c generators, they should share the power load on the system in proportion to their ratings. In addition they should also share the kva load in proportion to their kva ratings. Since the combined output of any number of synchronous generators in parallel is limited by the generator which first becomes fully loaded, the foregoing division of load gives the maximum combined output since all the generators become fully loaded at the same time. These conditions in regard to the division of the power load and the kva load are the same as for transformers in parallel.

To sum up, the successful operation of synchronous generators in parallel depends upon these two facts:

1. The output of a synchronous generator with fixed terminal voltage, fixed excitation, and fixed frequency cannot be changed without changing the power angle at which it operates.

2. The output of a prime mover which operates at fixed speed cannot be changed without changing its governor setting in such a way as to give it more steam at the same speed in a steam turbine or a steam engine or more water in a water wheel.

It follows that the kw output of a synchronous generator operating in parallel with a system of constant frequency and constant voltage cannot be changed by changing its excitation since this cannot change its speed. It must operate at the same frequency as the system with which it is in parallel. If the generator speed does not change, the governor on its prime mover cannot act to alter the generator output. A change in the excitation alone merely changes the power factor at which the synchronous generator operates without changing the power output.

The Effect of a Change in the Excitation of a Synchronous Generator Which Is in Parallel with Others. A synchronous generator operating in parallel with a system of constant frequency and constant terminal voltage is represented by the full lines in the vector diagram of Fig. 210. The terminal voltage V of the system and of the synchronous generator is assumed constant. Assume that the synchronous reactance of the generator is constant and that its resistance is negligible. Neglect any change in core loss. Under these conditions the internal power developed is a direct function of the component of the armature current in phase with the terminal voltage. Assume that the generator has non-salient poles. Then from Eq. (204), page 354,

$$P = \frac{VE'_a}{x_s} \sin \delta$$

Although the synchronous reactance changes somewhat with a change in excitation, this change affects merely the magnitude of the change in the power angle when the excitation is altered. If the generator has salient

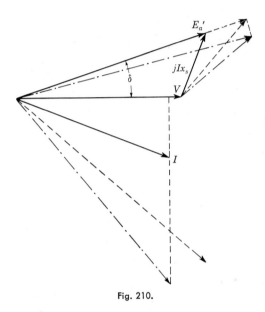

Fig. 210.

poles, the general conditions are the same as for a cylindrical-rotor generator.

Let the excitation voltage be increased by increasing the field current. Assume that this increase can take place before the generator can alter its speed momentarily to accommodate itself to the new conditions. The conditions existing immediately after E'_a is increased are shown by the lines of dashes in Fig. 210. Since according to the assumptions the angle δ has not changed, the power P has increased because of the increase in the voltage E'_a. The generator is now delivering more power than it is receiving from its prime mover and starts to slow down. E'_a swings in phase toward V, decreasing the power angle δ and the output P. E'_a continues to swing in phase with respect to V until equilibrium is reestablished between output and input. When this condition has been reached, the generator is delivering the same power output as before the excitation was increased, except for a small change in the losses, which are neglected in this discussion. Since the speed of the generator and its prime mover cannot change except momentarily, the governor on the prime mover cannot have acted. The only effect produced by the change in excitation is an increase in the reactive component of the armature current. The final conditions are shown by the dot-and-dash lines of Fig. 210.

The effect of the change in the reactive component of the armature current of a synchronous generator in parallel with others, caused by a change in excitation, is to neutralize the effect on the terminal voltage of the change in excitation. If the excitation is increased, a lagging reactive component is added to the armature current, and this lagging component current weakens the resultant field by the change in armature reaction produced by it. The leakage reactance drop caused by this reactive component of the armature current also decreases the terminal voltage. These two effects balance the effect of the increase in excitation on the terminal voltage. If the excitation is decreased, a leading reactive component is added to the armature current. The armature reaction and the leakage-reactance drop caused by this leading component current tend to increase the terminal voltage and to balance the effect of the change in excitation on the terminal voltage.

When there is an increase in the reactive component of the armature current of a synchronous generator caused by something that is done to the generator or to its prime mover, the other generators in parallel with it must carry an equal but opposite reactive current to balance this increase, since the load current of the system, which does not change, is equal to the vector sum of the currents delivered by the separate generators.

A synchronous generator in parallel with others cannot change its speed except momentarily and then only by an amount sufficient to cause the change in power angle necessary to maintain equilibrium between input to the generator and its output. There cannot be any permanent change in speed if synchronism is to be maintained.

The Effect of Giving More Power to a Synchronous Generator Which Is Operating in Parallel with Others. The initial conditions under which the synchronous generator is operating are shown by full lines in Fig. 211. The resistance drop is neglected. Let the tension of the governor spring be slightly decreased on the prime mover which drives the generator. This causes the governor to move to a new position and to admit more steam. The generator receives from its prime mover an amount of power greater than output plus losses. This unbalanced power tends to cause the generator to increase its speed. The vector E'_a momentarily has a higher frequency than the vector V, and E'_a swings ahead in phase

with respect to V, increasing the power angle δ.   The momentary change
in speed is too small to affect the magnitude of V.   Since the power angle
δ is increased, the power output P of the generator increases until equilibrium has been established between the power output of the generator
plus its losses and the input from the prime mover.   The new condition is
shown by the dot-and-dash lines
on Fig. 211.   The generator is
now delivering a larger output
than before the governor was
changed but at a different power
factor.   Since the resultant current output of any number of generators in parallel is equal to the
vector sum of the currents delivered by the individual generators,
the change in the reactive current
carried by any generator must be
balanced by an equal but opposite

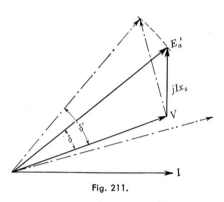

Fig. 211.

change in the reactive current carried by the other generators in parallel.
The original power factor can be restored by adjusting the excitation voltage E_a'.   Under the conditions illustrated in Fig. 211, the excitation voltage should be increased to restore the original power factor.

Up to the limit of its maximum power output, the output of a synchronous generator in parallel with others can be adjusted by changing the
governor on its prime mover.   The power factor at which it operates
can then be adjusted by changing its excitation.   The proper operating
conditions for synchronous generators in parallel are (1) for all generators
to carry loads which are proportional to their kva ratings and (2) for all
generators to operate at the same power factor, *viz.*, the power factor of
the load on the system.

Giving a synchronous generator in parallel with others more power by
changing the tension of the governor spring on its prime mover advances
the phase of its excitation voltage with respect to its terminal voltage,
increases its power angle, and consequently increases its power output.   Increasing the excitation of a synchronous generator in parallel
with others causes its excitation voltage to swing in the direction of lag
with respect to its terminal voltage and causes the current it delivers to
lag more, but changing the excitation alone cannot change the power
output of the generator.   With fixed terminal voltage and fixed excitation
voltage there is only one value of power for each power angle.   Any
change in the power angle must be accompanied by a corresponding
change in the power output.   With the power fixed, any change in either
the terminal voltage or the excitation voltage must be accompanied by a

change in the power angle in order that with the new voltage there shall be the same power. The load carried by a synchronous generator in parallel with others should not be changed without changing both the governor on its prime mover and its field excitation. The governor changes the kw load. The field excitation adjusts the power factor at which the generator operates.

The Parallel Operation of Synchronous Generators Is a Stable Condition. If a synchronous generator in parallel with others momentarily swings ahead of the phase position corresponding to the load it is carrying, its output is increased since its power angle is increased. Slight momentary changes in speed which produce momentary changes in the power angle are in general too small and of too short duration to cause the governor to act and to change the amount of power given to the generator. The generator therefore is developing more power than it is receiving from its prime mover, and it immediately starts to slow down and to return to its initial phase position and power angle. If the generator momentarily slows down, its power angle and therefore its power output decrease, but the input from its prime mover cannot have changed since the governor has not had time to act. The average speed of the generator must be constant if it is to remain in synchronism. The output of the generator is now less than the power it receives from its prime mover, and the generator speeds up until the initial phase position and power angle are restored.

A synchronous generator can momentarily swing beyond the power angle corresponding to maximum output and still be in a stable condition, provided the output developed at this new power angle is greater than the load the generator is carrying plus the losses, since there is then unbalanced power to cause the generator to swing back into its proper phase position with corresponding power angle.

Inductance Is Necessary in the Armatures of Synchronous Generators. The synchronizing action of synchronous generators in parallel is dependent on their armature inductive reactance. If enough capacitance should be inserted in the armature circuits of synchronous generators in parallel to make the resultant reactance of their armature circuits capacitive instead of inductive, they would be unstable and would not operate in parallel. If the armature reactance is capacitive, it is negative. When the reactance is negative, P in Eqs. (204) and (205), page 354, is negative for positive values of the power angle. Under this condition, if a synchronous generator in parallel with others should swing ahead of the proper phase position for the load it is carrying, it would develop less power. The power input from its prime mover would now be greater than the output of the generator plus its losses, and the generator would swing still further ahead of its proper phase position and its output would still further decrease. It would continue to swing ahead in phase until it dropped out of step. The reaction which would occur if a synchronous generator with capacitive reactance in its armature circuit should swing out of its proper phase position would be such as to make it swing still further out of phase instead of restoring it. Synchronous generators in parallel are stable on capacitive loads as the capacitance of the load is beyond the point of paralleling, is not in the armature circuits, and cannot influence the relation between the power developed by a synchronous generator and its power angle, as given by Eqs. (204) and (205).

An Example of Synchronous Generators in Parallel. Consider two equal non-salient-pole three-phase 60-cycle Y-connected 6,600-volt

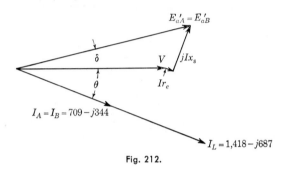

Fig. 212.

10,000-kva synchronous generators, each with an effective resistance of 0.063 ohm per phase and a synchronous reactance of 1.52 ohms per phase. Assume that the resistances and reactances of the generators are constant. These generators in parallel carry an inductive load of 18,000 kva at 0.9 power factor.

1. Find the excitation voltages of the generators and their power angles when they share the kw load equally and operate at the same power factor.

2. Assume that the power load on generator A is increased by adjusting the governors on the prime movers, keeping the frequency the same, until generator A carries two-thirds of the load on the system. The excitation of generator A is kept constant, but the excitation of generator B is adjusted to keep the terminal voltage of the system 6,600 volts. Under these new conditions find the excitation voltage of generator B and the new power angles.

The vector diagram for the first condition is shown in Fig. 212. This diagram is not drawn to scale but illustrates the conditions under which the generators are operating. Since the generators are carrying equal loads and operate at the same power factor, they must deliver equal currents.

$$V = \frac{6,600}{\sqrt{3}} = 3,810 \text{ volts per phase}$$

$$\text{Load current} = I_L = \frac{18,000,000}{\sqrt{3} \times 6,600} = 1,576 \text{ amp per phase}$$

$$\cos \theta = 0.9 \qquad \sin \theta = 0.4359$$

$$I_A = I_B = \frac{I_L}{2} = \frac{1,576}{2} (0.9 - j0.4359) = 709 - j344 \text{ amp per phase}$$

$$E'_a \text{ (for each generator)} = 3,810(1 + j0) + (709 - j344)(0.063 + j1.52)$$
$$= (3,810 + j0) + (568 + j1,056)$$
$$= 4,378 + j1,056 \text{ volts per phase}$$
$$|E'_a| = \sqrt{(4,378)^2 + (1,056)^2} = 4,504 \text{ volts per phase}$$
$$\sqrt{3} |E'_a| = \sqrt{3} \times 4,504 = 7,800 \text{ volts between terminals}$$
$$\tan \delta = \frac{1,056}{4,378} = 0.2413$$
$$\delta = 13.57 \text{ deg}$$

Neglecting the effect of armature resistance, the maximum power occurs at a power angle of 90 deg. If the terminal voltage and frequency

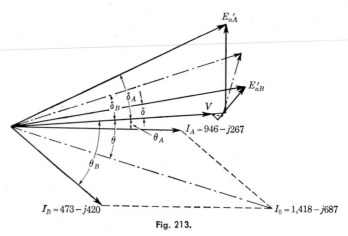

Fig. 213.

of one of the generators could be maintained constant, the power angle could increase $(90 - 13.57)$ deg before the generator would develop its maximum power. This maximum power is

$$3 \times \frac{E'_a V}{x_s} \sin \delta = 3 \times \frac{4,504 \times 3,810}{1.52} \times 1.00$$
$$= 33,870 \text{ kw}$$

and is greater than three times the rating of the generator. This neglects the change in synchronous reactance with the change in load and also neglects the effect of the armature resistance. The generator can swing beyond a power angle of 90 deg and still be stable, provided that at this greater angle the power output of the generator plus losses is larger than the power input from its prime mover, since under this condition the generator would swing back to its initial phase position.

The conditions when the load on generator A has been increased to two-thirds of the entire load on the system are shown in Fig. 213, which is not drawn to scale. The dot-and-dash lines on this figure show the initial conditions.

$$I_A \text{ (active)} = \tfrac{2}{3} \times 1{,}576 \times 0.9 = 946 \text{ amp per phase}$$
$$I_B \text{ (active)} = \tfrac{1}{3} \times 1{,}576 \times 0.9 = 473 \text{ amp per phase}$$

Generator A

$$4{,}504\underline{/\delta_A} = 3{,}810(1 + j0) + (946 + jI_r)(0.063 + j1.52)$$

where I_r is the reactive component of the armature current.

$$4{,}504\underline{/\delta_A} = 3{,}810 + (59.60 - 1.52I_r) + j(1{,}438 + 0.063I_r)$$
$$(4{,}504)^2 = (3{,}870 - 1.52I_r)^2 + (1{,}438 + 0.063I_r)^2$$
$$I_r = +5{,}273 \text{ or } -267 \text{ amp}$$

Use the smaller value. Then

$$I_A = 946 - j267 \text{ amp}$$
$$E'_a = 3{,}810(1 + j0) + (946 - j267)(0.063 + j1.52)$$
$$= 4{,}275 + j1.421$$
$$|E'_a| = \sqrt{(4{,}275)^2 + (1{,}421)^2} = 4{,}503 \text{ volts per phase}$$

which checks with the value 4,504 volts previously found for generator A.

$$\tan \delta_A = \frac{1{,}421}{4{,}275} = 0.3325$$
$$\delta_A = 18.39 \text{ deg}$$

Generator A has swung ahead in phase from its initial phase position by $(18.39 - 13.57) = 4.82$ deg

Generator B

$$I_B = I_L - I_A$$
$$= 1{,}576(0.900 - j0.4359) - (946 - j267)$$
$$= 472 - j420 \text{ amp per phase}$$
$$E'_a = 3{,}810(1 + j0) + (472 - j420)(0.063 + j1.52)$$
$$= 4{,}478 + j692 \text{ volts per phase}$$
$$E'_a = \sqrt{(4{,}478)^2 + (692)^2} = 4{,}531 \text{ volts per phase}$$
$$\tan \delta_B = \frac{692}{4{,}478} = 0.1545$$
$$\delta_B = 7.64 \text{ deg}$$

Generator B has swung behind its initial phase position by

$$(7.64 - 13.57) = -5.93$$

electrical degrees. Generator A has swung ahead of its initial phase position. The currents delivered by the two generators can be brought into phase without changing the terminal voltage of the system by increasing the excitation of generator A and decreasing the excitation of generator B.

The Synchronizing Power of a Synchronous Generator.

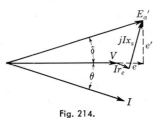

Fig. 214.

The synchronizing power of a synchronous generator can be defined as the change in the internal power with respect to a change in the power angle. It is the change in the internal power per unit change in the power angle. The synchronizing power of a cylindrical-rotor generator can be found in the following manner.

The internal power of a cylindrical-rotor generator as a function of its power angle can be found in the same manner as in Eq. (172), page 283. Refer to Fig. 214.

$$E_a' = V + Iz_s$$

where the letters have the same significance as when previously used.

$$I = \frac{E_a' - V}{z_s} = \frac{E_a' - V}{r_e + jx_s}$$

Take V as a reference axis and let

$$E_a' = e + je'$$

where e and e' are the components of E_a' with respect to the terminal voltage V.

$$I = \frac{e + je' - V}{r_e + jx_s}$$
$$= \frac{(er_e - Vr_e + e'x_s) + j(e'r_e - ex_s + Vx_s)}{r_e^2 + x_s^2}$$

The internal power is $E_a'I \cos \theta_I^{E_a'}$ and is equal to the product of the real parts of E_a' and I plus the product of the j parts of E_a' and I.

$$P' \text{ (internal)} = \frac{e(er_e - Vr_e + e'x_s)}{z_s^2} + \frac{e'(e'r_e - ex_s + Vx_s)}{z_s^2}$$
$$= \frac{r_e(e^2 + e'^2)}{z_s^2} + \frac{V(e'x_s - er_e)}{z_s^2}$$
$$= \frac{E_a'^2 r_e}{z_s^2} + \frac{VE_a'}{z_s^2}(x_s \sin \delta - r_e \cos \delta)$$

$$x_s \sin \delta - r_e \cos \delta = x_s \sin \delta - r_e \sin[\delta + (\pi/2)] = x_s \sin \delta - jr_e \sin \delta$$

This expression can be considered as two stationary vectors, shown in Fig. 215.

$$x_s \sin \delta - jr_e \sin \delta = z_s \sin \left(\delta - \tan^{-1} \frac{r_e}{x_s} \right)$$

$$P' = \frac{E_a'^2 r_e}{z_s^2} + \frac{VE_a'}{z_s} \sin \left(\delta - \tan^{-1} \frac{r_e}{x_s} \right) \qquad (206)$$

$$\frac{dP'}{d\delta} = \text{synchronizing power} = \frac{VE_a'}{z_s} \cos \left(\delta - \tan^{-1} \frac{r_s}{x_s} \right) \qquad (207)$$

Equation (207) shows why a system with quick acting automatic voltage regulators is more stable than one with voltage regulators of a slower type. If a sudden increase in the load on the system occurs, quick acting voltage regulators rapidly increase E_a' and tend to keep up the terminal voltage V of the system and to maintain sufficient synchronizing power to hold the generators in step. If the terminal voltage drops too far, because of any cause

Fig. 215.

such as a short circuit near the station, the synchronizing power may be reduced below that necessary to hold the generators in synchronism. If the terminal voltage becomes zero, the synchronizing power also becomes zero. The synchronizing power becomes zero when the angle $[\delta - \tan^{-1} (r_e/x_s)]$ becomes 90 deg. With fixed excitation and terminal voltage, the internal power of a cylindrical-rotor generator is a maximum when $[\delta - \tan^{-1} (r_e/x_s)]$ is 90 deg. Although the synchronizing power is zero for the angle corresponding to maximum internal power, it does not follow that a generator drops out of synchronism if it momentarily swings to an angle equal to or greater than $[\delta - \tan^{-1} (r_e/x_s)]$ between its excitation voltage and its terminal voltage. It is still stable and swings back provided the power it is receiving from its prime mover is less than the internal power it develops at the angle to which it has momentarily swung.

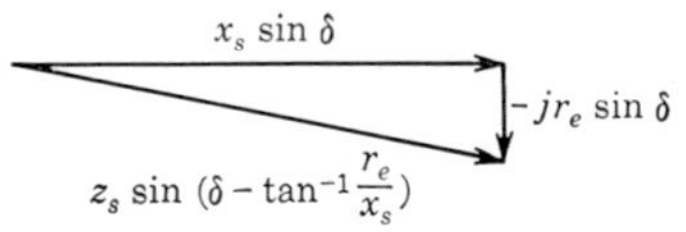

The Constants of Synchronous Generators Which Are to Operate in Parallel Need Not Be Inversely Proportional to Their Ratings. When transformers which have equal ratios of transformation are operated in parallel, all their primary voltages must be equal and in phase. All their secondary voltages must also be equal and in phase. The loads the transformers carry and the phase relations among the currents they deliver depend solely upon their constants. For this reason it is important that transformers which are to operate in parallel should have constants approximately inversely proportional to their ratings. The conditions for successful parallel operation of synchronous generators are not nearly so rigid since the excitation voltages, corresponding to the primary voltages in the case of transformers, do not have to be equal or in phase.   If the constants of the synchronous generators are not in the inverse ratios of their ratings it makes little difference, since the kw load can be divided among the generators in any desired ratio by means of the governors of the prime movers, and the armature currents can be brought into phase by adjusting the field excitations.

Chapter 27

Period of phase swinging or hunting. Damping. Governors. Effect of differences in the slopes of the prime-mover speed-load characteristics on the division of load among synchronous generators operating in parallel. Effect of changing the tension of the governor spring on the load carried by a synchronous generator in parallel with others.

Period of Phase Swinging or Hunting. If a synchronous generator operating in parallel with others momentarily changes its angular velocity owing to any cause, synchronizing action is developed which tends to restore the proper phase relations. The generator which swings ahead in phase has more load put on it and the generator which swings behind in phase is relieved of load, but the amounts of power they receive from their prime movers cannot change since the governors do not respond to momentary changes in speed which accompany momentary changes in phase. This causes the machines which lag to speed up and those which lead to slow down. Because of the inertia of the moving parts of the generators, they swing past the position of no synchronizing action. The synchronizing action then reverses and tends to pull the generators together again. This action is hunting and is the same as the hunting which takes place in synchronous motors. It would continue indefinitely if it were not for the damping action of the losses in the pole faces produced by hunting and by the damping action of the currents induced by hunting in the dampers. Since a change in load on a synchronous machine must be accompanied by a change in the power angle, such a machine is subject to hunting. With a properly designed synchronous generator, the amount of oscillation about the phase position corresponding to the load is small and of little importance under the usual operating conditions. It is only when the oscillation becomes extreme because of periodic changes in the load, in the terminal voltage, or in the frequency of the system that hunting is likely to become serious and to produce violent surges of current in the circuit or to cause the machines to drop out of synchronism. Violent hunting may be caused by periodic variations in the driving torque of a synchronous generator which is driven by a reciprocating engine, especially a gas engine, if there is insufficient flywheel effect to limit the variation of angular velocity to a small value.

The period of hunting of a synchronous generator can be found in the same way as that of a synchronous motor. A cylindrical-rotor generator is assumed and the effect of the damping action on the period of hunting is neglected. The variation in the angular velocity produced by hunting is equivalent to an oscillation superimposed on the constant angular velocity of the generator which is fixed by the frequency of the system. For this oscillation in speed, the rotor acts like a torsional pendulum and has a definite period of oscillation fixed by the constants of the generator, by its terminal and excitation voltages, by the combined moment of inertia of the generator and its prime mover, and by the damping. The effect of the damping is small and is neglected.

The time of oscillation of a torsional pendulum is

$$t = 2\pi \sqrt{\frac{J}{M}} = 2\pi \sqrt{\frac{\Sigma mr^2}{M}} \tag{208}$$

where $J = \Sigma mr^2$ is the moment of inertia of the pendulum and M is the restoring couple per unit of angular displacement from its position of equilibrium. In a synchronous generator, J is the moment of inertia of the generator and its prime mover. If the damping is neglected, M is proportional to the synchronizing power, which from Eq. (207), page 365, for a cylindrical-rotor machine, is

$$\frac{dP'}{d\delta} = \frac{VE'_a}{z_s} \cos \left(\delta - \tan^{-1} \frac{r_e}{x_s} \right) \tag{209}$$

If r_e, the armature effective resistance, is small compared with the synchronous reactance x_s, the expression for the synchronizing power reduces to

$$\frac{dP'}{d\delta} = \frac{VE'_a}{x_s} \cos \delta \tag{210}$$

For small values of the power angle δ, $\cos \delta$ changes slowly for moderate changes in δ and can be considered constant as a first approximation. Making this approximation and neglecting the effect of the damping, the natural period of hunting is

$$t = 2\pi \sqrt{\frac{Jx_s}{KVE'_a}} \text{ approximately} \tag{211}$$

where K is a constant.

The natural period of hunting of the cylindrical-rotor generator therefore varies directly as the square root of the product of the moment of inertia and the synchronous reactance and inversely as the square root of the product of the terminal and excitation voltages. The period is affected by the load and its power factor since the value of the power angle, included in the constant in Eq. (211), depends upon both. With fixed excitation and terminal voltages, the period of hunting increases with the load since the cosine of the power angle decreases with the load. If the natural period of oscillation of the generator as a torsional pendulum coincides with any periodic variation in the system, hunting may become extreme and the angular displacement caused by the hunting may become so great that the generator drops out of synchronism. When a generator which is driven by a reciprocating engine, especially by a gas engine, is to be paralleled with other generators, it is necessary to make the combined moments of inertia of the generator and its engine large in order to limit to a very small value the angular variation in the speed of the generator caused by the impulses of the pistons of the engine. It is also necessary to have the natural period of oscillation of the generator and its engine differ from any periodic variation in the load on the system. It must also differ from the natural period of the governor of the prime mover.

Damping. Synchronous generators which are to operate in parallel require a certain amount of damping, but as a rule the damping does not have to be so great as for synchronous motors unless the generators are to be subjected to violent fluctuations in load or their prime movers have very irregular driving torques. Since the damper of a generator does not serve also as a starting winding, as in a synchronous motor, less damping is required as a rule for synchronous generators than for synchronous motors. Dampers for synchronous generators therefore can be designed for damping only. For best damping, the resistance of a damper should be low. The frequency of the currents induced in a damper by hunting is very low. Its resistance for damping is therefore practically its d-c resistance.

Steam turbines have practically uniform driving torque. Because of their high speeds, synchronous generators driven by such prime movers have few poles, and consequently small angular variations in their speeds cause relatively small variations in their power angles. For example, with four poles, a variation in the angular velocity of 1 space degree during a revolution causes a variation of only 2 electrical degrees or a variation of only 2 deg in the power angle. With a slow-speed synchronous generator with 20 poles, an equal variation of 1 space degree in the angular velocity causes a variation of 10 deg in the power angle. Turbine generators, because of their high speeds and the resulting small number of poles and also because of the uniformity of the driving torques of their prime movers, require little damping other than the natural damping caused by eddy currents and hysteresis losses produced in their solid cylindrical rotors by hunting.

Synchronous generators driven by internal-combustion engines require strong damping, when operated in parallel, because of the irregularity of the driving torque of internal-combustion engines and their relatively low speed. The variation in the angular velocity of such an engine must be limited to a small amount by the use of a heavy flywheel and a suitable number of cylinders.

Hunting may be caused by any periodic variation in the circuit fed by the synchronous generators, as for example a periodic variation in the load, but it is seldom that the frequency of variation in a load is the same as the natural frequency of the generators.

Turbines for both water and steam have uniform turning moments. Turbine generators are therefore free from hunting caused by their prime movers.

Governors. Hunting may be caused by improperly designed governors. Governors for engines which are to drive synchronous generators must not be too sensitive and must be sufficiently damped to prevent overrunning. If ω_1 and ω_2 represent the maximum and minimum angular velocities of an engine during one revolution, the mean angular velocity is

$$\omega_m = \frac{\omega_1 + \omega_2}{2}$$

The variation in speed referred to the mean speed is

$$\sigma = \frac{\omega_1 - \omega_2}{\omega_m}$$

This is called the cyclic irregularity of engine speed. The cyclic irregularity of steam or water turbines is zero. With too sensitive a governor, the cyclic irregularity of engine speed sets up oscillations in the governor and causes hunting. To avoid this action, governors must be sufficiently damped so that they do not respond to the cyclic irregularity of speed.

Owing to the friction of the parts moved by the governor of an engine or turbine, there must be a certain change in speed of an engine or turbine before its governor acts. With an engine running at a certain mean speed ω_m, if the greatest and least speeds the engine can have without the governor acting are ω_1' and ω_2',

$$\Delta = \frac{\omega_1' - \omega_2'}{\omega_m}$$

is known as the coefficient of governor insensitiveness. The speeds ω_1', ω_2', and ω_m are plotted in Fig. 216 against the load.

The curves marked ω_1' and ω_2' give the maximum and minimum speeds for different loads at which the engine can operate without action of its governor. If the engine is a reciprocating engine, its speed varies by an amount equal to $0.5\sigma\omega_m$ above and below its mean speed. Therefore in order that its speed shall lie within the limits of maximum and minimum speed fixed by the lines marked ω_1' and ω_2', Fig. 216, its average speed during a revolution must be less than the speed represented by the line ω_1' and must be greater than the speed represented by the line ω_2' by $0.5\sigma\omega_m$. Subtracting $0.5\sigma\omega_m$ from the ordinates of the curve marked ω_1' and adding it to the ordinates of the curve marked ω_2' gives curves ω_1'' and ω_2''. These last two curves show the maximum and minimum average speeds corresponding to different loads at which the engine can operate without its governor acting. The more sensitive the governor, the closer

are the curves ω_1' and ω_2'. Their separation must always be greater than the cyclic irregularity of the engine; otherwise the governor acts because of the variation in speed during each revolution. A governor must always be damped sufficiently to make its coefficient of insensitiveness greater than the cyclic variation in the engine speed. If the prime mover is a turbine, there is practically no cyclic

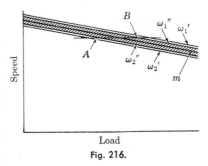

Fig. 216.

variation in its speed and the curves ω_1' and ω_1'' coincide as do also the curves ω_2' and ω_2''.

When any number of identical synchronous generators driven by identical engines are operated in parallel, they do not necessarily share the load equally. Let the horizontal line AB, Fig. 216, represent the speed of the system. The portion AB of this line lying between the curves ω_1'' and ω_2'' represents the possible loads corresponding to the assumed speed. The portion of the line AB included between the two curves ω_1'' and ω_2'' decreases as the slope of the speed-load curve increases. Consequently the greater the drop in speed for a given increase in load, the more uniform is the distribution of the load among the generators. A drooping speed characteristic is undesirable on account of the large change in frequency produced by change in load. A speed regulation of about 3 per cent usually gives satisfactory results.

Effect of Differences in the Slopes of the Prime-mover Speed-load Characteristics on the Division of Load among Synchronous Generators Operating in Parallel. To show the effect of the slopes of prime-mover speed-load characteristics on the distribution of load among synchronous generators in parallel, the slight variation in the distribution of load caused by lack of sensitiveness of the governors is neglected. Under this condition, the speed-load curve of a prime mover can be represented by a single line which corresponds to the mean-speed line marked ω_m in Fig. 216, page 371.

Figure 217 shows speed-load curves of two prime movers which are dissimilar. Assume that the synchronous generators have the same number of poles or, if the number of poles is different, that the prime mover speed is plotted in terms of generator frequency, as in Fig. 217.

The frequency must be the same for both generators. At the frequency f, both generators carry equal loads. As load is added to the system, the frequency drops. Both generators continue to operate at equal frequencies, but the frequency is lower than before. Let this new frequency be f'. Generator 1 is now carrying load L_1' which is greater than load L_2'.

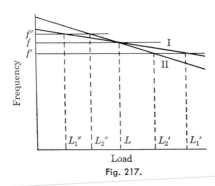

Fig. 217.

If the load on the system is decreased, the frequency rises to a new value f''. Generator 2 is now carrying the greater load. In general, as the load on a system is increased, the generators driven by the prime movers having the least drop to their frequency-load characteristics increase their loads faster than the generators driven by the prime movers with the most drooping characteristics. As the load is decreased the generators driven by the prime movers with the most drooping characteristics drop their loads slowest. In order that the generators shall divide the load equally, they must be driven by prime movers with identical frequency-load characteristics. If the generators are not of the same rating, the frequency-load characteristics of their prime movers should be identical when outputs are plotted in percentage of full-load output instead of in kilowatts.

The frequency-load characteristics of the prime movers should have considerable droop. Too flat characteristics exaggerate the effect of any slight differences which may exist among the slopes. Perfectly flat characteristics produce unstable operating conditions. The difference between the slopes of the characteristics shown in Figs. 218 and 219 is the same, but the actual slopes of the characteristics shown in Fig. 219 are greater.

The slopes of the characteristics shown in Fig. 219 are exaggerated in order to show more clearly the effect of the droop. It is seen by referring to Fig. 218 that, when the droop in the frequency-load characteristics is small, a slight difference in the slopes makes a large difference in the distribution of load carried by the generators for any given total load on the system. From Fig. 219 it is evident that, when the droop is large, the same difference between the slopes of the characteristics has a much less marked effect on the distribution of the load between the two generators. As has already been stated in another connection, the frequency-load characteristics of prime movers that are to drive synchronous generators in parallel should have droops of about 3 per cent in speed from no load to full load.

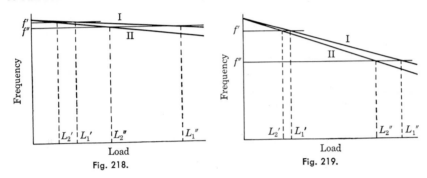

Fig. 218.

Fig. 219.

Effect of Changing the Tension of the Governor Spring on the Load Carried by a Synchronous Generator in Parallel with Others.

The method of adjusting the load carried by a synchronous generator in parallel with others has already been given on page 358, but what actually takes place may be made clearer by referring to Fig. 220, which gives the frequency-load characteristics for two prime movers.

The characteristics are I and II. At the frequency f the load carried by generator 1 is L_1 and the load carried by generator 2 is L_2. If it is desired

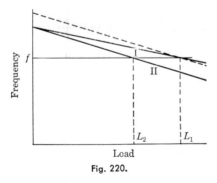

Fig. 220.

to make 2 increase its load, the tension of the governor spring of its prime mover is increased. This raises the frequency-load characteristic nearly parallel to itself to some new position shown by the dotted line. As the prime movers must still run at the same frequency, generator 2 must now carry more load. In the figure it carries the same load as the other generator.

Chapter 28

Effect of wave form on parallel operation of synchronous generators.

Effect of Wave Form on Parallel Operation of Synchronous Generators.

If two or more synchronous generators having dissimilar wave forms are paralleled, a voltage due to the unbalanced harmonics causes a circulatory current between generators. This circulatory current serves no useful purpose and increases the copper loss in the armatures. Suppose two three-phase generators having the following phase induced voltages are paralleled:

$$e_a = 3{,}900 \sin \omega t + 195 \sin 3\omega t + 117 \sin \left(5\omega t + \frac{\pi}{6} \right)$$

$$e_b = 3{,}900 \sin \omega t + 19.5 \sin \left(3\omega t + \frac{\pi}{6} \right) + 10 \sin \left(5\omega t - \frac{\pi}{6} \right)$$

The third harmonics cannot appear in the line voltages. If the generators are assumed not to have their neutrals connected, the third harmonics are without effect. If the individual loads and excitations are adjusted so that the voltages due to the fundamentals are equal and opposite on the series circuit of the generators, there is an unbalanced voltage acting in this circuit due to the fifth harmonics. This voltage has a maximum value

$$E_m = 117 \left(\cos \frac{\pi}{6} + j \sin \frac{\pi}{6} \right) - 10 \left(\cos \frac{\pi}{6} - j \sin \frac{\pi}{6} \right) = 112.3$$

This voltage causes a fifth-harmonic current to circulate in each phase of the armature. Since the phase order of the fifth harmonics in a three-phase balanced circuit is opposite to that of the fundamentals, the armature-reaction field caused by the fifth-harmonic currents revolves in a direction opposite to that of the field poles. The speed of this armature-reaction field with respect to the armature is five times synchronous speed. With respect to the field poles, this armature-reaction field revolves at six times synchronous speed. If seventh harmonics were present, since the seventh harmonics in a balanced three-phase circuit have the same phase order as the fundamentals, the armature-reaction field due to them would revolve at seven times synchronous speed with respect to the armature and at six times synchronous speed with respect to the field poles. Fifth and seventh harmonics in the armature cause a sixth harmonic in the pole flux. This harmonic in the pole flux is more or less damped out by the effect of a damper and by the effect of eddy-current and hysteresis losses in the field poles caused by the flux variation. Since the armature reaction produced by harmonics in the armature current is not fixed with respect to the field poles, there can be no such thing for harmonics as synchronous reactance. The armature reaction due to harmonics and its effect are complicated and too difficult to consider in this text.

Leakage reactance varies as the frequency. With the ordinary ratios of leakage reactance to synchronous reactance for the fundamentals of modern synchronous generators, it is probable that the leakage reactance of synchronous generators for the fifth and seventh harmonics and for harmonics of higher order is in general nearly as great as the synchronous reactance for the fundamental. For this reason the circulatory currents caused in synchronous generators in parallel by differences ordinarily found in their wave forms are not likely to be of importance.

If synchronous generators in parallel having the same wave form become displaced in phase, the resultant voltages acting to produce circulatory currents are exaggerated for the harmonics as compared with the fundamental, since a phase displacement of α electrical degrees for the fundamentals of the machines is equivalent to a phase displacement of 5α electrical degrees for the fifth harmonics and 7α electrical degrees for the seventh harmonics. Harmonic circulatory currents produced in this way are not likely to be troublesome except in three-phase Y-connected generators having pronounced third harmonics in their phase voltages and operated with interconnected neutrals.

If the synchronous generators are Y-connected and their neutrals are inter-connected, the effect of any third-harmonic component in the coil voltages cannot be neglected. This is the only condition under which ordinary differences in wave form are likely to cause trouble.

If Y-connected synchronous generators with like wave forms are put in parallel and their neutrals grounded, there is no current in the neutrals except that caused by an unbalanced load on the system, provided the phase voltages of the generators are equal and are in phase. If the phase voltages are not equal or become displaced in phase, there is a resultant third-harmonic voltage acting between lines and neutral which causes a triple-frequency current in each phase. This triple-frequency current is in addition to the circulatory current produced by the fundamental and harmonics of other frequencies than the third. The triple-frequency currents are all in phase and add directly in the neutral. Since the third-harmonic currents in the three phases are in phase, the armature reaction produced by them in a synchronous generator with non-salient poles and a distributed armature winding is approximately zero.   It may be represented as the sum of three equal space vectors differing by 120 deg in phase.   If the generators have salient poles, the triple-frequency currents produce some armature reaction, but it is relatively small.   Consequently the reactance for the triple-frequency currents is nearly equal to the leakage reactance for the third harmonics.   Suppose two equal three-phase Y-connected synchronous generators in parallel with interconnected neutrals become displaced in phase by α deg.   As a result of this phase displacement there is a triple-frequency current in each phase equal to

$$I_3 = \frac{2E_3' \sin (3\alpha/2)}{2 \sqrt{r^2 + x_3{}^2}}$$

where I_3 and E_3' are rms values. E_3' is the third-harmonic phase voltage. If the ratio of the synchronous reactance to the leakage reactance for the fundamental is assumed to be a, z_3 is approximately $\sqrt{r^2 + (3x_1/a)^2}$, or is approximately equal to $3z_1/a$.   If b is the ratio of the third-harmonic voltage to the fundamental voltage,

$$I_3 = \frac{a(bE_1') \sin (3\alpha/2)}{3z_1}$$

E_1'/z_1 is the steady-state or sustained short-circuit current I_{sc} for the fundamental voltage and is nearly equal to the steady-state short-circuit current of the generator.   Therefore

$$I_3 = \frac{abI_{sc} \sin (3\alpha/2)}{3}$$

If a is equal to 5 and b is equal to 0.10 and the generators are displaced by 20 electrical degrees, the triple-frequency current is

$$I_3 = \frac{(5)(0.1)(I_{sc})(\frac{1}{2})}{3} = 0.083 I_{sc}$$

If the short-circuit current of this machine is 150 per cent of the full-load current, the third-harmonic current in each phase corresponding to a phase displacement of 20 electrical degrees is 12 per cent of the full-load current.   The current in the neutral is three times the current per phase.   The trouble caused by this triple-frequency current is due not so much to the increase in the armature copper loss as to the tripping out of the circuit breakers of slow-speed generators which may have a much larger short-circuit ratio than that just given.   The displacement in phase may be produced by hunting, by a sudden change in load, or by careless synchronizing.

The trouble caused by the triple-frequency circulatory currents when

Y-connected synchronous generators with grounded neutrals are paralleled can be avoided in two ways: (1) by grounding each generator through a low resistance or reactance, and (2) by grounding only one of a group of generators in parallel. With reactance in the ground connections there is danger under some conditions of producing harmful oscillations. Grounding only one generator does away with the trouble caused by triple-frequency currents in the neutral and at the same time gives full protection to the system. The current which can be delivered on short circuit between any line conductor and neutral is limited to that which one generator can supply, and it affects only the grounded machine.

Chapter 29

A résumé of the conditions for parallel operation of synchronous generators. The difference between paralleling synchronous generators and d-c generators. Synchronizing devices. Connections for synchronizing single-phase synchronous generators. A special form of synchronizing transformer. Connections for synchronizing three-phase synchronous generators. Connections for synchronizing three-phase synchronous generators using synchronizing transformers. Lincoln synchronizer.

A Résumé of the Conditions for Parallel Operation of Synchronous Generators. Under ideal conditions for parallel operation of synchronous generators, the armature currents carried by the generators should be in phase when considered with respect to the parallel circuit and each generator should carry a current which is proportional to its rating. If the generators are free from hunting and have similar wave forms, these conditions can be fulfilled, as has been explained, by properly adjusting the power outputs of the prime movers and the excitations of the generators.

Synchronous generators which are to be operated in parallel should have (1) the same voltage rating, (2) the same frequency rating, and (3) approximately the same wave form.

It is desirable, but not necessary, that they should have synchronous impedances and effective resistances which are approximately inversely proportional to their current ratings.

The prime movers which drive the synchronous generators should have (1) the same frequency load characteristics, (2) drooping frequency load characteristics, and (3) constant angular velocity during each cycle.

In addition, the free period of oscillation of the governors of the prime movers and the mechanical period of oscillation of the system must be different from the frequency of any periodic variation in the prime-mover torques, the governors must not be too sensitive, and the natural electrical and mechanical frequencies of oscillation of the system must be different. If the synchronous generators are Y-connected and are to operate with their neutrals interconnected, their wave forms must not contain marked third harmonics.

The necessity for all the above conditions has been explained in preceding pages.

The Difference between Paralleling Synchronous Generators and D-C Generators. When paralleling a d-c generator with others, it is necessary to bring it up to rated speed, make its voltage approximately equal to the bus-bar voltage, and then close its main switch. It is necessary that the polarity of the incoming generator be the same as that of the bus bars. The polarity of a d-c generator does not change under normal conditions, and when a generator has been once correctly wired to the switch which connects it to the bus bars, it always builds up with the correct polarity unless some abnormal condition occurs to reverse its polarity. Such abnormal conditions are rare.

Paralleling of synchronous generators is somewhat more complicated, for in addition to having equal voltages, their frequencies must be the same. Moreover the polarity of synchronous generators changes with a periodicity equal to twice their frequency. For this reason some device is necessary for indicating not only when the frequencies are equal but also when the polarities are the same. It is necessary to "synchronize" the synchronous generator, which is being put in circuit, with those already operating. Since the relative speeds of d-c generators which operate in parallel are not fixed, the distribution of load among d-c generators in parallel can be controlled by merely adjusting their field excitations. With synchronous generators the conditions are different, since their relative speeds are fixed. Under ordinary conditions the relative field excitations of synchronous generators in parallel have practically nothing to do with the distribution of the power load among the machines. The relative field excitations control the power factors at which the synchronous generators operate. The portion of the total load on the system carried by any synchronous generator is controlled solely by the setting of the governor of its prime mover.

Synchronizing Devices. The simplest form of synchronizer is an incandescent lamp connected between the bus bars and the terminals of the synchronous generator to be synchronized in such a way as to indicate by its brilliancy when the generator is running in synchronism with other generators. Except with low-voltage synchronous generators, transformers must be used with the lamps.

Connections for Synchronizing Single-phase Synchronous Generators. Figure 221 shows the connections for synchronizing a low-voltage synchronous generator without the use of synchronizing transformers. Figure 222 shows the connections when such transformers are necessary. BB on these figures are the main bus bars. B_sB_s are synchronizing bus bars.

When the synchronous generators are synchronized without transformers, using the connections shown in Fig. 221, the maximum voltage impressed across the synchronizing lamp is equal to twice the voltage of the generators. The voltage across the lamp has this value when the generator which is being synchronized is exactly 180 deg out of phase with the bus-bar voltage. The lamp is dark when this generator is in phase with the bus-bar voltage. If the generator which is being synchronized is running a little fast or a little slow, the lamp flickers with a frequency equal to the difference between the frequencies of the voltages of the bus bars and the incoming generator. If $e_b = E_m \sin \omega t$ is the voltage of the bus bars and $e_g = E'_m \sin (\omega \pm \Delta\omega)t$ is the voltage of the

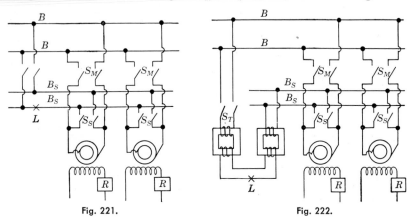

Fig. 221.   Fig. 222.

generator which is being synchronized,

$$e_L = e_b - e_q = E_m \sin \omega t - E'_m \sin (\omega \pm \Delta\omega)t$$

is the voltage across the lamp. If $E_m = E'_m$,

$$e_L = 2E_m \sin \frac{\pm\Delta\omega}{2} t \cos \left(\omega \pm \frac{\Delta\omega}{2} \right) t \qquad (212)$$

since $\sin a - \sin b = 2 \cos \frac{1}{2} (a + b) \sin \frac{1}{2} (a - b)$.

The sine term of Eq. (212) shows that the voltage impressed on the lamp has a frequency equal to the mean frequency of the voltage of the bus bars and the generator. Equation (212) also shows that the maximum value of each voltage wave is different from that immediately preceding it. Equation (212) if plotted gives a curve like that shown in the lower half of Fig. 223. The upper half of Fig. 223 shows the voltage waves of the generator and the bus bars.

The envelope of the lower curve shown in Fig. 223 has for its equation

$$2E_m \sin \frac{\pm\Delta\omega}{2} t$$

Therefore if the wave forms of the voltages of bus bars and generator are sinusoidal, the successive maximum values of the voltage across the lamp vary according to the sine law, and since the lamp is bright once in each half of the cosine wave, the lamp flickers Δf times per second where f is the frequency of the bus-bar voltage. If the frequency of the bus-bar voltage is 60 cycles and that of the generator is 60.3 cycles, Δ is $0.3/60 = 0.005$ and the lamp is bright once in every $3\frac{1}{3}$ sec. There is in general no difficulty in making this interval 9 or 10 sec or even greater.

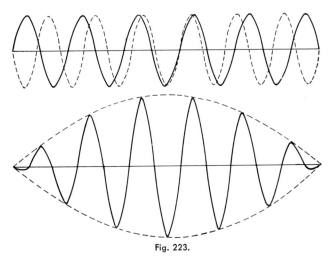

Fig. 223.

If the connections with transformers are used, the conditions are essentially the same as without transformers, except that the lamp can be either bright or dark for synchronism according to the way the secondaries of the transformers are connected. The maximum voltage which is impressed across the lamp is equal to twice the secondary voltage of the transformers. When transformers are used, it is usually better to connect them so as to make the lamp bright at synchronism, for it is easier to judge the instant of maximum brightness of the lamp than to estimate the middle point of its period of darkness. If the lamp is replaced by a voltmeter, the sensitiveness of the synchronizing device is greatly increased.

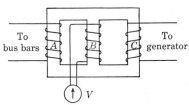

Fig. 224.

A Special Form of Synchronizing Transformer. The two transformers shown in Fig. 222 can be replaced by a single transformer with three separate windings shown in Fig. 224.

The branch B of the transformer core forms a return path for the fluxes produced by the exciting windings A and C. If the voltages impressed on these windings are in conjunction, the fluxes in A and C

add directly in the branch B of the core, and the voltage induced in the winding B, which is connected to the voltmeter or lamp, is a maximum. If the voltages impressed on the two exciting windings A and C are in opposition, the fluxes produced by these windings are in opposition in the branch B of the core and neutralize in that branch. Under this condition the voltage across the lamp or voltmeter is zero.

Connections for Synchronizing Three-phase Synchronous Generators. The connections for synchronizing two three-phase synchronous generators which are of low enough voltage not to require transformers for the synchronizing lamps are shown in Fig. 225.

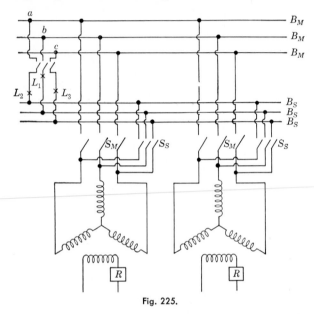

Fig. 225.

Closing switch S_s for the generator to be synchronized connects the generator to the synchronizing bus bars. When the generators are in conjunction, all three lamps L_1, L_2, L_3 are dark. If the voltage of the generators is too high to be taken directly by the lamps, the lamps can be replaced by suitable transformers with the lamps across their secondaries.

When synchronizing three-phase synchronous generators for the first time, it is necessary to synchronize at least two phases. If a single phase alone is synchronized, the other two phases may be 120 deg out of phase. After the connections have once been made and have been found to be correct, synchronizing one phase is sufficient. In spite of this, it is customary to provide synchronizing lamps for all three phases.

With the connections shown in Fig. 225, the lamps are all dark for synchronism. A better arrangement of lamps, known as the Siemens and

Halske arrangement, is obtained by interchanging the connections of two of the leads from the lamps at either the main bus bars or at the synchronizing bus bars, by interchanging the connections at a and c, for example, Fig. 225. With this arrangement of the synchronizing lamps, the lamp L_1 is dark at synchronism and the lamps L_2 and L_3 are equally bright but below the maximum brightness. If the generator which is being synchronized is running too fast or too slow, the lamps glow in rotation and indicate by the direction of this rotation whether the generator is running too fast or too slow.

It is highly desirable to know whether a synchronous generator which is being synchronized is running too fast or too slow, not only on account of the time which can often be saved by this knowledge, but also on account of the desirability of having the generator which is being synchronized run too fast rather than too slow when it is paralleled. If the incoming generator is running too fast when put in circuit, the synchronizing action which pulls it into step puts load on it and at the same time relieves the other generators of some of their load. If the incoming generator is put in circuit when running too slow, it takes motor power to pull it into step.

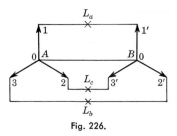

Fig. 226.

The effect of interchanging the connections of two of the synchronizing lamps can be seen by referring to Fig. 226. For simplifying the explanation of this arrangement of the lamps, the generators are assumed to be connected in Y with their neutrals interconnected. In Fig. 226, the vectors 1, 2, 3 and $1'$, $2'$, $3'$ represent the voltages of the generators. When the two generators are in phase, as they are shown in Fig. 226, lamp L_a is subjected to a voltage $V_{01} + V_{1'0} = 0$. The voltages impressed across the lamps L_c and L_b are $V_{02} + V_{3'0} = \sqrt{3}\, V_{02}$ and

$$V_{03} + V_{2'0} = \sqrt{3}\, V_{03}$$

If generator A is running too fast, the voltage across lamp L_b decreases until generator A has gained 120 deg in phase, when the voltages $V_{02'}$ and V_{03} are in phase. At this instant lamp L_b has a voltage $V_{03} - V_{02'} = 0$ impressed on it and is dark. The other lamps are equally bright. When the generator has gained another 120 deg, lamp L_c is dark. In other words, the lamps are dark in succession. If generator A is running more slowly than B, the order in which the lamps become dark reverses. The lamps can conveniently be placed on the corners of an equilateral triangle, showing by the rotation of their brilliancy whether the generator which is being synchronized is running too fast or too slow. The proper time to put the incoming generator on the line is when lamp L_a, Fig. 226, is dark and the other two lamps L_c and L_b are equally bright. The maximum voltage to which any lamp is subjected is twice the Y voltage of the generators. The neutral connection shown in Fig. 226 is not necessary and is not used in practice.

Connections for Synchronizing Three-phase Synchronous Generators Using Synchronizing Transformers. If transformers are required for synchronizing synchronous generators, as is practically always the case, it is customary to provide one set for each generator and one set to con-

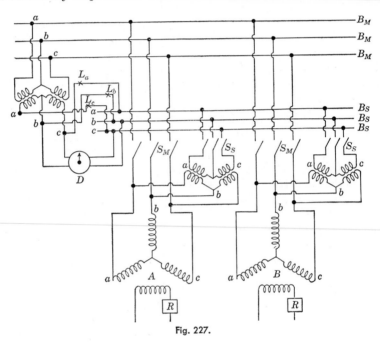

Fig. 227.

nect the synchronizing bus bars to the main bus bars. In this way the synchronizing bus bars, as well as all synchronizing switches and other synchronizing devices, are kept at low voltage. The synchronizing transformers are often connected in V to reduce the number required. The proper connections for operating three-phase synchronous generators in parallel are shown in Fig. 227.

The connections of the two synchronizing lamps L_a and L_c are reversed from their natural order. This reversed order has already been shown in Fig. 226. With this arrangement lamp L_b, Fig. 227, is dark at synchronism. If the sequence of the phases of the generators is a, b, c, a right-hand rotation of the brilliancy of the lamps indicates that the generator which is being synchronized is running too fast. This assumes that the direction of rotation of the generator is right-handed. D, Fig. 227, is some form of synchronizing device which indicates the exact point of synchronism more closely than the lamps.

Lincoln Synchronizer. The necessity for determining the point of synchronism more exactly than is possible by the use of incandescent lamps has led to the development of several forms of synchronizing device of which the Lincoln synchronizer is typical. The Lincoln synchronizer is a rotary-field synchronizer indicating by a pointer that moves over a graduated dial the exact difference in phase between the voltage of the generator which is being synchronized and the voltage of the bus bars. The direction of rotation of the pointer also indicates whether the generator is running too fast or too slow. The Lincoln synchronizer is in reality a small motor having a laminated field excited by the bus bars through a large noninductive resistance. The armature has two windings in space quadrature, excited from the generator to be synchronized, one through a noninductive resistance, the other through a reactance. Figure 228 shows the essential features of this device.

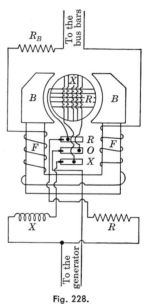

Fig. 228.

The armature takes up a position with the axis of the field produced by its windings coincident with the axis of the field produced by the poles BB, when the latter field is a maximum. The winding X on the armature is in series with a large reactance and carries a current which is practically in quadrature with the voltage of the generator. The winding R on the armature is in series with a large noninductive resistance and the current in it is nearly in phase with the generator voltage. The current in the field coils FF is nearly in phase with the voltage of the bus bars. If the voltage of the generator and the voltage of the bus bars are in phase, the currents in the field winding FF and in the armature winding R also are in phase. Under these conditions, the armature takes a position such that the magnetic fields due to these two windings coincide. The armature therefore rotates until the axis of the winding R is horizontal (Fig. 228). If the voltages of the generator and the bus bars are 180 deg out of phase, the axis of the winding R is again horizontal but is 180 deg from its former position. If the voltages are in quadrature, the axis of the winding X is horizontal. The position of the armature always indicates the difference of phase between the voltages of the generator and the bus bars. If the frequency of the generator differs from that of the bus bars, the armature rotates in one or the other direction according as the generator is running too fast or too slow.   The speed of its rotation is proportional to the difference of the frequencies of the voltages of the generator and the bus bars.   A pointer attached to the armature indicates on a graduated dial the exact difference in phase between the voltage of the generator which is being paralleled and the voltage of the bus bars.

PART FIVE

POLYPHASE INDUCTION MOTORS

Chapter 30

Asynchronous machines. Polyphase induction motor. Operation of the polyphase induction motor. Slip. Revolving magnetic field. Rotor blocked. Rotor free. Load is equivalent to a noninductive resistance on a transformer. Transformer diagram of a polyphase induction motor. Equivalent circuit of a polyphase induction motor. Polyphase induction regulator.

Asynchronous Machines. Up to this point only machines which operate at synchronous speed have been considered. There is another class known as asynchronous machines. As their name implies, these do not operate at synchronous speed. Their speed varies with the load. One distinguishing feature of commercial synchronous machines is that they require fields excited by direct current. An asynchronous machine requires no d-c excitation. Both parts of an asynchronous machine, armature and field, carry alternating current and are either connected in series, as in the series motor, or are inductively related, as in the induction motor. The induction motor and induction generator, the series motor and repulsion motor, and all forms of a-c commutator motors are included in the general class known as asynchronous machines. The induction motor is the most important and most widely used type of asynchronous motor. It has essentially the same speed and torque characteristics as a d-c shunt motor and is suitable for the same kind of work. Its ruggedness and ability to stand abuse make it a particularly desirable type of industrial motor.

Polyphase Induction Motor. The polyphase induction motor is equivalent to a static transformer operating on a noninductive load. It is a transformer with a secondary capable of rotating with respect to the primary. Although the secondary is usually the rotating part, the motor operates equally well if the secondary is fixed and the primary revolves. In what follows, the primary is assumed to be stationary and is referred to as the primary, stator, or field. The secondary, which rotates, is called the secondary or rotor. The terms primary and secondary are perfectly definite, meaning the part which receives power directly from the line and the part in which the current is produced by electromagnetic induction. The terms stator and rotor are not so definite, since their significance is not determined by the electrical connections but by the part which is stationary.

Operation of the Polyphase Induction Motor.   The stator winding of a polyphase induction motor is similar to the armature winding of a polyphase synchronous generator or synchronous motor.   The polyphase currents in the stator winding produce a rotating magnetic field which corresponds to armature reaction in the synchronous generator.   The

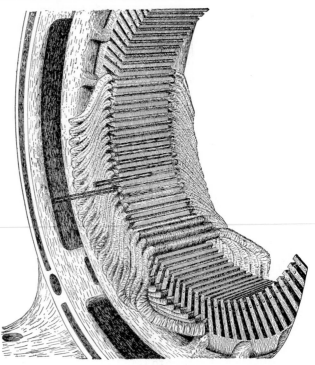

Fig. 229.

fundamental of this field revolves at synchronous speed with respect to the stator in the same way that the armature reaction of a synchronous machine revolves with respect to the armature.   With respect to the rotor the field revolves at a speed which is the difference between synchronous speed and the speed of the rotor.   This difference in speed is known as the slip.   A portion of the stator field of an induction motor with a few coils in place is shown in Fig. 229.

The rotor, which must have a polyphase winding with the same number of poles as the stator, has currents induced in it by the revolving magnetic field.   These currents cause the rotor to revolve in the same direction as the revolving magnetic field set up by the stator.   Although the rotor and stator windings must be wound to give the same number of poles, they need not be wound for the same number of phases. If it were not for rotational losses, synchronous speed would be reached at no load. Under load conditions, the difference between the speeds of the magnetic field and the rotor is just sufficient to cause enough current to be induced in the rotor to produce the torque required for the load and the rotational loss.

The speed of the revolving magnetic field depends upon the frequency and the number of poles for which the motor is wound and is independent of the number of phases. The only condition which must be fulfilled in regard to the number of phases is that the space relations of the windings for the different phases in electrical degrees must be the same as the time-phase relations of the currents they carry. Thus for a three-phase winding, the windings must be 120 electrical degrees apart, and for a four-phase winding, they must be 90 electrical degrees apart.

Slip. If f_1 and p are the impressed frequency and the number of poles, the speed of the revolving magnetic field and also the synchronous speed of the motor in revolutions per minute is

$$n_1 = \frac{2f_1}{p} 60 \tag{213}$$

If n_2 is the actual speed of the rotor and s is the slip expressed as a fraction of synchronous speed, then

$$s = \frac{n_1 - n_2}{n_1} \tag{214}$$

and

$$n_2 = n_1(1 - s) \tag{215}$$

Revolving Magnetic Field. Assume the rotor to be on open circuit. This corresponds to the condition in a static transformer when the secondary is open. The only mmfs acting are the mmfs produced by the primary windings.

The primary winding of an induction motor is distributed and is similar to the armature winding of a synchronous machine with the same number of phases and poles.

At any instant, the space distribution of the flux caused by any one phase is determined by the distribution of the winding and by the wave form of the impressed voltages which are assumed to be sinusoidal. The air gap of an induction motor is uniform and except for the presence of the slots does not affect the flux distribution. The space distribution of the flux set up by the stator becomes more nearly sinusoidal as the number of slots per phase is increased.   This distribution can be found by the method indicated on page 216 in the section on Synchronous Generators.   The time variation of the air-gap flux due to any one phase may or may not be sinusoidal depending upon the wave form of the impressed voltages.   If the impressed voltages are balanced and the space distribution of the flux produced by each stator phase is sinusoidal, the fundamentals of the time variation of the air-gap flux for all phases combine to produce a magnetic field which revolves at synchronous speed, is constant in magnitude, and is sinusoidal in its space distribution.

Actually, the space distribution of the air-gap flux is not sinusoidal though the harmonics existing under normal operating conditions are small.   They produce voltages of fundamental frequency in the stator winding as in the synchronous generator, and their effect is included in the leakage reactance of the stator winding.   At certain speeds and at starting, space harmonics in the flux may have a marked and undesirable effect on the torque of the motor.   For this reason, in designing an induction motor, care must be taken to diminish as much as possible any harmonics in the flux which would prevent the motor from readily attaining its normal operating speed.   Certain harmonics in the flux may cause objectionable vibration of the motor under the usual operating conditions.   In what follows the air-gap flux is assumed to be sinusoidal.

The flux due to any one phase is alternating with respect to the winding. As in the transformer, it induces a voltage in the winding which is equal to the voltage impressed on the phase minus the impedance drop due to the resistance and leakage reactance of the winding.   Except as this induced voltage is influenced by the impedance drop, it is of the same wave form as the impressed voltage and the magnetizing current must adjust itself to meet this condition.   If the impressed voltage is sinusoidal, the induced voltage is nearly sinusoidal, since the impedance drop is small.   If the impressed voltage contains harmonics, the induced voltage contains the same harmonics.

Figure 230 shows the developed stator of a three-phase induction motor. The dots represent inductors and the numbers indicate the phases to which the inductors belong.

The full line, the dotted line, and the dot-and-dash line show the fundamentals of the space distribution of the fluxes produced in the air gap by phases 1, 2, and 3.   Each distribution is shown for maximum current in the phase considered.   Consider a point b, situated α electrical degrees from the beginning of phase 1.   The flux density $\mathcal{B}_b$ at this point is

$$\mathcal{B}_b = \mathcal{B}_1 \sin \alpha + \mathcal{B}_2 \sin (\alpha - 120°) + \mathcal{B}_3 \sin (\alpha - 240°) \quad (216)$$

where $\mathcal{B}_1$, $\mathcal{B}_2$, and $\mathcal{B}_3$ are the flux densities at the centers of phases 1, 2, and 3.   If only the fundamental of the time variation of flux is considered, Eq. (216) can be written

$$\mathcal{B}_b = \mathcal{B}_m\,[\sin \alpha \sin \omega t + \sin(\alpha - 120°)\sin(\omega t - 120°)$$
$$+ \sin(\alpha - 240°)\sin(\omega t - 240°)]$$
$$= \tfrac{3}{2}\mathcal{B}_m \cos(\alpha - \omega t)$$
$$= \tfrac{3}{2}\mathcal{B}_m \sin\left(\omega t + \frac{\pi}{2} - \alpha\right) \tag{217}$$

Equation (217) shows that the flux density at any point b is sinusoidal with respect to time. It also shows that at any given time the space distribution of the air-gap flux is sinusoidal.

If $\alpha = \omega t$, Eq. (217),

$$\mathcal{B}_b = \tfrac{3}{2}\mathcal{B}_m \sin \frac{\pi}{2}$$
$$= \tfrac{3}{2}\mathcal{B}_m$$

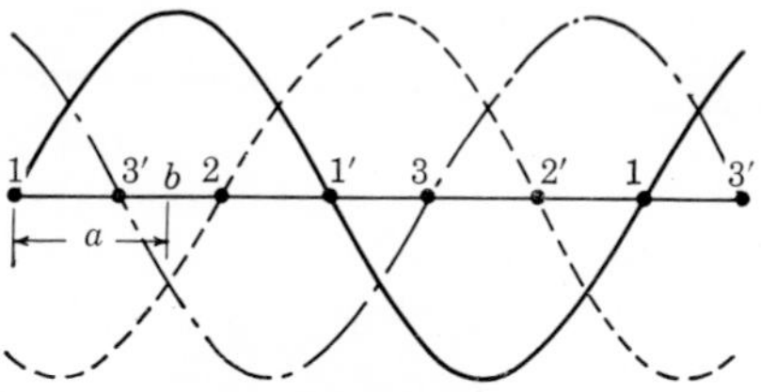

Fig. 230.

i.e., the magnetic field travels about the air gap at synchronous speed and has a constant value.

If w is the thickness of the stator core and A is the pole pitch, the fluxes φ_1, φ_2, and φ_3 through phases 1, 2, and 3 are at any instant of time t

$$\varphi_1 = \tfrac{3}{2}w\mathcal{B}_m \frac{A}{\pi} \int_{0°}^{\pi} \sin\left(\omega t + \frac{\pi}{2} - \alpha\right) d\alpha$$
$$= 3w\mathcal{B}_m \frac{A}{\pi} \sin \omega t \tag{218}$$
$$\varphi_2 = \tfrac{3}{2}w\mathcal{B}_m \frac{A}{\pi} \int_{120°}^{\pi+120°} \sin\left(\omega t + \frac{\pi}{2} - \alpha\right) d\alpha$$
$$= 3w\mathcal{B}_m \frac{A}{\pi} \sin(\omega t - 120°) \tag{219}$$
$$\varphi_3 = \tfrac{3}{2}w\mathcal{B}_m \frac{A}{\pi} \int_{240°}^{\pi+240°} \sin\left(\omega t + \frac{\pi}{2} - \alpha\right) d\alpha$$
$$= 3w\mathcal{B}_m \frac{A}{\pi} \sin(\omega t - 240°) \tag{220}$$

It is evident from Eqs. (218), (219), and (220) that the total flux through each phase is sinusoidal with respect to time and that the total fluxes linking the phases differ in time phase by 120 deg.

The maximum flux linking a full-pitch turn of the primary winding is then $\varphi_m = 3w\mathcal{B}_m(A/\pi)$. This induces a voltage in each phase of a distributed fractional-pitch primary winding as in the case of a synchronous generator, and this voltage E_1 equals $4.44\,N_1 f_1 \varphi_m k_{b_1} k_{p_1}$ where N_1, k_{b_1}, and

k_{p_1} are, respectively, the number of primary turns per phase, the breadth factor of the primary winding, and the pitch factor of the primary winding.

Rotor Blocked. If the rotor is blocked, the revolving magnetic field produces a voltage in each phase of both the rotor and the stator. This rotor voltage E_2 equals $4.44N_2f_1\varphi_mk_{b_2}k_{p_2}$ where N_2, k_{b_2}, and k_{p_2} are the rotor turns per phase, the breadth factor, and the pitch factor, respectively. If the rotor circuits are closed, polyphase currents are produced in them of the same frequency as the stator currents. These currents, like those in the stator, set up an mmf which rotates at synchronous speed and this mmf reacts on the stator to cause additional currents to flow in the stator windings. This reaction is similar to that which takes place between the primary and secondary of a transformer. Let the stator current per phase when the rotor is open be represented by I_n as in a transformer. This is the flux-producing component of the stator current. The additional stator current per phase I'_1 caused by the reaction of the rotor currents is called the load component of the stator current. Neglecting I_n, the total mmf produced by the stator and rotor currents must equal zero. Therefore [see Eq. (142), page 211, under Synchronous Generators]

$$0.90N_1I'_1k_{b_1}k_{p_1}m_1 + 0.90N_2I_2k_{b_2}k_{p_2}m_2 = 0$$

and

$$-\frac{I'_1}{I_2} = \frac{N_2k_{b_2}k_{p_2}m_2}{N_1k_{b_1}k_{p_1}m_1}$$

where I_2 is the rotor current per phase and m_2 and m_1 are the number of phases in the rotor and stator windings, respectively. If the number of phases in the rotor and stator are alike and N'_1 and N'_2 are the effective turns per phase in the stator and rotor windings, then

$$-\frac{I'_1}{I_2} = \frac{N'_2}{N'_1} = \frac{1}{a}$$

where a is the ratio of transformation.

The rotor current lags behind E_2, the induced voltage in the rotor winding, by an angle whose cosine equals $r_2/\sqrt{r_2{}^2 + x_2{}^2}$ where r_2 and x_2 are the rotor effective resistance and the rotor leakage reactance per phase at stator frequency. The vector diagram for a polyphase induction motor with rotor blocked is exactly the same as that for a short-circuited transformer. The magnetizing component of the stator current and the stator and rotor reactances x_1 and x_2 are larger for the motor because of the air gap between stator and rotor windings.

The rotor current, considered with respect to the revolving magnetic field, produces a torque which acts in the direction of rotation of the field, and if the rotor is free to revolve, it speeds up.

Rotor Free. When the rotor is blocked, the speed of the stator field with respect to the rotor inductors is proportional to the primary frequency. When the rotor revolves, the speed of the stator field with respect to the rotor inductors is equal to the difference between the speed of the field in space and the rotor speed. This relative speed is

$$n_s = n_1 - n_2$$

where n_1 is the speed of the stator field and n_2 is that of the rotor.

Replacing n_1 and n_2 by their values from Eqs. (213) and (215), page 389,

$$n_s = \frac{2f_1}{p} 60s$$

The frequency of the rotor currents corresponding to the speed n_s is

$$f_s = f_1 s$$

The rotor currents at a frequency $f_1 s$ produce a rotating mmf force in the rotor. This revolves at a speed

$$n_s = \frac{2f_1}{p} 60s$$

with respect to the rotor.

Since this mmf is revolving in the same direction as the rotor, its speed with respect to the stator is equal to its speed with respect to the rotor plus the speed of the rotor with respect to the stator, or to

$$n_s + n_2 = \frac{2f_1}{p} 60s + \frac{2f_1}{p} 60(1 - s)$$
$$= \frac{2f_1}{p} 60$$

Its speed with respect to the stator is the same as the speed of the stator field in space. The frequency of the rotor current considered with respect to the stator current is the impressed frequency f_1 of the circuit. The rotor current reacts on the stator current at this frequency whatever the speed of the rotor.

The resultant mmf causing the air-gap flux is equal to the vector sum of the mmfs of the stator and rotor currents. This is shown as $0.9N_1'I_n m_1$ on Fig. 231. The component mmf due to the load component of the primary current is equal and opposite to the mmf due to the rotor currents.

The flux φ corresponding to the flux producing or exciting component I_n of the stator current I_1 induces a voltage E_1 in the stator and a voltage E_2s in the rotor. E_2s has a frequency f_1s with respect to the rotor but a frequency f_1 with respect to the stator. The secondary current I_2 corresponding to the voltage E_2s is

$$I_2 = \frac{E_2s}{\sqrt{r_2{}^2 + x_2{}^2s^2}} \tag{221}$$

where x_2 is measured at primary frequency. At a frequency $f_2 = f_1s$, the secondary reactance is x_2s.

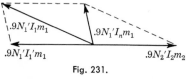

Fig. 231.

In Eq. (221), the current I_2, the voltage E_2s, the resistance r_2, and the reactance x_2s are at the rotor frequency $f_2 = f_1s$ and are the actual current, voltage, resistance, and reactance that exist in the rotor circuit. The current, voltage, resistance, and reactance in Eq. (221) can be referred to the stator frequency by dividing both numerator and denominator of the right-hand side of the equation by s. This change gives Eq. (222),

$$I_2 = \frac{E_2}{\sqrt{(r_2/s)^2 + x_2{}^2}} \tag{222}$$

This expression is used later with the vector diagram. E_2, r_2/s, and x_2 in Eq. (222) are all referred to the stator and are at stator or impressed frequency. I_2 is also at stator frequency when so referred. The resistance r_2/s is the apparent resistance of the rotor when referred to the stator.

If it were possible to observe the variation of the rotor current from a fixed point on the stator, this current would be seen to vary at a frequency f_1s due to the actual variation of the current in the rotor plus a superimposed variation $(1 - s)f_1$ due to the movement of the rotor conductors past the fixed point on the stator. The sum of these two variations is the apparent variation in the rotor current as seen from the stator. This is stator frequency f_1.

If the rotor is blocked and the resistance of each of its phases is increased to r_2/s by inserting resistances in its windings, the reaction of the rotor currents on the stator is the same as when the motor operates with a slip s and with no added resistances. The impressed voltages are assumed to be the same in both cases.

E_2 is the voltage which would be induced in the rotor by the flux φ if the rotor were blocked. It corresponds to the voltage induced in the secondary of a static transformer. E_2s is the actual voltage induced in the rotor when the rotor revolves with a slip s. The difference $E_2(1 - s) = E_R$ may be considered to be the voltage induced in the rotor due to its speed n_2. In other words $E_2(1 - s) = E_R$ is the rotational or armature voltage of the motor. E_R corresponds to the back emf in a d-c motor.

Load Is Equivalent to a Noninductive Resistance on a Transformer. The internal power developed per phase by any motor is equal to the

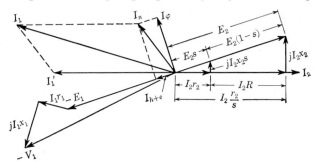

Fig. 232.

product of the current, the rotational voltage, and the cosine of the phase angle between the current and the rotational voltage. The internal power P'_2 developed by an induction motor is

$$P'_2 = E_2(1 - s)I_2 \frac{r_2}{\sqrt{r_2{}^2 + x_2{}^2 s^2}}$$

Replacing E_2 by its value from Eq. (221), page 394, gives

$$P'_2 = I_2{}^2 r_2 \left(\frac{1 - s}{s}\right) \tag{223}$$

Since $\dfrac{1 - s}{s}$ is a scalar quantity, $r_2 \left(\dfrac{1 - s}{s}\right)$ can be considered as a fictitious resistance R which depends upon the slip or upon the mechanical load.

$$P'_2 = I_2{}^2 R \tag{224}$$

The internal power developed by an induction motor is equivalent to a noninductive load on a transformer. $I_2 R = V_2$ corresponds to the secondary terminal voltage of the equivalent transformer.

Transformer Diagram of a Polyphase Induction Motor. The transformer diagram of a polyphase induction motor is given in Fig. 232. Everything in this diagram is per phase and is referred to the stator by use of the ratio of transformation.

The relative positions of the vectors on the secondary side of the diagram may be changed to correspond to their usual positions in the ordinary transformer diagram as indicated in Fig. 233. I_2R corresponds to the potential difference V_2 at the secondary terminals of a transformer.

In the vector diagram of Fig. 232, in the equivalent circuits which follow and in the equations developed in the next chapter, all quantities used are for an equivalent induction motor referred to the primary side. The equivalent induction motor is used to replace the actual motor for

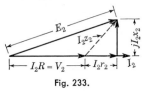

Fig. 233.

convenience in making calculations. This is a similar process to that used with transformers (see page 43). The equivalent induction motor is one having characteristics identical with those of the motor it replaces, but it has a ratio of transformation of one and therefore, for the equivalent induction motor, $E_1 = E_2$ and $I_1' = -I_2$.

Since all quantities are referred to the primary or stator, all primary values for the equivalent induction motor are the same as for the actual motor. Because of the polyphase windings, the reduction factors needed for referring secondary quantities to the primary are changed slightly from those used with transformers. They are influenced not only by the turns ratio but also by the coil pitch, the distribution of the windings, and the relative number of phases on the stator and rotor.

To give equal induced voltages in the primary and secondary of the equivalent motor, all secondary voltages of the actual motor must be multiplied by a ratio of transformation defined as follows:

$$a = \frac{N_1 k_{b1} k_{p1}}{N_2 k_{b2} k_{p2}} \tag{225}$$

Since the mmfs produced by polyphase currents in a polyphase winding combine for all phases, in order that the resultant mmf produced by the secondary currents of the equivalent induction motor shall be equal to that of the actual motor and also satisfy the relationship $I_1' = -I_2$, secondary currents of the actual motor must be multiplied by the following reduction factor:

$$\frac{m_2}{m_1 a} = \frac{m_2 N_2 k_{b2} k_{p2}}{m_1 N_1 k_{b1} k_{p1}} \tag{226}$$

where m_1 and m_2 are the numbers of phases in the primary and secondary windings. Since an impedance is equal to the voltage across it divided by the current through it, it follows that the reduction factor for impedances is equal to the quotient of the reduction factors for voltages and currents. Therefore in order to refer resistances, reactances, and impedances to the primary, they must be multiplied by

$$\frac{m_1 a^2}{m_2} = \frac{m_1}{m_2} \left(\frac{N_1 k_{b1} k_{p1}}{N_2 k_{b2} k_{p2}} \right)^2 \tag{227}$$

Equivalent Circuit of a Polyphase Induction Motor. The conditions of the vector diagram are the same as for the circuit shown in Fig. 234, which is the equivalent circuit of the induction motor. This circuit is the same as the equivalent circuit of a transformer which supplies power

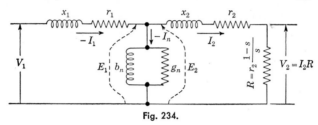

Fig. 234.

to a noninductive load R. Everything in the equivalent circuit is referred to the primary or stator. For example, r_2 on Fig. 234 is the actual secondary resistance per phase multiplied by the reduction factor given in Eq. (227). The susceptance and conductance b_n and g_n are such that

$$I_n = E_1(g_n - jb_n)$$

With the ordinary transformer, little error is introduced in the calculations based on the equivalent circuit if the portion of the circuit represented by b_n and g_n is placed directly across the impressed voltage. When

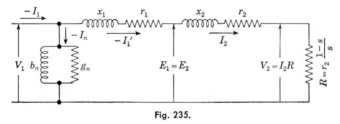

Fig. 235.

this change is made in the equivalent diagram of an induction motor, the error introduced is much greater, since the exciting current I_n of an induction motor is large compared with the load component I_1' of the stator current at full load. On account of the air gap, the primary leakage reactance x_1 of an induction motor is much larger than the leakage reactance of the primary winding of a transformer. The approximate equivalent circuit of an induction motor is given in Fig. 235. The use of this circuit introduces a nearly constant error of 2 or 3 per cent in the induced voltages E_1 and E_2 between no load and full load. Since the power and torque corresponding to any given slip vary as the square of E_2, the error in these quantities introduced by the use of the approximate circuit may be as high as 4 to 6 per cent.

Since everything in Fig. 235 is referred to the primary, $-I_1' = I_2$ and $I_1 = I_n + I_1'$ vectorially.

The calculation of the true equivalent circuit can be considerably simplified by dividing the impedance drop in the primary into two components, one produced by the exciting current I_n and the other by the load component I_1'. Under ordinary conditions, the drop due to the exciting current subtracts almost arithmetically from the impressed voltage and can be assumed constant for ordinary operating values of slip without introducing any substantial error in the value of the induced voltage E_1. From Fig. 234, according to these assumptions,

$$E_1 = V_1 - (-I_n)\sqrt{r_1{}^2 + x_1{}^2} - (-I_1')(r_1 + jx_1)$$
$$= V_1' - (-I_1')(r_1 + jx_1)$$

where V_1' is a constant voltage obtained by subtracting $I_n\sqrt{r_1{}^2 + x_1{}^2}$ arithmetically from V_1. The error in E_1 produced by this arithmetical subtraction is very small under ordinary load conditions.

Polyphase Induction Regulator. Any polyphase induction motor with a coil-wound rotor which is blocked can be used as a polyphase transformer in which the time-phase relation between the primary and secondary voltages of corresponding phases can be changed progressively merely by moving the rotor into different positions with respect to the stator. Such a device is in common use in laboratories as a phase-shifting transformer, but its chief commercial use is as an induction regulator for controlling the voltage of polyphase lines. When used as an induction regulator, it adds voltages of fixed magnitude but of adjustable phase to the Y voltages of the circuit.

If the rotor of an induction motor with a coil-wound rotor is blocked, with the centers of the rotor phase belts directly opposite the centers of corresponding stator phase belts, the maximum flux through the rotor and stator windings of like phases, due to the revolving magnetic field produced by the exciting currents in the stator windings, occurs at the same instant of time. Under this condition the primary and secondary induced voltages of like phases are in time phase and the ratio of their magnitudes is equal to the ratio of the number of effective primary and secondary turns per phase. The effective number of turns per phase is equal to the actual number of turns per phase multiplied by the pitch and breadth factors. If the rotor is turned through any angle, such as α electrical degrees, in the direction of rotation of the magnetic field, the maximum flux linking any rotor winding does not change in magnitude but it occurs $\alpha/2\pi f$ sec later. The secondary induced voltages are displaced by an angle α and lag by an angle α behind the corresponding primary induced voltages. By changing the relative position of the rotor and stator windings, any desired phase angle can be obtained between the voltages without altering their relative magnitude.

The connections for a three-phase induction regulator are given in Fig. 236. The primaries are connected in Y although they might equally well be connected in Δ. The primaries are shunted across the line of which the voltage is to be regulated and one secondary is inserted in each of the line conductors.

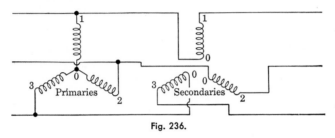

Fig. 236.

The ratio of transformation of the main transformer feeding the circuit should be such as to maintain the desired voltage at the load at times of average load. When the load is greater than average, the regulator is set to increase the line voltage and when the load is less than average it is set to decrease the line voltage. This permits the use of a regulator of minimum capacity. For a 20 per cent variation of voltage from 10 per cent above to 10 per cent below the average, a regulator of only 10 per cent of the kva capacity of the line is required.

Vector diagrams for one phase are shown in Fig. 237. From left to right these represent the condition of no change in voltage, an increase in

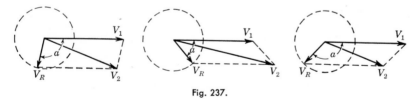

Fig. 237.

voltage (not maximum), and a decrease in voltage (not minimum). V_1 and V_2 are the line voltages to neutral on the primary and secondary sides of the regulator. V_R is the voltage added by the regulator.

The relative position of the rotor and stator windings is controlled by means of a worm and gear with the gear mounted on the rotor shaft. This device not only makes it possible to turn the rotor to any desired position, but it also serves to lock it so that it cannot turn under the influence of the torque developed between the rotor and stator when their windings carry currents.   The worm is usually mounted on the shaft of a small motor which is controlled either by hand or automatically.   Some device must be provided to limit the movement of the rotor of the regulator to 180 deg.

The difference between the action of single-phase and polyphase induction regulators should be noted.   The single-phase regulator adds a variable voltage which is either in phase or 180 deg out of phase with the line voltage.   The polyphase regulator adds a voltage fixed in magnitude but adjustable in phase.

Chapter 31

Analysis of the vector diagram. Internal torque. Maximum internal torque and the corresponding slip. The effect of reactance, resistance, impressed voltage, and frequency on the breakdown torque and the breakdown slip. Speed-torque curve. Stability. Starting torque. Fractional-pitch windings. Effect of shape of rotor slots on starting torque and slip. Low power factor of low-speed induction motors. The effect on the operation of a polyphase induction motor of unbalanced applied voltages.

Analysis of the Vector Diagram. Refer to the vector diagram of the induction motor (Fig. 232, page 395). The power input to the motor per phase is

$$P_1 = V_1 I_1 \cos \theta^{-V_1}_{I_1} \tag{228}$$

Resolving the impressed voltage drop $-V_1$ into its components,

$$P_1 = (-E_1 + I_1 x_1 + I_1 r_1) I_1 \cos \theta^{-V_1}_{I_1}$$
$$= E_1 I_1 \cos \theta^{-E_1}_{I_1} + (I_1 x_1) I_1 \cos \frac{\pi}{2} + (I_1 r_1) I_1 \cos 0$$
$$= E_1 I_1 \cos \theta^{-E_1}_{I_1} + 0 + \text{stator copper loss}$$

$(-E_1 + I_1 x_1 + I_1 r_1)$ as used in the foregoing expression for P_1 represents the magnitude of the vector sum of the three components $-E_1$, $I_1 x_1$, and $I_1 r_1$ of the voltage drop $-V_1$.

The expression for P_1 can be further expanded by replacing I_1 by its components.

$$P_1 = E_1(I'_1 + I_\varphi + I_{h+e}) \cos \theta^{-E_1}_{I_1} + 0 + \text{stator copper loss}$$
$$= E_1 I'_1 \cos \theta^{-E}_{I_1'} + E_1 I_\varphi \cos \frac{\pi}{2} + E_1 I_{h+e} \cos 0$$
$$+ 0 + \text{stator copper loss}$$
$$= E_1 I'_1 \cos \theta^{-E_1}_{I_1'} + 0 + \text{core loss} + 0 + \text{stator copper loss}$$

$(I'_1 + I_\varphi + I_{h+e})$ as used in the above expression for P_1 represents the magnitude of the vector sum of the three components I'_1, I_φ, and I_{h+e} of the current I_1.

$E_1 I_1' \cos \theta_{I_1'}^{-E_1}$ is the power transferred across the air gap to the rotor by electromagnetic induction and is the total rotor power P_2'.

$$P_2' = E_1 I_1' \cos \theta_{I_1'}^{-E_1} = E_2 I_2 \cos \theta_{I_2}^{E_2} \tag{229}$$

If E_2 is resolved into its components, the expression for P_2' becomes

$$P_2' = [I_2 r_2 + I_2 x_2 s + E_2(1 - s)]I_2 \cos \theta_{I_2}^{E_2}$$

$$= I_2{}^2 r_2 \cos 0 + I_2{}^2 x_2 s \cos \frac{\pi}{2} + E_2(1 - s)I_2 \cos \theta_{I_2}^{E_2(1-s)}$$

$$= \text{rotor copper loss} + 0 + \text{internal power}$$

$[I_2 r_2 + I_2 x_2 s + E_2(1 - s)]$ in the foregoing expression for P_2' represents the magnitude of the vector sum of the three components $I_2 r_2$, $I_2 x_2 s$, and $E_2(1 - s)$ of E_2.

The internal power P_2 per stator phase developed in the rotor is

$$P_2 = I_2 E_2(1 - s) \cos \theta_{I_2}^{E_2(1-s)} \tag{230}$$

Replacing I_2 by $\dfrac{E_2 s}{\sqrt{r_2{}^2 + x_2{}^2 s^2}}$ and $\cos \theta_{I_2}^{E_2 (1-s)}$ by $\dfrac{r_2}{\sqrt{r_2{}^2 + x_2{}^2 s^2}}$,

$$P_2 = \frac{E_2{}^2(1 - s)s r_2}{r_2{}^2 + x_2{}^2 s^2} \tag{231}$$

For any fixed slip the internal power developed by a polyphase induction motor varies as the square of the voltage E_2, the voltage induced in the rotor by the air-gap flux. E_1 does not differ greatly from V_1 under ordinary conditions of operation, since $I_1 z_1$ is not large, although much larger than in the transformer. E_2 is proportional to E_1 and is equal to E_1 when referred to the primary. E_2 is the voltage induced in the blocked rotor by the same flux that induces E_1 and differs from E_1 only on account of the difference in the effective number of rotor and stator turns. E_2 when referred to the primary by multiplying E_2 by the ratio of the effective number of stator to rotor turns is equal to E_1. For a fixed slip s the internal power developed by a polyphase induction motor is nearly proportional to the square of the impressed voltage. It must be remembered that this statement holds only as long as the primary impedance drop is negligible with respect to the primary impressed voltage. It is strictly true only if the voltage induced by the air-gap flux is proportional to the impressed voltage, a condition which never occurs in practice.

Internal Torque. The power developed by any motor is equal to its torque times the angular velocity of its rotor. Let T_2 be the internal torque per stator phase corresponding to the internal power P_2.

$$P_2 = 2\pi n_2 T_2$$

Replacing n_2 by its value from Eqs. (213) and (215), page 389, and leaving out the 60 in Eq. (213) in order to get the speed in revolutions per second,

$$P_2 = 2\pi \frac{2f_1}{p} (1 - s) T_2 \tag{232}$$

From Eq. (231),

$$T_2 = \frac{p}{4\pi f_1} \frac{E_2{}^2 s r_2}{r_2{}^2 + x_2{}^2 s^2} \tag{233}$$

The voltage E_2 in Eq. (233) corresponds to the voltage induced in the secondary of a static transformer by the mutual flux. Although E_2 for a transformer can be assumed constant for most purposes and independent of the load, E_2 of an induction motor cannot be so assumed. On account of the much larger leakage reactance of the induction motor, the induced voltage varies considerably from no load to full load. For this reason, E_2 in Eq. (233) should be replaced by the impressed voltage V_1, which is constant under ordinary operating conditions. An approximate value of V_1 in terms of E_2 can be obtained from the approximate equivalent circuit given in Fig. 235, page 397.

Neglecting the exciting current, from Fig. 235,

$$V_1 = I_2 \left[(r_1 + jx_1) + \left(r_2 + r_2 \frac{1 - s}{s} + jx_2 \right) \right]$$

$$V_1 s = I_2 \sqrt{(r_1 s + r_2)^2 + s^2 (x_1 + x_2)^2} \tag{234}$$

From Fig. 232, page 395,

$$E_2 s = I_2 \sqrt{r_2{}^2 + x_2{}^2 s^2}$$

Therefore

$$E_2{}^2 = V_1{}^2 \frac{r_2{}^2 + x_2{}^2 s^2}{(r_1 s + r_2)^2 + s^2 (x_1 + x_2)^2} \tag{235}$$

If $E_2{}^2$ from Eq. (235) is substituted in Eq. (233), the expression for internal torque becomes

$$T_2 = \frac{p V_1{}^2}{4\pi f_1} \frac{s r_2}{(r_1 s + r_2)^2 + s^2 (x_1 + x_2)^2} \tag{236}$$

In terms of the rotor current the internal torque is

$$T_2 = \frac{p}{4\pi f_1} I_2{}^2 \frac{r_2}{s} \tag{237}$$

The motor internal power can be obtained in terms of V_1 by combining Eqs. (231) and (235).

$$P_2 = \frac{V_1{}^2 (1 - s) s r_2}{(r_1 s + r_2)^2 + s^2 (x_1 + x_2)^2} \tag{238}$$

In terms of rotor current P_2 is

$$P_2 = I_2{}^2 r_2 \frac{1 - s}{s} \tag{239}$$

In all the preceding equations for the induction motor and also those which follow, the secondary constants r_2 and x_2 are assumed to be referred to the primary winding.

Equation (235), which gives the relation between E_2 and V_1, neglects the effect of the exciting current I_n on the primary impedance drop. For a transformer, the effect of I_n on the primary impedance drop is negligible in most cases on account of the small magnitude of the exciting current of a transformer.

For an induction motor the conditions are different. For the motor the exciting current I_n is large, often as much as 35 to 40 per cent of full-load current. The impedance of an induction motor is also larger than that of a transformer. For these reasons, the component I_n of the primary current of the motor cannot be neglected when finding the primary impedance drop without causing serious error.

Equations (236) and (238) for torque and power and all subsequent equations derived from them involve Eq. (235), in which I_n is neglected. Since V_1 appears in these equations as squared, any percentage error produced in V_1 by the neglect of I_n in Eq. (235) is exaggerated in P_2.

In all equations for torque and power, V_1 should be understood to be the actual stator impressed voltage per phase minus the part of the stator impedance drop per phase that is caused by the component I_n of the stator current. The error made by assuming that $V_1 - (-I_n)z_1$ vectorially is equal to $V_1 - I_n z_1$ arithmetically is negligible (see page 398). It is also sufficiently accurate in most cases to assume that $V_1 - (-I_n)z_1$ vectorially is equal to $V_1 - I_n x_1$ arithmetically.

When calculating torque or power for normal operating conditions from any of the equations involving V_1, V_1 should be replaced by $V_1' = V_1 - I_n x_1$ where the subtraction is made arithmetically. Under these conditions I_n can be assumed as constant and equal to the no-load current per phase at rated voltage.

If Eq. (238) or the corresponding equation for torque is solved for slip, an equation of the following form results:

$$as^2 - bs + c = 0$$

The solution of this equation is

$$s = \frac{b \pm \sqrt{b^2 - 4ac}}{2a}$$

For the smaller of the two values of slip given by the equation, which is the slip on the normal operating part of the speed-torque curve and is the slip ordinarily desired, the numerator of the equation is the difference between two terms that are so nearly equal in magnitude that little accuracy for s can be obtained. This difficulty can be avoided by rationalizing the equation by multiplying both its numerator and denominator by its numerator with the sign of the second term reversed. Rationalization puts the equation in the following form:

$$s = \frac{2c}{b \mp \sqrt{b^2 - 4ac}}$$

For the smaller of the two values of slip, the two terms in the denominator are now additive and a 10-in. or better a 20-in. slide rule gives sufficiently accurate results.

Equations (238) and (239) give the internal power per phase P_2 in watts if voltage and current are expressed in volts and amperes and resistances and reactances are expressed in ohms. To get the external power or pulley power of the motor the friction and windage loss must be subtracted from P_2 multiplied by the number of stator phases.

When voltage and current are expressed in volts and amperes and resistances and reactances are expressed in ohms, Eqs. (236) and (237) give the internal torque per phase in a unit which is in the same system of units as the watt but has no name and is not used. To convert the torque given by Eqs. (236) and (237) to pound-feet, it must be multiplied by $550/746 = 0.737$. To get the external torque or pulley torque, the friction and windage torque must be subtracted from the torque multiplied by the number of stator phases.

The stator resistance r_1 in the equations for power and torque is the effective resistance per phase at stator frequency. The rotor resistance r_2 is the effective resistance per phase at rotor frequency referred to the stator. For ordinary loads, the slip is small and the frequency of the rotor currents is consequently small. Under such conditions, the difference between the rotor effective resistance and the rotor ohmic resistance as measured with direct current is negligible. The rotor ohmic resistance should therefore be used when the slip is small, as it is under ordinary operating loads. At the instant of starting, the rotor current has stator frequency. Under this condition, the rotor effective resistance per phase at stator frequency and referred to the stator must be used. The reactances x_1 and x_2 are both at stator frequency.

Maximum Internal Torque and the Corresponding Slip. For any fixed stator frequency f_1 and impressed voltage V_1, the torque is a maximum when the last factor of Eq. (236), page 403, is a maximum. Therefore the slip at which the maximum torque occurs can be found as follows:

$$\frac{d}{ds}\left[\frac{sr_2}{(r_1s + r_2)^2 + s^2(x_1 + x_2)^2}\right] = 0$$

$$s = \frac{r_2}{\sqrt{r_1{}^2 + (x_1 + x_2)^2}} \tag{240}$$

Substituting this value of s in Eq. (236) gives for the maximum internal torque

$$T_m = \frac{pV_1{}^2}{4\pi f_1}\frac{1}{2[r_1 + \sqrt{r_1{}^2 + (x_1 + x_2)}]^2} \tag{241}$$

From Eqs. (240) and (241) it follows that the slip at which maximum internal torque occurs is directly proportional to the secondary or rotor resistance and that the maximum internal torque itself is independent of the rotor resistance. The effect of increasing the rotor resistance is to increase the slip at which maximum internal torque occurs without changing the value of that torque. Neither the maximum internal torque nor the slip at which it occurs is independent of the primary or stator resistance. Both torque and slip are decreased by increasing the primary resistance.

The Effect of Reactance, Resistance, Impressed Voltage, and Frequency on the Breakdown Torque and the Breakdown Slip. The maximum torque which can be developed by an induction motor is its breakdown torque, the torque at which the motor becomes unstable with increasing slip. However the motor is stable at any point of its speed-torque curve if the speed-torque curve of the load is steeper than the speed-torque curve of the motor at the point of operation.

The maximum torque and the slip at which maximum torque occurs depend upon the stator and rotor leakage reactances. Both maximum torque and the slip at which it occurs decrease with increasing leakage reactance, Eqs. (240) and (241). In order to have large breakdown torque, the leakage reactance of an induction motor must be small. Since the leakage reactance, like that of a transformer with an air gap between its primary and secondary windings, increases with the length of the air gap, the necessity for small reactance requires the use of a short air gap. A long air gap not only decreases the maximum torque by increasing the leakage reactance, but it also increases the reluctance of the magnetic circuit and increases the magnetizing current, thus lowering the power factor. For these reasons the air gap of an induction motor is always very small. Since the maximum torque developed by an induction motor varies as the square of the impressed voltage, Eq. (241), good voltage regulation is highly desirable on circuits from which induction motors are to be operated. Also, since both x_1 and x_2 are proportional to the primary frequency f_1, it is clear that induction motors are best suited for low frequencies. The effect of f_1 in the expression for maximum torque does not in itself show that the maximum torque differs for different frequencies, since when compared on the only rational basis, viz., the same speed, the ratio of p to f_1, Eq. (241), is constant. The reason high-frequency motors are less satisfactory than low-frequency motors is because of the effect of f_1 on the reactances x_1 and x_2.

Increasing the rotor resistance r_2 brings the maximum torque point toward 100 per cent slip but does not affect the maximum value of the internal torque, Eqs. (240) and (241). The external torque is slightly decreased by an increase in r_2 on account of the increase in the rotor core loss with an increase in slip.

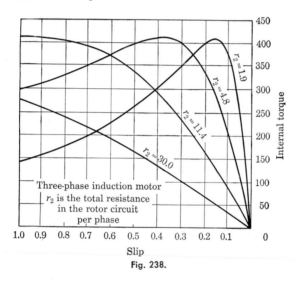

Fig. 238.

Speed-torque Curve. The speed-torque curve of a polyphase induction motor can be plotted from Eq. (236). Four such curves are plotted in Fig. 238. These curves are plotted against slip instead of speed.

Stability. The internal torque is zero at synchronous speed. The working part of any speed-torque curve is from the point of maximum torque to synchronous speed. Synchronous speed cannot be reached even at no load, since at synchronous speed no torque would be developed to balance the opposing torque caused by the rotational loss. The motor is stable for any slip between zero and that corresponding to maximum torque, since within these limits an increase in load causes the motor to slow down and develop more torque to carry the increase in load. Beyond the point of maximum torque, an increase in slip is accompanied by a decrease in torque. With ordinary loads, a motor is unstable when operating at a slip greater than that for maximum torque since under this condition an increase in load decreases the speed and causes the motor to develop less torque to carry the increase in load. However, as has just been stated, it is possible to have a motor stable with a slip greater than that corresponding to its maximum torque provided the speed-torque curve of the load is steeper than that of the motor. The ratio of the breakdown torque to the full-load torque is largely a question of design. For most motors the ratio is 2 or even greater.

Starting Torque. At starting, the slip is unity. Under this condition, Eq. (236), page 403, becomes

$$T_{st} = \frac{pV_1^2}{4\pi f_1} \frac{r_2}{(r_1 + r_2)^2 + (x_1 + x_2)^2} \tag{242}$$

T_{st} in Eq. (242) is the starting torque. Replacing V_1^2 by its value from Eq. (234), page 403, and remembering that $s = 1$ at starting,

$$T_{st} = \frac{p}{4\pi f_1} I_2^2 r_2 \tag{243}$$

That is, the starting torque is proportional to the copper loss in the secondary or rotor circuit and can be increased up to a certain maximum value by increasing the resistance r_2 of the rotor circuit. It makes no difference whether the increase in resistance is obtained by actually increasing the rotor resistance or by putting external resistance in series with the rotor windings. The starting torque is a maximum when the constants of the motor are such as to make the slip unity in Eq. (240), page 406. From Eq. (240) for maximum torque at starting,

$$r_2^2 = r_1^2 + (x_1 + x_2)^2 \tag{244}$$

By properly adjusting r_2 the maximum internal torque can be made to occur at starting, but if this value of resistance remains unchanged as the motor speeds up, the slip under normal running conditions is large and the efficiency is low. One curve in Fig. 238, page 407, is drawn for that value of r_2 which gives maximum torque at starting.

It is evident from Fig. 238 that the portion of the speed-torque curve between maximum torque and synchronous speed is approximately a straight line. Therefore when the maximum torque is made to occur at starting by increasing the rotor resistance, the slip at which full-load torque occurs is approximately equal to the ratio of full-load torque to maximum torque. Under this condition both the speed regulation and the efficiency are poor.

For best running conditions, r_2 should be as small as possible. For best starting torque, it should be large. In any motor, a compromise must be made between these two requirements. A higher rotor resistance under starting conditions than when running with small slip under load is automatically attained because of the decrease in rotor frequency as the motor speeds up. This change in frequency causes the rotor resistance to change from effective resistance at full stator frequency, when the slip is unity at the instant of starting, to practically d-c resistance when the slip is small under normal load conditions. By making use of this decrease in rotor resistance as the motor speeds up, it is possible by proper design of the rotor to obtain good speed regulation under load as well as a high starting torque. When large starting torque is required, the rotor resistance can be temporarily increased by inserting resistance in the rotor circuit. The resistance is cut out when the motor is up to speed. This is considered more in detail when different types of rotors are discussed in Chap. 32.

Fractional-pitch Windings. In order to obtain good operating characteristics, it is desirable to make the reactance of induction motors low. See Eqs. (236), (241), and (242). For this reason fractional-pitch windings are generally used for both stator and rotor since such windings reduce the leakage reactance. The decrease in the reactance increases the maximum torque and also increases the slip at which maximum torque occurs, Eqs. (240) and (241). The change in slip, however, is of minor importance compared with the change in maximum torque. By decreasing the effective number of turns (effective number equals the actual number multiplied by the pitch and breadth factors), a fractional-pitch winding increases the magnetizing current and thus reduces the power factor for a given load. A fractional-pitch winding also reduces the harmonics in the air-gap flux and thus improves the operation of the motor. The effect of harmonics on the operation of an induction motor is considered later.

Effect of Shape of Rotor Slots on Starting Torque and Slip. By proper shaping of the rotor slots and also of the inductors much can be accomplished in increasing the starting torque without sacrificing good speed regulation. If deep, narrow rotor slots with low-resistance inductors and end connections are used, the rotor resistance at standstill can be made several times greater than the resistance under normal running conditions.

The increase in the effective resistance at standstill is due in part to the local losses set up by the slot leakage flux, but the chief cause of the increase is the tendency of the slot leakage flux to force the current toward the top of the inductors. If the inductors are considered to be divided into horizontal elements similar to those shown in Fig. 150, page 228, the linkages with these elements due to the slot leakage flux increase in passing from the top to the bottom of an inductor, causing the leakage reactance of the lower elements to be higher than the leakage reactance of those above.   As a result, the current is not distributed uniformly over the cross section of the inductors but is forced toward their upper portions, producing an increase in the effective resistance.   The effect is similar to the ordinary skin effect in conductors but is much more marked in the motor because of the presence of the iron which surrounds the inductors.   The reactance of these elements and the apparent increase in the resistance is dependent upon the frequency.   At starting, the frequency of the rotor current is f_1, the stator frequency.   At any slip s, it reduces to $f_1 s$, and at full load, it has 2 to 8 per cent of its starting value according to the size and type of the motor.   Because of the decrease in the local losses and in the skin effect with decreasing frequency, the resistance of the rotor when running can be much less than at starting. The effect of the slot leakage flux on the distribution of the current over the cross section of the rotor inductors is used in the double-squirrel-cage type of rotor.   This type of rotor is considered later.

Low Power Factor of Low-speed Induction Motors.   The power factor of low-speed induction motors is inherently lower than the power factor of high-speed induction motors of similar design for the same frequency and output.   Consider as an example two motors for the same frequency, the same impressed voltage, and the same output.   Let the speed of one motor be twice that of the other.   Let T, I, Z, σ, β, and L represent torque, rotor current, total number of rotor inductors, rotor radius, average air-gap flux density, and axial length of the rotor inductors.   Let subscript 2 indicate the low-speed motor and subscript 1 the high-speed motor.   Assume that both motors are designed for the same average flux density and that the number of rotor inductors per inch of rotor periphery is the same for both.   Also assume that the ratio of axial lengths of rotor iron and the ratio of radii of the rotors are the same and equal to q.   Let p_2 and p_1 be the number of poles where $p_2/p_1 = 2$.   Then

$$\frac{T_2}{T_1} = \frac{KZ_2 I_2 \sigma_2 L_2 \beta_2}{KZ_1 I_1 \sigma_1 L_1 \beta_1}$$

where K is a constant of proportionality.

For the same output, I_1 and I_2 are equal and β_1 and β_2 are assumed equal.   For equal output T_2/T_1 must equal 2 since the speeds are in the ratio of 1:2.

$$\frac{T_2}{T_1} = 2 = \frac{(Z_1 q)(\sigma_1 q)(L_1 q)}{Z_1 \sigma_1 L_1} = q^3$$

$$q = \sqrt[3]{2} = 1.26$$

Let E_2 and E_1 be the voltages induced in the rotors referred to stator frequency.   These voltages must be approximately equal since the impressed voltages are assumed to be equal.

$$\frac{E_2}{E_1} = 1 = \frac{2.22(Z_2 k_2) f \varphi_2}{2.22(Z_1 k_1) f \varphi_1}$$

where k_2 and k_1 are the products of the pitch and breadth factors for the two motors. These products are assumed to be equal. φ_2 and φ_1 are the fluxes per pole.

$$1 = \frac{(Z_1 q)(\varphi_1 q^2 \frac{1}{2})}{Z_1 \varphi_1}$$

$$q^3 = 2 \qquad \text{and} \qquad q = 1.26$$

which checks the value of q obtained from consideration of the torques and justifies the assumption of an equal number of rotor inductors per inch of periphery of the rotors.

$$\beta = \frac{\left(\dfrac{Z}{2} k\right) I_\varphi}{\displaystyle\sum \dfrac{l}{\mu}}$$

where I_φ is the magnetizing current. As an approximation assume that the effective length of the air gaps of the two motors is the same and that the reluctance of the iron portions of the magnetic circuits is negligible as compared with the reluctance of the air gaps. Then

$$\frac{\beta_2}{\beta_1} = 1 = \frac{\left(\dfrac{Z_2}{2}\dfrac{p_1}{p_2} k\right) I_{\varphi 2}}{\left(\dfrac{Z_1}{2} k\right) I_{\varphi 1}}$$

$$= \frac{(Z_1 \frac{1}{2} q) I_{\varphi 2}}{Z_1 I_{\varphi 1}}$$

$$I_{\varphi 2} = I_{\varphi 1}\frac{2}{q} = I_{\varphi 1}\frac{2}{1.26} = 1.58 I_{\varphi 1}$$

The magnetizing current of the lower-speed motor is 58 per cent greater than that of the higher-speed motor, but since the outputs are equal, the load components of the primary currents are equal for equal outputs. The power factor of the lower-speed motor is therefore less than that of the higher-speed motor. Assume that the power factor of the lower-speed motor is 0.90 at full load. Its magnetizing current is

$$\sin (\cos^{-1} 0.90) = 0.44$$

of its full-load current. The power factor of the lower-speed motor is approximately

$$\cos \left(\tan^{-1} \frac{0.44 \times 1.58}{0.90} \right) = 0.79$$

By modifying the design and the dimensions of the lower-speed motor its power factor can be improved, but it is inherently less than the power factor of the higher-speed motor. Increasing the radius of the rotor of the lower-speed motor with respect to its axial length increases its power factor. If the axial length of the rotors is the same for the two motors, the radius of the rotor of the lower-speed motor has to be 1.41 times that of the higher speed motor for the same output. Under this condition the power factor of the lower-speed motor is 0.82. It can be shown that the power factor of a low-frequency motor is inherently less than the power factor of a high-frequency motor for the same output and speed. This last statement assumes that the two frequencies are within the range of ordinary commercial frequencies.

The Effect on the Operation of a Polyphase Induction Motor of Unbalanced Applied Voltages. In the following discussion, the flux distribution in the air gap due to each phase is assumed to be sinusoidal. In other words, the harmonics in the air-gap flux which are caused by the distribution of the windings are assumed to be negligible. These harmonics are in general small and their effect under operating conditions can usually be neglected. The effect of harmonics on the operation of a polyphase induction motor is considered in Chap. 32. The impressed voltages are assumed to be sinusoidal. All of the equations which have already been developed for the polyphase induction motor are based on the above assumptions. A three-phase motor is assumed.

When balanced sinusoidal voltages are impressed on the stator of a polyphase induction motor which has the usual symmetrical windings, a magnetic field is set up which is constant in magnitude and rotates at a constant speed fixed by the number of poles on the motor and the frequency of the applied voltages. This field is sinusoidal in its space distribution and can be represented by a rotating vector which is equal in length to the maximum value of the flux density in the air gap and rotates at a constant speed fixed by the frequency of the impressed voltages and the number of poles on the motor. The end of the vector which represents the field traces out a circle, and the field is said to be circular. The direction of rotation of the vector is fixed by the phase order or phase sequence of the applied voltages. If the sequence of the applied voltages is reversed, the direction of rotation of the vector reverses.

If the applied voltages are unbalanced, they can be resolved into positive-sequence and negative-sequence components. There can be no zero-sequence components since the vector sum of the line voltages must be zero. Under this condition, the rotating field in the air gap is no longer of fixed magnitude and it does not rotate at constant speed. It can be considered to be the resultant at each instant of two circular fields produced by the positive-sequence and the negative-sequence impressed voltages. Each of these component fields rotates at synchronous speed with respect to the stator but they rotate in opposite directions. Except when the two fields are equal in magnitude, which occurs only for single-phase operation, the resultant of the two oppositely rotating vectors which represent the fields traces out an ellipse and the field is said to be elliptical. The limit of unbalance occurs under single-phase conditions. In this case the two oppositely rotating fields are equal in magnitude and the resultant of the two revolving vectors which represent them is a vector which varies sinusoidally along a fixed axis. This condition can be considered as the limiting case of an ellipse when the minor axis becomes zero. Under the ordinary degree of unbalance which occurs in the line voltages of a three-phase circuit in practice, the magnitude of the negative-sequence voltages is not large and the negatively rotating component field is small. Under badly unbalanced conditions it can be large.

The constants which determine the operating characteristics of an induction motor are the stator resistance and leakage reactance and the rotor resistance and leakage reactance referred to the stator. These are functions of the applied frequency, the slip or both and for any fixed frequency and given slip they are actually constant. The motor can therefore be treated as a linear circuit for both positive-sequence and negative-sequence components in unbalanced impressed voltages. For the positive-sequence components of voltage and current the slip is s, the actual slip of the motor. This is small for ordinary load conditions. For the negative-sequence components, the slip is $(2 - s)$. This is nearly equal to 2.

When the impressed voltages are unbalanced, the resultant torque and power are found for any slip s by calculating the torque and power for the positive-sequence and for the negative-sequence component voltages separately by the formulas already developed for balanced conditions, using s for the slip for the positive-sequence components and $(2 - s)$ for the slip for the negative-sequence components. For the positive-sequence components, V_1' is found by subtracting the stator impedance drop per phase due to the positive-sequence exciting current from the positive-sequence phase voltage. For the negative-sequence components, V_1' is practically equal to the actual negative-sequence phase voltage since the exciting current is negligible for the usual magnitudes of the negative-sequence voltages and the corresponding large slip, $(2 - s) = 2$(approximately).

Since the windings are symmetrical, the constants of all three phases are identical for each group of components. The copper loss for the entire motor is

$$3[(I_1^+)^2 r_1^+ + (I_2^+)^2 r_2^+] + 3[(I_1^-)^2 r_1^- + (I_2^-)^2 r_2^-]*$$

* R. R. Lawrence, "Principles of Alternating Currents," 2d ed., p. 391.

The currents are per phase. The subscripts 1 and 2 refer to primary and secondary. The resistances $r_1{}^+$ and $r_1{}^-$ are the primary positive-sequence resistance per phase and the primary negative-sequence resistance per phase and are equal to each other and to the primary effective resistance per phase at stator frequency. The resistances $r_2{}^+$ and $r_2{}^-$ are the secondary positive-sequence resistance and the secondary negative-sequence resistance per phase, but they are not equal. The resistance $r_2{}^+$ is the rotor effective resistance per phase at a frequency f_1s. For ordinary values of operating slip this is for all essential purposes equal to the d-c resistance. The resistance $r_2{}^-$ is the rotor effective resistance per phase at a frequency $f_1(2 - s)$. This is very nearly equal to the rotor effective resistance per phase at double stator frequency. The relative magnitude of positive-sequence and negative-sequence currents is determined by the relative magnitude of positive-sequence and negative-sequence components in the applied voltages and by the corresponding slips s and $(2 - s)$.

Although torque and power produced by negative-sequence components in the applied voltages are in general small compared with torque and power produced by positive-sequence components, the copper loss caused by negative-sequence components may be large. The load component I_1' of the stator current in any phase is equal in magnitude to the secondary phase current referred to the stator and is given by Eq. (234), page 403.

$$I_1' = \frac{V_1's}{\sqrt{(r_1s + r_2)^2 + s^2(x_1 + x_2)^2}} = \frac{V_1'}{\sqrt{\left(r_1 + \dfrac{r_2}{s}\right)^2 + (x_1 + x_2)^2}}$$

Because of the large difference in the value of slip for the positive-sequence and the negative-sequence components of I_1' and I_2, the negative-sequence components of these currents may be larger than the positive-sequence components if the impressed voltages are badly out of balance. For normal operating conditions the rotor resistance for the positive-sequence currents is the rotor effective resistance at very low frequency and is nearly equal to the rotor d-c resistance. For the negative-sequence rotor currents, the rotor resistance is the rotor effective resistance at nearly double stator frequency and is much larger than the rotor positive-sequence resistance. Unbalanced impressed voltages may produce a marked increase in the copper loss for a given output. The relative magnitude of positive-sequence and negative-sequence component currents is determined by the relative value of the corresponding components in the impressed voltages and of the slips, s and $(2 - s)$. Under single-phase operation, with one stator line open, the currents in two of the phases are equal in magnitude and opposite in phase and the current in the third phase is zero. This statement assumes that the motor is Y-connected or that the actual motor has been replaced by an equivalent Y-connected motor. If the currents are resolved into positive-sequence and negative-sequence components, it is found that in the stator these components are equal in magnitude. In the rotor, the positive-sequence currents are slightly smaller than the negative-sequence currents because of the difference between the magnetizing currents in the stator corresponding to these components. If the impressed voltages are badly out of balance, the total copper loss for a given output can be reduced by disconnecting one line terminal and operating the motor single phase.

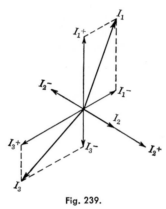

Fig. 239.

The copper losses in the stator phases of a motor operating with unbalanced impressed voltages are not equal. The frequency of positive-sequence and negative-sequence component currents in each phase of the motor is the same. Hence the current in any stator phase is equal to the vector sum of the positive-sequence and negative-sequence component currents in that phase. That the currents in the three stator phases are not equal is evident from Fig. 239. This figure is drawn for conditions where the negative-sequence currents have one-half the magnitudes of the positive-sequence currents, and in phase 1 the negative-sequence current lags the positive-sequence current by 60 deg. The resultant currents I_1 and I_3 in phases 1 and 3 are equal in magnitude and each is approximately 1.44 times as great as the current I_2, so that the copper losses in phases 1 and 3 are slightly more than double the copper loss in phase 2. Since the frequencies $f_1 s$ and $f_1(2 - s)$ of the positive-sequence and the negative-sequence rotor currents are not the same except when the motor is at rest, the rotor currents cannot be combined vectorially under operating conditions. Under operating conditions, the average copper losses in the rotor phases are equal.

Chapter 32

Rotors, number of rotor and stator slots, air gap. Coil-wound rotors. Squirrel-cage rotors. Advantages and disadvantages of the two types of rotors. Double-squirrel-cage rotor. Different types of rotor slots and the corresponding starting characteristics. The effect of harmonics and rotor and stator slots on the speed-torque characteristics of a polyphase induction motor. Harmonics in the time variation of the impressed voltages. Harmonics in the space distribution of the air-gap flux. The effect of slot harmonics in the air-gap flux.

Rotors, Number of Rotor and Stator Slots, Air Gap. Two distinct types of rotors are used in induction motors, the coil-wound and the squirrel-cage types. Each of these possesses distinct advantages. Each has slots which are usually partially closed. Very open slots are undesirable as they materially increase the effective length of the air gap. This increases the magnetizing current and hence decreases the power factor. On the other hand, completely closed slots are undesirable as they decrease the reluctance of the path of the leakage flux and consequently increase the stator and rotor leakage reactances, thus decreasing the maximum torque developed by the motor. Magnetic wedges are sometimes used to hold the coils in the slots. Such wedges give the effect of closed or partially closed slots and decrease the effective length of the air gap. An induction motor always has a very short air gap. For this reason, it should be provided with bearings which minimize the effect of wear and the danger of the rotor striking the stator.

The number of slots in the rotor must be different from the number in the stator. In order to prevent a periodic variation in the reluctance of the magnetic circuit of the motor, the ratio of these numbers must not be an integer. Moreover, if the rotor and stator have the same number of slots, there is a tendency for the rotor at starting to lock in the position which makes the reluctance of the magnetic circuit a minimum.

Coil-wound Rotors. The windings of coil-wound rotors are similar to those of synchronous generators. The rotor must be wound for the same number of poles as the stator. The rotor must have polyphase windings, but the number of phases need not be the same as for the stator although in practice it usually is the same. Either mesh or star connection may be used.   Y connection is preferable to Δ connection not only for the rotor but also for the stator as it gives a better slot factor than the Δ connection.   The terminals of the rotor winding are brought out to slip rings mounted on the shaft.   These slip rings may be short-circuited for normal running conditions, or may be connected through suitable resistances for starting or for varying the speed.   Since the current in the rotor

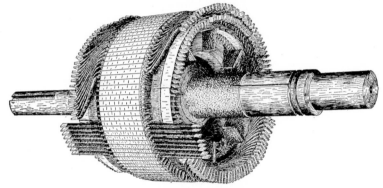

Fig. 240.

is induced, the operation of the motor is not influenced by the voltage for which the rotor is wound.   The best voltage for a rotor is usually that which makes the cost of construction a minimum.   A coil-wound rotor is shown in Fig. 240.

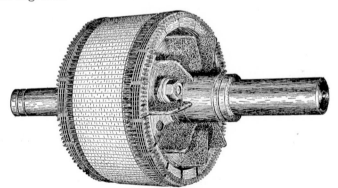

Fig. 241.

Squirrel-cage Rotors.   The windings of squirrel-cage rotors may consist of solid copper inductors of either circular or rectangular cross section, placed in the rotor slots, usually without insulation, and short-circuited by copper end rings or straps to which the inductors are welded or brazed.   A rotor of this type is shown in Fig. 241.

Many motors, however, with squirrel-cage rotors are built with the rotor squirrel-cage winding, including both bars and end rings, of aluminum which is cast directly in the rotor.   The end rings can be turned down to give the desired rotor resistance.   Such a squirrel-cage winding has no joints to work loose and cannot be injured by a temperature below that of the melting point of aluminum.

Skewed Slots.   Some induction motors of the squirrel-cage type have rotor slots which, instead of being parallel to the shaft, are skewed at an angle with respect to the parallel position.   With skewed slots, the revolving flux encounters an air gap of more constant reluctance resulting in a more uniform torque and a quieter motor.   Skew, however, also increases the rotor resistance and influences the effective ratio of transformation.   The ratio of the voltage produced in a skewed inductor to the voltage that would be produced if the inductor were parallel to the shaft is known

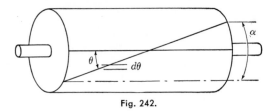

Fig. 242.

as the skew factor.   Although both are not used on the same machine, Fig. 242 shows a skewed and an unskewed inductor, the skewed inductor crossing the unskewed at its center.   The angle of skew is α and is measured in electrical radians or degrees.   The revolving flux cuts across the entire length of the unskewed inductor at the same instant.   It however cuts one end of the skewed inductor first and gradually proceeds across its length.   Let E_m be the maximum voltage produced per unit angle of the skewed inductor.   At the time when the maximum amount of flux is cutting the center of this inductor, the voltage in the portion of the inductor included in the differential angle $d\theta$, displaced by an angle θ from the center, equals $E_m \cos \theta \, d\theta$.   The voltage at this time produced in the entire length of the skewed inductor is then equal to

$$\int_{-\alpha/2}^{+\alpha/2} E_m \cos \theta \, d\theta = 2E_m \sin \frac{\alpha}{2}$$

If the entire length of the inductor was cut by maximum flux, the voltage would be $E_m\alpha$.   This is the case in the unskewed inductor.   The skew factor k_s is then given by the following relationship:

$$k_s = \frac{2E_m \sin (\alpha/2)}{E_m\alpha} = \frac{\sin (\alpha/2)}{\alpha/2} \tag{245}$$

If skewed inductors are used, the secondary turns N_2 in formulas (225), (226), and (227) must be multiplied by k_s.

Advantages and Disadvantages of the Two Types of Rotors. The chief advantage possessed by the coil-wound rotor is the possibility it offers of having its resistance varied by inserting resistances between slip rings. This increase in resistance can be used to increase the starting torque or to vary the speed. The chief disadvantages of this type of rotor are that it costs more, has slightly higher resistance when referred to the stator, and is less rugged than the squirrel-cage type, which is extremely rugged.

By properly shaping the slots of a squirrel-cage rotor much can be accomplished to control the starting and operating characteristics of the motor. The frequency of the rotor current at the instant of starting is stator frequency. Under running conditions, the frequency of the rotor current is very low. Therefore, at the instant of starting, the rotor resistance is its effective resistance at stator frequency, and under normal operating conditions, its resistance is for all essential purposes the d-c resistance. This change in the rotor resistance between starting conditions and running conditions can be utilized to give a motor good starting characteristics as well as good running characteristics. For good starting characteristics, high rotor resistance is necessary. For good operating characteristics, low rotor resistance is necessary if low slip and high efficiency are to be secured. The decrease in rotor resistance from starting conditions to running conditions is much greater in the double-squirrel-cage rotor than in the ordinary squirrel-cage rotor.

For low starting current, the motor reactance should be high, but for high maximum torque the motor reactance should be low. A small air gap and nearly closed slots, needed for low magnetizing current and high operating power factor, decrease the reluctance of the leakage-flux paths of a motor and increase its reactance. The design of a satisfactory motor necessitates a compromise between these opposing factors, but much can be done to obtain satisfactory starting and operating characteristics by shaping the rotor slots and by using rotor squirrel-cage windings of the proper resistance.

Double-squirrel-cage Rotor. The change of resistance of the squirrel-cage winding with frequency can be much increased by the use of what is known as a double-squirrel-cage rotor winding. This consists of two squirrel-cage windings placed in the same slots one above the other. The upper winding is a high-resistance low-reactance winding. The lower winding is a low-resistance high-reactance winding. The high reactance of the lower winding is obtained by constricting the portion of the slots between the two windings. Figure 243 shows a diagram of a typical slot for a double-squirrel-cage rotor with high-resistance and low-resistance bars H and L.   Constricting the portion of the slot between the high-resistance and low-resistance bars increases the leakage flux produced per ampere in the lower bar, thus increasing its reactance.   At the instant of starting, the frequency of the currents in the rotor bars is stator frequency.   Under this condition, the reactance of the lower bar is much higher than that of the upper bar.   The division of the rotor current between the two squirrel cages is determined by their relative impedance, and as their reactances at stator frequency are high compared with their resistances, the division of rotor current between them at starting is determined chiefly by the relative reactance of the two squirrel cages.   At

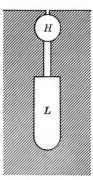

Fig. 243.

starting, most of the rotor current is carried by the upper high-resistance squirrel cage and the motor starts with the characteristics of a motor with a high-resistance rotor.   As the motor speeds up, the frequency of the rotor current decreases and is very low when full speed is attained.   Under this condition, the reactances of the two squirrel-cage windings are low compared with their resistances.   The division of current between them is determined chiefly by their relative resistance, and most of the rotor current is carried by the low-resistance squirrel cage, the motor operating like a motor with a low-resistance rotor.   By properly shaping the rotor slots and by adjusting the relative resistance of the two squirrel cages, it is possible to have high starting torque with moderate starting current and also to have small slip and small rotor loss under running conditions.

Different Types of Rotor Slots and the Corresponding Starting Characteristics.   The general arrangement and shape of the slots used for the rotors of polyphase induction motors with squirrel-cage windings are shown in Fig. 244.   By the use of these types of slots, much can be accomplished in controlling the starting and operating characteristics of motors.

Slots similar to those shown in Fig. 244a are used for a general-purpose motor.   Such a motor comes under class A of the National Electrical Manufacturers' Association and in the four-pole size should develop a normal starting torque specified as equal to 150 per cent of full-load torque and also should have a breakdown or maximum torque of not less than 200 per cent of full-load torque when rated voltage is applied.   The lower-speed motors develop somewhat smaller starting torques.   All motors of class A take large starting currents which may be seven times full-load current.   Except in small sizes, these motors must be started on reduced voltage.

The slots shown in Fig. 244b are for high-reactance rotors and are used for double-squirrel-cage windings. A motor with slots similar to those shown in this figure should develop normal starting torque with relatively low starting current and can be started on full voltage.

The slots shown in Fig. 244c are used for a motor which is to develop relatively high starting torque with relatively low starting current. The low starting current is obtained by the high inherent resistance produced by the size and shape of the slots.

The slots shown in Fig. 244d are for high-resistance motors with single-squirrel-cage windings. A motor with such slots develops high starting torque with moderate current which is obtained by the use of a high-

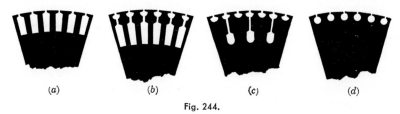

(a)   (b)   (c)   (d)

Fig. 244.

resistance rotor which sacrifices small operating slip. A high-resistance rotor is used in a motor for operating electric elevators, electric cranes, hoists, etc., or for flywheel sets. Such a set consists of an induction motor coupled directly to a d-c generator and has a heavy flywheel mounted directly on the shaft. Flywheel sets are used to smooth out violent fluctuations in the load applied to the d-c generator by utilizing the kinetic energy of the flywheel. In order to utilize this energy without passing on to the a-c line the fluctuations in the d-c load, the induction motor must have a very drooping speed-torque characteristic such as is obtained by the use of a high-resistance rotor.

The Effect of Harmonics and Rotor and Stator Slots on the Speed-torque Characteristics of a Polyphase Induction Motor. There are two classes of harmonics that must be considered, time harmonics in the time variation of the impressed voltages and space harmonics in the space distribution of the air-gap flux caused by the distribution of the motor windings. The effect of the two classes of harmonics on the torque characteristics of a motor is considered separately from the effect of the slots.

Harmonics in the Time Variation of the Impressed Voltages. If the effect of saturation is neglected so that the superposition theory can be applied, the effect of any harmonic, such as the fifth, in the time variation of the impressed voltages is the same as that produced by impressing sinusoidal voltages of five times the frequency of the fundamental voltages. The number of poles is the same as for the fundamentals in the impressed voltages, but as the frequency is five times as great as that of the fundamentals, the revolving field produced by the fifth harmonics revolves five times as fast with respect to the stator as the revolving field due to the fundamentals. In general, the speed of the revolving field produced by the nth harmonics in the impressed voltages is n times the speed of the field set up by the fundamentals. There are no third harmonics in the time variation of the air-gap flux of a three-phase machine since third harmonics do not exist in the line voltages of a balanced three-phase system. Similar statements apply to all harmonics which are multiples of the third.

In a three-phase system the phase order for the fifth, seventh, eleventh, thirteenth, seventeenth, nineteenth, etc., harmonics alternates with respect to that of the fundamentals. The phase order of the fifth, eleventh, seventeenth, etc., harmonics is opposite to that of the fundamentals. The phase order of the seventh, thirteenth, nineteenth, etc., harmonics is the same as that of the fundamentals. Therefore in a three-phase machine, the fifth-harmonic component of the flux caused by the fifth harmonics in the impressed voltages revolves five times as fast as the fundamental revolving field and revolves in the opposite direction. The seventh-harmonic component field revolves seven times as fast as the fundamental field and in the same direction. The discussion which follows is limited to a three-phase motor, but with some modification in regard to the harmonics present and their phase orders, it applies to any polyphase machine.

Consider the fifth harmonics in the impressed voltages. The slip of a motor is the difference between synchronous and actual speed divided by synchronous speed. Since the revolving field set up by the fifth harmonics in the impressed voltages revolves five times as fast with respect to the stator as the field set up by the fundamentals and revolves in the opposite direction, the slip of the motor with respect to the fifth-harmonic field is

$$s_5 = \frac{-5\left(\dfrac{2f_1}{p}\right) - \dfrac{2f_1}{p}(1 - s_1)}{-5\left(\dfrac{2f_1}{p}\right)} = \frac{6 - s_1}{5} \qquad (246)$$

For the seventh-harmonic field the slip is

$$s_7 = \frac{7\left(\dfrac{2f_1}{p}\right) - \dfrac{2f_1}{p}(1 - s_1)}{7\left(\dfrac{2f_1}{p}\right)} = \frac{6 + s_1}{7} \qquad (247)$$

where f_1, p, and s_1 are the frequency of the fundamentals in the impressed voltages, the number of poles for which the motor is wound, and the actual slip of the motor, *i.e.*, its slip with respect to the fundamental revolving field.

The equation for the maximum torque of a polyphase induction motor per phase [Eq. (241), page 406] is

$$T_m = \frac{pV_1^2}{4\pi f_1} \frac{1}{2[r_1 + \sqrt{r_1^2 + (x_1 + x_2)^2}]}$$

The resistance r_1 is the effective resistance of the stator at stator frequency, which is $5f_1$ for the fifth harmonic. Because of the greater frequency, the effective resistance r_1' for the fifth harmonic is much greater than the corresponding resistance for the fundamental. If the effect on the reactances due to change in distribution of the currents in the conductors with change in frequency is neglected, the term $(x_1 + x_2)$ is five times as large as at fundamental frequency. The maximum torque developed by fifth harmonics in the impressed voltages is

$$T_{m5} = \frac{p(kV_1)^2}{4\pi(5f_1)} \frac{1}{2[(r_1') + \sqrt{(r_1')^2 + (5x_1 + 5x_2)^2}]} \tag{248}$$

where k is the relative magnitude of fifth harmonics and fundamentals in the applied voltages. If $k = 0.1$ and there is a 10 per cent fifth harmonic, the maximum torque due to the term $p(kV_1)^2/4\pi(5f_1)$ in the equation for maximum torque is $(0.1)^2/5 = 0.002$ as large as for the fundamentals and the maximum torque is reduced to a negligible value by the term involving the reactances. A similar statement applies to the torques produced by other harmonics in the impressed voltages. The maximum torque produced by any harmonic of a magnitude likely to occur is too small to have any significant effect on the shape of the speed-torque curve of a motor. The harmonics in the impressed voltages cause harmonics in the currents taken by the motor, but because of the high impedance of the motor for the harmonic currents, the load components of the stator harmonic currents are very small. The harmonic components in the magnetizing currents are also very small. The phase impressed voltage is equal to the voltage induced by the mutual flux plus the primary leakage impedance drop. If this drop is negligible, the impressed voltage and the voltage induced by the mutual flux are equal. Since for a sinusoidal induced voltage $E = 4.44Nk_bk_p\varphi_m f$, with a 10 per cent fifth harmonic, φ_{m5} is $\dfrac{0.1}{5} \times \dfrac{k_{p1}k_{b1}}{k_{p5}k_{b5}} \times \varphi_{m1}$. With a $\frac{5}{6}$ pitch winding, the ratio of the products of the pitch and breadth factors for the fifth harmonic and fundamental is about 0.07. Under this condition, φ_{m5} is about $0.28\varphi_{m1}$. The impedance drop in the primary winding is not negligible for the harmonic except near synchronous speed. Since the fifth harmonic slip, Eq. (246), is equal to $(6 - s_1)/5$, it follows that for values of s_1 between 1 and 0, between standstill and synchronous speed, s_5 varies between 1 and $\frac{6}{5}$. Under these conditions, if the leakage impedances of stator and rotor are equal, the impedance drop in the stator is about one-half the impressed voltage. According to these assumptions, the fifth-harmonic flux between the limits $s_1 = 1$ and $s_1 = 0$ is of the order of magnitude of

$$\frac{0.1}{5} \times \frac{1}{0.07} \times \frac{1}{2} = 0.14$$

of the fundamental flux. For such a magnitude of the fifth-harmonic flux the fifth-harmonic component of the corresponding magnetizing current does not produce any significant effect on the operating current of the motor.

Harmonics in the Space Distribution of the Air-gap Flux. Sinusoidal impressed voltages are assumed, and under this assumption the speed-torque curve of a polyphase induction motor for a sinusoidal distribution of the air-gap flux is a smooth curve. In many cases the actual speed-torque curve is not smooth at low speeds and may indeed be rough. This roughness is due to the effect of odd harmonics in the air-gap flux which are caused by the distribution of the motor windings and by the variation in the permeance of the air gap caused by the slots.

Since the space distribution of the magnetomotive force of an armature coil can be expressed in a Fourier series of the form $A(\sin x + \frac{1}{3} \sin 3x + \frac{1}{5} \sin 5x + \dots)$, as was done in discussing armature reaction and leakage reactance of a synchronous generator, the number of poles produced in the air-gap flux by the distribution of the armature winding is equal to the number of poles for which the motor is wound multiplied by the order of the harmonic. For example, there are five times as many poles for the fifth harmonic produced in the air-gap flux by the distribution of the armature winding as for the fundamental. The time variation of all the harmonics in the air-gap flux is the same as for the fundamental since they are all produced by the same current. Therefore with sinusoidal impressed voltages the revolving harmonic component fields in the air-gap flux, due to the windings, revolve more slowly than the fundamental field. For example, the fifth harmonic revolves one-fifth as fast as the fundamental, the seventh harmonic one-seventh as fast. There are no third harmonics in the air-gap flux of a three-phase machine under balanced conditions since the third harmonics in the flux distributions due to the three phases of such a machine are in space phase and differ in time phase by 120 deg. They therefore cancel. Similar statements apply to all harmonics which are multiples of three.

All the harmonic component fields which exist in the distribution of the air-gap flux of a three-phase machine rotate at speeds $1/q$ times the speed of the fundamental field and the direction of their rotation with respect to the stator alternates. The fifth, eleventh, seventeenth, etc., rotate in the opposite direction to the fundamental field, and the seventh, thirteenth, nineteenth, etc., rotate in the same direction. The magnitudes of the components cannot be large. According to the Fourier analysis, the magnitude of the mmf of a single coil for the fifth harmonic is only one-fifth as large as for the fundamental, and only one-seventh as large for the seventh harmonic. For the whole winding these values are much reduced by the pitch and breadth factors for the fifth and seventh harmonics. For example, if the winding pitch is $\frac{5}{6}$, the ratio of the products of pitch and breadth factors for the fifth harmonic and the fundamental for a three-phase winding is

$$\frac{k_{p5}k_{b5}}{k_{p1}k_{b1}} = \frac{(0.26 \times 0.26)}{(0.95 \times 0.95)} = 0.07$$

The slip of the rotor with respect to any one of the space-harmonic fields is equal to the difference between the speed of the harmonic field with respect to the stator and the speed of the rotor with respect to the stator divided by the speed of the harmonic field. Since the fifth-harmonic field has five times as many poles as the fundamental field and revolves in the opposite direction, the slip of the rotor with respect to the fifth-harmonic field is

$$s_5 = \frac{-\dfrac{2f_1}{5p_1} - \dfrac{2f_1}{p_1}(1 - s_1)}{-\dfrac{2f_1}{5p_1}} = 6 - 5s_1$$

where p_1 is the number of poles for which the motor is wound, s_1 is the slip of the rotor with respect to the fundamental field, and f_1 is the stator frequency. The frequency of the currents in the rotor inductors induced by the fifth-harmonic rotating field is

$$f_2 \text{ (for the fifth harmonic)} = f_1 s_5 = (6 - 5s_1)f_1$$

Since the seventh-harmonic field rotates in the same direction as the fundamental field, the slip of the rotor with respect to the seventh-harmonic field is

$$s_7 = \frac{+\dfrac{2f_1}{7p_1} - \dfrac{2f_1}{p_1}(1 - s_1)}{+\dfrac{2f_1}{7p_1}} = -6 + 7s_1$$

The frequency produced in the rotor inductors by the seventh-harmonic field is

$$f_2 \text{ (for the seventh harmonic)} = f_1 s_7 = (-6 + 7s_1)f_1$$

If the slip s_1 of the motor is 5 per cent, the rotor frequencies induced by the fifth-harmonic and the seventh-harmonic fields, assuming a fundamental frequency of 60 cycles, are $(6 - 5 \times 0.05)60 = 345$ cycles and $(-6 + 7 \times 0.05)60 = 339$ cycles.

Consider the speed of the rotor field with respect to the stator due to currents induced in the rotor by any harmonic stator field, such as the seventh. The speed of the rotor field with respect to the stator is equal to the speed of the rotor field with respect to the rotor plus the speed of the rotor with respect to the stator. For the seventh-harmonic field,

$$\frac{2f_7}{p_7} + \frac{2f_1}{p_1}(1 - s_1) = \frac{2}{7p_1}f_1(-6 + 7s_1) + \frac{2f_1}{p_1}(1 - s_1) = \frac{2f_1}{7p_1}$$

This is equal to the speed of the seventh-harmonic component mmf due to the stator currents. Since the two fields rotate at the same speed, they can react to produce induction-motor action. The seventh-harmonic mmf due to the stator currents can therefore produce induction-motor action. Synchronous speed for the seventh harmonic is one-seventh the synchronous speed for the fundamental. Since the resultant field revolves one-seventh as fast as the fundamental field but has seven times as many poles, it produces fundamental frequency in the stator windings. Similar conditions exist for any of the space harmonics in the stator field.

The currents of any given frequency in the rotor inductors produce rectangular mmfs similar to those produced by the stator currents. These rectangular mmfs can be resolved into a series of space harmonics by Fourier analysis as is done for the stator mmfs. These space harmonics give rise to component fields which revolve at $-\frac{1}{5}$, $+\frac{1}{7}$, $-\frac{1}{11}$, etc., of the speed of the fundamental field produced by the rotor currents of a given frequency. Therefore each of the component rotating fields due to the stator currents produces a series of right-handed and left-handed component rotor fields, but only the rotor fields produced by the fundamentals of the rotor mmfs can react to produce induction-motor action with the harmonic components of the stator field producing them.

Since there is induction-motor action corresponding to each space harmonic in the stator mmf (third harmonics and multiples of thirds excepted in a three-phase machine), a polyphase induction motor can be considered equivalent to a number of motors mechanically coupled, with different numbers of poles and connected in series in such a way that the phase order of their applied voltages alternates. For example, consider induction-motor action due to the fundamental and the fifth and seventh harmonics in the flux distribution. The motor is equivalent to three mechanically coupled motors with numbers of poles p_1, $5p_1$, $7p_1$. The seventh harmonic produces a dip in the speed-torque curve at one-seventh synchronous speed. There is a dip in the speed-torque curve due to the fifth harmonic, but because the phase order of the fifth harmonic is opposite to that of the fundamental, this dip occurs at a slip which is greater than unity. It occurs at a slip $(1 + \frac{1}{5}) = \frac{6}{5}$. The existence of dips in the speed-torque curve of an induction motor has been known for a long time. Certain motors show a succession of dips at low speeds which make their speed-torque curves rough. The existence

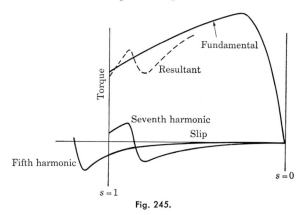

Fig. 245.

of a marked dip in the speed-torque curve of a motor causes a decrease in torque with an increase in speed as illustrated in Fig. 245, and if large enough, a dip may cause a motor to hang or "crawl" at a subsynchronous speed. Crawling is usually accompanied by considerable noise and in some cases by noticeable vibration. Because of the small magnitude of the stator mmfs corresponding to the space harmonic fields, the induction-motor torques corresponding to them must be small. The harmonic torques shown in Fig. 245 are much exaggerated in magnitude. This explanation of the "crawl" of an induction motor is only partial since the induction-motor action caused by space harmonics must be small. The effect of the induction-motor action due to space harmonics can be diminished by skewing the rotor slots. The effect of the stator and rotor slots plays an important part in producing irregularities and roughness in the speed-torque curve of a poorly designed motor.

The Effect of Slot Harmonics in the Air-gap Flux. The slots play an important part in producing irregularities in the speed-torque curve of an induction motor and in causing subsynchronous speeds, noise, and vibration. The effect of the slots is due to the variation in the permeance of the air gap as the motor revolves. This variation in the air-gap permeance causes ripples in the air-gap flux. These ripples can be resolved into right-handed and left-handed rotating fields that may introduce large irregularities in the torque curve of a motor and may cause it to be noisy and to vibrate badly. Under certain conditions revolving fields produced by rotor and stator mmfs may interlock at certain speeds and produce subsynchronous torque, as illustrated in Fig. 246, where the vertical line represents the subsynchronous torque

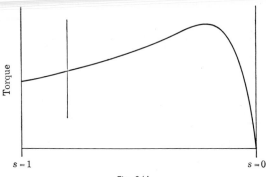

Fig. 246.

condition. There are certain relations between the number of rotor and stator slots which must be avoided if a motor is to operate quietly and is not to show a tendency to crawl. In certain cases the windings of a motor are made slightly unsymmetrical to diminish the effect of the slots on the operation of the motor.

The effect of slots on the operation of an induction motor is too complicated to consider in detail here. Discussions of the effect of space harmonics and of slots on the operation of induction motors can be found in scientific journals.[1]

---

[1] G. Kron, "Induction Motor Slot Combinations," *AIEE Transactions*, Vol. 50, p. 757, 1931.

E. E. Dreese, "Synchronous Motor Effects in Induction Machines," *AIEE Transactions*, Vol. 49, p. 1033, 1930.

Chapter 33

Methods of starting polyphase induction motors. Methods of varying the speed of polyphase induction motors. Methods of controlling the speed of large polyphase induction motors. Scherbius system of speed control. The brush-shifting polyphase induction motor.

Methods of Starting Polyphase Induction Motors. Referring to the vector diagram of a polyphase induction motor (Fig. 232, page 395), the stator current is

$$I_1 = I_n + I'_1$$
$$= I_n - I_2$$

or

$$-I_1 = -I_n + I_2 \tag{249}$$

From either the vector diagram or the equivalent circuit of Fig. 234, page 397, the following relationship may be obtained:

$$V_1 = - I_1(r_1 + jx_1) + I_2(r_2 + R + jx_2)$$
$$= - I_n(r_1 + jx_1) + I_2(r_1 + jx_1) + I_2(r_2 + R + jx_2)$$
$$= - I_n(r_1 + jx_1) + I_2[r_1 + r_2 + R + j(x_1 + x_2)]$$

But $V_1 + I_n(r_1 + jx_1) = V'_1$ is nearly equal to $V_1 - I_n z_1$ numerically as explained on pages 398 and 404. Making this substitution and solving, the magnitude of the secondary current referred to the primary is

$$I_2 = \frac{V'_1}{\sqrt{(r_1 + r_2 + R)^2 + (x_1 - x_2)^2}} = \frac{V'_1}{\sqrt{\left(r_1 + \dfrac{r_2}{s}\right)^2 + (x_1 + x_2)^2}} \tag{250}$$

If the stator and rotor constants are assumed approximately equal when referred to the stator or primary, the exciting current I_n when the slip is unity is only about half as large as it is when the motor is running with normal slip, assuming the same impressed voltage. Consequently the starting current taken by a motor which has no resistance added to its rotor circuit can be considered approximately equal to the secondary current referred to the primary. Therefore, from Eq. (250),

$$I_{st} = \frac{V_1}{\sqrt{(r_1 + r_2)^2 + (x_1 + x_2)^2}} \tag{251}$$

The approximate power factor corresponding to this is

$$pf_{st} = \frac{r_1 + r_2}{\sqrt{(r_1 + r_2)^2 + (x_1 + x_2)^2}} \qquad (252)$$

The reactances are usually three or four times the resistances. Therefore the power factor at starting is low if no resistance is added to the rotor circuit. It must be remembered that Eqs. (251) and (252) can be applied only when no resistance is added to the rotor circuit since only under this condition is the exciting current negligible.

The starting torque, Eq. (243), page 408, is

$$T_{st} = \frac{p}{4\pi f_1} I_2{}^2 r_2 \qquad (253)$$

It is proportional to the copper loss in the rotor circuit and can be increased to the maximum torque of the motor by increasing the resistance of the rotor circuit.

Small motors can be started by connecting them directly to the line, but when started in this way they draw large currents at low power factor. Power companies usually require that the starting currents of larger motors be limited in order to prevent a troublesome dip in the voltage resulting from large low-power-factor starting currents. This limitation of starting current is usually accomplished by (1) reducing the impressed voltage, (2) using motors of the line-start type, or (3) inserting resistance in the rotor circuits.

The reduced voltage for starting is usually obtained by means of a compensator which includes an autotransformer and suitable switches. In the starting position the secondary voltage of the transformer is applied to the motor. This voltage is often about 50 per cent of the voltage of the lines. The current drawn by the motor is reduced in direct proportion to the voltage reduction. The current taken from the lines is still further reduced by the ratio of transformation of the transformer. When 50 per cent taps are used on the transformer the line current is approximately one-quarter of the value obtained with full voltage starting. After the motor is brought up to speed on this reduced voltage, it is then switched to full line voltage and the transformer is disconnected from the circuit. As shown by Eq. (242), page 408, the starting torque is proportional to the square of the impressed voltage. Reducing the voltage at starting, therefore, reduces considerably the starting torque.

Line-start motors are of three general types: (1) high-resistance rotor, (2) high-reactance rotor, and (3) double-squirrel-cage rotor.

The resistance and reactance of the rotor is dependent on the size and shape of the rotor conductors. When cast aluminum conductors are used, this is controlled by the size and position of the slots as illustrated in Fig. 244, page 421. Small open slots near the surface of the rotor give high resistance and low reactance. Larger slots, more deeply imbedded in the iron, give reduced resistance but larger reactance. If the slots are closed at the rotor surface, this increases considerably the leakage reactance.

The high-resistance rotor gives a large starting torque, often about 300 per cent of full-load torque, at a good starting power factor. Because of the high resistance the slip at full load is likely to be as great as 0.10 and the efficiency is poor because of the large copper loss. This type of motor is often used on such applications as punch presses, cranes, and winches.

While the starting current can be reduced by using a high-reactance rotor, this type of motor has limited application because of the reduced starting torque and the poorer power factor during starting and also under running conditions.

The double-squirrel-cage rotor however is capable of combining excellent starting conditions with good speed regulation and good efficiency, under running conditions. This result is obtained because this type of rotor may have a high resistance at starting and a low resistance at running speeds as explained on page 419.

For starting motors having coil-wound rotors, Y-connected variable resistances are connected to the rotor slip rings. This adds an external resistance in series with the windings of each phase of the rotor. When starting, the external resistance is set at its maximum value and full line voltage is applied to the stator. After the motor is up to speed, the external resistance may be gradually reduced to zero, thereby shorting the rotor windings at the slip rings. If the resistance units are designed to carry the full-load rotor current continuously, these same resistances may be used for varying the speed of the motor as explained on page 433.

When this method of starting is employed, a motor can be brought up to speed under any load which requires a torque not exceeding the maximum torque of the motor. The current required to develop a given torque when starting with resistance in the rotor circuit is the same as that required to develop the same torque under running conditions. The torque per ampere is a characteristic constant of the induction motor when operating on the stable part of its speed-torque curve, the part between synchronous speed and maximum torque. If full-load torque is required, the current is equal to the normal full-load current of the motor. Equation (236), page 403, for torque can be written

$$T_2 = \frac{pV_1^2}{4\pi f_1} \frac{r_2/s}{\left(r_1 + \dfrac{r_2}{s}\right)^2 + (x_1 + x_2)^2} \tag{254}$$

From Fig. 235, page 397,

$$I_1 = V_1 \left\{ \left[g_n + \frac{r_1 + \dfrac{r_2}{s}}{\left(r_1 + \dfrac{r_2}{s} \right)^2 + (x_1 + x_2)^2} \right] \right.$$

$$\left. - j \left[b_n + \frac{x_1 + x_2}{\left(r_1 + \dfrac{r_2}{s} \right)^2 + (x_1 + x_2)^2} \right] \right\} \quad (255)$$

$$I_1 = V_1(G - jB) \quad (256)$$

At starting, $s = 1$. Therefore if r_2 plus a resistance r_2' inserted in each phase of the rotor circuit at starting is made equal to r_2 divided by the slip at full load, both the torque developed by the motor and the current it takes are the same as at full load except for the negligible effect of the rotor-core loss when starting. From Eq. (256) the approximate power factor is

$$\text{pf} = \frac{G}{\sqrt{G^2 + B^2}}$$

This is also the same as at full load when $r_2 + r_2'$ is made equal to r_2/s, where s is the slip at full load.

If it is desired that the motor develop its maximum torque at starting, $r_2 + r_2'$ must be made equal to $\sqrt{r_1^2 + (x_1 + x_2)^2}$, Eq. (240), page 406. This, however, also gives a large starting current. If $r_2 + r_2'$ is made equal to r_2/s where s is now the slip at 150 per cent of full load, the starting current is limited to 150 per cent of full-load current and the starting torque is 150 per cent of full-load torque. Moreover, since this value of r_2' is greater than that required for maximum torque, by gradually reducing the external resistance until the motor starts, any starting torque up to the maximum torque of the motor may be obtained. A motor with a coil-wound rotor is therefore an excellent motor to use when high starting torque is desired.

Methods of Varying the Speed of Polyphase Induction Motors. The exact speed of a polyphase induction motor in revolutions per minute is given by Eq. (215), page 389. There are five ways by which the speed of a polyphase induction motor can be changed:

1. By inserting resistance in the rotor circuit
2. By using a stator winding which can be connected for different numbers of poles
3. By varying the frequency
4. By concatenation, series or cascade connection for two or more motors
5. By inserting voltages in the rotor circuits

By Resistance. This method of controlling the speed of an induction motor requires a coil-wound rotor with slip rings. Rotors are usually star-connected. For normal speed, the slip rings are short-circuited. During starting and also when the speed is to be reduced below normal speed, the slip rings are connected through suitable resistances. The slip of an induction motor can be found by

$$s = \frac{r_2}{r_1{}^2 + (x_1 + x_2)^2} \left\{ -\left(r_1 - \frac{K}{2T_2}\right) + \sqrt{\left(r_1 - \frac{K}{2T_2}\right)^2 - [r_1{}^2 + (x_1 + x_2)^2]} \right\} \quad (257)$$

where $K = pV_1{}^2/4\pi f_1$, Eq. (236), page 403. For given impressed voltage and frequency, K is constant. For fixed internal torque T_2 and constant impressed voltage and frequency, the slip of an induction motor varies directly as the rotor resistance r_2. Consequently the speed of an induction motor can be varied by inserting resistance in its rotor circuits.

According to Eq. (229), page 402, the power transferred across the air gap to the rotor is

$$P_2' = E_2 I_2 \cos \theta_{I_2}^{E_2}$$

But

$$I_2 = \frac{E_2 s}{\sqrt{r_2{}^2 + x_2{}^2 s^2}} \qquad \text{and} \qquad \cos \theta_{I_2}^{E_2} = \frac{r_2}{\sqrt{r_2{}^2 + x_2{}^2 s^2}}$$

Hence

$$P_2' = \frac{I_2{}^2 r_2}{s}$$

and

$$s = \frac{I_2{}^2 r_2}{P_2'} \quad (258)$$

The slip is equal to the ratio of the copper loss in the rotor circuit to the total power received by the rotor from the stator, *i.e.*, the copper loss in the rotor circuit is proportional to the slip. If the slip is 25 per cent, the electrical efficiency of the rotor is 75 per cent. If the slip is 50 per cent, the rotor efficiency is 50 per cent. If the slip is increased to 75 per cent, the efficiency of the rotor is reduced to 25 per cent. The percentage decrease in the rotor efficiency is proportional to the slip.

Although the resistance method of controlling the speed is simple and often convenient, it is not economical and the drop in speed obtained by this method is dependent upon the load. A motor delivering full-load torque, which has its speed decreased to 50 per cent of its synchronous speed by adding resistance to the rotor circuit, speeds up to nearly normal speed when the load is removed. The speed regulation of a motor with resistance added to its rotor circuit is poor.

It has been shown that the maximum internal torque developed by a polyphase induction motor is independent of the resistance of the rotor circuit. Adding resistance changes the slip at which this maximum torque occurs and at the same time lowers the efficiency. Adding resistance to the rotor circuit of a polyphase induction motor has much the same effect as adding resistance to the armature circuit of a d-c shunt motor. The resistance method of speed control is used with motors for operating electric cranes, electric hoists, elevators, etc.

By Changing Poles. The speed of an induction motor is directly proportional to the frequency and inversely proportional to the number of poles for which the stator is wound. If induction motors which operate at the same frequency are to run at different synchronous speeds, they must have different numbers of poles. Induction-motor windings can be arranged so as to be connected for two different numbers of poles in the ratio 2:1. By the use of two independent windings four speeds can be obtained. Unless squirrel-cage rotors are used with such motors, the general arrangement of the rotor winding must be similar to that of the stator and its connections must be changed whenever the connections of the stator are changed in order that the rotor and stator shall have the same number of poles. On account of the additional slip rings and the complication involved in arranging the rotor windings for pole changing, squirrel-cage rotors are generally used for multispeed motors unless resistances are to be used in the rotor circuits to increase the starting torque or to control the speed.

Multispeed induction motors are used to some extent on electric locomotives. The locomotives on some of the Italian State Railroads and on the Norfolk & Western Railroad in this country are of this type. Two-speed motors are used for ship propulsion where two efficient running speeds are required. Multispeed motors for driving machine tools and for other purposes can be obtained from many of the companies manufacturing electrical machinery.

The difficulties in the design of a satisfactory multispeed motor are due to the change in the effective number of turns per phase, and consequently in the flux density, and to the change in the coil pitch when the connections are altered to change the number of poles. There are several practical ways to change the number of poles, but all of these, if the voltage is kept constant, involve changes in the flux density and magnetizing current which may, in some cases, be as high as 100 per cent and involve a change in the blocked current which is even greater. As a result the power factor and breakdown torque may be quite different for the different connections. The design of multispeed motors must be more or less of a compromise.

In the practical design of two-speed induction motors, the speed ratio with a single winding is 2:1. In these motors the coils are generally of such a pitch as to give full pitch for the connection producing the greater number of poles. Consequently when connected for the smaller number of poles the pitch is one-half. The connections are made in such a manner that half of the poles are consequent poles when connected for the greater number of poles. The smaller number of poles is obtained by conducting the current to the center point of the windings of each phase. To keep the flux density nearly constant, a change can be made from Δ to Y or from series to parallel connection. If for example for the larger number of poles the connections are series Δ and for the smaller number of poles parallel Y, the flux densities are not markedly different for the two speeds.

Figure 247 illustrates for one phase how the number of poles can be changed in the ratio 2:1 by changing the connections of the winding.

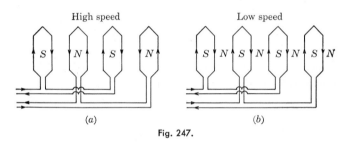

(a)   (b)

Fig. 247.

The winding for each phase consists of two sections, and the change in the number of poles is accomplished by reversing the current in one section relatively to the other. The arrows on the two parts of Fig. 247 indicate the relative direction of the current in the two sections of the winding. For high speed, one section produces north poles and the other produces south poles. For low speed, all sections produce south poles. The north poles which must necessarily exist between south poles are consequent poles. If the coils are full-pitch coils for the low-speed connection, their pitch is one-half for the high-speed connection. The corresponding pitch factors are 1.00 and 0.707. If a pitch of ⅓ is used for the low-speed connection, the pitch is ⅔ for the high-speed connection. In this case each pitch factor is 0.866.

The stator windings of a two-speed three-phase motor consist of one or more groups of six identical equally spaced sections, with two sections in each group for each phase. The angle between adjacent sections is 60 electrical degrees for the high-speed connections and 120 electrical degrees for the low-speed connections. There are several ways by which the current can be reversed in the adjacent sections of each phase to change the number of poles.

The relative characteristics of the motor for the two speeds and the flux densities at which it operates depend on the manner in which the reversal of current in the adjacent sections of each phase is accomplished. The ratio of the blocked currents for the two connections is inversely proportional to the ratio of the blocked equivalent impedances of the motor for the two connections [Eq. (251), page 429]. The ratio of the maximum or pull-out torques is very nearly equal to the inverse ratio of these impedances [Eq. (241), page 406]. The winding pitch and the method adopted for changing the connections determine the ratio of the impedances and depend on the operating characteristics desired at the two speeds. Two-speed induction motors are generally either for

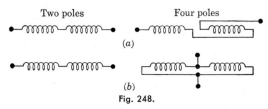

Fig. 248.

constant power or constant torque at the two speeds. The constant-torque motors are more common.

Figure 248 illustrates two methods for changing the connections of one phase. In Fig. 248a, there are the same number of series turns in each phase between terminals for both connections. In Fig. 248b there are twice as many series turns in each phase between terminals for the low-speed connection as for the high-speed connection. In a three-phase motor the connections can be changed from Δ to Y or vice versa. The change from Δ to Y changes the effective number of turns per volt. The ratio of the number of turns per volt of the phase voltage for the two connections is $\sqrt{3}:1$. Two methods by which the number of poles of a motor can be changed in the ratio 2:1 are illustrated in Fig. 249. The connections of Fig. 249a change series Y for the four-pole connection to parallel Y for the eight-pole connection. The connections of Fig. 249b change series Δ for the four-pole connection to parallel Y for the eight-pole connection.

By Varying the Frequency. The speed of an induction motor is directly proportional to the frequency impressed on the stator. By varying this frequency, the speed can be varied. This method for varying the speed has the objection of requiring a separate synchronous generator for each motor, and for this reason it is applicable only in special cases.

Since an induction motor is a transformer, the flux at any fixed voltage varies inversely as the applied frequency. In order to prevent the increase in flux density with its attendant increase in core loss, magnetizing current, and magnetic leakage when the frequency is lowered, the voltage impressed on the motor must be lowered in proportion to the frequency. This does not involve any difficulty if a separate generator is used to furnish power to the motor since the voltage of a synchronous generator varies in direct proportion to the frequency when the field excitation is kept constant. This method of speed control can be used when induction motors are used for ship propulsion since the generators furnish power only to the motors. Separate generators are used for lighting and for auxiliary apparatus. When the voltage and frequency

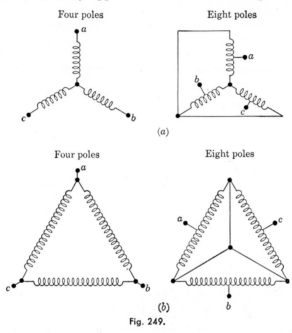

Fig. 249.

impressed on the stator of an induction motor are decreased in the same proportion, the starting torque is increased up to the point where the increase in the term

$$\frac{r_2}{(r_1 + r_2)^2 + (x_1 + x_2)^2}$$

in Eq. (242), page 408, is balanced by the decrease in the term $pV_1{}^2/4\pi f$. The resistances are smaller than the reactances at normal frequency and decrease somewhat with a decrease in frequency. If the change in the distribution of the current over the cross section of rotor and stator conductors is neglected, rotor and stator reactances decrease in proportion to frequency. Actually they decrease a little more rapidly than the frequency because of the change in current distribution with frequency. If the ratio of frequency to impressed voltage is constant, the torque for any small slip varies very nearly in proportion to speed since the term $s^2(x_1 + x_2)^2$ in Eq. (236), page 403, is negligible when s is small.

The torque of an induction motor however is proportional to the product of the strength of the rotating magnetic field, the rotor current, and the rotor power factor. When the voltage and frequency are reduced in the same ratio, the strength of the rotating magnetic field remains nearly constant. The torque for a given rotor current therefore would increase somewhat when the speed was decreased because of the increase in the rotor power factor.

By Concatenation. Concatenation, cascade or series connection for induction motors gives much the same effect as series connection for d-c series motors. In both cases, if the current taken from the mains is equal to the full-load current of one motor, approximately twice the full-load torque of one motor at approximately one-half full-load speed results.

Motors which are to be connected in concatenation should have wound rotors and their ratios of transformation should preferably be unity. The rotors must be rigidly coupled. The stator of one motor is connected to the mains and its rotor is connected to the stator of the second motor. The rotor of the second motor is either short-circuited or connected through resistance. The resistance is used either during starting or when intermediate speeds are required.

If the ratios of transformation are not unity, the motors can still be operated in concatenation provided they have equal ratios of transformation, although the rotors then must be electrically connected as well as mechanically coupled. The stator of one motor must be connected to the mains and the stator of the other motor short-circuited.

Let p_1 and p_2 be the number of poles and s_1 and s_2 be the slip for the two motors. If f_1 is the frequency of the voltage impressed on the first motor, the frequency f_2 of the current in the primary of the second motor is

$$f_2 = f_1 s_1$$

Therefore synchronous speed in revolutions per second for motor 2 is

$$\frac{2f_1 s_1}{p_2}$$

Its actual speed is

$$\frac{2f_1 s_1}{p_2} (1 - s_2)$$

The speed of motor 1 is

$$\frac{2f_1}{p_1} (1 - s_1)$$

Since the motors are rigidly coupled they must run at the same speed. Hence

$$\frac{2f_1 s_1}{p_2} (1 - s_2) = \frac{2f_1}{p_1} (1 - s_1)$$

and

$$s_1 = \frac{p_2}{p_1 - s_2 p_1 + p_2}$$

As the rotor of the second motor is short-circuited, s_2 is small and the term $s_2 p_1$ can be neglected, giving

$$s_1 = \frac{p_2}{p_1 + p_2} \text{ approximately} \qquad (259)$$

The speed of the system is the same as the speed of the first motor or

$$\frac{2f_1}{p_1} (1 - s_1) = \frac{2f_1}{p_1} \left(1 - \frac{p_2}{p_1 + p_2}\right) = \frac{2f_1}{p_1} \frac{p_1}{p_1 + p_2} \qquad (260)$$

If p_1 and p_2 are equal, the speed of the system is equal to one-half the synchronous speed of either motor. The use of two identical motors in parallel and in concatenation gives two efficient running speeds, full speed with the motors in parallel and half speed with the motors in concatenation. When the motors are in concatenation, other speeds can be obtained by the use of resistance in the rotor of the second motor. When the motors are in parallel, other speeds can be obtained by the use of resistance in both rotors. The use of two identical motors gives essentially a constant-torque system since approximately twice the full-load torque of one motor can be obtained at any speed without exceeding full-load current in either motor.

If motors having a different number of poles are used, three running speeds can be obtained, but two of these speeds make use of only one motor at a time. The full torque of the system is available only when the motors are in concatenation. The three speeds are obtained by the use of (1) motor 1 alone, (2) motor 2 alone and (3) motors 1 and 2 in concatenation.

For example, let the motors have 8 and 12 poles and let the frequency be 25 cycles. Then $p_1 = 8$, $p_2 = 12$, and $f_1 = 25$. The speeds obtainable in revolutions per minute are

(1) Motor 1 alone:

$$\text{Speed} = \frac{2(25)}{8} 60 = 375 \text{ rpm}$$

(2) Motor 2 alone:

$$\text{Speed} = \frac{2(25)}{12} 60 = 250 \text{ rpm}$$

(3) Motors 1 and 2 in concatenation:

with motor 1 connected to the mains,

$$\text{Speed} = \frac{2(25)}{8}\, 60 \left(\frac{8}{8 + 12}\right) = 150 \text{ rpm}$$

with motor 2 connected to the mains,

$$\text{Speed} = \frac{2(25)}{12}\, 60 \left(\frac{12}{12 + 8}\right) = 150 \text{ rpm}$$

In concatenation, it makes no difference so far as the speed of the system is concerned which motor is connected to the mains.

Consider two identical Y-connected motors in concatenation. The connection of motors having a ratio of transformation of unity are shown

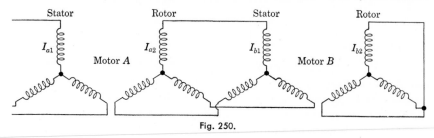

Fig. 250.

in Fig. 250. If the ratios of transformation are not unity but are equal, the rotors must be connected electrically and the stator of the second motor must be short-circuited. Let the resistances of the stator and rotor windings per phase be r_1 and r_2, and let x_1 and x_2 be the reactances per phase of the stator and rotor at a frequency f_{a1}, the frequency impressed on the stator of the first motor. The subscripts a and b are used with currents, voltages, etc., when necessary to indicate to which motor the quantities belong. The subscripts a and b are not needed with resistances and reactances as the motors are identical. Each motor is assumed to have p poles.

The speed of the first motor is

$$n_a = \frac{2f_{a1}}{p}\, (1 - s_a)$$

Since the motors are mechanically coupled, this is also the speed n_b of the second motor. The frequency of the currents and voltages in the rotor windings of the first motor is

$$f_{a2} = f_{a1} s_a$$

This is also the frequency of the currents in the stator windings of the second motor since the stator of the second motor is electrically connected to the rotor of the first motor. The speed of the rotating field due to the stator currents in the second motor is

$$\frac{2f_{b1}}{p} = \frac{2f_{a1}s_a}{p}$$

with respect to the stator. The slip of the second motor is

$$s_b = \frac{\text{speed of stator field of second motor} - \text{speed of rotor of second motor}}{\text{speed of stator field of second motor}}$$

$$s_b = \frac{\dfrac{2f_{a1}s_a}{p} - \dfrac{2f_{a1}}{p}(1 - s_a)}{\dfrac{2f_{a1}s_a}{p}}$$

$$= \frac{2s_a - 1}{s_a}$$

The equivalent circuit of the first motor is given in Fig. 251. Everything in this figure is referred to the stator. V_{a2} is the actual phase

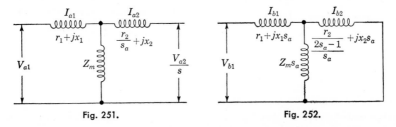

Fig. 251.   Fig. 252.

voltage across the rotor terminals of the first motor and is equal to the phase voltage V_{b1} impressed on the stator terminals of the second motor. Since in the equivalent circuit the rotor resistance and the rotor reactance are referred to stator frequency, V_{a2} must also be referred to stator frequency by dividing it by the slip s_a before it can be used in the equivalent circuit.

The equivalent circuit of the second motor with its rotor short-circuited is given in Fig. 252. Everything on this figure is referred to the stator frequency of the second motor, i.e., to the frequency $f_{b1} = f_{a2}$. For example, x_1 is the reactance of the stator windings of both motors at frequency f_{a1}. At frequency $f_{b1} = f_{a2}$ the reactance is x_1s_a. If the impressed voltage V_{b1} and the impedances are divided by the slip s_a, the current in the motor remains unchanged. This division of the impressed voltage and the impedances in Fig. 252 by s_a refers everything in the equivalent circuit of the second motor to the frequency of the equivalent circuit of the first motor. The two equivalent circuits can then be joined as in Fig. 253 to give the equivalent circuit of two identical motors in concatenation.

The speed-torque curve of the two motors in concatenation can be obtained by superimposing the speed-torque curves of the two motors as obtained from the equivalent circuit of Fig. 253. The first motor operates with a resistance in the rotor circuit equal to $\left(\dfrac{r_2}{s_a} + \dfrac{r_1}{s_a}\right)$ in series with the resistance of the two impedances Z_m and $\left(\dfrac{-r_2}{1 - 2s_a} + jx_2\right)$ in parallel.

The torque developed by a polyphase induction motor per phase is given by $(p/4\pi f)I_2{}^2 r_2$ where r_2 is the total resistance per phase of the rotor circuit. When the slip s_a of the first of the motors in concatenation is 0.5, $i.e.$, the system is operating at one-half synchronous speed, the phase

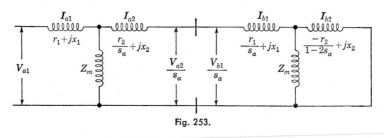

Fig. 253.

resistance $-r_2/(1 - 2s_a)$ of the rotor circuit of the second motor is infinite. Under this condition I_{b2} is zero and the torque developed by the second motor is zero. When $s_a = 0.5$, the resistance of the rotor circuit of the second motor is infinite, the rotor current of the first motor is very small, and the torque developed by this motor is nearly zero. For a slip less than 0.5, the resistance $-r_2/(1 - 2s_a)$ of the rotor of the second motor is negative and the torque of this motor $T_b = \dfrac{p}{4\pi f_b} I_{b2}{}^2 \dfrac{-r_2}{1 - 2s_a}$ is negative. For a slip greater than 0.5, $-r_2/(1 - 2s_a)$ is positive and the torque of the second motor is positive. For a value of s_a greater than 0.5, the resistance of the rotor circuit of the first motor is positive and the torque is positive. The slip-torque curve of each motor and of the motors in concatenation is given in Fig. 254.

If two identical motors connected in concatenation are started at rest, they come up to approximately one-half synchronous speed and operate stably on the part of the speed-torque curve marked ab in Fig. 254. If the speed of the motors is increased in some way to nearly synchronous speed, they operate stably at nearly synchronous speed on the part of the speed-torque curve marked cd in Fig. 254.

It can be shown by the use of symmetrical phase components that the speed-torque curve of a polyphase induction motor which has a single-phase rotor is similar to that of two identical motors in concatenation.[1] A polyphase motor with a three-phase rotor and one rotor phase open is equivalent to a polyphase induction motor with a single-phase rotor.

By Inserting Voltages in the Rotor Circuits. If a voltage is added to the armature circuit of a d-c shunt motor, the speed of the motor is changed. If the voltage inserted assists the voltage impressed on the armature, the speed of the motor is increased. The motor receives power from the circuit on which it operates and additional power from the source of the added voltage. The armature is doubly fed, *i.e.*, it receives power from two independent sources. If the voltage inserted opposes

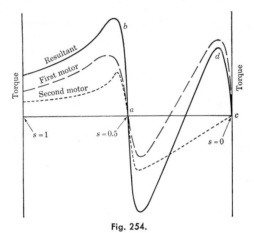

Fig. 254.

that impressed on the armature, the speed of the motor is decreased and power is given to the source of auxiliary voltage. In both cases the speed of the motor is independent of the load except as it is affected by the normal resistance drop in the armature and by the small effect of armature reaction. At any given armature current, a voltage inserted in the armature in opposition to the armature voltage has the same effect on the speed as a resistance producing an equal voltage drop. However with resistance the decrease in speed is accompanied by a loss in power (the I^2r loss in the resistance), but with the inserted voltage the power corresponding to the reduction in speed is not lost but is given to the auxiliary source of voltage. When the speed is raised by adding the auxiliary voltage, the effect on the speed at any current is equivalent to that produced by a negative resistance. When the speed of a d-c shunt motor is changed by adding a voltage to its armature circuit, the speed regulation of the motor is not appreciably affected. Voltages added to the rotor circuits of an induction motor in the proper phase and fre-

[1] Wagner and Evans, "Symmetrical Components," p. 358, 1933.

quency have the same effect as a voltage added to the armature circuit of
a shunt motor.   Moreover, by shifting the phase of the voltages which are
inserted in the rotor circuits, the power factor at which the motor operates
can be controlled.   If the speed is reduced by inserting voltages in the
rotor circuits of an induction motor, the effect on the speed is the same
as would be produced by inserting resistances, except that the change in
speed is nearly independent of the load and the power which would be
lost in the resistances is returned to the circuit supplying the inserted
voltages.   If the speed is raised, the effect produced is equivalent to that
caused by inserting negative resistances.   In this case as before the speed
is nearly independent of the load
and power is supplied to the armature circuits by the voltages which
are inserted.   This power corresponds to the power necessary to

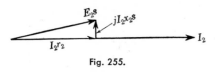

Fig. 255.

produce the increase in speed.   The addition of voltages at rotor frequency and in the proper phase is the basis of power-factor control of
induction motors.   It is also the underlying principle involved in the
operation of adjustable-speed induction motors.

If the voltages added to the rotor circuit of an induction motor are
approximately in phase with or in phase opposition to the voltages E_2s
induced in the rotor windings by the relative motion of the rotor conductors with respect to the revolving field, the speed is changed and at
fixed torque there is little effect on the power factor at which the motor
operates.   If the voltages inserted are approximately in quadrature with
the voltages induced in the rotor conductors by their relative motion
with respect to the revolving field, the power factor at which the motor
operates is changed but there is little effect on the speed and torque.

Refer to Fig. 255, the vector diagram of one phase of the rotor of a
polyphase induction motor operating with a slip s.   The letters in this
diagram have the usual significance.   The rotor current is

$$I_2 = \frac{E_2s}{\sqrt{r_2{}^2 + x_2{}^2s^2}}$$

If a voltage E_2' approximately opposite in phase to E_2s is added to the
rotor circuit, the slip s and the rotor frequency $f_2 = f_1s$ are increased.
If the new slip is s', the new rotor frequency is f_1s'.   The new rotor voltage E_2s' is greater than the corresponding voltage E_2s before the voltage
E_2' is inserted.   The new rotor reactance x_2s' is greater than x_2s since the
rotor frequency f_1s' is greater.   To simplify the explanation, assume that
the air-gap flux is constant and therefore E_2 is constant.   This is equivalent to assuming that the change in the stator impedance drop caused by
change in the stator current is negligible.   The phase angle of the rotor

current with respect to the rotor voltage is greater than before because of the increase in the rotor reactance from x_2s to x_2s'. Let the voltage E_2' added to the rotor circuit be of such magnitude and phase with respect to E_2s that the magnitude of the current I_2 is unchanged and I_2 is in its original position with respect to E_2s. To produce this condition, E_2' is not exactly opposite to E_2s since the rotor reactance is increased by the increase in the rotor frequency. The following relations must hold to fulfill these conditions:

$$I_2 = I_2' = \frac{E_2s}{r_2 + jx_2s} = \frac{E_2' + E_2s'}{r_2' + jx_2s'}$$

The vector diagram with the added voltage E_2' which makes $s' = 2s$ and keeps the current I_2 unchanged in magnitude and phase is given in Fig. 256, r_2 being assumed equal to r_2' for simplicity.

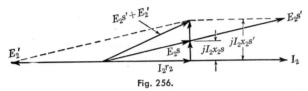

Fig. 256.

The torque of the motor is proportional to the product of the current, flux, and cosine of the angle between the current and the voltage E_2s induced by the air-gap flux. Since the air-gap flux is assumed to be constant and I_2 is unchanged in magnitude and makes the same phase angle with E_2s as before, the torque is unchanged. Since I_2 has a component opposite in phase to E_2', I_2 corresponds to power delivered to the circuit supplying voltage E_2'. So far as the effect of E_2' on the speed is concerned, it is the same as inserting a resistance in the rotor circuit, except that the voltage E_2' is not influenced by current in the same way as the drop due to a resistance. Consequently the speed does not change greatly with a change in load. Also there is no power loss due to the decrease in speed as there is with resistance, since the power corresponding to E_2' and I_2 is returned to the circuit supplying the voltage E_2' instead of being absorbed as an I^2r loss in resistance.

Consider the case where the added voltage E_2' is inserted to assist the voltage induced in the rotor circuit by the air-gap flux. This condition is shown in Fig. 257. In this figure the magnitude of the voltage E_2' and its phase are such as to make $s = -s'$ and to keep the current I_2 unchanged in magnitude and phase with respect to the voltage E_2s induced in the windings by the air-gap flux. Since s' is equal to s in magnitude but opposite to s in sign, $x_2s' = -x_2s$. If the current I_2 is kept constant and fixed in phase, the added voltage E_2' is in phase with I_2. In this as in the previous case the torque is unchanged. E_2', which is a voltage rise, is in phase with the current I_2. The power represented by the voltage E_2' in conjunction with the current I_2 is power delivered to the rotor to produce the increase in speed. The rotor receives power by induction from the stator and also by conduction from the source of power which supplies the voltage E_2'. The rotor is doubly fed in that it receives power from two independent sources. By inserting a voltage E_2' in each of the rotor circuits of a polyphase induction motor, the speed

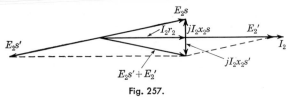

Fig. 257.

can be adjusted and for each value of E_2' the speed changes little with change in load.

The power factor of an induction motor can be changed by inserting in each of the rotor circuits a voltage E_2' of the proper magnitude and phase. If the magnitude and phase of E_2' are such as to swing I_2 ahead of the position with respect to E_2s, which existed before the voltage was added, and at the same time are such as to keep the component of the current in

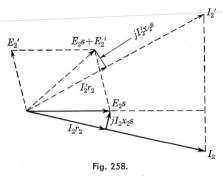

Fig. 258.

phase with E_2s unchanged in magnitude, the torque is unchanged, but the stator current, which must have a component opposite in phase to I_2, is swung toward lead, and the motor can be made to operate either at unity power factor or with a leading current. The vector diagram for one phase of the rotor for this condition is given in Fig. 258. In this figure I_2 is the current before the voltage is added. I_2' is the current after the voltage is added. The effects of speed control and power-factor adjustment can be combined by inserting the voltage E_2' of the proper magnitude and phase. Figure 259 is the complete vector diagram of an induction motor showing power-factor control. The full lines represent the conditions before the voltages are added to the rotor circuits. The conditions after adding the voltages which produce nearly unity power factor are shown by lines of dashes. In this figure the air-gap voltage is assumed constant. This assumption makes the impressed voltages slightly different for the two conditions. In the diagram impedance drops are much larger than the actual drops for the sake of clearness.

The voltages for speed control or for power-factor compensation can be added at the slip rings of a wound rotor. The frequency of the source of power from which the voltages are obtained must be equal to the rotor frequency, which is the slip frequency, and must therefore vary with the slip. The additional equipment needed to supply these voltages is expensive, and consequently this method of speed control is generally limited to motors of large size such as those used for the rolls in steel mills. It is sometimes more economical to convert alternating current to direct current and then to use adjustable-speed d-c motors.

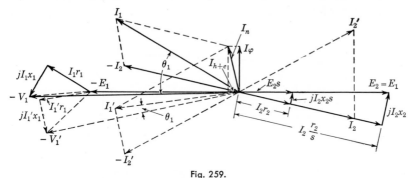

Fig. 259.

Methods of Controlling the Speed of Large Polyphase Induction Motors.

The methods of controlling the speed of large polyphase induction motors fall into two classes, constant-torque systems and constant-power systems. These methods of speed control can be used with any wound-rotor polyphase induction motor, but the expense is not warranted except for large motors where it is more than balanced by economies gained by their use.

In the constant-power system of speed control, neglecting losses, all power taken from the line ultimately appears at the shaft of the driving motor, coming partly through an auxiliary motor connected to the shaft of the main motor. In the constant-torque drive, the power corresponding to the increase in slip is returned to the power line. As this power is not returned to the shaft of the main motor, it is not necessary to have the auxiliary or regulating motor connected to the shaft of the main motor. Therefore the auxiliary or regulating motor can be a high-speed machine. The two systems for controlling the speed depend on the insertion of voltages in the rotor circuits of the main motor. Several methods have been devised for obtaining the required voltages but only the Scherbius system is described here. Those especially interested in the speed control of mill motors should consult technical publications.[1]

[1] K. A. Pauly, "Some Methods of Obtaining Adjustable Speed with Electrically Driven Rolling Mills," *General Electric Review*, Vol. 24, p. 422, 1921.

L. A. Umansky, "Adjustable Speed Drives for Rolling Mills," *Iron and Steel Engineering*, September, 1924.

Scherbius System of Speed Control.   The diagram of connections for the single-range constant-torque Scherbius system is shown in Fig. 260. The single-range system of speed control gives speeds below synchronism. The double-range system gives speeds both below and above synchronism. M is the main motor which is a wound-rotor induction motor with the terminals of the rotor winding brought out to slip rings. R is the regulating machine which provides the voltages inserted in the rotor circuits of the main motor to control its speed. I is a polyphase induction machine which is mechanically coupled to the regulating machine

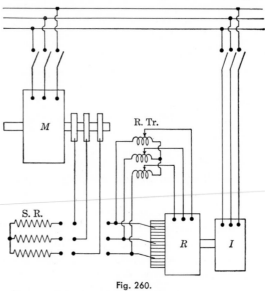

Fig. 260.

and serves to return to the power mains the power absorbed by the regulating machine.   The regulating machine can be mechanically coupled to the shaft of the main motor.   In this case the power absorbed by the regulating machine is returned to the shaft of the main motor.   This gives a constant-power system.   If the regulating motor is not connected to the shaft, the system is a constant-torque system.

In the double-range system, I furnishes the power to the regulating machine which must be delivered to the main motor in order to drive it above synchronous speed.   S.R. is the starting resistance for the main motor.   R.Tr. is the regulating transformer which controls the magnitude of the voltages inserted in the rotor circuits of the main motor by the regulating machine.

The regulating machine has an armature similar to the armature of a d-c generator and a stator or field similar to that of a three-phase induction motor.   If the stator of R is excited with three-phase voltages of frequency f_1, the frequency of the currents induced in the rotor conductors is f_1s where s is the slip of the rotor with respect to the rotating field set up by the stator currents. If brushes are placed 120 deg apart on the commutator, the frequency appearing at any brush on the commutator is due to the actual frequency f_1s of the currents in the rotor conductors plus a frequency $f_1(1 - s)$ due to the movement of the conductors past the brush. The frequency appearing at the brush is the sum of the two frequencies and is $f_1(s) + f_1(1 - s) = f_1$, which is entirely independent of the speed at which the machine is driven. If three brushes are placed

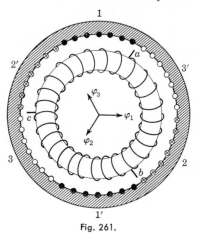

Fig. 261.

120 deg apart on the commutator (a two-pole field assumed), three-phase voltages of frequency f_1 appear between pairs of brushes. As in a d-c generator, the electrical power developed does not come from the power used for field excitation. It comes only from the mechanical power applied to the shaft. If the brushes are placed as in Fig. 261, the voltages appearing between the pairs of brushes a-b, b-c, and c-a are in phase with the voltages impressed on the stator windings or in phase opposition to them, depending upon the direction of rotation of the armature and the phase order of the voltages impressed on the stator. The above statement neglects resistance and leakage-reactance drops in the machine. A displacement of the brushes from the positions shown in Fig. 261 changes the phase of the voltages appearing between them.

Interpoles placed at the commutating zones with their windings connected in series with the brush circuits are provided to aid commutation. The interpoles are omitted in Fig. 261. A gramme-ring armature is shown in the figure. Actually, the ordinary drum-type armature winding is used. A two-pole machine is shown for simplicity. The phase belts for the stator windings are 1-1', 2-2', 3-3'. The vectors representing the directions of the fluxes produced by the three phases are marked φ_1, φ_2, φ_3.

The brushes of the regulating machine are connected to the slip rings of the main motor. The stator of the regulating machine is excited from the slip rings of the main motor through the regulating transformer marked R.Tr. by means of which the voltage applied to the slip rings of the main motor can be varied. If the voltages of the regulating machine are opposite to those induced in the rotor circuits of the main motor by its own slip, the main motor operates below synchronism.   In this case the regulating machine operates as a motor and delivers power to the induction machine I which operates as an induction generator.   Under this condition all the power corresponding to the decrease in speed of the main motor except that due to the losses in the regulating machine and in the induction generator is returned to the circuit from which the system operates.   The system just described cannot operate at synchronous speed since at that speed the slip-ring voltages of the main motor are zero and there is no excitation for the regulating machine.   However, the system operates equally well above synchronous speed if the voltages applied to the slip rings of the main motor are reversed and some device is provided for exciting the regulating machine at synchronous speed.

Above synchronous speed the regulating machine operates as a generator and receives mechanical power from the induction machine I which now operates as a motor.   For a given range in speed, the double-range system requires control apparatus about half as large as for the single-range system, since the change in speed from synchronism need be only half as great for the same speed range.   For example, to produce a 50 per cent change in speed based on the higher speed requires a change in speed from one-third above synchronous speed to one-third below synchronous speed and necessitates a regulating machine of one-third the capacity of the main motor.   For the same change in speed, the synchronous speed of the main motor for a single-range system must be one-third greater than the synchronous speed of the main motor for the double-range system, and the regulating machine for the single-range system must drop the speed of the main motor to one-half its synchronous speed.

In order to have the speed in the double-range system pass through synchronous speed, it is necessary to feed d-c excitation to the slip rings of the main motor at synchronous speed in order to have it operate as a synchronous motor at that speed and develop torque.   For this purpose a special machine is used which can excite the regulating machine with direct current at synchronous speed of the main motor.   The regulating machine then excites the main motor with direct current.   The special machine is known as an ohmic-drop exciter since the voltage it develops at synchronous speed has to overcome only ohmic resistance drops.

When an ohmic-drop exciter is driven at synchronous speed, d-c voltage appears at the commutator brushes.   Neglecting the voltage drops in the armature windings, the ratio of the d-c voltage to the alternating voltage impressed on the slip rings is fixed by the number of phases and is not influenced by the speed or the excitation at which the ohmic-drop exciter operates since both voltages are generated in the same winding. The frequency appearing across the d-c brushes on the commutator can be considered to be made up of two frequencies, f_1 and $f_1(1 - s)$. The frequency f_1 is the frequency of the current in the armature inductors and is the same as the frequency impressed on the slip rings. The other frequency $f_1(1 - s)$ is due to the rotation of the armature inductors past the brushes where s is the slip of the machine defined in the same way as for an induction motor. The actual frequency appearing at the d-c brushes is $f_1 - f_1(1 - s) = f_1 s$. If the machine operates at synchronous speed, $f_1 s$ is zero and d-c voltage appears at the brushes. If the machine

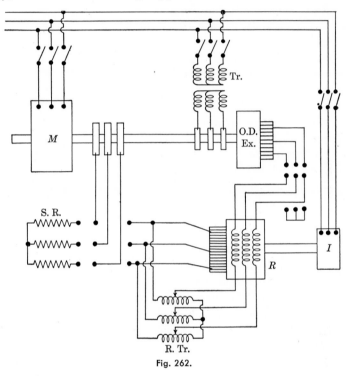

Fig. 262.

is driven mechanically at some other speed, with slip s, the frequency of the voltage at the commutator is $f_1 s$. This gives the value of the frequency both when s is positive and when it is negative.

The ohmic-drop exciter has an armature like that of a d-c machine with the addition of taps brought out from the windings and connected to slip rings as in a synchronous converter (see Chap. 44). These slip rings are connected to the a-c source through a transformer. Instead of the usual poles it has a laminated iron ring pressed on over the armature to complete the magnetic circuit and to decrease the magnetizing components of the alternating currents which produce the flux. This ring rotates with the armature. There are no stationary parts except brushes and supports for the bearings. Neglecting the voltage drops, the voltages appearing at the commutator of the machine are fixed in magnitude by the magnitude of the alternating voltages impressed on the slip rings, and the frequency is fixed by the slip.

The diagram of connections for the double-range system of speed control is given in Fig. 262 and shows the addition of the ohmic-drop exciter (O.D. Ex.). The ohmic-drop exciter is wound for the same number of poles as the main motor and is coupled directly to the shaft of the main motor. As it operates at the same speed as the main motor, the frequency at its commutator must be f_1s, or the same as the frequency at the slip rings of the main motor. The ohmic-drop exciter has brushes on its commutator 120 electrical degrees apart. At synchronous speed the frequency between brushes is zero, but the voltage between them is not zero since it is fixed not by speed but by the alternating voltage impressed on the slip rings.

The double-throw switch shown just below the ohmic-drop exciter in Fig. 262 is closed in the downward position for speeds below synchronous speed. The equipment then functions as in the single-range system. With the switch closed in the upward position, the ohmic-drop exciter reverses the voltage applied to the stator of the regulating machine. This causes the regulating machine to supply to the rotor of the main motor a voltage aiding its own E_2s and therefore causes an increase in speed. At exactly synchronous speed the ohmic-drop exciter supplies a d-c excitation to the regulating machine which causes this machine to introduce into the rotor of the main motor the direct current needed to carry it through synchronous speed.

The Brush-shifting Polyphase Induction Motor. The brush-shifting induction motor is a motor which, in effect, incorporates the main motor and the regulating machine of the Scherbius system of speed control into a single machine. A schematic diagram of this motor is shown in Fig. 263. For simplicity the diagram is drawn for a two-pole machine. Although this motor has been made in larger sizes, the brush-shifting type of induction motor is used principally in sizes from 5 to 50 hp. It is similar to an induction motor with a coil-wound rotor except that the primary winding is wound on the rotor and the secondary winding on the stator. Were the secondary shorted or connected to the usual resistances used for starting and speed control, this motor would have essentially the characteristics of the coil-wound rotor type of induction motor.

The three-phase currents, in the primary windings of the brush-shifting motor produce, in the rotor, a magnetic field which rotates at synchronous speed. This, like the rotating field of any polyphase induction motor, reacts with the currents produced in the secondary winding, which is now the stator winding, to produce torque. This torque causes the rotor to rotate in a direction opposite to that of its rotating magnetic field. The rotating field induces voltages of primary frequency in the auxiliary winding. The rotating field however rotates past the brushes on the commutator at a speed equal to the synchronous speed minus the actual speed of rotation of the rotor. The voltages appearing at these brushes are therefore at slip frequency and are of the same frequency as the voltages produced in the stator windings.

If the stator windings are connected to the commutator brushes as indicated in Fig. 263, this provides a means of accomplishing power-factor adjustment and speed control. How this is done is indicated in Figs. 256, 257, and 258 on pages 445 and 446. The voltage E_2s is now the

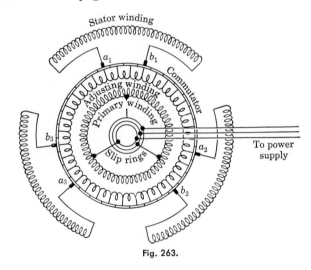

Fig. 263.

voltage induced in each phase of the stator winding; the voltage E_2' is the voltage across the commutator brushes to which the stator winding is connected. The phase relation of E_2' with respect to E_2s is varied by moving all brushes equally around the commutator; the magnitude of E_2' is varied by changing the separation between corresponding a and b brushes.

If the brushes are located on the commutator so as to give a voltage E_2' which is in opposition to E_2s, a reduction in speed is thereby produced. Moving the a and b brushes closer together reduces the magnitude of E_2' and so causes an increase in speed, until when both brushes are on the same commutator segment; the stator winding then being shorted, normal induction motor speed is obtained. If the brushes are now moved past each other so that the relative positions of the a and b brushes have been reversed, an E_2' aiding E_2s will be obtained, causing a further increase in speed. Further separation of the a and b brushes will now give speeds considerably in excess of synchronous speed. This becomes, therefore, a double-range constant-torque method of speed control.

Chapter 34

Performance tests for polyphase induction motors. Losses. Measurement of stray-load loss. Motor characteristics from equivalent circuits. Determination of the constants of the equivalent circuit from test data. Calculation of performance characteristics from the equivalent circuit. The calculation of the performance of a polyphase induction motor from the equations for power, torque, etc., and the vector diagram.

Performance Tests for Polyphase Induction Motors. The efficiency and other characteristics of polyphase induction motors can best be determined by load tests using either a brake or a dynamometer for measuring the motor output. This is, however, not always convenient and in the case of large motors may not be possible. Methods, therefore, have been developed for determining these characteristics from measurements of the motor constants and losses. When possible the motor is loaded in some convenient manner in order that the input, power factor, and slip may be determined from direct measurements. Where this is not convenient, the characteristics may be approximated by calculations involving the equivalent circuit of the motor or by the formulas given in the preceding chapters of this text.

Losses. The losses in an induction motor are as follows: (1) primary copper loss, (2) secondary copper loss, (3) core loss, (4) friction and windage loss, and (5) stray-load loss.

When it is possible to load the motor and measure its input and slip but not possible to measure directly the output, the efficiency is obtained by subtracting these losses from the input and dividing the result by the input. The primary copper loss is usually obtained by measuring the ohmic resistance per phase with direct current and multiplying this resistance by the square of the primary phase current and the number of phases. Although the effective resistance at line frequency should be used for this purpose, any method of measuring the effective resistance is likely to involve errors as large as those involved in using the more easily obtainable ohmic value.

The Standards of the ASA as approved in March, 1943, recommends the use of the AIEE Test Code for Polyphase Induction Machines (No. 500, August, 1937). This test code specifies that whenever the slip is accurately determinable, the secondary I^2R loss be taken as the product of the measured primary input minus the primary copper loss and the slip. This is very closely in accordance with the formula (258) given on page 433.

The friction and windage loss plus the core loss may be obtained by running the motor at no load with normal frequency and voltage applied. The input is then equal to the no-load primary copper loss, the friction and windage loss, the core loss, and the no-load secondary copper loss. This last loss is small and is usually neglected. Although it is possible to separate the core loss and the friction and windage loss it is sufficient in most cases to determine the sum of the two. This combined loss is generally assumed to remain constant at all values of load.

The stray-load loss includes all losses not otherwise accounted for. These losses take account of the high-frequency rotor I^2r loss and the increase in the core loss caused by the change in the magnetic fields when the motor is loaded. They also include the increase in the eddy-current losses in the conductors and structural parts of the motor under load conditions. Reference to Harmonics in the Space Distribution of the Air-gap Flux, page 424, will indicate how the high-frequency rotor currents are produced. The following method of measuring the stray-load loss conforms with the recommendations in the AIEE Test Code.

Measurement of Stray-load Loss. The direct measurement of the stray-load loss involves two tests: a d-c excitation test and a test to find the blocked-rotor torque with balanced polyphase voltages at rated frequency.

In the d-c excitation test, two stator terminals are connected to a d-c source, the third terminal being left open. A variable series resistance may be used for adjusting the current value. (For motors of the wound-rotor type, the direct current is applied to the rotor and the stator terminals are short-circuited.) The motor is then driven at synchronous speed, using either a dynamometer or a calibrated motor, and the mechanical driving power is measured for a series of values of d-c excitation. Under these conditions, the currents of fundamental frequency induced in the conductors of the short-circuited member create an mmf which very nearly balances the mmf of the d-c excitation and there is little core loss due to the mutual flux of fundamental frequency. The leakage and harmonic fields are nearly the same as those existing when the motor is operating at a load current equivalent to the d-c excitation and produce losses which are nearly equal to those existing when the motor is operating at this load current.

The direct current in each of the two phases of a Y-connected machine produces an mmf equal to $I_{dc}N_{ph}$ where N_{ph} is the number of turns per phase, but these two mmfs are acting along axes displaced by 60 electrical degrees. The total mmf due to the d-c current will therefore equal

$\sqrt{3}\,I_{dc}N_{ph}$. With balanced polyphase currents of normal frequency, the mmf equals $\frac{3}{2}\,\sqrt{2}I_{ac}N_{ph}$ where I_{ac} is the rms value of the phase alternating current. If equal magnetic fields are to be produced by the a-c and d-c currents, $\sqrt{3}I_{dc}N_{ph}$ must equal $\frac{3}{2}\,\sqrt{2}I_{ac}N_{ph}$. The d-c current which is equivalent to a given a-c load current is therefore

$$I_{dc} = \frac{\sqrt{3}}{\sqrt{2}}\,I_{ac} = 1.225I_{ac} \tag{261}$$

Since the distribution of the winding and the use of fractional-pitch coils affect both mmfs similarly, these effects are omitted in the foregoing analysis. It can be shown that Eq. (261) holds equally well for a Δ-connected machine, and it may therefore be applied to any three-phase motor.

Let W equal the mechanical power, expressed in watts, required to drive the motor in the d-c excitation test. This mechanical power equals the sum of the friction and windage loss, a small fundamental frequency core loss, the I^2R loss in the shorted member, and the stray-load loss. If however a d-c excitation equivalent to the no-load current of the motor is used, the stray-load loss is essentially eliminated. Let W_0 be the mechanical driving power at this value of d-c excitation. Then $W - W_o$ equals the stray-load loss plus the difference in secondary I^2R loss.

The blocked-rotor test is made to determine this secondary I^2R loss. With reduced balanced polyphase voltages of rated frequency applied to the motor, the blocked-rotor torque is measured for a range of currents. The blocked-rotor torque equals the starting torque as given by Eq. (243), page 408. In pound-feet for n phases this torque $T_b = (p/4\pi f) \times I_2{}^2r_2n\,^{550}\!/_{746}$. Therefore $I_2{}^2r_2n = T_b \times 4\pi(f/p)^{746}\!/_{550}$. The synchronous speed of the motor

$$n_s = \frac{f \times 120}{p} \text{ in rpm and } \frac{f}{p} = \frac{n_s}{120}$$

$$I_2{}^2r_2n = T_b n_s \frac{4\pi \times 746}{120 \times 550} = 0.142T_b n_s \tag{262}$$

The d-c current for which W of the preceding paragraph is determined is equivalent to an a-c load current in accordance with Eq. (261). If T_b is measured for a blocked-rotor current equal to this a-c load current and T_o is the blocked-rotor torque[1] for a blocked-rotor current equal to the no-load current of the motor, then the stray loss for the given value of a-c load current is

$$P_{sl} = (W - W_o) - 0.142n_s(T_b - T_o) \tag{263}$$

[1] For detailed explanation of measurement of blocked-rotor torque, see AIEE Test Code for Polyphase Induction Machines (No. 500, August, 1937).

This calculation is illustrated by the curves of Figs. 264 and 265. Figure 264 is plotted from data taken in a d-c excitation test and Fig. 265 from data from a blocked-rotor test.

Motor Characteristics from Equivalent Circuits. Although an equivalent circuit similar to that used for transformers may be drawn for an induction motor as shown in Fig. 234, page 397, its use for determining the characteristics of a motor is complicated by several factors noted below.

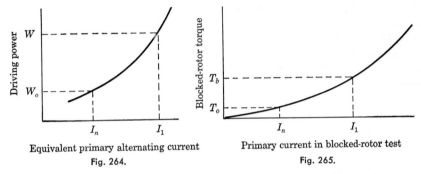

Equivalent primary alternating current

Fig. 264.

Primary current in blocked-rotor test

Fig. 265.

The frequency of the secondary currents of an induction motor is a function of the slip and for usual load conditions is much lower than the primary frequency. Also, the I_n component of the primary current is an appreciable part of the full-load current of the motor. A third factor is that, while the effect of a load on a motor may be taken into account by using a resistance $R = r_2 (1 - s)/s$ as shown in the equivalent circuit, the power absorbed in this resistance is the internal power per phase and not the shaft power of the motor. This internal power multiplied by the

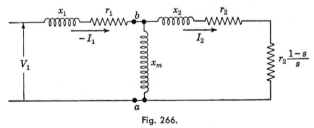

Fig. 266.

number of phases is greater than the shaft power by the power absorbed in the friction and windage and stray-load losses of the motor. These complicating factors make it desirable to introduce certain approximations justified by experience in order to avoid unduly laborious tests. The method which follows is essentially that given by the AIEE in its Test Code for Polyphase Induction Motors previously referred to.

The somewhat simplified equivalent circuit of Fig. 266 is used. In this equivalent circuit the conductance g_n, which absorbs the fundamental frequency core loss, is omitted and the shaft power is obtained by subtracting from the internal power this core loss in addition to the friction and windage and stray-load losses. An advantage of this change is that the friction and windage and core losses are thus combined and do not have to be separated.

Determination of the Constants of the Equivalent Circuit from Test Data. The resistance r_1 of the equivalent circuit of Fig. 266 is taken as the primary (stator) ohmic resistance per phase and is found from d-c measurements. To find r_2, x_1, and x_2 a blocked rotor test at rated current is used. If P_b, I_b, and V_b are the total power input, the line current, and the line voltage measured with the rotor blocked, then for a three-phase machine with Y-connected primary

$$r_2 = \frac{P_b}{3I_b{}^2} - r_1$$

and

$$x_0 = x_1 + x_2 = \frac{f}{f_t}\sqrt{\left(\frac{V_b}{3I_b}\right)^2 - \left(\frac{P_b}{3I_b{}^2}\right)^2}$$

where f/f_t is the ratio of rated frequency to the frequency used in the test. Because the rotor frequency under normal load conditions is low, the AIEE Test Code recommends that the blocked rotor test be made at 15 cycles although a 25-cycle test may be used if the values found from this test do not differ appreciably from those obtained from a test at normal frequency.

The reactance x_0 is split on an empirical basis between x_1 and x_2 as follows:

$$x_1 = 0.5x_0 \qquad x_2 = 0.5x_0 \text{ for class A motors}$$
$$x_1 = 0.4x_0 \qquad x_2 = 0.6x_0 \text{ for class B motors}$$
$$x_1 = 0.3x_0 \qquad x_2 = 0.7x_0 \text{ for class C motors}$$
$$x_1 = 0.5x_0 \qquad x_2 = 0.5x_0 \text{ for class D motors}$$
$$x_1 = 0.5x_0 \qquad x_2 = 0.5x_0 \text{ for wound-rotor motors}$$

Class A motors are normal starting torque, normal starting current squirrel-cage motors and are frequently referred to as general-purpose motors.

Class B motors are normal starting torque, low starting current squirrel-cage motors.

Class C motors are high starting torque, low starting current squirrel-cage motors.

Class D motors are high-slip (high-resistance rotor) squirrel-cage motors.

The magnetizing reactance x_m is determined from a no-load test at rated voltage and frequency. If P_n, I_n, and V_n are, respectively, the total power input, the line current, and the line voltage measured in this no-load test, then for a Y-connected machine

$$x_m = \frac{V_n}{\sqrt{3}\,I_n} - x_1$$

Although the first term on the right-hand side of this equation is not wholly reactive, the discrepancy is negligible. The sum of the core loss and the friction and windage loss is also determined from this no-load test and equals the power input P_n minus the primary copper loss.

Calculation of Performance Characteristics from the Equivalent Circuit. As an example of the calculation of the performance characteristics of a polyphase induction motor from the equivalent circuit, a 1,000-hp 2,200-volt Y-connected 25-cycle 12-pole squirrel-cage induction motor is used. This is a general-purpose motor with copper bars in rotor. Test data obtained from no-load and blocked-rotor runs and the measured stator resistance are as follows:

Ohmic resistance of stator between terminals at 25°C = 0.210 ohm:

> At no load and a temperature of 25°C:
> Stator line voltage............ 2,200 volts
> Stator line current........... 75.1 amp
> Total stator input........... 15.2 kw

With the rotor blocked at approximately full-load current at a temperature of 25°C and a frequency of 15 cycles:

> Stator line voltage.............. 185 volts
> Stator line current.............. 250 amp
> Total stator input.............. 38 kw

Stray load loss = 7.0 kw

A slip of 0.02 and an operating temperature of 75°C are assumed. The stator resistance at this operating temperature is

$$r_1 \text{ at } 75°\text{C} = \frac{0.210}{2} \times \frac{234.5 + 75}{234.5 + 25} = 0.1252 \text{ ohm}$$

The rotor resistance referred to the stator as found from the blocked-rotor test is

$$r_2 \text{ at } 25°\text{C} = \frac{38,000}{3 \times (250)^2} - \frac{0.210}{2} = 0.0977 \text{ ohm}$$

$$r_2 \text{ at } 75°\text{C} = 0.0977 \times \frac{234.5 + 75}{234.5 + 25} = 0.1165 \text{ ohm}$$

The combined reactance of the stator and rotor is

$$x_0 = \frac{25}{15} \sqrt{\left(\frac{185}{\sqrt{3} \times 250}\right)^2 - \left(\frac{38,000}{3 \times (250)^2}\right)^2} = 0.626 \text{ ohm}$$

and

$$x_1 = x_2 = 0.5x_0 = 0.313 \text{ ohm}$$

The magnetizing reactance as found from the no-load test is

$$x_m = \frac{2,200}{\sqrt{3} \times 75.1} - 0.313 = 16.60 \text{ ohms}$$

$$b_m = \frac{1}{16.60} = 0.0602 \text{ ohm}$$

The core loss plus the friction and windage loss equals

$$P_c + P_{f+w} = 15,200 - 3 \times (75.1)^2 \times 0.105 = 13,420 \text{ watts}$$

Refer to the equivalent circuit of Fig. 266.

$$z_2 = r_2 + r_2 \frac{1-s}{s} + jx_2 = \frac{r_2}{s} + jx_2$$

$$= \frac{0.1165}{0.02} + j0.313$$

$$= 5.825 + j0.313$$

$$y_2 = \frac{1}{z_2} = g_2 - jb_2 = \frac{1}{5.825 + j0.313} = 0.1712 - j0.0092$$

The total admittance to the right of ab is

$$g_2 - jb_2 - jb_m = 0.1712 - j0.0092 - j0.0602$$
$$= 0.1712 - j0.0694$$

The corresponding impedance is

$$z_{ab} = \frac{1}{0.1712 - j0.0694} = 5.02 + j2.04$$

The primary impedance is

$$r_1 + jx_1 = 0.1252 + j0.313$$

The total impedance of the equivalent circuit is

$$z_e = z_{ab} + z_1 = 5.02 + j2.04 + 0.1252 + j0.313$$
$$= 5.15 + j2.35$$

The primary current for the assumed slip of 0.02 is

$$I_1 = \frac{2,200}{\sqrt{3} \times \sqrt{(5.15)^2 + (2.35)^2}} = \frac{2,200}{\sqrt{3} \times 5.66} = 224$$

The primary power factor is

$$pf = \frac{r_e}{z_e} = \frac{5.15}{5.66} = 0.910$$

The power input to the primary equals

$$3P_1 = \frac{\sqrt{3} \times 2{,}200 \times 224 \times 0.910}{1{,}000} = 777 \text{ kw}$$

The secondary current referred to the primary is

$$I_2 = I_1 \frac{z_{ab}}{z_2} = 224 \frac{5.41}{5.83} = 208 \text{ amp}$$

The internal power is

$$3P_2 = I_2{}^2 r_2 \frac{1-s}{s} n = (208)^2 \times 0.1165 \times \frac{0.98}{0.02} \times 3$$
$$= 743{,}000 \text{ watts or } 743 \text{ kw}$$

The shaft output equals

$$P_0 = 3P_2 - P_c - P_{f+w} - \text{stray-load loss}$$
$$= 743 - 13.4 - 7.0 = 723 \text{ kw or } 970 \text{ hp}$$

The efficiency is $723/777 = 0.93$ or 93 per cent.

The shaft torque is $5{,}250 \times \dfrac{970}{250 \times 0.98} = 20{,}800$ lb-ft

The Calculation of the Performance of a Polyphase Induction Motor from the Equations for Power, Torque, etc., and the Vector Diagram. To apply the equivalent-circuit method of calculating the performance of an induction motor, the slip may be assumed as in the preceding example. A value of slip found by taking r_2/s equal to the quotient of the rated voltage per phase and the rated phase current will generally yield a motor output near full load. When the performance is to be calculated from the equations and vector diagram, the slip corresponding to the desired load condition can be found by solving Eq. (238), page 403, as is explained on page 404.

From page 459, the phase voltage, phase currents, phase resistances, and phase reactances of the 1,000-hp motor used in the equivalent-circuit calculations are

$V_1 = 1{,}270$ volts per phase
$I_n = 75.1$ amp per phase
$r_1 = 0.1252$ ohm at 75°C
$r_2 = 0.1165$ ohm at 75°C
$x_1 = 0.313$ ohm
$x_2 = 0.313$ ohm
$z_1 = \sqrt{(0.1252)^2 + (0.313)^2} = 0.337$ ohm
$x_1 + x_2 = 0.626$ ohm

All values are referred to the stator.

In order to compare the results obtained from the equations and vector diagram with those derived from the equivalent circuit, the computations are made for a slip of 0.02.

From page 404,

$$V_1' = V_1 - I_n x_1 \text{ arithmetically (approximately)}$$
$$= 1{,}270 - 75.1 \times 0.313$$
$$= 1{,}246$$

From Eq. (238), page 403,

$$P_2 = \frac{V_1'^2(1-s)sr_2}{(r_1 s + r_2)^2 + s^2(x_1 + x_2)^2}$$

in which V_1' is used for V_1 (see page 404)

$$(r_1 s + r_2)^2 = (0.1252 \times 0.02 + 0.1165)^2$$
$$= 0.01416$$
$$s^2(x_1 + x_2)^2 = (0.02 \times 0.626)^2$$
$$= 0.000157$$
$$3P_2 = 3 \times \frac{(1{,}246)^2 \times 0.98 \times 0.02 \times 0.1165}{0.01432}$$
$$= 743{,}000 \text{ watts}$$

If the I_{h+e} component of the no-load current is omitted in calculating the primary current and instead the core loss is combined with the friction and windage and stray-load losses and subtracted from the internal power in calculating the shaft power, little error is introduced. In fact, this is the same approximation as was used in the equivalent circuit calculations in accordance with AIEE recommendations (see page 458).

The shaft power then equals

$$P_0 = 3P_2 - P_c - P_{f+w} - \text{stray-load loss}$$
$$= 743 - 13.4 - 7.0 = 723 \text{ kw or 970 hp.}$$

From Eq. (234), page 403,

$$I_2 = \frac{V_1' s}{\sqrt{(r_1 s + r_2)^2 + s^2(x_1 + x_2)^2}}$$
$$= \frac{1{,}246 \times 0.02}{\sqrt{0.01432}} = 208 \text{ amp}$$

The no-load power factor is

$$\frac{15.2 \times 1{,}000}{\sqrt{3} \times 2{,}200 \times 75.1} = 0.0531$$
$$I_\varphi = I_n' \sqrt{1 - (\text{no-load power factor})^2}$$

where I'_n is the no-load current of the motor at rated voltage.

$$I_\varphi = 75.1 \sqrt{1 - (0.0531)^2}$$
$$= 75.0 \text{ amp}$$

Refer to the vector diagram given in Fig. 232, page 395. Use E_2 as an axis of reference and neglect the I_{h+e} component of the current.

$$I_1 = -I_2 \left(\frac{r_2}{\sqrt{r_2{}^2 + x_2{}^2 s^2}} - j \frac{x_2 s}{\sqrt{r_2{}^2 + x_2{}^2 s^2}} \right) + jI_\varphi$$

$$= -208 \left[\frac{0.1165}{\sqrt{(0.1165)^2 + (0.313 \times 0.02)^2}} \right.$$
$$\left. -j \frac{0.313 \times 0.02}{\sqrt{(0.1165)^2 + (0.313 \times 0.02)^2}} \right] + j75.0$$

$$= -208 + j86.2$$

In magnitude
$$I_1 = \sqrt{(208)^2 + (86.2)^2} = 225 \text{ amp}$$

Primary copper loss $= 3 \times (225)^2 \times 0.1252$	$= 19,020$ watts
Secondary copper loss $= 3 \times (208)^2 \times 0.1165$	$= 15,140$ watts
Core loss plus friction and windage loss	$= 13,420$ watts
Stray-load loss	$= 7,000$ watts
Total losses	$= 54,580$ watts

$$\text{Efficiency} = \frac{\text{output}}{\text{output} + \text{losses}} \times 100$$
$$= \frac{723}{723 + 54.6} \times 100$$
$$= 93.0 \text{ per cent}$$

$$\text{Power factor} = \frac{\text{input}}{3V_1 I_1}$$
$$= \frac{(723 - 54.6) \times 1,000}{3 \times 1,270 \times 225} = 0.91 \text{ or } 91 \text{ per cent}$$

The torque can be found from the power as on page 461 or it can be found directly from the equations for torque given on page 403. It should be remembered that the equations give internal torque. To get external or shaft torque, the torque corresponding to the sum of friction and windage, core loss, and stray-load loss must be subtracted. The equations give the torque in units corresponding to the units substituted in the equations. If volts, amperes, and ohms are used, the torque is expressed in a unit for which there is no name but which is equal to 10^7 dyne-cm. To reduce torque to pound-feet, it is necessary to multi-

stator comes entirely from the synchronous machines and is the exciting current I_n of the vector diagram of Fig. 232, page 395. The core loss is supplied by the synchronous machines. The mechanical power required to drive the rotor at synchronous speed is equal to the friction and windage loss.

2. Below synchronous speed there is rotor current. To balance the demagnetizing action of this current there must be an equivalent component current in the stator circuit. Under this condition only motor power can be developed.

3. Above synchronous speed the current in the rotor reverses in direction as does also the component current in the stator required to balance the demagnetizing action of the rotor current. At a speed above synchronism generator action occurs, but power is not delivered to the external circuit until the current in the stator, which balances the demagnetizing effect of the rotor current, has a component equal and opposite to the current I_{h+e} required to supply the core loss. At the slip at which this particular condition occurs, the generator supplies its own core loss. Its external output is zero. At larger slip, power is delivered to the load.

Power Factor of the Induction Generator. The only current to produce generator power in an induction generator is that component of the primary current which is equal and opposite to the rotor current. A 1:1 ratio of transformation between the rotor and stator is assumed. The power factor of this component current with respect to the primary induced voltage is fixed by the rotor constants and by the slip.

$$\cos \theta_{I_2}^{E_2} = \frac{r_2}{\sqrt{r_2{}^2 + x_2{}^2 s^2}}$$

Since the slip is small, $x_2{}^2 s^2$ is small compared with $r_2{}^2$ and $\cos \theta_{I_2}^{E_2}$ is nearly unity. The load component of the primary current, the I_1' of the usual transformer diagram, is therefore nearly in phase with the primary induced voltage. Neglecting the magnetizing current and the phase displacement of the terminal voltage due to the resistance and reactance drops in the primary windings, the primary current is very nearly in phase with the terminal voltage. This is the basis of the common but incorrect statement that an induction generator can deliver power only at unity power factor. The magnetizing current is not negligible, and the power factor in consequence of this may differ considerably from unity. The correct statement is that an induction generator can deliver power only at leading power factor. The power factor, in large machines, usually is over 90 per cent at full load, but at no load or at small loads it may be very low. The quadrature component of the current, mainly magnetizing, changes little with the load.

Phase Relation between Rotor Current Referred to the Stator and Rotor Induced Voltage E_2. The current in the rotor of an induction machine is given by

$$I_2 = \frac{E_2 s}{r_2 + j x_2 s}$$

Rationalizing this by multiplying both numerator and denominator by $r_2 - j x_2 s$,

$$I_2 = E_2 \left(\frac{r_2 s}{r_2{}^2 + x_2{}^2 s^2} - j \frac{x_2 s^2}{r_2{}^2 + x_2{}^2 s^2} \right) \qquad (264)$$

Below synchronous speed s is positive and the expression for I_2 takes the form $I_2 = A - jB$ which represents a lagging current with respect to $E_2 s$. Above synchronous speed s is negative, the real part of the Eq. (264) reverses its sign, the sign of the j part remains unchanged, and the expression for the rotor current is $I_2 = -A - jB$. This represents a leading current with respect to $-E_2$.

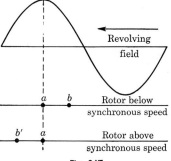

Fig. 267.

The current I_2 in the rotor cannot actually lead the voltage in the rotor which causes it, since the rotor circuit is inductive. It is only when this current is considered with respect to the stator that it has this apparent phase relation. The reason for the apparent phase relation is the reversal of the relative direction of motion of the revolving magnetic field and the rotor when the slip changes sign. This can be seen by referring to Fig. 267. Let the magnetic field move to the left, as shown by the arrow. Consider the voltage induced in an inductor a on the rotor. This voltage has its maximum value when the inductor is at the point of maximum flux density of the stator field, in position a.

Below synchronous speed the rotor moves to the right relatively to the field, and since the rotor circuit is inductive, the inductor moves to some position as b before the current in it reaches its maximum value. Above synchronous speed, the rotor moves faster than the field and moves to the left with respect to the field. In this case the inductor a moves to some such position as b' before the current in it reaches its maximum value. In both cases the rotor when considered with respect to the stator moves in the same direction as the field, *i.e.*, from right to left. Therefore if the voltage and current in the rotor are observed from any fixed point on the stator, the voltage is seen to pass through its maximum value before the current passes through its maximum value when the rotor is below synchronous speed and after the current passes through its maximum value when the rotor is above synchronous speed.

Vector Diagram of the Induction Generator. The vector diagram of the induction generator is given in Fig. 268. I_1 is the total stator current and $\cos \theta_1$ is the stator power factor, which is calculated from the readings of instruments placed in the mains leading from the generator. The angle θ_1 cannot be an angle of lag under any conditions. For fixed terminal voltage V_1, current I_1, frequency f, and slip there can be but one value of θ_1 and this must be an angle of lead. The induction generator has its power factor fixed by its constants and its slip and not by the power factor of the load.

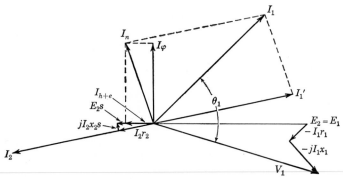

Fig. 268.

Voltage, Magnetizing Current, and Function of Synchronous Apparatus in Parallel with an Induction Generator. The voltage of the induction generator depends upon the magnetizing component I_φ of the primary current I_1, and unless the load calls for a component equal to I_φ, the generator loses its voltage. The function of synchronous apparatus which must be operated in parallel with an induction generator is to adjust the power factor of the load on the induction generator to that at which it can deliver the required power.

With respect to the synchronous apparatus in parallel with the induction generator, the magnetizing current I_φ, which leads when referred to the terminal voltage of the induction generator, is a lagging current when referred to the terminal voltage of the synchronous apparatus. The synchronous apparatus must not only supply whatever lagging current is called for by the load but must also supply a lagging current equal to the leading magnetizing current of the induction generator. For this reason the use of induction generators is limited to systems which inherently have high power factors. The synchronous apparatus can be synchronous generators, synchronous motors, or synchronous converters. The induction-generator action of induction motors is seldom used except when induction motors are used for heavy traction where steep grades are encountered. On the downgrades, the motors speed up above synchronism and automatically become induction generators and give regenerative braking.

Use of a Condenser Instead of a Synchronous Generator in Parallel with an Induction Generator. It is possible to operate an induction generator without synchronous apparatus in parallel with it provided a suitable condenser is connected across its terminals, but this method of operation has no practical importance on account of the size and cost of the condenser required, as well as on account of the very drooping voltage characteristics of such a system.

Although it is often stated that an induction generator is not self-exciting, this statement is not strictly true. If sufficient capacitance is placed across the terminals of an induction generator, it generally builds up as a synchronous generator because of the residual magnetism in the rotor from previous operation. After the voltage starts to build up, it builds up rapidly and usually the capacitance must be reduced to prevent excessive voltage. There is a critical capacitive reactance, similar to the critical resistance in the field circuit of a d-c shunt generator, which must be less than a certain value in order for the induction generator to build up. This is illustrated in Fig. 269. If a line is drawn (Fig. 269) with voltage across the condenser as ordinates, this voltage is the same as the voltage across the terminals of the induction generator, and with condenser current as ordinates, the induction generator does not build up unless this line intersects the saturation curve of the machine.

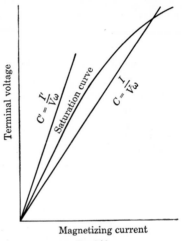

Fig. 269.

Voltage, Frequency, and Load of the Induction Generator.

Case I. In Parallel with a Synchronous Generator. The voltage of the induction generator is equal to the voltage impressed across its terminals by the synchronous generator with which it is connected. The magnetizing current of the induction machine automatically adjusts itself to give this voltage. The frequency is determined by the frequency of the magnetizing current and is the same as the frequency of the synchronous generator. The load is fixed by the rotor current which depends on the slip.

Case II. In Parallel with a Synchronous Motor or a Synchronous Converter. As in Case I, the voltage of the induction generator is determined by the terminal voltage of the synchronous motor or of the converter. The initial excitation must come from a synchronous generator or from a synchronous motor or from a converter driven as a generator. The frequency is fixed by the speed of the rotor and by the load and is

$$f = \frac{p}{2} \frac{n}{60(1-s)}$$

where p and n are the number of poles and the speed in revolutions per minute. It should be remembered that s is negative for generator action. The induction generator carries the entire load and in addition supplies all the losses of the synchronous machine. The synchronous machine supplies sufficient quadrature current to adjust the power factor of the load on the system to that corresponding to the inherent power factor of the induction generator for that load. The slip is fixed by the load. The voltage regulation of the system is similar to the voltage regulation of a synchronous generator. The voltage at any given load is fixed by the constants of the induction generator, the excitation of the synchronous machine, and the power factor of the circuit external to the induction generator.

Short-circuit Current of the Induction Generator. Since an induction generator depends for its excitation upon the synchronous apparatus which is in parallel with it, the current it can supply on short circuit depends upon the drop in voltage produced at the terminals of the synchronous apparatus. There is a transient short-circuit current which can be large, but its duration is very short.[1] If the terminal voltage becomes zero, the steady short-circuit current is zero. Little current is supplied on partial short circuit since the maximum power an induction generator can deliver at any fixed slip and frequency is proportional to the square of its terminal voltage, Eq. (238), page 403. The inability to back up a short circuit greatly reduces the resulting damage and permits the use of smaller and less expensive circuit breakers than could be used safely if the whole capacity of the system were in synchronous generators.

Hunting of the Induction Generator. An induction generator is free from hunting since it does not operate at synchronous speed. Any change in load must be accompanied by an actual change in speed instead of by a small angular displacement as with a synchronous generator. The irregularities in the angular velocity of prime movers during a single revolution are so small as to produce only insignificant changes in load.

Advantages and Disadvantages of the Induction Generator. Most of the advantages and disadvantages possessed by an induction generator

[1] This assumes that all phases are short-circuited.

are obvious from what has already been said.   The following are the
advantages.   The rotating part is rugged, the machine does not back up a
short circuit and is free from hunting, the construction of its rotating part
makes it well suited to high speeds, no synchronizing is necessary, and its
voltage and frequency are automatically controlled by the voltage and
frequency of the synchronous machines which must operate in parallel
with it.

The following are the disadvantages.   The power factor is fixed by
its slip and not by the power factor of the load, it is therefore necessary
to operate synchronous apparatus in parallel with it, this synchronous
apparatus must operate with reactive lagging current in addition to
that demanded by the load on the system and consequently must operate
at less than load power factor.

Polyphase Induction Motors Used as Selsyn Motors.   The word
selsyn is an abbreviation of self-synchronizing.   It is applied to devices

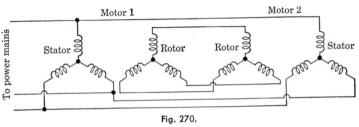

Fig. 270.

which are connected electrically, are fed by alternating current, and
are arranged so that angular displacement of the rotating element
of one is reproduced by the rotating element of the other.   Many
of these devices develop little torque and are used for signaling or for
indicating the position of objects at a distance.   One of the earliest
applications of such a device in a large way was in connection with the
operation of the locks of the Panama Canal.   The operator has a miniature model of the canal locks, and the movements of the parts of the
model are made to correspond to the movements of the parts of the
actual locks by the use of selsyns.

In other applications the selsyns must develop considerable torque.
Two identical three-phase induction motors with wound rotors can
be made to serve as a power selsyn system capable of developing considerable torque.   The stators of the two motors are connected together
electrically and are connected to a three-phase power system.   The
rotor slip rings are connected electrically in such a way that the voltages
of the rotors are in opposition.   This condition is shown diagrammatically
in Fig. 270.   Under these conditions the voltages induced in the rotors
by the stator rotating fields are in opposition, no currents flow in the

rotor circuits, and no torque is developed by either rotor. Let the rotor of motor 1 be displaced by a small angle α. There is now a resultant voltage $2V \sin \alpha/2$ acting in each rotor circuit where V is the voltage induced in each rotor phase by the rotating magnetic field. The currents caused by these resultant voltages produce torques in the two motors which act to restore the rotors to phase opposition. The result is that one rotor follows the movement of the other. The phase displacement between the two rotors depends upon the torque against which the driven rotor turns, on the constants of the motors, and on their speed of rotation. By suitable design the phase displacement in space degrees can be made small, especially with multipolar motors. Power selsyn

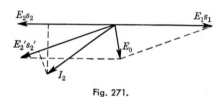

Fig. 271.

motors are used to hold the two parts of a certain apparatus in line, to keep two shafts revolving at the same speed, etc. Selsyn motors are used to some extent in the manufacture of paper, for the operation of certain types of draw-bridges, etc. When considerable torque is required, d-c motors furnish the main power and selsyn motors mounted on the shafts of the main motors serve to equalize their speed. In this case the selsyn motors merely take care of differences in torque and can therefore be made much smaller than if they had to furnish the entire driving torque.

Consider two machines connected as in Fig. 270, rotating at equal speeds and with their rotors initially so connected that the rotor voltages are in opposition. Let $E_1 s_1$ and $E_2 s_2$ (Fig. 271) represent these rotor voltages for motor 1 and motor 2, respectively. If f is the frequency of the source, the rotor voltages will have frequencies $f s_1$ and $f s_2$, respectively. Under the initial conditions assumed, the resultant voltage acting in the rotor circuits is zero, and there will be no rotor current and no torque produced by either motor. Should the speed of motor 2 for any reason become less than that of motor 1, s_2 becomes greater than s_1, and $E_2 s_2$ will change its phase relation with respect to $E_1 s_1$ in the direction indicated by $E_2' s_2'$, producing a resultant voltage E_0 and setting up a current I_2. From Eq. (230), page 402, if $\cos \theta_{I_2}^{E_2(1-s)}$ is taken as equal to $\cos \theta_{I_2}^{E_2 s}$, then motor power is developed when this angle is less than 90 deg. Referring again to Fig. 271, motor 2 develops motor power proportional to the cosine of the angle between $E_2' s_2'$ and I_2 while the motor power of motor 1 is negative (motor must be driven) since the cosine of the angle between $E_1 s_1$ and I_2 is negative. If therefore these induction machines are coupled to d-c motors, the d-c motor coupled to motor 2 will speed up because of the aid it receives from motor 2 while the d-c motor coupled to motor 1 will have the additional load of driving motor 1 added to its initial load which will tend to reduce its speed. The result is an equalization of the speed of the two d-c motors.

Since the secondary voltages of power selsyns are proportional to slip, very little secondary current or torque is produced when they are operated near synchronous speed. This limits the practical operation of power selsyns to slips of 0.33 or greater. Because of the large slip the secondary leakage reactance $x_2 s$ is large and therefore the secondary current I_2 lags behind the resultant secondary voltage E_0 by a large angle as shown in Fig. 271. The large values of slip so increase the rotor core loss above that usually encountered in the rotors of induction motors that rotors of motors for selsyns must be especially designed to take care of this increase in rotor core loss. The initial condition imposed may be of some importance when full three-phase excitation is applied when the rotors are in relative positions that give initially high resultant voltages in the rotor circuits, because of the difference in phase of these voltages. Under this condition they may overshoot their normal corresponding positions. To prevent this the selsyns may first be excited with single phase while at standstill before the three-phase power is applied.

Position Selsyns. Position selsyns or synchros, as they are sometimes called, are single-phase devices of small size. They are often used for indicating at some remote point the positions of some item of equipment. They may also be used as a part of a servomechanism to control the position of some much larger device and for other control purposes. These synchros are of four general types: (1) transmitter selsyns, (2) receiver selsyns, (3) differential selsyns, and (4) transformer selsyns. The stators of all four types are similar to the stators of polyphase induction motors. The three distributed stator windings are displaced from each other by the usual 120 deg and may be connected either in Y or Δ. The rotors differ according to the particular type but in all cases are assembled from laminations. The rotors of transmitter and receiver selsyns have salient poles and a single-phase winding. The receiver often has some kind of mechanical damping device attached to the rotor to reduce oscillations. Differential and transformer selsyns have cylindrical rotors. The differential selsyn has a winding on the rotor similar to that on the stator, *i.e.*, three separate distributed windings displaced by 120 deg and connected in Y or Δ. The transformer selsyn has a single distributed rotor winding.

Figures 272, 273, and 274 show three of the common methods of connecting position selsyns. Fig. 272 shows a transmitter and a receiver selsyn. With this connection the rotor of the receiver selsyn tends to assume a position identical with that of the rotor of the transmitter. If the transmitter is turned to a new position, the receiver will follow. For the circuit of Fig. 273, two motions may be applied, one to each of the transmitter selsyns, the sum or difference of these motions, depending upon the relative connections, being indicated on the differential selsyn.   In Fig. 274, with the two rotor windings located as in the diagram, no voltage is produced in the rotor of the transformer selsyn.

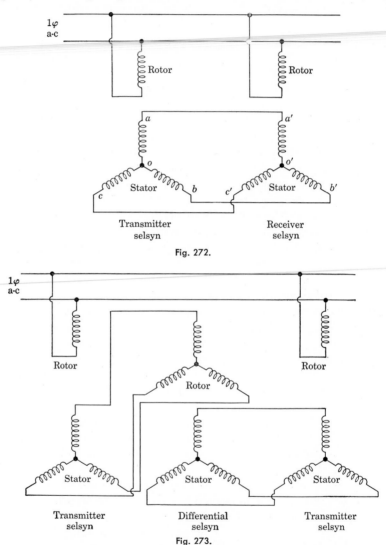

Fig. 272.

Fig. 273.

If however the rotor of the transmitter selsyn is moved to some new position, a voltage is produced in the rotor winding of the transformer selsyn.   The magnitude of this voltage is dependent upon the difference in relative positions of the two rotors.   This so-called "error voltage" may then be applied to some amplifier system for positioning some larger piece of equipment.

In the circuit of Fig. 272, with the rotors in the relative positions shown, all of the flux which crosses the air gap from each rotor links the respective stator windings oa and $o'a'$ producing voltages E_{oa} and $E_{o'a'}$ which are

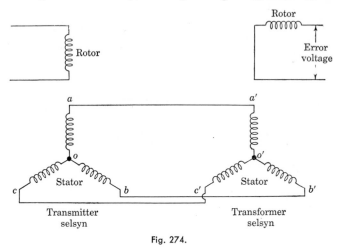

Fig. 274.

equal and in phase. If the flux distribution is assumed to be sinusoidal, only one-half of the air-gap flux links the b and c windings. The voltages E_{bo}, E_{co}, $E_{b'o'}$ and $E_{c'o'}$ are therefore each equal to one-half of E_{oa} and are in phase with E_{oa}. Since the voltage around any closed circuit in the stator is zero, no current flows in the stator windings and no torque is produced. If now the rotor of the transmitter selsyn is moved through an angle of

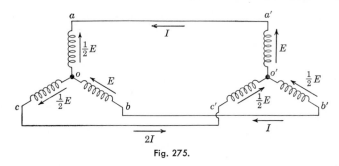

Fig. 275.

60 deg in a counterclockwise direction while the rotor of the receiver selsyn is held in its initial position, in the transmitter selsyn all the air-gap flux from the rotor will link winding bo while only one-half of this flux links windings oa and oc. The relative magnitudes and directions of the stator voltages and currents will then be as shown in Fig. 275.

Figure 276 represents the receiver selsyn, the small circles indicating the arrangement of the stator windings. The dots and crosses within the circles indicate the voltage directions when the flux is increasing in a downward direction. The current directions for the same period of time are determined by reference to Fig. 275 and are indicated by the dots and

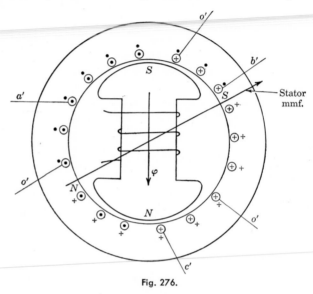

Fig. 276.

dashes just outside the small circles, the current in winding $c'o'$ being twice as large as the currents in the other two windings. These currents react with the rotor flux to produce torque tending to turn the rotor counterclockwise. If the rotor is free to turn, it will rotate in this counterclockwise direction until it occupies a position corresponding to the position of the rotor of the transmitter selsyn at which point the rotor currents and torque will again be zero.

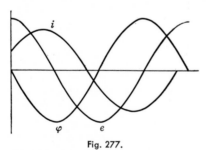

Fig. 277.

The above explanation applies strictly during the quarter cycle when the rotor flux is increasing in a downward direction. During the next quarter cycle, while the flux is still downward, the voltages reverse in direction but the currents, because of the reactance of the windings, continue in the direction shown for a considerable portion of this quarter cycle, producing counterclockwise torque. This is illustrated in Fig. 277, where e and i may apply to any one of the three windings provided a proper scale of magnitudes is used. When both flux and current reverse, the torque is still counterclockwise and so a counterclockwise torque is produced during a major portion of each cycle.

Frequency Converters. Frequency converters are used for changing from one frequency to another or for interconnecting two power lines of different frequencies. Two synchronous machines mounted on a common shaft can be used for a frequency converter provided the ratio of the number of poles is equal to the ratio of the two frequencies. For example, to change from 25 to 60 cycles the ratio of the number of poles must be $^{25}\!/_{60} = {}^{5}\!/_{12}$. The minimum number of poles that can be used for this transformation is 10 for the 25-cycle machine and 24 for the 60-cycle machine. Such a change in frequency is necessary when residential lighting is to be furnished from a 25-cycle source of power since 25 cycles is unsatisfactory for lighting. The two sides of a frequency converter consisting of two synchronous machines, *i.e.*, the two sides of a synchronous-synchronous frequency changer, are independent and the excitation of either machine can be changed without affecting the other. When two or more frequency converters are operated in parallel, it is necessary to control the phase relation between the voltages on both sides in order to parallel an unloaded converter with one that is already loaded and to adjust the loads carried by the converters when operating in parallel. For this purpose the frame of one of the synchronous machines of each converter is mounted in such a way that the frame can be rotated slightly with respect to the frame of the other machine.

An induction motor with wound rotor can be used as a frequency changer by operating the rotor at the proper speed. The frequency in the rotor of a polyphase induction motor is equal to the stator frequency multiplied by the slip, that is, $f_2 = f_1 s$. By driving the rotor at the proper speed a desired frequency can be obtained from the rotor slip rings. For example, if the stator of a 60-cycle wound-rotor induction motor is connected to a 60-cycle source of power and its rotor is driven at $^{35}\!/_{60}$ of synchronous speed, the frequency at the slip rings of the rotor is 25 cycles, and if the ratio of transformation between stator and rotor windings is unity, the voltage across the rotor slip rings is $^{25}\!/_{60}$ of the stator voltage. This statement in regard to the voltages neglects resistance and leakage reactance drops in the motor windings. Neglecting the losses, $(60 - 35)/60 = {}^{25}\!/_{60}$ of the power received by the stator of the motor appears as electrical power at the rotor terminals. The remaining $^{35}\!/_{60}$ appears as mechanical power at the shaft and drives as generator the machine connected to the shaft of the induction motor. If this machine is a synchronous machine with the proper number of poles, its stator can be connected to the 25-cycle circuit. With this arrangement $(60 - 35)/60 = {}^{25}\!/_{60}$ of the power delivered to the 25-cycle circuit comes from transformer action between the stator and rotor of the induction machine and ${}^{35}\!/_{60}$ is due to the synchronous machine driven as a synchronous 25-cycle generator by the mechanical power developed by the induction motor. The kva capacity of the synchronous machine in this case need be only ${}^{35}\!/_{60}$ as great as that of the induction machine. This statement neglects the losses and the exciting current of the induction machine. The transformation can equally well be from 25 cycles to 60 cycles. In this case ${}^{25}\!/_{60}$ of the 60-cycle power is from transformer action between rotor and stator of the induction machine and ${}^{35}\!/_{60}$ comes from mechanical power delivered to the shaft of the induction

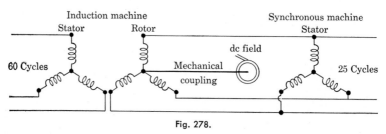

Fig. 278.

machine by the synchronous machine which now operates as a synchronous motor. If two or more induction-synchronous frequency converters are in parallel, it is necessary that the stator of one machine of each converter be mounted so that it can be rotated slightly in order to control the phase relation between voltages of the converter and of the circuit with which it is to be paralleled.

The total machine capacity involved in an induction-synchronous frequency converter is less than that required for a synchronous-synchronous frequency converter. For example, neglecting losses in the machines, to convert 1,000 kva by a synchronous-synchronous frequency converter requires two 1,000-kva synchronous machines. If the frequency change is from 25 to 60 cycles or vice versa, the capacity of the induction machine of an induction-synchronous frequency converter is 1,000 kva but the kva capacity of the synchronous machine is only ${}^{35}\!/_{60} \times 1,000 = 583$ kva. This statement neglects the exciting current of the induction machine.

If the rotor of the induction machine is driven at a slip of 2, the rotor frequency is twice the stator frequency and, neglecting losses, half the power on the slip-ring side is from transformer action between stator and rotor of the induction machine and half is from mechanical power required to drive the rotor at slip 2.

When an induction-synchronous frequency converter is used to tie together two power systems of different frequencies, the direction of flow of power can be controlled by changing the power angles of the systems in the same way as when two power systems of the same frequency are interconnected. The schematic diagram of connections of an induction-synchronous frequency converter which changes from 25 to 60 cycles or vice versa is given in Fig. 278.

Both the synchronous-synchronous frequency converter and the induction-synchronous frequency converter have a fixed ratio of frequencies, and any tendency of two systems connected through a frequency converter to depart from this ratio, which might be due to a sudden change in the load carried by either system or to a short circuit, may cause serious disturbance in the operation of the converter.

PART SIX

SINGLE-PHASE INDUCTION MOTORS

Chapter 36

Single-phase induction motor. Windings. Methods of treating a single-phase induction motor. The determination of the operating characteristics of a single-phase induction motor by the use of symmetrical-phase components. The equivalent circuit of a single-phase induction motor. The rotating field of a single-phase induction motor. Measurement of the core loss of a single-phase induction motor. An example of the calculation of the torque, power output, current, and power factor of a single-phase induction motor.

Single-phase Induction Motor. The running characteristics of a single-phase induction motor are fairly satisfactory, but the motor is not so good as a polyphase motor since it possesses no starting torque. It is also heavier and has a lower efficiency and lower power factor than a polyphase motor for the same speed and output. The greater weight for a given output is not an inherent peculiarity of the single-phase induction motor alone but is characteristic of any single-phase motor or generator.

A polyphase induction motor has a starting torque which can be increased up to a certain limiting value by putting resistance in the rotor circuit. No amount of resistance inserted in the rotor of a single-phase induction motor can give it an initial starting torque. It must be started by some form of auxiliary device and must attain considerable speed before it develops sufficient torque to overcome its own friction and windage. The direction of its rotation depends merely upon the direction in which it is started. Once started, it operates as well in one direction as in the other. The greater weight and the absence of a starting torque with the consequent necessity for some form of auxiliary starting device are the chief factors which limit the use of single-phase induction motors to small sizes. Although millions of fractional-horsepower single-phase induction motors are in use for desk fans, small ventilating fans, electrical refrigerators, etc., they are seldom built in sizes of over 5 or 10 hp and in general they are not used in sizes as large as this except when three-phase power is not available or the amount of power involved does not warrant running the three wires necessary for three-phase power. Although single-phase induction motors are always small and many are fractional-horsepower motors, the total volume of their yearly sales in dollars is very large and comparable to that of heavy electrical machinery.

Windings. The general features of construction of single-phase and polyphase induction motors are similar. The essential difference between the motors is in the windings and in the necessity for some form of starting device for the single-phase motor. The stator of a single-phase motor always has a distributed single-phase winding usually of fractional pitch. The rotor is generally of the squirrel-cage type except when the auxiliary starting torque is obtained by converting the motor into a repulsion motor for coming up to speed. If the repulsion-motor action is not used for starting, the stator must have an auxiliary starting winding or its equivalent, in addition to the regular winding. By subdividing the main winding it is possible to make a part serve as the auxiliary winding.

Methods of Treating a Single-phase Induction Motor. There are two chief ways of treating the single-phase induction motor: (1) by considering the motor as a three-phase Y-connected induction motor operating single phase with one line terminal open and resolving the currents into their symmetrical-phase components; (2) by the quadrature-field theory in which a quadrature field in both space and time quadrature with the stator field is assumed to be produced, when the motor is in operation, by a component current generated in the rotor by its rotation in the field produced by the stator winding. Both methods give the same operating characteristics, but for the ordinary single-phase induction motor, the method involving symmetrical-phase components is the simpler.

The Determination of the Operating Characteristics of a Single-phase Induction Motor by the Use of Symmetrical-phase Components. In order to apply this method, the motor must have a symmetrical rotor winding, such as a symmetrical polyphase winding or a squirrel-cage winding.

For any unbalanced three-phase circuit the following relations hold[1]:

$$V_1{}^+ = I_1{}^+ z_1{}^0 + I_1{}^- z_1{}^- + I_1{}^0 z_1{}^+ \tag{265}$$

$$V_1{}^- = I_1{}^- z_1{}^0 + I_1{}^+ z_1{}^+ + I_1{}^0 z_1{}^- \tag{266}$$

$$V_1{}^0 = I_1{}^0 z_1{}^0 + I_1{}^+ z_1{}^- + I_1{}^- z_1{}^+ \tag{267}$$

where

$$z_1{}^+ = \tfrac{1}{3}(z_1 + az_2 + a^2 z_3) \tag{268}$$

$$z_1{}^- = \tfrac{1}{3}(z_1 + a^2 z_2 + az_3) \tag{269}$$

$$z_1{}^0 = \tfrac{1}{3}(z_1 + z_2 + z_3) \tag{270}$$

and z_1, z_2, z_3 are the complex impedances of the phases and a is an operator which rotates a vector to which it is applied through 120 deg in a counterclockwise or positive direction when V_1 leads V_2 and in a clockwise or

[1] R. R. Lawrence, "Principles of Alternating Currents," 2d ed., p. 392.

negative direction when V_1 lags V_2. V_1, V_2, V_3 are the phase vector voltages of the circuit. The I's are the corresponding vector currents. The plus, minus, and zero exponents are used to indicate positive-sequence, negative-sequence, and zero-sequence components.

In a polyphase induction motor there is no neutral and the equivalent impedances of the phases referred to the stator are identical for a given slip s of the rotor with respect to the rotating field caused by the positive-sequence voltages. The equivalent impedances of the phases are also identical for a slip $(2 - s)$ of the rotor with respect to the rotating field caused by the negative-sequence voltages. Therefore for each of the slips, s and $(2 - s)$, z^+ and z^- are equal to zero for each phase. Since the positive-sequence voltages and the negative-sequence voltages each form a balanced system and the motor constants are balanced, the operating characteristics of the motor operating either with unbalanced applied voltages or single phase can be found by calculating its performance due to the positive-sequence components and the negative-sequence components separately, using the ordinary equations for a three-phase induction motor, and then combining the results.

From the equation already developed in Chap. 31 for a polyphase induction motor, for slip s

$$z_+{}^0 = \left(r_1 + \frac{r_2}{s}\right) + j(x_1 + x_2) \tag{271}$$

and for slip $(2 - s)$

$$z_-{}^0 = \left(r_1 + \frac{r'_2}{2 - s}\right) + j(x_1 + x_2) \tag{272}$$

where $z_+{}^0$ and $z_-{}^0$ are the equivalent phase zero-sequence impedances of the motor referred to the stator for the positive-sequence and the negative-sequence component voltages. As in the polyphase motor, x_2 and r_2 are the actual rotor phase reactance at stator frequency and the actual rotor phase resistance at rotor frequency, each referred to the stator by multiplying the actual values by the square of the ratio of transformation between stator and rotor.

Consider first a three-phase Y-connected motor operating with one phase open. If Δ-connected, the motor can be replaced by the equivalent Y-connected motor. Refer everything to the stator. I_2, referred to the stator, is equal to $-I'_1$ of the ordinary vector diagram for the polyphase induction motor. If line 1 is opened the phase currents are

$$I_{01} = 0 + j0$$
$$I_{02} = I + j0$$
$$I_{03} = -I + j0$$

and if $I_{01}{}^+$ is assumed to lead $I_{02}{}^+$

$$I_{01}{}^+ = \frac{1}{3}(I_{01} + I_{02}\underline{/+120°} + I_{03}\underline{/-120°})$$

$$= \frac{1}{3}(0 + I\underline{/+120°} - I\underline{/-120°})$$

$$= \frac{1}{\sqrt{3}}(0 + jI)$$

$$I_{01}{}^- = \frac{1}{3}(I_{01} + I_{02}\underline{/-120°} + I_{03}\underline{/+120°})$$

$$= \frac{1}{3}(0 + I\underline{/-120°} - I\underline{/+120°})$$

$$= \frac{1}{\sqrt{3}}(0 - jI)$$

$$I_{01}{}^0 = \frac{1}{3}(I_{01} + I_{02} + I_{03})$$

$$= 0 + j0$$

The vectors for the positive-sequence and the negative-sequence currents for the three phases are shown in Fig. 279. The zero-sequence components are not shown as they are zero.

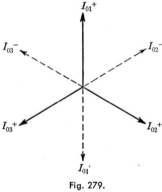

Fig. 279.

In the stator the positive-sequence and the negative-sequence currents are equal in magnitude. The rotor currents referred to the stator are equal to the stator currents minus the exciting currents. Since for normal operating conditions the slip s for the positive-sequence currents is small, the positive-sequence exciting currents are large. Under similar conditions, the slip for the negative-sequence currents is large and is equal to $(2 - s)$ which is nearly 2. The negative-sequence exciting currents are therefore small. Consider phase 01 in which the resultant of the positive and negative sequence currents is zero. In phase 01 the positive-sequence and negative-sequence currents are equal in magnitude and opposite in phase since their vector sum must be zero. In what follows the subscripts 1 and 2 are used to indicate stator and rotor currents. The subscript 01 on voltages is used to indicate voltages in phase 01.

$$V_{01}{}^+ = I_1{}^+(r_1 + jx_1) + I_2{}^+\left(\frac{r_2}{s} + jx_2\right) \tag{273}$$

$$V_{01}{}^- = I_1{}^-(r_1 + jx_1) + I_2{}^-\left(\frac{r_2'}{2 - s} + jx_2\right) \tag{274}$$

The resistance r_2 is the rotor effective resistance at a frequency fs

where f is the stator frequency. Since fs is very small under normal operating conditions, r_2 is nearly equal to the rotor ohmic resistance. The resistance r_2' is the rotor effective resistance at a frequency $f(2 - s)$, nearly $2f$ under operating conditions, and r_2' is much greater than the ohmic resistance.

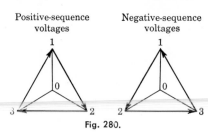

Positive-sequence voltages

Negative-sequence voltages

Fig. 280.

Figure 280 gives the positive-sequence and the negative-sequence line and phase voltages. From this figure,

$$V_{23}^+ = j\sqrt{3}V_{01}^+ \qquad V_{23}^- = -j\sqrt{3}V_{01}^-$$
$$V_{23} = V_{23}^+ + V_{23}^- = j\sqrt{3}(V_{01}^+ - V_{01}^-)$$

$$\frac{1}{j\sqrt{3}}V_{23} = (V_{01}^+ - V_{01}^-)$$

$$= \left[I_1^+(r_1 + jx_1) + I_2^+\left(\frac{r_2}{s} + jx_2\right)\right]$$
$$- \left[I_1^-(r_1 + jx_1) + I_2^-\left(\frac{r_2'}{2 - s} + jx_2\right)\right] \quad (275)$$

In order to avoid double exponents, I_2^+ is used in Eq. (275) for $I_1'^+$ where $I_1'^+$ is equal to $-I_2^+$ multiplied by the reciprocal of the ratio of transformation between stator and rotor. The resistance and reactance r_2 and x_2 are assumed to be referred to the stator by multiplying the actual values by the square of the ratio of transformation between stator and rotor. In all the equations which follow, I_2^+ and I_2^- are used for $I_1'^+$ and $I_1'^-$.

$$I_1^+ = I_2^+ + I_n^+ \tag{276}$$
$$I_1^- = -I_1^+ = -(I_2^+ + I_n^+) \tag{277}$$
$$I_2^+ = I_1^+ - I_n^+ \tag{278}$$
$$I_2^- = I_1^- - I_n^- \tag{279}$$
$$= -I_1^+ - I_n^-$$
$$= -I_1^+ \text{ approximately under operating conditions} \tag{280}$$

$$\frac{1}{j\sqrt{3}}V_{23} = \left[I_n^+(r_1 + jx_1) + I_2^+(r_1 + jx_1) + I_2^+\left(\frac{r_2}{s} + jx_2\right)\right]$$
$$+ \left\{[I_n^+(r_1 + jx_1) + I_2^+(r_1 + jx_1)]\right.$$
$$\left. + \left[I_n^+\left(\frac{r_2'}{2 - s} + jx_2\right) + I_2^+\left(\frac{r_2'}{2 - s} + jx_2\right)\right]\right\}$$
$$= I_n^+\left[2(r_1 + jx_1) + \left(\frac{r_2'}{2 - s} + jx_2\right)\right]$$
$$+ I_2^+\left[2(r_1 + jx_1) + \left(\frac{r^2}{s} + \frac{r_2'}{2 - s} + j2x_2\right)\right] \quad (281)$$

$$\frac{1}{j\sqrt{3}} V_{23} = I_n^+ \left[\left(2r_1 + \frac{r_2'}{2-s} \right) + j(2x_1 + x_2) \right]$$
$$+ I_2^+ \left[\left(2r_1 + \frac{r_2}{s} + \frac{r_2'}{2-s} \right) + j2(x_1 + x_2) \right] \quad (282)$$

The magnitude of the term $I_n^+ \left[\left(2r_1 + \dfrac{r_2'}{2-s} \right) + j(2x_1 + x_2) \right]$ can be subtracted arithmetically from the magnitude of the term $(1/\sqrt{3})V_{23}$ to give $(1/\sqrt{3})V_{23}'$ producing only a negligible error for the same reason that $I_n x_1$ can be subtracted arith-

metically from V_1 to get V_1' in the polyphase induction motor operating with balanced impressed voltages (see page 404). The vector diagram shown in Fig. 281 should make clear the justification of the arithmetical subtraction.

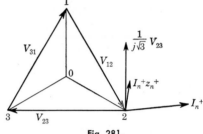

Fig. 281.

Refer to Fig. 281. I_n^+ for phase 1 is nearly 90 deg behind the voltage drop V_{01} which is nearly equal to V_{01}^+ since V_{01}^- is small on account of the large slip for the negative-sequence components under operating conditions. The voltage drop

$$I_n^+ \left[\left(2r_1 + \frac{r_2'}{2-s} \right) + j(2x_1 + x_2) \right] = I_n^+ z_n^+$$

is nearly 90 deg ahead of I_n since the part of this drop due to resistance is small compared with the part due to reactance. The error introduced by subtracting $I_n^+ z_n^+$ arithmetically from $(1/j\sqrt{3}) V_{23}$ is negligible

$$I_2^+ = \frac{\dfrac{1}{\sqrt{3}} V_{23}'}{\sqrt{ \left(2r_1 + \dfrac{r_2}{s} + \dfrac{r_2'}{2-s} \right)^2 + 4(x_1 + x_2)^2 }} \quad (283)$$

The vector diagram for the positive-sequence components is the same as the vector diagram of the polyphase induction motor for balanced impressed voltages, given in Fig. 282.

$$I_1^+ = I_1'^+ + I_n^+ \quad (284)$$

In Eq. (284) and in Fig. 282, $I_1'^+$ is the positive-sequence power component of the primary current and is the same as $-I_2^+$ used in other equations.

Since the phase orders of the positive-sequence and negative-sequence currents are opposite, the torques due to these currents are opposite.

Therefore
$$T = T^+ - T^-$$

From Eq. (237), page 403, and the magnitudes of the currents $I_2{}^+$ and $I_2{}^-$ from Eqs. (283) and (280), the torque can be determined. The

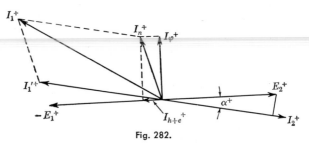

Fig. 282.

magnitude of $I_1{}^+$ which determines the magnitude of $I_2{}^-$ can be calculated from Eq. (284) and the vector diagram in Fig. 282.

$$\text{Internal torque} = T = 3\frac{p}{4\pi f}\left[(I_2{}^+)^2\frac{r_2}{s} - (I_2{}^-)^2\frac{r_2'}{2-s}\right] \quad (285)$$

$$\text{Pulley torque} = T' = T - \text{(friction and windage torque)} \quad (286)$$

$$\text{Pulley power} = 2\pi \text{ (speed in revolutions per second) } T''$$

$$= 2\pi\frac{2f}{p}(1-s)T' \quad (287)$$

where p is the number of poles and f is the frequency in cycles per second. Equation (285) gives the torque in terms of a unit which has no name but is in the same system of units as the watt. To convert the torque given by Eq. (285) to pound-feet it must be multiplied by $550/746 = 0.737$.

Efficiency =

$$\frac{\text{output}}{\text{output} + \text{copper loss} + \text{core loss} + \text{friction and windage loss}} \quad (288)$$

$$\text{Copper loss} = 3\{[(I_1{}^+)^2 + (I_1{}^-)^2]r_1 + (I_2{}^+)^2r_2 + (I_2{}^-)^2r_2'\}* \quad (289)$$

$$\text{Power factor} = \frac{\text{input}}{V_{23}I_{23}} \quad (290)$$

Referring to Fig. 279 it is evident that each of the currents in the phases which carry currents is equal in magnitude to

$$\sqrt{3}\,I_1{}^+ = I_{23}$$

A single-phase induction motor can be treated as a Y-connected three-phase motor, operating with one line terminal open, by using one-half the single-phase resistance and reactance as the resistance and

* R. R. Lawrence, "Principles of Alternating Currents," 2d ed., p. 391.

reactance of the three-phase motor, provided the single-phase motor has a symmetrical polyphase rotor winding or its equivalent.

The positive-sequence and negative-sequence torque curves and the resultant torque curve of a single-phase induction motor or of a Y-connected induction motor operating with one terminal open are given in Fig. 283. A slip of zero for the positive-sequence torque corresponds to a slip of 200 per cent for the negative-sequence torque, and a slip of zero for the negative-sequence torque corresponds to a slip of 200 per cent for the positive-sequence torque. The component torque curves in

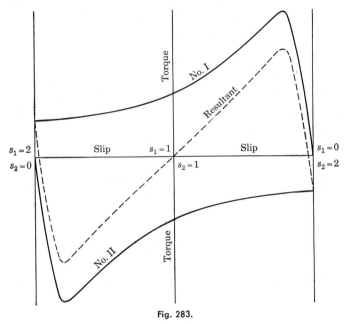

Fig. 283.

Fig. 283 are marked No. I and No. II. The direction of rotation of the motor determines which of the two curves represents the positive-sequence torque. The motor operates equally well in both directions. The direction in which it revolves depends upon the direction in which it is started. The sum of the two component torques at any slip is the resultant torque. At standstill the slip for both torque curves is the same, $s = 1$. Under this condition the two component torques are equal and opposite and the resultant torque is zero. There is no inherent starting torque such as is possessed by the polyphase induction motor. A single-phase induction motor requires some form of auxiliary starting device. The methods of starting single-phase induction motors are discussed later.

If the motor is started in either direction and is brought up to such a speed that the resultant torque is greater than that required for load plus losses, the motor gains speed until the stable part of the speed-torque curve is reached, corresponding to the direction of rotation in which the motor is started. A peculiarity of the single-phase induction motor is that its internal torque becomes zero at a speed slightly below synchronous speed. A single-phase induction motor could never reach synchronous speed even if its rotational losses were zero.

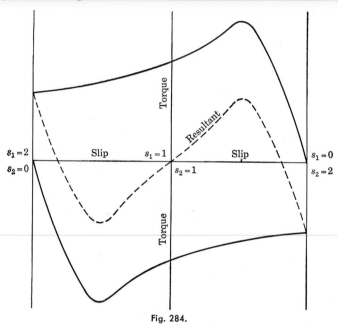

Fig. 284.

Adding resistance to the rotor of a single-phase induction motor increases its slip and also decreases its maximum internal torque, whereas the maximum internal torque developed by a polyphase motor is independent of rotor resistance [Eq. (241), page 406. A comparison of Figs. 283 and 284 shows the effect on the torque of adding resistance to the rotor of a single-phase induction motor. Figure 284 shows the torque curves with resistance added to the rotor circuit.

The Equivalent Circuit of a Single-phase Induction Motor. The equivalent circuit of a single-phase induction motor is given in Fig. 285. This circuit also applies to a three-phase Y-connected induction motor operating with one line terminal open. The resistances, reactances, currents, and voltage have the same significance as in the preceding equations for the motor.

The Rotating Field of a Single-phase Induction Motor. Except when at standstill, the single-phase induction motor has a revolving magnetic field which can be considered as the resultant of two oppositely rotating fields produced by the magnetizing components of the positive-sequence and the negative-sequence currents in the stator windings of the equivalent three-phase motor. The positive-sequence and negative-sequence stator ampere-turns are equal in magnitude since the positive-sequence and negative-sequence stator currents are equal in magnitude (Fig. 279,

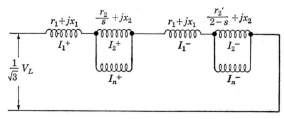

Fig. 285.

page 483). The magnetizing components of these two groups of currents and the component fluxes produced by them are the same as would be produced under similar conditions of slip in a three-phase motor operating under balanced conditions. In a three-phase motor under balanced conditions, the magnetizing components of the stator currents at full load are of the order of magnitude of 30 to 40 per cent of the full-load stator currents. At standstill, when the slip is unity, the magnetizing components are only 3 to 4 per cent of the full-load stator currents, and with slips greater than unity, they are even smaller.

Each of two oppositely rotating fields produced by the positive-sequence and the negative-sequence magnetizing currents can be represented by a revolving vector which rotates at synchronous speed with respect to the stator and has a fixed length for a given slip. When the motor is at standstill, these two revolving vectors are equal in magnitude and combine to give a resultant vector which varies in length along a fixed axis which is the axis of the stator winding (Fig. 286a). At a value of slip other than unity, the two rotating vectors are unequal in magnitude and combine to give a revolving vector which varies in magnitude as it revolves. The end of this vector traces an ellipse. The resultant field is elliptical under this condition, with its

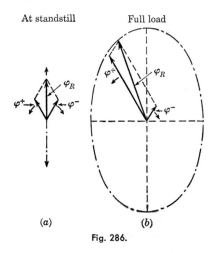

Fig. 286.

major axis along the axis of the stator winding (Fig. 286b).   As the slip decreases, the ratio of the positive-sequence field to the negative-sequence field increases and the resultant field approaches a circular field.   Under ordinary operating conditions the resultant field is elliptical.   It varies in magnitude as it revolves, with its maximum value along the axis of the stator winding and its minimum value at right angles to the axis of the stator winding.   Although the resultant field makes one revolution in each cycle, its speed of rotation varies somewhat during a revolution.

If a three-phase motor is operated on rated line voltage with one line terminal open, the core loss in the stator is somewhat less than when the motor operates with balanced rated line voltages, because the maximum flux density in the stator decreases as the field revolves from a maximum along the axis of the stator winding to a minimum at right angles to this axis.   With balanced impressed voltages, the maximum flux density in the stator does not vary as the field revolves.   With balanced impressed voltages, the rotor core loss is small under load conditions because of the small slip of the rotor with respect to the revolving field.   This is also true under the conditions of single-phase operation in regard to the core loss produced in the rotor by the positive-sequence flux.   Although the flux produced by the negative-sequence magnetizing currents is small under operating conditions, the rotor core loss produced by this flux is not negligible because the rotor slip with respect to the negative-sequence flux is $(2 - s)$, which is nearly 2 for operating values of slip.   Although the stator core loss of a three-phase induction motor, operating with fixed impressed voltage and frequency, with one line terminal open, is slightly less than when it operates polyphase, this decrease in stator core loss is balanced in part by the increase in the rotor core loss due to the double-frequency negative-sequence flux in the rotor.   For this reason, the total core loss of the polyphase motor when operating with one line terminal open is probably not greatly different than when operating with balanced impressed voltages.   A similar statement applies to the single-phase motor when replaced by the equivalent three-phase motor used to determine the single-phase operating characteristics by the use of symmetrical-phase components.

Measurement of the Core Loss of a Single-phase Induction Motor. The approximate core loss of a single-phase induction motor can be found by measuring the stator input at no load with rated impressed voltage and frequency and subtracting from this input the friction and windage loss and the stator and rotor copper losses.

$$P_c = VI \cos \theta_I^V - P_{f+w} - 3\{[(I_1{}^+)^2 + (I_1{}^-)^2]r_1 + (I_2{}^+)^2 r_2 + (I_2{}^-)^2 r_2'\}$$

V is the rated impressed voltage and I is the no-load stator current.

The resistance r_1 is the stator effective resistance per phase of the equivalent three-phase motor at stator frequency, and r_2 and r_2' are the rotor effective resistances per phase at approximately zero frequency and approximately double frequency.

Since the copper loss in the rotor at no load due to the positive-sequence rotor current is negligible and $I_2{}^-$ equals $I_1{}^+$ in magnitude,

$$P_c = VI \cos \theta_I^V - P_{f+w} - 3 \left\{ \left[\left(\frac{I_1}{\sqrt{3}} \right)^2 + \left(\frac{I_1}{\sqrt{3}} \right)^2 \right] r_1 + \left(\frac{I_1}{\sqrt{3}} \right)^2 r_2' \right\}$$

$$= VI \cos \theta_I^V - P_{f+w} - I_1{}^2 (2r_1 + r_2') \qquad (291)$$

An Example of the Calculation of the Torque, Power Output, Current, and Power Factor of a Single-phase Induction Motor. A 25-hp 230-volt 60-cycle six-pole single-phase induction motor is used. This is an unusually large single-phase motor. The no-load and the blocked or locked data and other data are given.

No load				Locked rotor			
Volts	Frequency	Amperes	Watts	Volts	Frequency	Amperes	Watts
220	60	36.8	938	62.0	60	108.0	2,650

Friction and windage loss.................... 375 watts
Ohmic resistance between stator terminals.... 0.1261 ohm
Ratio of effective resistance to ohmic resistance:
Stator at 60 cycles......................... 1.1
Rotor:
    At 60 cycles............................. 1.2
    At 120 cycles............................ 1.8

The stator and rotor reactances are assumed to be equal when referred to the stator. The operating conditions for a slip of 2.0 per cent are calculated.

$$r_{eq} \text{ (equivalent)} = \frac{2,650}{2 \times (108.0)^2} = 0.1136 \text{ ohm per phase at 60 cycles}$$

$$z_{eq} \text{ (equivalent)} = \frac{62}{2 \times 108.0} = 0.287 \text{ ohm per phase at 60 cycles}$$

$$x_{eq} \text{ (equivalent)} = \sqrt{(0.287)^2 - (0.1136)^2} = 0.2636 \text{ ohm per phase at 60 cycles}$$

Equivalent Three-phase Constants

$$r_1 \text{ (effective at 60 cycles)} = \frac{0.1261}{2} \times 1.1 = 0.06936 \text{ ohm per phase}$$

$$r_2 \text{ (effective at 60 cycles)} = 0.1136 - 0.06936 = 0.0442 \text{ ohm per phase}$$

$$r_2 \text{ (ohmic)} = \frac{0.0442}{1.2} = 0.03683 \text{ ohm per phase}$$

$$r_2' \text{ (effective at 120 cycles)} = 0.03683 \times 1.8 = 0.0663 \text{ ohm per phase}$$

$$x_1 = x_2 \text{ (referred to stator)} = \frac{0.2636}{2} = 0.1318 \text{ ohm per phase}$$

From Eq. (282), page 485,

$$z^+ = z^- = \left(2r_1 + \frac{r_2}{s} + \frac{r_2'}{2-s}\right) + j2(x_1 + x_2)$$

$$z_n{}^+ = \left(2r_1 + \frac{r_2'}{2-s}\right) + j(2x_1 + x_2)$$

$$z^+ = z^- = \left(2 \times 0.06936 + \frac{0.03683}{0.02} + \frac{0.0663}{1.98}\right) + j2(0.2636)$$

$$= 2.014 + j0.5272 \text{ vector ohms}$$

$$= \sqrt{(2.014)^2 + (0.5272)^2} = 2.082 \text{ ohms}$$

$$z_n{}^+ = \left(2 \times 0.06936 + \frac{0.0663}{1.98}\right) + j(2 \times 0.1318 + 0.1318)$$

$$= 0.1722 + j0.3954 \text{ vector ohm}$$

$$= \sqrt{(0.1722)^2 + (0.3954)^2} = 0.431 \text{ ohm}$$

$$\frac{V_L}{\sqrt{3}} = \frac{220}{\sqrt{3}} = 127.1 \text{ volts}$$

$$\frac{V_L'}{\sqrt{3}} = \frac{V_L}{\sqrt{3}} - I_n z_n{}^+ = 127.1 - \frac{36.8}{\sqrt{3}} \times 0.431 \text{ (arithmetically)}$$

$$= 117.9 \text{ volts}$$

$$I_2{}^+ = \frac{117.9}{2.082} = 56.6 \text{ amp}$$

Consider the positive-sequence components. The vector diagram for these components is given in Fig. 287.

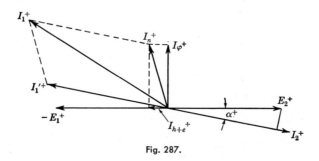

Fig. 287.

$$\text{No-load power factor} = \frac{938}{220 \times 36.8} = 0.1156$$

$$\cos \theta_n{}^+ = 0.1156 \qquad \sin \theta_n{}^+ = 0.993$$

$$\text{Core loss} = 928 - 375 - (36.8)^2(2 \times 0.06936 + 0.0663)$$

$$= 285 \text{ watts}$$

$$\text{Single-phase } I_n = -\frac{285}{220} + j36.8 \times 0.993$$

$$= -1.29 + j36.5 \text{ vector amperes}$$

Neglecting the negative-sequence magnetizing current,

$$I_n{}^+ = \frac{1}{\sqrt{3}}(-1.29 + j36.5) = -0.748 + j21.1 \text{ vector amperes}$$

$$\cos \alpha^+ = \frac{r_2}{\sqrt{(r_2)^2 + (x_2 s)^2}} = \frac{0.03683}{\sqrt{(0.03683)^2 + (0.1318 \times 0.02)^2}} = 0.997$$

$$\sin \alpha^+ = \frac{x_2 s}{\sqrt{(r_2{}^2) + (x_2 s)^2}} = \frac{0.1318 \times 0.02}{\sqrt{(0.03683)^2 + (0.1318 \times 0.02)^2}} = 0.0714$$

$$I_1{}^+ = I_1'^+ + I_n{}^+ = -56.6(0.997 - j0.0714) + (-0.748 + j21.1)$$
$$= -57.15 + j25.1 \text{ vector amperes}$$
$$= \sqrt{(-57.15)^2 + (25.1)^2} = 63.25 \text{ amp}$$

If the magnetizing current for the negative-sequence components is neglected, $I_2{}^- = I_1{}^+$ in magnitude if referred to the stator. Making this approximation,

$$T = T^+ - T^-$$

$$= 3\frac{p}{4\pi f}\left[(I_2{}^+)^2 \frac{r_2}{s} - (I_2{}^-)^2 \frac{r_2'}{2-s}\right]\frac{550}{746}$$

$$= 3\frac{6}{2 \times 377}\left[(56.6)^2 \frac{0.03683}{0.02} - (63.25)^2 \frac{0.0663}{1.98}\right]\frac{550}{746}$$

$$= 0.02386(5{,}900 - 134)\frac{550}{746}$$

$$= 101.5 \text{ lb-ft internal torque}$$

$$P = \text{pulley output} = 2\pi \text{ speed } T = 2\pi\frac{2f}{p}(1-s)T - P_{f+w}$$

$$= 2\pi\frac{120}{6}(1.00 - 0.02)101.5 - 375 \times \frac{550}{746}$$

$$= 12{,}230 \text{ ft-lb per sec}$$

$$= \frac{12{,}230}{550} = 22.2 \text{ hp}$$

$$\text{Copper loss} = 3[(I_1{}^+)^2 r_1 + (I_1{}^-)^2 r_1 + (I_2{}^+)^2 r_2 + (I_2{}^-)^2 r_2']$$
$$= 3[2(63.25)^2 0.0694 + (56.6)^2 0.0368 + (63.25)^2 0.0663]$$
$$= 3(555 + 118 + 265)$$
$$= 2{,}814 \text{ watts}$$

Copper loss	2,814 watts
Core loss	285 watts
Friction and windage loss	375 watts
Total losses	3,474 watts

$$\text{Efficiency} = \frac{22.2 \times 746}{22.2 \times 746 + 3{,}474} = 0.827 \text{ or } 82.7 \text{ per cent}$$

$$I_1 = \sqrt{3} I_1{}^+ = \sqrt{3} \times 63.25 = 109.5 \text{ amp}$$

$$\text{Power factor} = \frac{22.2 \times 746 + 3{,}474}{220 \times 109.5} = 0.831$$

Chapter 37

Quadrature field of the single-phase induction motor.   Revolving field of the single-phase induction motor.   Explanation of the operation of the single-phase induction motor.   Comparison of the losses in single-phase and polyphase induction motors.   Vector diagram of the single-phase induction motor based on the quadrature-field theory.   Generator action of the single-phase induction motor.

Quadrature Field of the Single-phase Induction Motor.   In this chapter the single-phase induction motor is considered from the standpoint of the quadrature-field theory.   At any speed other than zero a single-phase induction motor has a revolving magnetic field, which can be considered as produced by two component fields which are in space quadrature and in time quadrature.   One of these component fields is due to the stator winding and for a given impressed voltage would be constant if it were not for the change in stator impedance drop with change in load. The other component field is due to the current in the rotor produced by its rotation in the stator field.   This second or quadrature field varies in magnitude with the speed.   It is zero at zero speed and would be equal to the stator field at synchronous speed, if it were not for the resistance and leakage-reactance drops in the rotor winding.

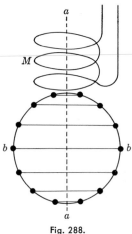

Fig. 288.

The trace of the extremity of the vector which represents the revolving field produced by these two quadrature component fields is nearly circular at synchronous speed.   Both below and above synchronous speed it is elliptical, with the major axis of the ellipse along the axis of the stator field below synchronous speed and at right angles to the axis of the stator field above synchronous speed.

Figure 288 represents diagrammatically a single-phase induction motor with a squirrel-cage rotor. M is the stator winding which is distributed in the actual motor.

Assume that the axis aa of the stator field is vertical. The variation in the stator flux induces voltages in the inductors of the squirrel-cage rotor. These voltages act in opposite directions on opposite sides of the axis aa. So far as these voltages are concerned, the rotor inductors can be paired off to form a series of closed coils, as indicated by the horizontal lines of Fig. 288. The voltages induced in these coils by a variation in the stator field produce currents in the coils, and these currents react on the stator winding just as the current in the secondary of a short-circuited static transformer reacts on the primary. So far as concerns the effect of these currents on the stator winding, the

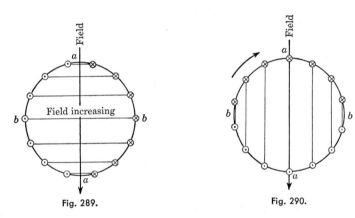

Fig. 289.   Fig. 290.

rotor winding can be replaced by a single concentrated winding with its axis coincident with the axis of the stator winding and with its sides at bb.

When the rotor turns, two component voltages are induced in the inductors by the stator field. One voltage is caused by the transformer action of the stator field and is the same as the voltage induced in the rotor when at standstill. The other voltage is induced by the rotation of the rotor inductors through the stator field. The first is a transformer voltage, the second a speed voltage. The voltages induced in the inductors by the transformer action are in the same direction in all inductors on the same side of the axis aa of the stator field (Fig. 289). Therefore the axis of the rotor for these voltages is vertical in Fig. 289 and any current the voltages cause reacts on the stator field. The rotor winding so far as the effect of this current is concerned acts like the closed secondary of a static transformer and has the same effect as an equivalent number of turns concentrated at bb.

The application of the right-hand rule shows that the voltages induced in the rotor inductors by their movement through the stator field act in one direction in all inductors above the horizontal axis (Fig. 290) and in the opposite direction in all inductors below the horizontal axis. The axis of the rotor for these voltages is horizontal in Fig. 290 and any currents they cause react on the stator along a horizontal axis. So far as the effect of these currents on the stator is concerned, the rotor winding can be replaced by a single concentrated winding having its axis at right angles to the axis of the stator field and its sides at aa.

Since the axis for the speed voltages is at right angles to the stator field, any currents these voltages produce can have no electromagnetic reaction on the stator winding.

The axes of the rotor for the two component voltages induced in it by the stator field are at right angles, i.e., in space quadrature. Both component voltages are produced in the same inductors by the same flux. The transformer voltage is in time quadrature with the stator flux. The speed voltage is produced by a movement of the inductors across the stator field at a speed which is constant for a fixed load. The speed voltage therefore must be directly proportional to the stator field at every instant. The two component voltages therefore are in time quadrature.

The component of the rotor current which can be considered to be produced by the rotation of the rotor in the stator field gives rise to a field which has its axis along bb, i.e., in space quadrature with the axis of the stator field. Since there is no winding on the stator or on any other part of the motor upon which this current can react electromagnetically, the rotor must act, so far as this current and the axis bb are concerned, like a reactor. The rotor is equivalent to a single winding on a magnetic circuit with two air gaps. Its reactance for the axis bb is high and the current producing the quadrature field must lag nearly 90 deg behind the speed voltage. Since the maximum of the speed voltage coincides in time with the maximum of the stator flux, the current producing the quadrature field is nearly in time quadrature with the stator field. Therefore the quadrature field, which is in phase with the magnetizing component of this current, must be in time quadrature with the stator field.

The two component currents in the rotor are considered separately and are referred to the stator as is the rotor current of the polyphase induction motor. All voltages induced in the rotor are also referred to the stator. The actual current in a rotor inductor is the vector sum of the two component currents. The component currents in the rotor when referred to the stator are designated by the letter I with a subscript, a or b, to indicate the axis along which they react. For example, I_b is the component current producing an armature magnetic axis coinciding with the axis bb. It is the component current producing the quadrature field. Three subscripts must be used with the component voltages: one, a or b, to indicate the rotor axis to which it belongs;

a second, M or Q, to indicate whether it is produced by the main or stator field or by the quadrature field; a third, T or S, to indicate whether a transformer or a speed voltage is indicated. For example, E_{bMS} is the speed voltage induced in the rotor by its rotation in the stator field M. It produces in the rotor a component current I_b which reacts along the axis bb.

If φ_M is the stator field,

$$E_{aMT} = KN\varphi_M f \qquad (292)$$

and

$$E_{bMS} = KN\varphi_M n \qquad (293)$$

where K = a constant

N = effective number of turns on the rotor referred to the stator

f = frequency

n = speed in revolutions per second multiplied by the number of pairs of poles

At synchronous speed f and n are equal. Hence E_{bMS} and E_{aMT} must be equal at synchronous speed. This assumes a sinusoidal time variation and a sinusoidal space distribution of the flux.

The quadrature field φ_Q also produces two voltages in the rotor, a speed voltage and a transformer voltage. These voltages are

$$E_{aQS} = KN\varphi_Q n \qquad (294)$$

and

$$E_{bQT} = KN\varphi_Q f \qquad (295)$$

These voltages are also equal at synchronous speed. Since the rotor is short-circuited, the component current I_b at any speed has such a value that the vector sum of E_{bMS} and E_{bQT} equals the impedance drop in the rotor.

The transformer voltage E_{bQT} is a reactance voltage produced by the quadrature flux φ_Q.

$$E_{bMS} + E_{bQT} = I_b(r + jx) \qquad (296)$$

Since voltages and current are referred to the stator, r and x must also be referred to the stator. The r's for the two equivalent windings which have replaced the rotor winding must be equal. The x's must also be equal.

From Eqs. (293), (295), and (296),

$$KN\varphi_M n + KN\varphi_Q f = I_b(r + jx)$$

Under operating load conditions, the impedance drop $I_b(r + jx)$ is small as compared with the voltages E_{bMS} and E_{bQT}. Neglecting this drop,

$$-\frac{\varphi_M}{\varphi_Q} = \frac{f}{n}$$

Therefore if the rotor impedance drop $I_b(r + jx)$ is neglected, the main and quadrature fields are equal at synchronous speed since f and n are equal at that speed. Actually at synchronous speed φ_Q is slightly less than φ_M on account of the rotor impedance drop. Below synchronous speed φ_Q is less than φ_M since φ_Q varies with the speed. The two fields are always in space quadrature and in time quadrature. The two fields are shown in Fig. 291.

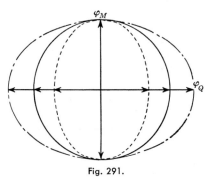

Fig. 291.

Revolving Field of the Single-phase Induction Motor. The two fields φ_M and φ_Q are in space quadrature and in time quadrature and combine to produce a revolving magnetic field. At synchronous speed these two component fields are sensibly equal and the end of the vector which represents their resultant traces a circle. Below synchronous speed the quadrature field is less than the main or stator field and the trace is an ellipse with the major axis along the axis of the stator field, shown dotted (Fig. 291). Above synchronous speed the trace, shown by the dot-and-dash line (Fig. 291), is still an ellipse but its major axis is at right angles to the axis of the stator field. The resultant field is constant and revolves at constant speed at synchronous speed only. When the motor is at standstill, the quadrature field is zero and the dotted ellipse becomes a straight line.

Explanation of the Operation of the Single-phase Induction Motor. For a motor to produce torque, the axis of the magnetic field due to its armature current must not be in space phase with the axis of the air-gap flux. The air-gap flux must not be in time quadrature with the armature current. The first of these conditions is made evident by considering a d-c motor. The magnetic axis of the armature winding of a d-c motor coincides with the brush axis. If the brushes of such a motor are moved so that the armature axis coincides with the field axis, the torque is zero. Under this condition, the two quarters of the armature winding on the same side of the brush axis produce torques in opposite directions and their effects neutralize. If the axes of the armature and of the field are in space quadrature, the torques produced by all armature inductors act in the same direction. The average torque throughout a cycle is zero unless the armature current and the air-gap flux have components in time phase.

At standstill, a single-phase induction motor has a rotor current produced by the transformer action of the stator flux. For this current the magnetic axis of the rotor coincides with the axis of the stator field. The resultant air-gap flux is in space phase with this rotor axis. The torque is zero. When the motor is running, the field φ_Q, due to the component produced in the rotor by rotation, has its axis in space quadrature with the stator field. This field φ_Q therefore develops torque with the current produced in the rotor by the transformer action of the stator flux, provided φ_Q and this current are not in time quadrature.

So far as the component current produced by the speed voltage is concerned, the rotor has a magnetic axis at right angles to the stator field. This component current therefore can develop torque with the stator field. Since the magnetic axis for this component current is at right angles to the stator field, the current cannot react on the stator and cannot cause any change in the stator current. Therefore the power developed by this component current cannot be directly supplied by the stator and it must be a part of the power developed by the rotation of the armature. This current can be considered as made up of a magnetizing component for the quadrature field and a power component supplying the core loss of that field. This current involves power developed and corresponds to generator action. Since there is no reaction between the stator and this current, the power must be supplied through an equivalent current in the rotor producing an equal motor action. The axis of the field due to this motor current must coincide with the axis of the stator. Hence it reacts on the stator to produce an equivalent current in the stator winding. The component current producing the quadrature field and supplying the core losses due to this field thus has its equivalent current in the stator.

A single-phase induction motor is in a sense a motor generator, since both motor and generator power are developed in its rotor. In addition to motor power, it must develop sufficient generator power to supply the exciting current for the quadrature field. So far as the axis aa is concerned, the rotor acts as a motor. With respect to the axis bb, the rotor acts as a generator. Under load conditions, the generator action is small compared with the motor action. Above synchronous speed, only generator action is developed with respect to both axes.

Although the single-phase induction motor has a revolving magnetic field, the simple explanation of the operation of a polyphase motor, based on its revolving magnetic field, cannot be applied to a single-phase induction motor, since the quadrature field is not produced by a stator winding. Also there is no winding on the stator inductively related to the current which produces the quadrature field and hence no reaction on such a winding.

The complete analysis of the operation of the single-phase induction motor is not simple. The action of such a motor based on the quadrature-field theory can best be understood by a study of its vector diagram. On account of the motor-generator action in the rotor, this diagram is somewhat complicated.

Comparison of the Losses in Single-phase and Polyphase Induction Motors. The losses of a single-phase induction motor are greater than those of a polyphase motor of the same speed and rating. This is due in part to the inherently greater losses of any single-phase motor or generator and in part to conditions which are peculiar to the single-phase induction motor alone.

To secure a satisfactory flux distribution and also to distribute the stator copper loss over as much of the stator surface as possible, it is necessary to use a greater phase spread for the stator winding in a single-phase motor than in the stator winding of a polyphase motor.

With a phase spread of 60 deg, the stator winding of a three-phase induction motor covers the entire stator surface. For the stator winding of a single-phase motor to cover the entire stator surface, a phase spread of 180 deg is necessary. A spread of 180 deg is not used in practice on account of the differential action of the end turns and the resulting low breadth factor of the winding.

Most induction motors have fractional-pitch windings. Assuming a three-phase motor with a $\frac{5}{6}$ pitch and a 60-deg phase spread, the total reduction factor, the breadth factor multiplied by the pitch factor, is 0.92 [see Eqs. (133) and (139), pages 184 and 191]. The total reduction factor for a single-phase motor, using two phases of a three-phase Y-connected winding with a $\frac{5}{6}$ pitch, is about 0.80. For the same mmf the single-phase motor requires approximately 15 per cent more ampere-turns than the three-phase motor for the same rotor current. This results in a corresponding increase in the stator copper loss or in an increase in the amount of copper required for the stator winding. The stator copper loss of a single-phase motor is in general greater than for a three-phase motor. This greater stator copper loss is not peculiar to the single-phase induction motor but exists in any single-phase motor or generator as compared with one of the polyphase type.

The frequency of the rotor current of a polyphase motor is $f_1 s$ where f_1 is the stator frequency and s is the slip. This frequency $f_1 s$ is low, especially for large motors, which at full load may have a slip as low as 2 per cent. The local losses in the rotor, caused by the rotor leakage flux, are negligible on account of this low frequency.

The current in each rotor inductor of a single-phase motor is the vector sum of two component currents, one due to the transformer action of the stator and the other due to the speed voltage produced by the stator field. Both of these component currents have stator frequency when referred to the stator. When referred to the rotor one has a frequency $f_2 = f_1 s$, the other a frequency $f_2' = (2 - s)f_1$. For ordinary values of slip the first of these frequencies is very low, the second is nearly twice stator frequency. The effect of the double-frequency component current is to increase substantially the rotor copper loss. There is no corresponding double-frequency component in the rotor current of a polyphase induction motor.

Another factor which tends to increase the copper loss of a single-phase motor is the magnetizing current for the quadrature field. This magnetizing current is carried by the rotor winding. No such magnetizing current exists in the rotor winding of a polyphase motor. The stator of a single-phase motor carries the magnetizing current for the main field and an equivalent of the magnetizing current for the quadrature field.

In a polyphase motor, the core loss produced in the rotor by the revolving field, neglecting the effect of harmonics in the stator flux, is due to a flux which has a frequency with respect to the rotor equal to stator frequency times slip. For the small values of slip at which an induction motor usually operates, this loss is generally negligible. A single-phase induction motor also has a revolving field. This field is not constant in magnitude except at synchronous speed. In general it can be resolved into two components, one, a revolving field with a constant strength equal to the maximum value of the quadrature field, the other, an oscillating field with its axis along the stator axis. The latter component has a maximum value equal to the difference between the maximum values of stator and quadrature fields. The oscillating component produces a flux variation of two frequencies in the rotor. If φ_m is the maximum value of the oscillating component and s is the slip, the oscillating flux in the rotor is

$$\varphi_m \sin \omega t \sin (1 - s)\omega t = \tfrac{1}{2}\varphi_m \cos s\omega t - \tfrac{1}{2}\varphi_m \cos (2 - s)\omega t$$

The frequency of the first component of the flux is stator frequency times slip, but that of the second component is twice stator frequency minus stator frequency times slip. The second component has therefore nearly twice stator frequency. The rotor core loss due to the first component is negligible but the rotor core loss due to the second component cannot be neglected. This second component tends to make the core loss per unit volume in the rotor of a single-phase induction motor slightly greater than the core loss per unit volume of the rotor of a polyphase induction motor.

The chief factor which makes the core loss of a single-phase motor greater than that of a polyphase motor of the same speed and rating is the greater amount of iron necessary in both rotor and stator of the single-phase motor.   For the same inductor copper loss, a three-phase generator or motor when operated single phase can deliver only about 60 per cent of its rated three-phase output.   If the motor or generator is rewound for single-phase operation only, its single-phase rating can be somewhat increased, but the rating is still below the polyphase rating.   It follows that a single-phase induction motor must have much more iron than a three-phase motor of the same speed and output and must have a greater core loss.

The losses of a single-phase induction motor are inherently greater than those of a polyphase motor of the same rating and speed, and thus the efficiencies of single-phase induction motors are less than those of corresponding polyphase motors.   The single-phase motor is heavier since it requires much more iron.

The power factors of single-phase induction motors are inherently less than the power factors of polyphase induction motors.   As the output of a single-phase induction motor is only about 60 per cent of the output of a polyphase motor of the same weight, the power component of the stator current of a single-phase induction motor is about equal to the power component of the current in one phase of a two-phase motor of the same weight.   The stator winding of a single-phase motor carries the magnetizing current for both the main field and the quadrature field and is about twice as great as the magnetizing current of one phase of the two-phase motor.   The ratio of magnetizing current to power current in the stator of a single-phase motor therefore is much greater than the ratio of the same currents for a two-phase motor or in general for a polyphase motor.   It follows that the power factor of a single-phase motor is less than the power factor of a polyphase motor of the same speed and output.

Vector Diagram of the Single-phase Induction Motor Based on the Quadrature-field Theory.   For convenience in reference, the main or stator field is represented by the poles MM and the quadrature field caused by the rotor is represented by the poles QQ, Fig. 292.   The actual motor has a distributed winding on the stator and a squirrel-cage winding or its equivalent on the rotor.   All vectors are referred to the stator.

At Synchronous Speed.   Consider first the conditions existing in the motor when it is driven up to synchronous speed by some outside source of power.   Assume that the direction of rotation of the rotor is clockwise. In the rotor there are two voltages to consider with respect to each of the two axes, aa and bb, a transformer voltage and a speed voltage.

A transformer voltage is 90 deg behind the flux producing it.   Some convention must be adopted for determining the time-phase relation between speed voltage and the flux producing it.   The maximum values occur at the same time but may not be of the same sign.

The positive direction of the stator flux can be arbitrarily assumed. This, with the direction of rotation of the rotor, fixes the positive direction of the quadrature flux. The revolving magnetic field of the motor must progress in the direction of rotation of the rotor. If upward fluxes (Fig. 292) are assumed to be positive for the poles MM, a clockwise rotation of the rotor makes the direction from left to right positive for the flux of the quadrature field QQ.

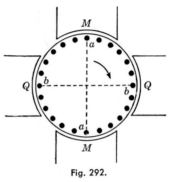

Fig. 292.

A current in a coil is positive if it produces a positive flux. A speed voltage is positive if the current due to this voltage produces a positive flux.

The time-phase vector diagram for a single-phase induction motor at synchronous speed is given in Fig. 293, in which the stator flux is drawn positive.

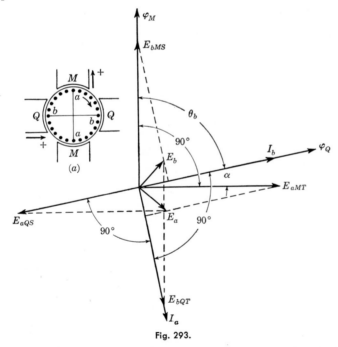

Fig. 293.

The rotation of the rotor in the stator field MM induces a component voltage E_{bMS}. Applying Fleming's right-hand rule, it is evident that this acts outward in all inductors above the horizontal axis and inward in all inductors below the horizontal axis. According to the convention adopted for determining the sign of a speed voltage, this voltage is positive. The component current I_b set up by this voltage reacts along the quadrature axis. It is largely a magnetizing current and lags nearly 90 deg behind the voltage producing it. To avoid unnecessary confusion in the diagram, the quadrature flux φ_Q caused by I_b is assumed to be in phase with I_b. Actually φ_Q is in phase with the magnetizing component of I_b and therefore lags slightly behind I_b. The flux φ_Q is a little less than the flux φ_M on account of the resistance and leakage-reactance drops in the rotor (page 498).

The flux φ_Q produces a transformer voltage E_{bQT} in the rotor inductors. The flux φ_M also produces a transformer voltage E_{aMT} but the axis of the rotor for E_{aMT} is in space quadrature with its axis for E_{bQT}. Because of the rotation of the rotor in the field φ_Q a speed voltage E_{aQS} is induced. According to the right-hand rule and the convention adopted, this voltage is negative.

The voltages E_{bMS} and E_{aMT} are caused by the same flux, and since the speed times $p/2$ and the stator frequency are equal at synchronous speed, the two voltages are equal at synchronous speed (page 497). Similarly the voltages E_{bQT} and E_{aQS} are equal at synchronous speed but are slightly less than the voltages E_{bMS} and E_{aMT}. This difference in magnitude results from the slight difference in magnitude of the fields φ_M and φ_Q caused by the rotor leakage-impedance drop.

The current I_b is equal to the resultant voltage E_b divided by the leakage impedance of the rotor. Similarly the current I_a is equal to the resultant voltage E_a divided by the leakage impedance of the rotor. E_{aMT} is equal to E_{bMS} and is 90 time degrees behind it. Also E_{aQS} is equal to E_{bQT} and is 90 time degrees behind it. Therefore E_b and E_a are equal and are in time quadrature. Since the resistance and leakage reactance of the rotor along the two axes aa and bb are equal, the currents produced by the voltages E_b and E_a are also equal. These currents are in time quadrature since the voltages E_b and E_a producing them are in time quadrature.

Referring to Fig. 293a, it is evident that the current I_b can produce torque with the flux φ_M only and the current I_a can produce torque with the flux φ_Q only. The power developed by a motor or a generator is always equal to the product of the projection of the rotor current on the speed voltage. From Fig. 293 it is evident that I_b has a positive projection on its speed voltage E_{bMS} and produces generator action. The current I_a is in time quadrature with its speed voltage E_{aQS} and produces neither motor nor generator action. To bring the motor up to synchronous speed, mechanical power must be supplied to the shaft. This power must be greater than the friction and windage loss by an amount equal to the product of the voltage E_{bMS} and the active component of the current I_b with respect to E_{bMS}. This mechanical power which must be supplied is equal to the core loss due to the quadrature field plus a small copper loss in the rotor.

The current I_a reacts on the stator by transformer action and produces an equivalent current in the stator winding. The current I_b which is the magnetizing current for the field QQ cannot react on the stator. Because of the rotation of the rotor, however, an equal component current I_a appears in the rotor and this component does react on the stator. The magnetizing current of the quadrature field is thus transferred to the stator winding owing to the rotation of the rotor. In addition to this transferred current, the stator carries magnetizing and core-loss currents for its own field. The stator consequently carries a component current which is equal to the sum of the magnetizing currents for the main and quadrature fields.

It should be clear that a single-phase induction motor cannot reach synchronous speed even at no load. To have it reach synchronous speed, sufficient power must be supplied to meet the core loss of the quadrature field even if the friction and windage and copper losses are zero. On the other hand, a polyphase induction motor runs at synchronous speed at no load if it has no friction and windage losses. If the harmonics in the rotating magnetic field are neglected, the rotor core loss at synchronous speed is zero. At synchronous speed therefore the internal rotor power is zero and the slip also is zero. See Eq. (238), page 403.

Below Synchronous Speed. If the mechanical power which brings the motor up to synchronous speed is removed, the motor slows down and the speed voltage E_{bMS} decreases. The current I_b also decreases, as well as the flux φ_Q. The voltage E_{aQS} decreases more rapidly than the speed because of the decrease in the flux φ_Q with speed. The voltage E_{aQS} is approximately proportional to the square of the speed.

If the reluctance of the magnetic circuit for the quadrature field is assumed constant, the flux φ_Q and the current I_b are proportional to each other and

$$E_{bQT} = KN\varphi_Q f$$

can be written

$$E_{bQT} = K'NI_b f$$

K' and N are constant, and if f is assumed constant

$$E_{bQT} = -I_b X$$

where X is the equivalent reactance of the rotor winding with respect to the axis bb. The minus sign is used because E_{bQT} is a voltage rise and $I_b X$ is a voltage drop.

E_{bQT} is thus a reactance voltage

$$E_{bMS} + E_{bQT} = I_b(r + jx)$$
$$E_{bMS} - I_bX = I_b(r + jx)$$
$$I_b = \frac{E_{bMS}}{r + j(x + X)} \tag{297}$$
$$\cos \theta_b = \frac{r}{\sqrt{r^2 + (x + X)^2}}$$

Therefore $\cos \theta_b$ is constant, since r and x are sensibly constant and X is assumed constant. For ordinary variations in speed, such as are produced by a change in load, X is nearly constant. Since $\cos \theta_b$ is constant for moderate changes in speed, the current I_b, Fig. 293, decreases in magnitude but does not change in direction as the motor is loaded. As the motor slows down, the generator action of the current I_b decreases. This decrease results from the decrease in the core loss of the quadrature field due to the decrease in φ_Q.

The conditions for the other component current I_a are different. The voltage E_{aMT} is constant, assuming φ_M to be constant, but the voltage E_{aQS} varies as the square of the speed. The decrease in speed causes E_a, Fig. 293, to rotate counterclockwise, carrying with it the current I_a which at the same time increases. The current I_a now has a negative projection on its speed voltage E_{aQS}. This negative projection represents motor action. If there were no copper loss and no friction and windage loss, the motor would run at such a speed below synchronism that the motor power developed by I_a would be equal to the generator power developed by I_b. This generator power supplies the core loss of the quadrature field.

If the load is now applied, the motor slows down, until at a certain small slip the motor action of I_a is enough greater than the friction and windage loss plus the generator action of I_b to cause a net motor power sufficient to enable the motor to carry its load. In general the change in the slip of a single-phase motor for a given change in load is less than for a corresponding polyphase motor. The speed voltage E_{aQS} of the single-phase motor varies with the square of the speed, but the speed voltage $E_2(1 - s)$ of a polyphase motor varies as the first power of the speed. The slip of a single-phase induction motor can be shown to be nearly proportional to the square root of the rotor copper loss. The slip of a polyphase induction motor has been shown to be proportional to the first power of the rotor copper loss.

It should be noted that the component current I_a, whose magnetic axis lies along the stator field, is the power current of the motor. The other component current I_b is merely the magnetizing current for the quadrature field. The current I_a in conjunction with this quadrature field produces the power output of the single-phase induction motor.

In the vector diagram of Fig. 294 are shown the main flux φ_M, the rotor voltage E_{AMT}, the rotor current I_a, and the stator currents and voltages. It indicates how the stator current and power factor are dependent on the component current I_a of the rotor. The component current I_b of the rotor cannot react on the stator winding directly since the axis of the field φ_Q produced by it is in space quadrature with the axis of the stator winding.

In Fig. 293, φ_M and φ_Q are fluxes, in the two axes, which cross the air gap between the rotor and stator. There will be in the rotor in addition to these fluxes the leakage fluxes produced by the currents I_a and I_b, which

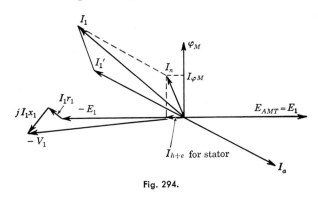

Fig. 294.

fluxes do not cross the air gap. These leakage fluxes produce transformer and speed voltages in the rotor windings. The transformer voltages are treated as Ix drops and in Fig. 293 are the components of E_a and E_b at right angles to the currents I_a and I_b, respectively. The speed voltages produced by these leakage fluxes are small and are omitted in the diagram in the interest of simplicity. In a strict analysis these should be added. The speed voltage produced by the leakage flux due to the current I_a will be in phase with I_a and will act around the b axis; the speed voltage produced by the leakage flux due to the current I_b will be in phase with I_b and will act around the a axis.

Generator Action of the Single-phase Induction Motor. If driven above synchronous speed, the single-phase induction motor acts as a generator. If such a motor is speeded up from below synchronous speed, the net internal motor power decreases and becomes zero at a small positive slip below synchronism. At this slip, the internal motor power developed is just equal to the core loss caused by the quadrature field. Above the speed corresponding to this slip, the mechanical power applied to the pully begins to supply the core loss of the quadrature field. At synchronous speed no internal motor power is developed and all the core loss of the quadrature field is supplied by the mechanical power. Above synchronous speed generator action is developed along the axis aa and the mechanical power begins to supply the core loss caused by the stator field MM. At some small negative slip above synchronism, both core losses are supplied by the mechanical power. Above this slip net generator power is developed and power is delivered to the circuit to which the motor is connected. Like a polyphase induction generator, the single-phase induction generator delivers only leading current at a power factor which is fixed by its constants and by the slip but not by the power factor of the load.

Chapter 38

Methods of starting single-phase induction motors.   The capacitor motor.
The repulsion-start single-phase induction motor.   Speed control of the
single-phase capacitor motor.

Methods of Starting Single-phase Induction Motors.   It is obvious
from Fig. 283, page 487, that single-phase induction motors have no starting torque.   A single-phase induction motor requires some form of starting device to bring it up to such a speed that the torque developed by the
motor is sufficient to overcome the opposing torque due to losses and the
load.   The methods of starting single-phase induction motors can be
grouped under the following four general heads:

1. By some mechanical device
2. By the use of shaded poles
3. By the creation of a rotating field by using an auxiliary winding
which may or may not be cut out when the motor is up to speed
4. By converting the motor into a repulsion motor while the motor is
coming up to speed

Method (1).   It is obvious from the speed-torque curve of a single-
phase induction motor (Fig. 283, page 487) that the torque of such a
motor increases rapidly with the speed.   If the rotor of such a motor is
rotated by hand for very small motors or by a d-c motor for larger motors,
the induction motor can be made to develop sufficient torque to bring
itself up to speed.   This method of starting single-phase induction
motors is obviously of no practical importance.

Method (2).   This is the simplest method of producing a starting
torque.   It depends upon the production of a periodic shift in the pole
flux across the pole faces by means of shading coils which cover a portion
of the poles of the motor.   The shaded-pole motor is constructed with
laminated salient poles.   About one-half the face of each pole is surrounded by a low-resistance short-circuited winding.

The arrangement of a four-pole motor of this type is shown in Fig. 295.
The short-circuited shading coils are shown as *S*.   The effect of the
currents induced in the short-circuited shading coils is to oppose in the
half of the poles they surround the change in the flux produced by the
main stator winding.   The result is that the flux rises to a maximum

in the unshaded portions before it reaches its maximum in the shaded portions of the poles. The effect is a progressive shift in the field from the unshaded to the shaded portions of the poles. The rotor rotates in the same direction as the shift in the field. The shading coils are left in circuit after the motor has reached speed as the loss in them is small and of no importance in the small motors for which this method of starting is used.

Method (3). This method makes use of an auxiliary winding on the stator in space quadrature with the main stator winding. The rotor is of the usual squirrel-cage type. The two stator windings are connected to the mains in parallel. A phase displacement between the currents in the stator windings is produced either by a difference between their constants or by inserting resistance, reactance, capacitance, or a combination of these in series with one of the windings.

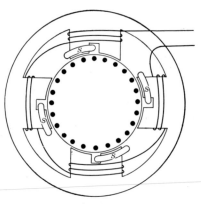

Fig. 295.

When a capacitance is used, the motor is called a capacitor or capacitor-start motor; otherwise it is known as a "split-phase" type of single-phase induction motor. The split-phase type of motor may have a resistance or an inductance connected in series with one of the windings or in some cases both may be used, *i.e.*, a resistance in series with one winding and an inductance in series with the other. In any case the added resistance and/or inductance are cut out after the motor is started. This may be accomplished either by removing them from the circuit or, where one is used in series with the auxiliary winding, a centrifugally operated switch automatically disconnects the auxiliary winding at about three-fourths of synchronous speed.

A more common arrangement, particularly for motors in the fractional-horsepower sizes, is to use a smaller gauge wire for the auxiliary winding than for the main winding or to use wire of higher specific resistance. Also fewer turns are used in the auxiliary winding, thus giving it a smaller leakage reactance but a higher resistance than the main winding. The auxiliary winding must be disconnected after starting as it would overheat if left in the circuit continuously. This is accomplished by means of a centrifugal switch which opens when the speed reaches 75 to 80 per cent of synchronous speed. A phase displacement of 20 or 30 deg between the currents in the two windings is obtained at starting. A vector diagram showing these starting currents is given in

Fig. 296. These currents produce a rotating magnetic field of varying magnitude. If the ampere-turns produced by the currents in the two windings were equal and the currents were displaced by a 90 deg phase angle, a starting condition similar to that existing in a two-phase motor would result. The torque produced varies very nearly as the sine of the angle between the two currents. The usual split-phase motor has a starting torque considerably less than its full-load torque and a line current at starting which is several times its rated current.

The advantages of this method of producing starting torque are its simplicity and its low cost. Its disadvantages are the large starting current, the small starting torque, and the rapid heating of the motor if it fails to start quickly or if it is started frequently.

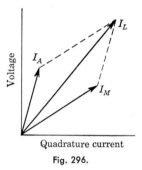

Quadrature current

Fig. 296.

When high starting torque is necessary as in starting pumps, compressors, etc., the starting torque of a split-phase motor can be increased by the use of a mechanical centrifugally operated clutch between the rotor and its shaft. In this case the rotor core is free to revolve without turning the shaft until a certain speed is reached which is in the neighborhood of 70 per cent of synchronous speed. In order to ensure smooth action the clutch is lined with a material similar to that used for the lining of automobile brakes. The operating speed for the switch which cuts out the auxiliary winding is much lower than for a motor without a clutch and is usually about 35 per cent of synchronous speed. Such a motor can develop a starting torque as high as 260 per cent of full-load torque.

When capacitance is inserted in series with the auxiliary winding of a split-phase motor, it is called a capacitor motor. If the auxiliary winding is then cut out when the motor speeds up, the motor is called a capacitor-start motor. By using the proper value of capacitance, the current in the starting winding can be made to lead the voltage across it. A greater phase displacement between the currents in the two windings and therefore a greater starting torque per ampere of line current can be obtained with capacitor-start motors than with other split-phase motors. A typical vector diagram of the starting currents of a capacitor-start motor is given in Fig. 297. Starting torques of over 200 per cent of full-load torque are usually obtained with the capacitor-start motor. This is made possible by the use of intermittent-duty electrolytic capacitors which have relatively high capacitance for their size and cost. A ¼-hp 110-volt motor would use a 233- to 281-μf capacitor. Because these capacitors cannot be left permanently in an a-c circuit, their use necessitates disconnecting the auxiliary winding after starting.

Another reason for disconnecting the auxiliary winding at about 75 per cent of synchronous speed is the rapid increase in the voltage across the capacitor as the speed is increased above this value. This voltage can be reduced by using a smaller capacitance, but this reduces the starting torque. When the auxiliary winding is left in the circuit during running as well as starting conditions, oil- or pyranol-insulated foil paper capacitors are used. These are bulkier and more expensive than electrolytic capacitors. A ¼-hp 110-volt motor of this type would use a capacitance of about 8 μf and would have a starting torque of less than 50 per cent of full-load torque. The operating power factor and other operating characteristics of the motor are improved by leaving the auxiliary winding with its capacitor in the circuit.

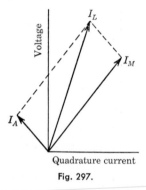

Quadrature current

Fig. 297.

The Capacitor Motor. According to the definition of the ASA, a capacitor motor is "a single-phase induction motor with a main winding arranged for direct connection to a source of power and an auxiliary winding connected in series with a capacitor." The auxiliary winding is not cut out when the motor is up to speed. The amount of capacitance in series with the auxiliary winding can be changed between starting and running conditions by means of a transformer which changes the voltage on the capacitance or by an actual change in the amount of capacitance.

Capacitor motors were built early in 1900 but did not come into commercial use because of the lack of reliable capacitors at reasonable cost. Because of the great improvement in commercial capacitors and the reduction in their cost, capacitor motors are now common and are replacing the repulsion-start single-phase induction motors (see page 513) when high starting torque is required for any purpose such as the operation of pumps or of air compressors. In order to secure the best starting torque and also the best operating characteristics under load, it is necessary to use two values of capacitance, one for starting and one for running. A capacitor motor can be built with a single value of capacitance permanently connected in series with the auxiliary or starting winding. When a single capacitance is used, its size is a compromise between that best for starting and that best for running. When a single capacitor is used, the motor is called a single-value capacitor motor. Although the single-value capacitor motor does not have such good characteristics as one using two values of capacitance, it is cheaper and is satisfactory for many purposes.

A motor which has a starting capacitance and a running capacitance is called a two-value capacitor motor. By the use of proper values of capacitance the currents in the two windings can be made to be approximately in quadrature at the instant of starting and also at full load. At standstill the reactance of the motor is the equivalent reactance of the windings and of the rotor circuit as a short-circuited transformer. As the motor speeds up the impedance increases. The impedance of the auxiliary winding with its series capacitance decreases as the motor speeds up. If the capacitive reactance which gives the best operating conditions at full load is used for starting, the current

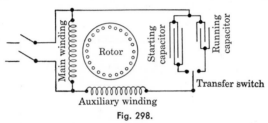

Fig. 298.

in the auxiliary winding at starting is too small. Therefore to secure the best starting and running conditions the capacitance must be reduced when the motor comes up to speed. The capacitance in series with the auxiliary winding can be changed, or a single capacitance can be used and its effect can be changed by means of an autotransformer. In either case the change is made by a centrifugally controlled switch. The two arrangements for the two-value capacitor motor are shown in Figs. 298 and 299. Keeping the capacitance in circuit during the normal operation

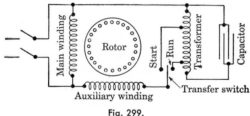

Fig. 299.

of the motor increases the breakdown torque, improves the efficiency, and makes the motor more quiet.

The Repulsion-start Single-phase Induction Motor. When high starting torque is required, a single-phase induction motor, when provided with a suitable rotor winding, can be converted into a repulsion motor for starting. A repulsion motor (Chap. 42) has essentially the same torque characteristics as a d-c series motor and can be brought up to speed with large torque and moderate current. The repulsion-start method of bringing single-phase induction motors up to speed was in commercial use in America as early as 1896. It has been extensively used when high starting torque and low starting current are required, but the capacitor motor is rapidly replacing the repulsion-start motor in fractional-horse-power sizes. The repulsion-start motor is suitable for starting pumps, compressors, etc. According to the ASA, four-pole 60-cycle repulsion-start motors up to ¾ hp should develop the following torques in per cent of full-load torque: starting, 350; pull-up, 200; breakdown, 200. The corresponding specifications for high-torque capacitor motors are: starting, 275 to 300 depending on the size; pull-up, 200; breakdown, 200.

A motor which is started as a repulsion motor has a drum-wound

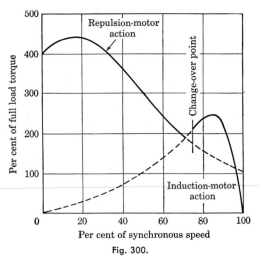

Fig. 300.

rotor similar to the armature of a d-c motor and is provided with brushes short-circuiting the rotor along its electrical diameter. The axis of the brushes is slightly displaced from the axis of the stator winding. A motor comes up to speed as a repulsion motor with torque characteristics similar to those of a d-c series motor. When a speed of about 75 per cent of synchronous speed is reached, a centrifugal governor forces a ring against the back of the commutator, thus short-circuiting each coil of the rotor winding, and at the same time it lifts the brushes off the com-mutator. The rotor then acts like a squirrel-cage rotor. Typical torque curves of a repulsion-start motor are shown in Fig. 300.

Speed Control of the Single-phase Capacitor Motor. The speed of the single-phase capacitor motor can be varied by changing the impressed voltage, provided the speed-torque curve of the load is of the proper shape with respect to that of the motor to give stable operating conditions. The speed-torque curve of a single-value capacitor motor is similar to that of a polyphase induction motor and the effect on the curve of changing the impressed voltage is much the same as in the polyphase motor. With fixed impressed frequency, the maximum torque, *i.e.*, the breakdown torque, varies as the square of the impressed voltage and the speed at which it occurs is independent of the voltage. Two typical speed-torque curves A and B of a single-value capacitor motor for two voltages are given in Fig. 301. Curve A is for rated voltage. Curve B is for one-half rated voltage. Curve C is a typical speed-torque curve of a propellor-type fan such as is used for a desk fan or for a ventilating fan. The stable operating speed of the motor with the fan as a load is

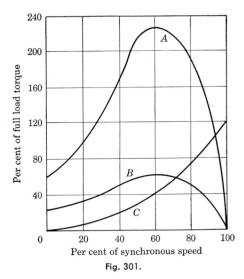

Fig. 301.

where the torque curve of the fan intersects the torque curve of the motor. For the conditions in Fig. 301, the low speed is about 70 per cent of the high speed. In order to be stable at any point of intersection of the torque curves for the motor and the load, the torque of the load at the point of intersection must increase faster than the torque of the motor for an increase in speed and must decrease faster for a decrease in speed. These conditions are fulfilled with a fan of the propeller type as a load because of the steep speed-torque curve of such a fan. Since the power required to operate the fan drops off more rapidly with speed than the output of the motor, there is no danger of overheating the motor when operating with the fan at reduced speed.

The change in voltage which is necessary to produce the desired change in speed can be obtained in a small motor by the use of a resistor or a reactor in series with the motor or in larger motors by the use of a transformer. The effect of an increase in the impressed voltage can be obtained by increasing the voltage impressed per turn on a portion of the winding, provided the change does not cause either excessive flux density or excessive exciting current. The necessary change in voltage per turn can be accomplished by bringing out a tap from the main winding and impressing rated voltage on a portion of the winding. If rated voltage is impressed on half the winding, the voltage per turn is doubled. The flux is also doubled and the effect on the speed-torque curve is the same as that produced by doubling the impressed voltage on the entire winding. If the number of turns in the main winding is altered, some provision must be made to change the voltage impressed across the auxiliary winding and its series capacitance in order to maintain the proper relation between the magnitudes of the currents in the

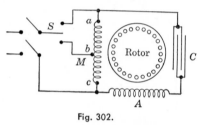

Fig. 302.

two windings. One method of accomplishing this is shown in Fig. 302. The main winding M has a tap b. By means of the switch S either the whole of the main winding or the portion between b and c can be used. When the switch S is connected to the tap b, the portions of the main winding between b and c and between a and b act like a step-up auto-transformer and give an increase in the voltage impressed across the auxiliary winding and its capacitor. For example, if the tap b is in the middle of the main winding, when the switch S is connected to b the voltage impressed per turn on the portion of the main winding used is doubled. At the same time double voltage is impressed on the auxiliary winding and its capacitor. When this method of controlling the speed is employed, the motor must be designed to prevent too high a voltage on the capacitance.

When the speed of a capacitor motor is varied by changing the impressed voltage, the motor does not operate at low speeds on the usual operating part of its speed-torque curve, i.e., the portion between maximum torque and synchronous speed, but as has been stated it is stable if the load has the proper speed-torque characteristic.

Chapter 39

The Induction Motor as a Phase Converter.   An induction motor can be used as a phase converter to change the number of phases of a system. Except when the transformation is to or from single phase it can be made more simply, much more economically, and with less unbalancing of the voltages by the use of static transformers.   When transformers are used for phase conversion, they produce no unbalancing with balanced loads provided their grouping is symmetrical.   Even the unsymmetrical arrangements which are in common use, such as the T and V connections, produce only slight unbalancing of the voltages.   The induction motor when used as a phase converter always produces some unbalancing of the voltages and this unbalancing may be large.

The principal case where the induction phase converter is of importance is for the conversion from single-phase power to polyphase power.   Such a conversion cannot be made by transformers.   One important application of phase conversion is for the operation of electric locomotives, using three-phase induction motors, from a single-phase source of power.   When only two efficient operating speeds are necessary, three-phase induction motors can be used to advantage for operating locomotives on long heavy grades.   The installation of a phase converter on a locomotive makes it possible to operate the locomotive from a single-phase trolley.   This gives the simplicity of single-phase line construction combined with advantages gained by the use of polyphase induction motors for the motive power.   The chief advantages of this method of operation are the ruggedness of the polyphase induction motor, the absence of commutation difficulties, and the possibility of electric braking and power regeneration on downgrades.   The main disadvantage is the limitation of efficient speed control.

An induction motor which has a symmetrical polyphase winding and is operated from a polyphase line with balanced voltages develops a revolving magnetic field constant in magnitude and rotating at constant speed.   It is evident that, if the motor has a second group of symmetrical windings on its stator with a number of phases different from that of the main windings, balanced polyphase voltages are induced in the second windings by the revolving magnetic field. If the load is balanced, the terminal voltages are balanced, but if the load is unbalanced, the terminal voltages are somewhat unbalanced because of the differences among the impedance drops in the windings. Such a device can be used to transform from polyphase to single phase or from single phase to polyphase. It can also be used to transform from one single-phase system to another single-phase system with any desired phase angle between the voltages of the two systems.

When a phase converter is used to transform from single phase to polyphase, the polyphase voltages cannot be balanced even at no load

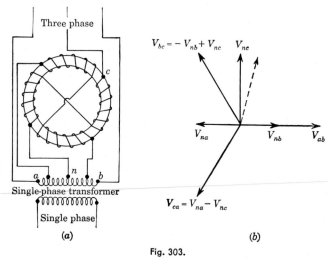

Fig. 303.

since the single-phase induction motor has an elliptical rotating field with the major axis of the ellipse along the axis of the single-phase winding. The fluxes linking the different phases of the polyphase winding therefore cannot be equal. To minimize the ellipticity of the rotating field, the phase converter should be designed with low resistance and low leakage-reactance drops.

When a phase converter is used to transform single-phase power to three-phase power, it generally has two windings in space quadrature. One of these windings is used as the primary or motor winding. The other is connected to the center of the secondary of the transformer which feeds the motor winding of the converter. The connections are similar to those on the three-phase side of Scott-connected transformers which transform from two phase to three phase. Neglecting the voltage drops in the converter and the transformer, the ratio of transformation between the quadrature winding of the converter and its motor winding should be $\sqrt{3}/2 = 0.866$. Instead of using a ratio of transformation of

0.866 between the two windings of the converter, the same result can be obtained by using a ratio of transformation of unity and connecting the motor winding to 0.866 tap on the secondary of the transformer which feeds the motor winding. A diagram showing the second method of connecting the converter to transform from single phase to three phase

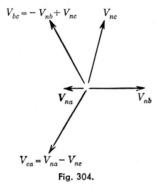

Fig. 304.

is given in Fig. 303a. A vector diagram of the voltages, neglecting the voltage drops, is given in Fig. 303b. Under load, the vector V_{nc} decreases in magnitude and shifts in phase in the direction of lag. The position of the vector V_{nc} under load is shown dotted in Fig. 303b. Compensation can be made for this shift in phase by moving the point n toward the point a, as is shown in Fig. 304.

A synchronous motor can be used as a phase converter. By changing the field excitation of the synchronous motor the power factor can be controlled.

PART SEVEN

SERIES AND REPULSION MOTORS

Chapter 40

Types of single-phase commutator motors with series characteristics. Starting. Doubly fed motors. Diagrams of connections for singly fed and doubly fed series and repulsion motors.

Types of Single-phase Commutator Motors with Series Characteristics. Single-phase commutator motors with series characteristics can be divided into two general classes: (1) series motors, and (2) repulsion motors

The chief distinction between these two types is in the way the armature receives power. The armature current of series motors is received by conduction from the line. The armature current of repulsion motors is received by induction from a winding on the stator.

For speeds greater than zero and less than about 1.4 synchronous speed, the commutation of repulsion motors is inherently better than that of series motors. In other respects the operating characteristics of the two types are similar. Each type requires three windings or their equivalent:

1. An exciting winding for the exciting or torque-producing field, the field which in conjunction with the armature current produces torque

2. An energy or armature winding for producing motor power

3. A compensating winding which compensates for armature reaction

In some types of motors, for example the simple repulsion motor, one winding can be made to serve as both compensating winding and exciting winding. From the way the armatures of the two types receive current, the motors might be called conductive series motors and inductive series motors. There are many modifications of the two general types of motors having series characteristics.

Starting. Series and repulsion motors can be started with resistance in series like a d-c series motor, but when variable speed is desired, it is customary to start motors of any considerable size on reduced voltage from a transformer. This transformer can also be used to vary the speed by changing the voltage impressed on the motor.

Doubly Fed Motors. In addition to singly fed series motors and repulsion motors, there is a class known as doubly fed motors and certain types of this class are sometimes called series-repulsion motors.

Doubly fed motors can be of either the series or the repulsion type. The armature of the doubly fed type receives power by induction from a winding of the stator and in addition receives power by conduction from the line. The power from the line comes from a transformer connected across the mains from which the motor is operated and the transformer serves for starting and for varying the speed of the motor.

Diagrams of Connections for Singly Fed and Doubly Fed Series and Repulsion Motors. The connections for a simple or singly fed series motor, a singly fed repulsion motor, a doubly fed series motor, and a

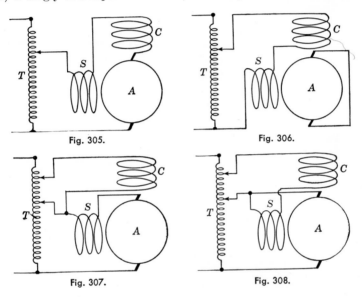

Fig. 305.   Fig. 306.

Fig. 307.   Fig. 308.

doubly fed repulsion motor are shown in Figs. 305, 306, 307, and 308. In these figures,

S = series winding producing the torque field
C = compensating winding
A = armature
T = speed-regulating and starting transformer

In the repulsion motor, a single distributed winding is made to serve for both series and compensating windings, Figs. 306 and 308. This is accomplished by shifting the brush axis out of coincidence with the axis of a single stator winding. The mmf of this single stator winding can then be resolved into two component mmfs at right angles to each other, one along the brush axis, the other at right angles to the brush axis. The first corresponds to the mmf of the compensating winding, the second to the mmf of the series winding. By changing the position of the brush axis with respect to the axis of the stator winding, the speed and the other operating characteristics of the motor can be changed.   The direction of rotation is determined by the direction in which the brushes are displaced from the axis of the stator winding.

To serve its purpose of neutralizing armature reaction, the compensating winding of a series or a repulsion motor should be distributed, since the mmf of a distributed armature winding can be neutralized only by a similarly distributed compensating winding.

Chapter 41

Singly fed series motor. Overcompensation and undercompensation. Starting and speed control. Commutation. Problems involved in the commutation of an a-c series motor. Shunted interpoles. General statements concerning a-c series motors. Universal fractional-horsepower motors.

Singly Fed Series Motor. Since the same current flows in the armature and field of a series motor, both the armature and field currents must reverse at the same time. Therefore a d-c series motor develops torque if supplied with alternating current, but owing to the high inductance of the armature and field windings as well as the large losses, little torque or power is developed and the power factor is low. Destructive sparking also occurs because of the transformer action in the armature coils short-circuited by the brushes during commutation.

The torque developed by a motor is proportional to the product of the armature ampere-turns and the component of the air-gap flux in time phase with the armature current. In a series motor, the flux is nearly in time phase with the current. A required torque can be obtained by using either a strong field and a few armature turns or a weak field and many armature turns. So far as torque is concerned, the ratio of armature turns to field turns can be either large or small. In order to minimize the effect of armature reaction and to reduce the cost of construction, the usual design of a series motor has a strong field and a relatively weak armature. If the motor is to operate on alternating current, the design must have a weak field and a relatively strong armature.

At constant frequency, the reactance of a coil varies as the square of the number of turns. By using a weak field and a strong armature, a field of few turns and an armature of many turns, the field reactance can be greatly reduced. This reduction in field turns however necessitates a corresponding increase in the armature ampere-turns and consequently in the armature reactance. Nothing would be gained by using a weak field and a strong armature if it were not possible to compensate for the armature reactance due to the cross flux produced by the armature ampere-turns.

The reactance drop in a series motor depends on f, N_f^2, and N_a^2, where f is the frequency, N_f the number of field turns, and N_a the number of armature turns. In order to make the reactance small, the frequency must be low. The design of a satisfactory 60-cycle series motor is not practical except in small sizes such as universal fractional-horsepower motors. On account of commutation difficulties caused by transformer action in the armature coils short-circuited during commutation, the series motor is essentially a low-voltage motor. This is not because of the increase in motor reactance with an increase in voltage. If the voltage is doubled, the number of turns in all windings must also be doubled. The reactance is quadrupled. The current for the same output is halved. The percentage reactance drop remains unchanged. Consequently the reactance is not the factor which limits the voltage for which an a-c series motor can be designed. The maximum operating voltage for which an a-c series motor can be designed is about 400 volts. Satisfactory a-c series motors cannot be designed for over 25 cycles, except in small sizes. A frequency of 25 cycles is standard in North America for a-c series motors used for railway electrification. In European practice a frequency as low as $16\frac{2}{3}$ is common.

The internal power developed by a motor is equal to the product of its armature current and the component of the armature voltage of rotation in phase with this current. In the series motor, the voltage of rotation is in time phase with the air-gap flux, since it is produced by the rotation of the armature inductors in the alternating field and not by transformer action of the field on the armature winding. Since the speed is constant for a given load, the voltage for a given load is directly proportional to the flux. Whenever the flux is a maximum, the voltage is a maximum. For a given current, the power developed is proportional to the voltage of rotation which is equal to the total voltage impressed on the motor minus the total impedance drop in both armature and field windings. If the total reactance is large, the voltage of rotation is small and little power is developed. The power factor is also low.

$$E_a = K \times \text{speed} \times \varphi \times N_a$$

where E_a, K, φ, and N_a are the voltage of rotation, a constant, the air-gap flux, and the number of armature turns.

An increase in the flux requires an increase in the number of turns on the field and hence an increase in field reactance. Increasing N_a increases armature reactance. Armature reactance is due to flux set up by armature reaction. If the brushes are set in the neutral plane, this flux acts along the brush axis and at right angles to the axis of the main field. It serves no useful purpose and can be suppressed without affecting the torque developed by the motor. The reactance of the main field cannot be compensated without destroying the field flux and therefore the motor torque.

It is necessary to design an a-c series motor with few field turns and many armature turns and then to eliminate the cross field due to armature reaction by a suitable compensating winding surrounding the armature. This compensating winding is placed with its magnetic axis parallel to the magnetic axis of the armature, and for complete compensation it must have the same number of effective turns as the armature and must be connected in series with the armature so as to oppose the armature reaction. The compensating winding should be distributed in order to compensate for the armature reaction of the distributed armature winding. It is placed in slots in the pole faces. Since it is not practicable to carry the conductors into the space between the poles, perfect compensation is not possible. Although the presence of an uncompensated zone between the poles is undesirable from the standpoint of commutation, the effect of this zone is not serious since the cross flux produced in the zone by the armature conductors is small and the omission of the compensating winding from this zone has relatively little effect on either the resultant reactance or the commutation of the motor. In all large a-c series traction motors the effect on commutation of armature reaction in the uncompensated zone is taken care of by the use of special interpoles.

Let E_a, N_a, N_f, x_a, x_f, and f be voltage of rotation, number of armature turns, number of field turns, armature reactance, field reactance, and frequency.

$$E_a \text{ varies as } N_a \varphi$$
$$x_a \text{ varies as } N_a{}^2 f$$
$$x_f \text{ varies as } N_f{}^2 f$$

Reducing f improves the power factor. N_a should be large with respect to N_f and the reactance drop due to the armature current in the N_a armature turns should be compensated. It is impossible to compensate leakage reactance. In order to have sufficient flux with few field turns, the air gap must be short. The magnetic circuits should be made short by the use of multipolar construction and should be operated at relatively low flux density. Since the armature reaction is compensated, the presence of a short air gap is not objectionable except for mechanical reasons. The length of the magnetic circuit of a multipolar motor is essentially shorter than that of a bipolar motor, and for the same total field mmf a larger flux can be obtained. The field reactance of a multipolar motor is less than that of a bipolar motor.

Figure 309 shows the arrangement of conductors for the armature and the compensating field of a four-pole motor without interpoles. Some traction motors have as many as 12 to 18 poles. The compensating coils must be connected in such a way that the current flows in the same direction in all the conductors under the same pole. The compensating winding can be connected in series with the armature, in which case the motor is said to be conductively compensated. It can be short-circuited,

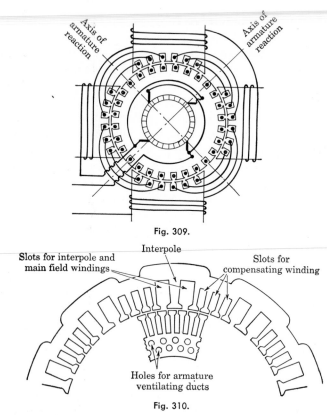

Fig. 309.

Fig. 310.

in which case the motor is inductively compensated and the compensating winding acts like the secondary of a short-circuited transformer for which the armature winding is the primary. Conductive compensation must be used if a motor is to operate on both direct current and alternating current.

Figure 310 shows the arrangement of field and armature slots used for a-c series compensated traction motors with interpoles manufactured by the General Electric Company.

Overcompensation and Undercompensation. The maximum power factor for any load is obtained when the compensation is perfect for the cross flux due to the armature current. Either overcompensation or undercompensation increases the resultant reactance drop in the armature and in the compensating field.

If the motor is overcompensated, the mmf of the compensating winding is greater than the mmf of the armature and the resultant mmf causes a flux in the same direction as the flux which would be produced by the compensating winding alone. Since this flux links both the armature and the compensating winding and is 180 deg from the flux which the armature alone would produce, the reactance drop in the armature due to it lags behind the armature current by 90 deg. The reactance drop in the compensating winding due to this resultant flux leads the current in that winding by 90 deg. It is larger than the drop in the

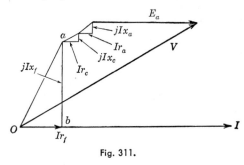

Fig. 311.

armature, since for overcompensation the effective turns on the compensating field must be greater than the effective turns on the armature. Therefore the net drop due to both armature winding and compensating winding leads the current by 90 deg. If the motor is undercompensated, there is an unbalanced mmf which sends a flux in the direction of the armature mmf. In this case the drop through the armature is greater. The resultant drop due to the cross flux caused by armature and compensating field windings is 90 deg ahead of the current, whether the flux is caused by undercompensation or overcompensation. Overcompensation cannot produce a capacitance effect. Hence the power factor of the motor cannot be increased by either overcompensation or undercompensation. Overcompensation or undercompensation gives a flux in the commutating zone and affects commutation. Overcompensation can be made to assist the reversal of the current in the armature coils short-circuited during commutation, but it cannot reduce the currents induced in these coils by the transformer action of the main field.

The vector diagram of Fig. 311 shows the effect of the reactances of the motor windings on the power factor. The subscripts f, a, and c refer to the main field, the armature, and the compensating field, respectively. The diagram is somewhat approximate in that it neglects the effect of the core loss and assumes the flux to be in time phase with the armature current. Perfect compensation is assumed and therefore Ix_a and Ix_c are small since these are then due entirely to the leakage fluxes. E_a is the voltage drop induced in the armature by rotation and is in phase with the flux. V is the voltage drop across the armature terminals.

Starting and Speed Control. An a-c motor can be started and its speed controlled by means of series resistance, like a d-c series motor. The reduction of voltage across its terminals for starting and the variation in the impressed voltage for controlling its speed can be obtained more economically than with resistance by using taps on the transformer or compensator from which the motor receives power. This method on account of its greater economy is often employed for large motors.

Commutation. The greatest difficulty in the design of an a-c series motor is due to the transformer voltage produced in the armature coils when short-circuited by the brushes during commutation. The armature coils which are short-circuited by the brushes link the full flux from the field and act like the short-circuited secondary of a transformer for which the field coils of the motor are the primary. Large currents causing excessive heating are produced in these short-circuited armature coils unless some device is used to limit their magnitude. Since these currents must be interrupted when the coils move under the brushes, bad sparking occurs. In addition to causing both excessive heating and sparking, the transformer currents in the coils short-circuited by the brushes react on the field and reduce the flux and therefore the torque developed by the motor for a given current. In order to make the operation of a series motor satisfactory commercially, some means must be adopted to reduce these short-circuited currents. They are most troublesome during starting when the field flux is usually greater than under normal running conditions. The time during which any armature coil remains short-circuited is also a maximum during starting.

In the first successful a-c series traction motors the sparking caused by the interruption of the transformer currents in the armature coils short-circuited by the brushes during commutation was held within reasonable limits by reducing these currents by inserting resistance leads between the commutator bars and the armature coils as shown diagrammatically in Fig. 312. Two of these leads are in parallel with respect to the external circuit but are in series with respect to the short-circuited coils. By reducing the flux per pole by multipolar construction and by using many commutator bars with but one armature turn between adjacent bars, the amount of resistance in the resistance leads necessary for satisfactory commutation is reduced to a minimum.

Although these resistance leads increase the armature resistance as measured between brushes, they actually diminish the total copper

loss in the armature on account of the large decrease they produce in
the copper loss in the short-circuited coils. The efficiency of the motor
is improved by their use and by diminishing sparking they make the
operation of the motor commercially satisfactory. These resistance
leads are laid in the bottom of the slots containing the armature coils.
Because of the limitation placed on the operation of the a-c series motor
by the transformer voltage in the armature coils short-circuited during
commutation, such a motor is essentially a low-frequency low-voltage
motor.

The difficulties involved in the use of resistance leads are the necessity for adequate space for them and the need to prevent their overheating and burning out. When a motor is running, any one resistance
lead is in circuit only a small part of the time. If only enough cooling

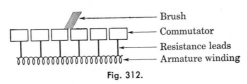

Fig. 312.

surface is provided to keep the leads cool when the motor is running,
they are liable to burn out when the motor is at rest and carrying current, or when the armature is revolving at slow speed as when starting.
Another difficulty caused by the short-circuit currents in the coils undergoing commutation is the unequal current density in the trailing and
leading edges of a brush. Since the pole flux is approximately in phase
with the armature current, the voltage produced by this flux in a coil
short-circuited by a brush is approximately in quadrature with the
armature current. Because of the presence of the resistance leads
between the armature coils and the commutator bars the current in the
coil caused by this voltage does not lag the armature current by a large
angle. Therefore the armature coils during commutation carry component currents which are in phase with the load current on one edge of
the brushes and in opposition to the load current on the other edge.
This causes a higher current density in one edge of a brush than in the
other edge. This difference may be sufficient to result in glowing at one
edge of the brushes at times of heavy load. In spite of these difficulties,
the early a-c series motor could be made to operate fairly satisfactorily.
Alternating-current series motors as at present designed often have
resistance leads.

Problems Involved in the Commutation of an A-C Series Motor.
There are two distinct commutation problems in the design of a satisfactory a-c series motor. One of these is the same as in a d-c motor and
concerns the reversal of the current in the armature coils undergoing

commutation. The other problem concerns the current caused by the voltage induced by transformer action of the main field in an armature coil short-circuited by a brush during commutation. No such voltage exists in d-c motors, but similar voltages exist in all types of a-c commutator motors. This problem is more difficult to handle successfully in an a-c series motor than in an a-c constant-speed commutator motor because a series motor does not operate at a fixed speed for a given armature current.

The time during which commutation takes place in an armature coil is so brief compared with the time of one cycle of the alternating current that the current in the line changes little in magnitude during the time required for commutation, and so far as commutation is concerned the line current can be considered constant. Commutation takes place under two limiting conditions, when the current is a maximum and when the current is zero. The limiting condition of maximum current is more severe on the commutation of a given motor when operating on alternating current than when operating on direct current, although the average condition of commutation is probably about the same in both cases. As an example of the duration of the commutation period, consider a motor with a commutator peripheral speed of 6,000 fpm. The brushes for an a-c series motor must be narrow so as not to short-circuit more than one or two armature coils at any instant. Assume that the brush thickness is 0.375 in. and that the thickness of the mica between commutator bars is 0.030 in. Under these conditions the time of commutation is

$$\frac{0.375 - 0.030}{(6,000 \times 12)/60} = \frac{1}{3,500} \text{ sec}$$

If the motor operates on 25 cycles, the time of commutation in terms of the time of one cycle is

$$\frac{1/3,500}{\frac{1}{25}} = \frac{1}{140}$$

of a cycle or about 2.6 electrical degrees. Assuming a sinusoidal current, the current during this time changes about 0.1 per cent if commutation takes place when the current has its maximum value. If the current has one-half its maximum value, the change is about 8 per cent. The reversal of the current in an armature coil during commutation is accomplished in the d-c motor by using suitable interpoles and by keeping the reactance voltage of the armature coils low. Low reactance voltage is obtained in the a-c series motor by using armature coils of a single turn arranged so that the coil sides in any slot do not commutate at the same time. Short pitch windings are also used in order to have the armature currents in the two sides of coils in the same slot reverse at different times.

Any armature coil which is undergoing commutation is linked by the full field flux (Fig. 313) and acts like a short-circuited secondary of a transformer having the main field winding as primary. The main flux induces a voltage in the short-circuited armature coil which is in quadrature with the flux and leads that flux, assuming that the voltage is a voltage drop. This transformer voltage in any armature coil is proportional to the number of turns in the coil, to the main flux, and to the frequency.

The current produced by transformer action in an armature coil must be interrupted when the commutator bars connected to the coil leave a brush. This produces severe sparking unless the current is limited by means of resistance leads (page 528) or by neutralizing the transformer voltage causing this current. The transformer voltage caus-

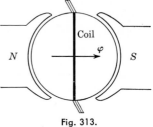

Fig. 313.

ing the current can be neutralized by introducing a speed voltage opposite to the transformer voltage in the armature coils undergoing commutation. A neutralizing voltage can be obtained by the use of an auxiliary flux in the interpolar space which is in quadrature with the main pole flux. The cutting of this flux by the armature coils short-circuited by the brushes induces a voltage in the coils in time phase with the flux. Its magnitude is given by

$$e = (\text{velocity})\beta L$$

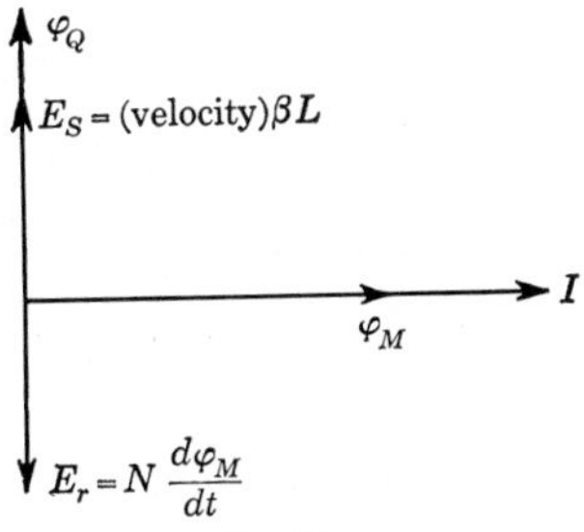

where the velocity is the peripheral speed of the armature conductors, and β and L are the flux density of the auxiliary field at the armature conductors and the embedded length of the conductors. How this transformer voltage is neutralized is illustrated in the vector diagram shown in Fig. 314, where φ_M is the flux in the main poles and φ_Q is the quadrature auxiliary flux in the commutating zone.

Fig. 314.

Since the transformer voltage is proportional to frequency and is independent of the speed of the motor, and the speed voltage is proportional to the speed of the motor and is independent of frequency, complete neutralization of the transformer voltage can take place only at a definite speed and a definite frequency for given values of the fluxes. There can be no neutralization when the speed is zero, as when starting.

Shunted Interpoles. One method for obtaining the auxiliary quadrature flux is by double feeding as illustrated in Fig. 307, page 521. This was used in many of the older type motors. A more modern and better method employs interpoles shunted with noninductive resistances which cause the interpole current and the interpole flux to lag the current taken by the motor. The component of the interpole flux which is in phase with the motor current has the same phase as the flux which would be produced by an unshunted interpole and can be made to take care of the ordinary commutation, *i.e.*, to produce the reversal of the current in the coil undergoing commutation. The quadrature lagging component of the interpole flux can be made to generate the speed voltage necessary to neutralize the transformer voltage in the short-circuited coil. Since

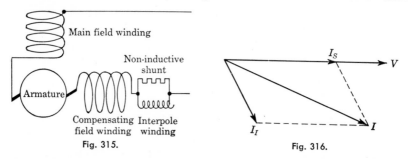

Fig. 315.   Fig. 316.

with shunted interpoles the quadrature component of the interpole flux is proportional to the main field flux except for the effect of saturation, nearly complete neutralization of transformer voltage can be obtained for a fixed speed over a wide range of current.

The diagram of connections for the shunted-interpole series motor is given in Fig. 315. The vector diagram for the interpole and its shunt is given in Fig. 316. The letters on this diagram have the following significance:

V = voltage drop across the interpole in parallel with its noninductive shunt

I_S = current in the noninductive shunt

I_I = current in the interpole winding

I = current taken by the motor

When the range of speed within which the motor is to operate is not great, satisfactory neutralization of the transformer voltage can be obtained through the operating range of speed with a single setting of the interpole shunt. The shunt is generally set to give the best commutation at somewhat more than half the maximum operating speed. When a greater range of speed is required, two or three interpole shunt settings may be necessary in order to neutralize the transformer voltage sufficiently throughout the operating range of speed in order to obtain satisfactory commutation. The necessary change of interpole shunt can be made automatic with the change in applied voltage necessary to give the required range of speed.   If the interpole shunt is adjusted for the low-operating speed range, it is necessary to insert reactance in series with the noninductive shunt when passing to the next higher speed range.   At the same time the resistance of the shunt must be reduced in order to keep the in-phase ampere-turns on the interpoles the same for a fixed line current.   It is also necessary to reduce the quadrature ampere-turns to correspond to the new range of speed.

When a-c series traction motors, usually built for 300 volts, are operated on 600-volt d-c railway circuits, two motors are connected in series and are operated as a unit.   Starting and speed control are handled in much the same way as for d-c series traction motors.

When an a-c series motor operates on direct current, the current divides between the interpole winding and its noninductive shunt inversely as their resistances instead of inversely as their impedances.   In order to maintain the proper commutating flux for d-c operation, it is necessary to reduce the shunt resistance.

General Statements Concerning A-C Series Motors.   As was stated, the a-c series motor is necessarily a low-voltage low-frequency motor with many poles.   Low voltage makes it possible to use single-turn armature coils and thus to reduce the transformer voltage in the armature coils short-circuited by the brushes during commutation.   Low frequency also reduces this voltage.   Many poles are necessary in order to make it possible to keep the flux per pole low and thus still further reduce the transformer voltage.   If a series motor is to operate without sparking, the transformer voltage must be reduced to a minimum consistent with the limitations of design.   Although this transformer voltage in a coil short-circuited by a brush during commutation can be more or less neutralized under running conditions, it cannot be neutralized at stand-still, since the neutralizing voltage, which is a speed voltage, is zero except when the motor is running.   The low voltage at which a-c series motors must operate makes it possible to use lap-wound armatures. Increasing the number of poles decreases the reluctance of the magnetic circuit of the motor and decreases the number of turns required in the main field winding for a given pole flux.   This cuts down the series reactance of the motor and increases the power factor.   The use of partially closed slots decreases the effective length of the air gap which further decreases the reluctance of the magnetic circuit.   Partially closed slots are undesirable from the standpoint of the reversal of the current in an armature coil undergoing commutation, but this reversal can still be accomplished satisfactorily by the use of properly designed shunted interpoles, short-pitch single-turn armature coils, and multiple coils per slot.   In Europe frequencies as low as $16\frac{2}{3}$ cycles are used for a-c traction motors. Such low frequencies call for heavier and more expensive transformers and generators.

Universal Fractional-horsepower Motors. The universal motor is a small series motor with essentially the same operating characteristics on alternating current and on direct current. Such motors are built in fractional-horsepower sizes in very large numbers and are usually designed for operating speeds of 3,500 rpm or even higher. In common with all series motors they have no-load speeds which are very high and are limited chiefly by the friction and windage losses and may be as high as 8,000 to 10,000 rpm. For low speeds, universal motors are provided with built-in reduction gears. When a universal motor with essentially constant speed characteristics is desired, a governor is used. Universal motors are used for portable drills, saws, home moving-picture projectors, small fans, vacuum cleaners, sewing machines, and other household appliances. They are built for voltages up to about 250 volts and frequencies as high as 60 cycles. Under normal operating conditions, their efficiencies are as high as 60 per cent and their power factors may reach 80 per cent.

Because of the small size of universal motors, the short-circuit transformer current in the armature coils undergoing commutation is small and does not present a serious problem even on 60 cycles if high-resistance brushes are used to assist in commutation and to limit the short-circuit current.

There are two general types of universal motors, compensated and uncompensated. Although the compensated type has more nearly the same operating characteristics on alternating current and direct current, it is more expensive and its construction is less simple than that of the uncompensated type. The uncompensated type is used for lower horsepower ratings and higher speeds than the compensated type. The compensated motor is of the distributed-field type. It has two windings which are at right angles to each other but by displacing the brushes slightly from the neutral axis a single winding is sometimes made to serve as both. The two-winding universal motor is really a miniature railway motor without interpoles. Universal motors in common with a-c series traction motors are designed with relatively weak fields. The uncompensated motor is of the salient-pole type and is more common than the compensated motor, especially in small sizes. The magnetic circuits of both types of motors must be laminated throughout.

A motor increases in speed until the back speed voltage is equal to the voltage impressed on the motor minus the total voltage drop in the motor. This drop is the impedance drop for a-c operation and the resistance drop for d-c operation. For fixed current and fixed impressed voltage, the speed voltage of any motor is less on alternating current than on direct current. The speed voltage in any motor is proportional to the product of speed and flux. For the same impressed voltage, the speed voltage is lower on alternating current than on direct current because of the greater voltage drop in the motor. In general a series motor runs more slowly on alternating current than on direct current. However the effective field is weaker on alternating current than on direct current, resulting from higher maximum flux density. This tends to make the motor run faster on alternating current than on direct current. The two effects of impedance drop and varying permeability of the mag-

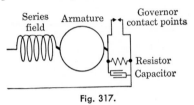

Fig. 317.

netic circuit influence the speed in opposite directions, and it is possible for a motor to operate faster on alternating current than on direct current. At light loads the effect of varying permeability on the flux and therefore on the speed is small.

When constant speed characteristics are required, a built-in governor is used to insert a small amount of resistance in series with the motor when its speed exceeds that for which the governor is set. It is customary to keep the resistance as low as possible. Instead of attempting to keep the speed constant at all loads, the resistance is usually made as low as possible without having the speed reach too high a value at light loads. A small capacitor is connected across the break in the auxiliary resistance circuit to prevent sparking at the contact points. The current handled is small and the governors cause little trouble. The diagram of connections of a motor with a governor is shown in Fig. 317.

Chapter 42

Singly Fed Repulsion Motor. The connections for a conductively compensated singly fed series motor are shown in Fig. 318. *MM* is the main field winding and *CC* is the compensating winding. The armature winding is not shown. In practice the compensating winding is placed in slots in the pole faces.

The connections for an inductively compensated singly fed series motor are shown in Fig. 319. The armature and compensating windings act as primary and secondary windings of a short-circuited trans-

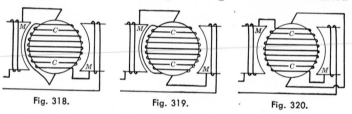

Fig. 318.   Fig. 319.   Fig. 320.

former. The operation of the motor is the same as with conductive compensation, if the magnetic circuit for the cross field is sufficiently good to prevent large magnetic leakage between the compensating and armature windings.

Instead of short-circuiting the compensating winding, this winding and the main winding can be connected in series and the armature short-circuited. This scheme of connections is shown in Fig. 320. The current in the armature is obtained by induction. The motor operates satisfactorily provided the magnetic circuit for the armature winding and compensating winding is sufficiently good to permit satisfactory transformer action.

The connections shown in Fig. 320 are those of a repulsion motor. The necessity for a good magnetic circuit along the axis of the main winding and also along the axis of the armature and the axis of the compensating winding requires non-salient poles and a uniform air gap. With non-salient poles both the main and compensating field windings must be distributed and are placed in slots in the stator. There is no particular object in using two independent windings for these two fields, and in the simple repulsion motor they are combined in a single uniformly distributed winding. The effect of two windings is obtained by displacing the brush axis from the axis of the single stator winding. The mmf of this single stator winding can then be resolved into two components, one along the brush axis, corresponding to the mmf of the compensating winding, the other at right angles to the brush axis, corresponding to the mmf of the main field winding.

The connections for a repulsion motor are shown diagrammatically in Fig. 321, in which salient poles are indicated for simplicity.

F represents the direction and magnitude of the mmf of the stator winding. This mmf is resolved into two components, T and S. The

component T, along the brush axis, corresponds to the compensating field and produces current in the armature by transformer action. It is called the transformer field. The component S, at right angles to the brush axis, corresponds to the main field of Fig. 320. It produces torque in conjunction with the current induced in the armature by the component T.

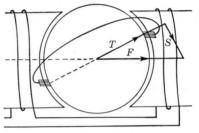

Fig. 321.

The component S cannot produce any voltage between brushes by transformer action. When the armature revolves, a speed voltage is induced in the armature between brushes by the field S. The field S is therefore called the speed field. It is also called the torque field, since in conjunction with the armature current it produces torque. The voltage induced in the armature by the field S is the speed voltage of the motor.

The relative magnitude of the two component fields due to T and S is determined by the position of the brush axis with respect to the axis of the stator winding. Increasing the displacement of the brush axis from the axis of the stator winding decreases the transformer field and increases the speed field and causes the motor to slow down. Reversing the direction of the displacement of the brushes reverses the speed field S without changing the direction of the transformer field T with respect to the armature and hence reverses the direction of rotation of the motor. The characteristics of the motor as well as its direction of rotation depend upon the position of the brush axis with respect to the axis of the stator winding.

The speed of the repulsion motor can be varied by changing the impressed voltage, as in the series motor, or by changing the position of the brush axis.

The series motor with either conductive or inductive compensation has no field along the brush axis, since the mmfs of the armature and the compensating field neutralize along this axis. Except at starting, the repulsion motor has a strong field along the brush axis. This is the essential difference between the two types of motor and is the cause of the chief difference in their operating characteristics.

The armature current of the repulsion motor results from transformer action between the armature and the component of the stator mmf which lies along the brush axis. The armature mmf and this component of the stator mmf act like the mmf in the two windings of a transformer. The load on this equivalent transformer is equal to the armature current multiplied by the speed voltage generated in the armature by its rotation in the speed field. Neglecting the magnetizing current in the stator winding for the transformer field, the armature current is proportional to the stator current and therefore to the current producing the speed field. Hence the repulsion motor has the torque and speed characteristics of the series motor. However it differs materially from the simple series motor in commutation because of the

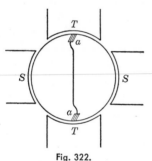

Fig. 322.

transformer voltage induced by the compensating field C (Fig. 308, page 521) in the armature coils short-circuited during commutation. This difference is due to the transformer field, which induces a speed voltage in the armature coils short-circuited by the brushes and neutralizes at synchronous speed the transformer voltage produced in these coils by the speed field.

Motor at Rest. In discussing the repulsion motor, the two component fields are replaced by two separate fields in space quadrature, one the torque-producing or speed field SS, the other the current-producing or transformer field TT, Fig. 322. The short-circuited armature brushes are shown as aa.

With the motor at rest, the component field TT in conjunction with the armature forms a short-circuited transformer. The voltage across TT is the equivalent impedance drop of this transformer.

The flux φ_s due to SS cannot produce any transformer voltage in the armature between brushes since the axis of SS is perpendicular to the brush axis. The total voltage impressed across the motor is the vector sum of the impedance drops due to the component field SS and to the short-circuited transformer formed by the component field TT and the armature. The conditions are equivalent to a short-circuited transformer in series with an impedance.

Since there cannot be any mutual induction between the field SS and the armature, the reactance of the field SS is high. The power absorbed by it is merely the copper loss in the winding and the core loss due to the flux φ_S. The resistance drop should be small compared with the total reactance drop. Consequently the magnetizing current and the flux φ_S of the field SS are nearly in time phase with the current taken by the motor. Since the component field TT and the armature act together like a short-circuited transformer, the current I_a in the armature must be very nearly in time-phase opposition to the current in the field TT. Therefore the armature current I_a and the flux φ_S must be nearly in time-phase opposition, and since the brush axis is at right angles to the axis of the flux φ_S, torque is developed and the motor speeds up. At starting, nearly all the drop in voltage through the motor is in the component field SS. Changing the brush position alters the relative number of ampere-turns in the two component fields SS and TT. The further the brush axis is moved from the axis of the stator winding, the greater is the field SS and the smaller is the field TT. For a given current, this increases the voltage drop across the motor at the instant of starting.

Motor Running. When the motor speeds up, a speed voltage E_s is produced across the brushes aa by the rotation of the armature in the component field SS. This voltage corresponds to the speed voltage of the series motor. It is in time phase with the flux φ_S, and since the stator current and φ_S are nearly in time phase, this voltage is nearly in time phase with the stator current. Neglecting the exciting current in the transformer formed by the stator winding and the armature winding, the armature current I_a is in time opposition to the stator current. Therefore E_s is nearly in time-phase opposition to the armature current. Hence the load on the transformer formed by TT and the armature is nearly noninductive. The motor under load, then, is equivalent to a loaded transformer in series with an impedance. The poles TT with the armature form the transformer. The poles SS with their winding form the impedance.

The current I_a is nearly in time-phase opposition to the stator current. If the sign of I_a is reversed in order to refer the armature impedance drop to the stator, the equation for the speed of the repulsion motor becomes

$$n = \frac{V - Iz - I_a z_a}{k\varphi_S} \tag{298}$$

This is also the equation for the speed of a simple series motor.

For any fixed current, an increase in the displacement of the brushes from the axis of the stator field increases φ_S and decreases the speed as well as the power factor. The effect is much the same as that produced by adding turns to the main field winding of a series motor.

Commutation. So far as commutation is affected by the transformer voltage induced in an armature coil undergoing commutation, the repulsion motor has an important advantage over the simple series motor. Each type of motor has a transformer voltage induced in the short-circuited armature coils. This voltage is due to the main field in the series motor and to the component SS of the stator field in the repulsion motor. At starting there is no voltage to oppose this voltage in either type of motor, and therefore there is no choice between the two types so far as commutation is concerned at starting. As the repulsion motor speeds up, however, a second voltage is induced in its short-circuited armature coils by the rotation of the armature in the component field TT. This speed voltage is nearly in time-phase opposition to the transformer voltage and is equal to it at synchronous speed. At synchronous speed therefore the commutation of the repulsion motor is perfect so far as the transformer action in the short-circuited armature coils is concerned. Between standstill and about 1.4 synchronous speed the conditions for commutation are better in the repulsion motor than in the series motor.

Comparison of Series and Repulsion Motors. There is little difference between the speed, torque, efficiency, and current curves of a simple series motor and those of a simple repulsion motor, but the commutation of a repulsion motor is inherently better between zero and 1.4 times synchronous speed. This is about the only factor in favor of motors of the simple repulsion type. This superiority in commutation is most marked near synchronous speed, and for this reason repulsion motors are designed so that their normal running speed is near synchronism. Above 1.4 times synchronous speed the commutation of the repulsion motor is inherently worse than that of the series motor. The repulsion motor has a distributed field winding without salient poles, and for this reason it requires for the same field strength a greater number of ampere-turns on its field than the series motor. Moreover the armature current is derived from transformer action. This necessitates a good magnetic circuit along the axis of the armature brushes as well as along the axis of the component field SS. For this reason a repulsion motor must be somewhat heavier than a series motor of the same speed and output. The power factor of the simple repulsion motor is less than the power factor of the series motor, since in addition to the reactive drop in the component field winding SS, Fig. 322, page 538, corresponding to the reactive drop in the main field winding of the series motor, the component field winding TT in the repulsion motor carries a quadrature current which produces the flux φ_T. There is no corresponding component in the series motor. One minor advantage of the repulsion motor is that there is no electrical connection between its armature and field windings. For this reason the field can be wound to receive power directly from high-voltage mains without the use of a transformer. This is of little practical advantage, since motors of the repulsion and series types are almost always used under conditions requiring variable speed.   Except with very small motors, economy calls for the use of a transformer for obtaining the variable voltage necessary for speed control.   If a transformer is required, there is no special advantage in being able to wind the stator for high voltage.   The series motor with shunted interpoles is superior to the repulsion motor.

At operating speeds, the fields φ_T and φ_S are in space quadrature and very nearly in time quadrature.   At synchronous speed they have been shown to be equal.   Therefore at synchronous speed the repulsion motor has a uniformly rotating field of constant strength.   At speeds other than synchronism the repulsion motor has an elliptical revolving field. Due to this rotating field, the rotor core loss at synchronous speed is small.

PART EIGHT

SYNCHRONOUS CONVERTERS

Chapter 43

Means for converting alternating current into direct current.

Means for Converting Alternating Current into Direct Current.
Alternating current can be converted into direct current by the use of
(1) mechanical rectifiers, (2) electrolytic rectifiers, (3) oxide-film rectifiers, (4) selenium rectifiers, (5) hot-cathode rectifiers, (6) multianode
mercury-arc rectifiers, (7) single-anode mercury-arc rectifiers, (8) motor
generators, or (9) synchronous converters.

1. *Mechanical Rectifiers.* On account of sparking, mechanical rectifiers are generally limited to small currents and low voltages and are of
little importance. To operate with minimum sparking, the current
must be reversed while passing through its zero value. The point of
zero current does not occur at the point of zero voltage however except
when the circuit is noninductive. In general the brushes short-circuit the
commutator when there is voltage between its two parts unless this is
prevented by providing an insulated segment between each two adjacent
live commutator segments.

Although mechanical rectifiers with polarized vibrating contacts have
been successfully used for charging small low-voltage storage batteries
where the charging current is small, such rectifiers have been largely
replaced by other types.

2. *Electrolytic Rectifiers.* An electrolytic rectifier depends on the
rectifying action of a lead plate and an aluminum plate immersed in a
solution of sodium bicarbonate or ammonium phosphate. Such an
arrangement allows current to pass from the solution to the aluminum
plate but not in the opposite direction. A reversal of current causes a
thin insulating film of aluminum oxide to form instantly on the aluminum
plate. This film acts as an insulator up to about 150 volts and prevents
the flow of current. Electrolytic rectifiers can be used for small currents
at low voltages. Their efficiency is low and they have been superseded
by rectifiers of the hot-cathode and oxide-film types.

3. *Oxide-film Rectifiers.* These rectifiers depend on the rectifying
action of a layer of copper oxide on a sheet of copper. Electrons can pass
easily from the copper to the copper oxide but not from the copper oxide
to the copper. Rectifiers of this type have a variety of applications in

circuits involving low currents and low voltages. Units of this type may be connected in series or in parallel for higher voltages or currents. They are simple and inexpensive and have an efficiency as high as 70 per cent under good conditions.

4. *Selenium Rectifiers.* Another dry-type rectifier with applications similar to those of the copper-oxide rectifier is the selenium rectifier. Selenium cells are formed by coating an iron or aluminum plate with a thin layer of selenium and heat-treating. A metal or alloy of low melting point is then sprayed onto the selenium and a thin film is formed between the selenium and the sprayed metal by electrochemical means. For a given voltage the electron flow is much greater in the direction from the sprayed metal to the selenium than in the reversed direction. This means that with an alternating emf applied, current flows largely in the direction from the selenium to the sprayed metal surface.

5. *Hot-cathode Rectifiers.* A hot-cathode rectifier consists of a hot tungsten cathode and a relatively cool anode enclosed in a glass or metal tube which either contains a gas at low pressure or is evacuated. These rectifiers depend for their action on the emission of electrons from the hot cathode. With suitable transformers, the gas-filled type are extensively used for charging small storage batteries.

6. *Multianode Mercury-arc Rectifiers.* A mercury-arc rectifier consists of two or more anodes with a mercury pool as cathode in an evacuated vessel of either glass or steel. The mercury pool serves as a source of electrons when an arc is established and maintained at the cathode. The glass-tube rectifier has been used for many years in charging storage batteries at charging rates up to about 50 amp. The glass-tube mercury-arc rectifiers have also been used in street lighting with series arc lamps which require unidirectional current and are operated from constant-current transformers. For street lighting the glass-tube rectifiers operate up to 5,000 or 6,000 volts with the usual current for series arc lamps.

Mercury-arc rectifiers with steel tanks are built in large sizes to operate on polyphase circuits. They have special airtight seals for the terminals and are operated with pumps to maintain the high vacuums (0.005 to 0.0001 mm of mercury) necessary for power rectification. Such rectifiers are built for large continuous ratings and operate 6 phase or 12 phase from 3-phase circuits through transformers. Many of these rectifiers are used for supplying 600 to 700 volts for the operation of street railways and 2,000 to 3,000 volts for locomotives on main-line railroads. The efficiency of large high-voltage rectifiers even at light loads is high and their cost of maintenance is low.

7. *Single-anode mercury-arc rectifiers.* Mercury-arc rectifiers having more than one anode are known as multianode rectifiers. A more recent development is the single-anode mercury arc rectifier requiring

special means of initiating and maintaining the arc. These single-anode rectifiers are now often used as a supply for welding equipment and, when several units are connected in polyphase arrangements, are competing with the multianode type for supplying power to railways and for other purposes.

8. *Motor Generators.* A motor generator consists of an a-c motor, of either the synchronous or induction type, which is coupled directly to a d-c generator. The chief advantage of motor generators is that their a-c and d-c sides are entirely independent. The relative merits of motor generators and synchronous converters are considered later.

9. *Synchronous Converters.* In a d-c generator the induced voltages are alternating and are rectified by means of a commutator. If taps are brought out to slip rings from equidistant points in the armature winding of a two-pole d-c generator, alternating current can be taken from the slip rings. The number of phases depends upon the number of taps and except for single phase is equal to the number of slip rings. If a multipolar d-c generator is used, there are as many taps per slip ring as there are pairs of poles, assuming a lap-wound armature such as would be used ordinarily for a synchronous converter of considerable size and voltage rating. A machine tapped in this way can be operated as a d-c motor or generator, as an a-c synchronous motor or generator or as a synchronous converter to change alternating current into direct current or to change direct current into alternating current. In most cases, synchronous converters are used to convert from alternating current to direct current. When a converter is used to change alternating current to direct current, it is called a direct converter but it is customary to drop the word "direct" and to speak of it merely as a converter or a synchronous converter. When used to convert direct current to alternating current, it is called an inverted converter.

Although any d-c generator provided with suitable taps and slip rings should be capable of operating as a converter, its proportions and design make its operation as a converter in general unsatisfactory if not impossible.

With relation to the external circuits, a synchronous converter on its a-c side has the characteristics of a synchronous motor and on its d-c side those of a d-c generator. In regard to its internal characteristics, it is radically different from either a synchronous motor or a d-c generator.

From one point of view the armature of a converter may be considered to carry the difference between the direct current delivered as generator and the alternating current received as motor. As a result the armature reaction and armature copper loss are neither those of a synchronous motor nor those of a d-c generator. Except in a single-phase converter the average armature copper loss is less than would be due to either the alternating current or the direct current alone. It follows that the output of a converter, the single-phase type excepted, is greater than the output of the same machine as a generator. This accounts for the large commutators found on synchronous converters.

The induced voltage of a d-c generator depends upon the number of inductors between adjacent brushes, the flux per pole, and the speed. The induced voltage of a synchronous generator depends upon the number of inductors between adjacent taps, the flux per pole, the speed, and the distribution of flux in the air gap. Since the a-c and the d-c induced voltages in a converter are induced by the same magnetic field and in the same armature winding, it follows that the ratio of the two induced voltages of a converter is fixed for any given flux distribution. Under ordinary conditions of operation, the distribution of the air-gap flux does not change. The synchronous converter is a machine of fixed voltage ratio.

Chapter 44

Voltage ratio of an n-phase synchronous converter.   Current relations.

Voltage Ratio of an n-phase Synchronous Converter. Let Fig. 323 represent a two-pole synchronous converter. The d-c brushes are assumed to be in the neutral plane at a and b. R is the direction of the resultant field.

Assume that the distribution of the air-gap flux is such as to produce a sine wave of induced voltage.   The voltage induced in an inductor, as at g, is

$$e = E_m \cos \Delta$$

where E_m is the maximum value of the voltage induced in a single induc-

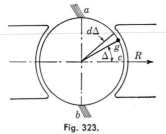

Fig. 323.

tor.   The position angle of the inductor g with reference to the resultant field R is Δ. The d-c voltage is

$$E_{dc} = \sum_{-\pi/2}^{+\pi/2} E_m \cos \Delta$$

$$E_{dc} = \int_{-\pi/2}^{+\pi/2} E_m \frac{Z}{\pi} \cos \Delta \, d\Delta$$

$$= 2 \frac{Z}{\pi} E_m$$

where Z is the number of inductors in series between brushes.

The maximum value of the voltage induced in the portion of the armature winding between any two adjacent taps occurs when the inductor at the center of that portion of the armature winding lies on the field axis R.

If there are n slip rings, the angle between two adjacent taps is $2(\pi/n)$ electrical radians.   If there are more than two poles, n is the number of taps per pair of poles.

The maximum a-c voltage between two adjacent taps is

$$E'_{ac} = \int_{-\frac{1}{2}\left(\frac{2\pi}{n}\right)}^{+\frac{1}{2}\left(\frac{2\pi}{n}\right)} E_m \frac{Z}{\pi} \cos \Delta \, d\Delta$$

$$= 2E_m \frac{Z}{\pi} \sin \frac{\pi}{n}$$

Since a sine wave of voltage is assumed, the effective or rms a-c voltage is

$$E_{ac} = \frac{2E_m(Z/\pi) \sin (\pi/n)}{\sqrt{2}}$$

The ratio of the voltages on the two sides of the converter is

$$\frac{E_{ac}}{E_{dc}} = \frac{[2E_m(Z/\pi) \sin (\pi/n)]/\sqrt{2}}{2(Z/\pi)E_m}$$

$$= \frac{1}{\sqrt{2}} \sin \frac{\pi}{n} \tag{299}$$

The actual voltage ratio of a synchronous converter differs slightly from that given by Eq. (299) on account of the effect of the distribution of the air-gap flux. The ratio of the terminal voltages under load is slightly influenced by the voltage drops in the armature winding. These drops are difficult to calculate on account of the peculiar wave form of the armature current (see Fig. 326 and Fig. 327, page 551). They are small and are usually neglected. The distribution of the air-gap flux is determined chiefly by the ratio of pole arc to pole pitch and by the shape of the pole shoes. Armature reaction plays little part in determining the distribution of the air-gap flux, since the distorting components of armature reaction caused by the a-c and the d-c components of the armature current of a synchronous converter are nearly equal and opposite under steady operating conditions and nearly cancel (see pages 559, 560, and 561). These components do not cancel when there is hunting.

The ratio of effective a-c voltage to d-c voltage for converters with different numbers of taps is given in Table 8. The ratios of voltages in this table are for a sinusoidal voltage on the a-c side. The d-c brushes are assumed to be in the neutral plane. It should be noted that the ratio of the maximum a-c voltage to the d-c voltage of a single-phase converter is unity.

TABLE 8

No. of taps per pair of poles	Number of phases	$\dfrac{E_{ac}}{E_{dc}}$
2	1	0.707
3	3	0.612
4	4	0.500
6	6	0.354
12	12	0.183

Current Relations.   Let

n = number of slip rings
I'_{ac} = coil alternating current
V_{ac} = phase a-c terminal voltage
pf = power factor
p = number of poles
I_{dc} = total direct current
V_{dc} = d-c terminal voltage
η = ratio of armature d-c power to armature a-c power

Since the input multiplied by the efficiency must be equal to the output, the following relation must hold for a two-pole converter between the power input on the a-c side and the power output on the d-c side:

$$(\text{pf})\eta n V_{ac} I'_{ac} = V_{dc} I_{dc} \tag{300}$$

If the converter is multipolar, there usually are as many parallel paths through the armature for each phase as there are pairs of poles. For such a converter Eq. (300) is

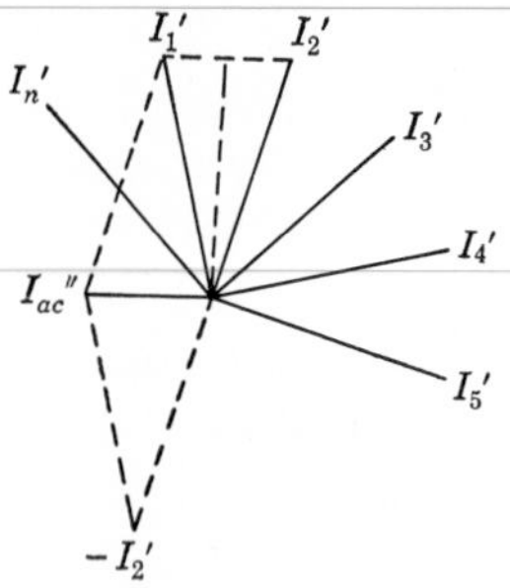

$$\frac{p}{2}(\text{pf})\eta n V_{ac} I'_{ac} = V_{dc} I_{dc}$$

and

Fig. 324.

$$\frac{(p/2)I'_{ac}}{I_{dc}} = \frac{1}{n(\text{pf})\eta}\frac{V_{dc}}{V_{ac}} \tag{301}$$

A converter must be mesh-connected since the armature winding of a converter is a d-c winding which has taps brought out for alternating current. All ordinary d-c windings are closed-circuit windings. A Y connection for the converter would necessitate an open-circuit armature winding.

Let the vectors I'_1, I'_2, I'_3, . . . and I'_n, Fig. 324, represent the coil currents of an n-phase mesh-connected converter with p poles.

The line current I''_{ac} per pair of poles, from the junction of phase 1 and phase 2, for balanced conditions is

$$I''_{ac} = 2I'_1 \sin\frac{\pi}{n} = 2I'_{ac}\sin\frac{\pi}{n}$$

In general the total line current is equal to the coil current multiplied by $2\sin \pi/n$ and by the number of pairs of poles.

Replacing the coil current in Eq. (301) by the total line current I_{ac},

$$\frac{I_{ac}}{I_{dc}} = 2\sin\left(\frac{\pi}{n}\right)\frac{1}{n(\text{pf})\eta}\frac{V_{dc}}{V_{ac}} \tag{302}$$

V_{ac}/V_{dc} is nearly equal to the ratio of the induced voltages and as an approximation may be assumed equal to this ratio

$$\frac{I_{ac}}{I_{dc}} = 2 \sin\left(\frac{\pi}{n}\right) \frac{1}{n(\text{pf})\eta} \frac{1}{\frac{1}{\sqrt{2}} \sin\frac{\pi}{n}} = \frac{2\sqrt{2}}{n(\text{pf})\eta} \tag{303}$$

Equation (303) shows that the ratio of line currents in converters is inversely proportional to the number of slip rings or inversely proportional to the number of phases except for single phase. Table 9 gives the ratio of currents on the two sides of converters with different numbers of phases. An efficiency of 100 per cent and unity power factor are assumed. The effect of armature reactance on the ratio of the voltages is neglected.

TABLE 9

No. of taps per pair of poles	Number of phases	$\dfrac{I_{ac}}{I_{dc}}$
2	1	1.41
3	3	0.943
4	4	0.707
6	6	0.471
12	12	0.236

It is evident from Table 9 that a three-phase converter having an efficiency of 94.3 per cent and operating at 100 per cent power factor has equal currents per terminal on its two sides.

Chapter 45

Copper losses of a synchronous converter. Induction heating. Inductor heating of an n-phase synchronous converter with a uniformly distributed armature winding. Relative outputs of a synchronous converter operated as a synchronous converter and as a generator. Efficiency.

Copper Losses of a Synchronous Converter. The output of all commutating machines is limited by commutation and by the heating produced by the losses. Most of the difficulties of commutation in d-c motors and generators are due to field distortion produced by armature reaction. Under steady operating conditions, polyphase synchronous converters are almost entirely free from field distortion. Since motor and generator currents are opposite when considered with respect to the induced voltage, the current carried by an armature inductor of a synchronous converter is the difference between the a-c and d-c components in the inductor. The average copper loss produced by the resultant current in the inductors is less than would be produced by either component alone, except for a single-phase synchronous converter.

The average copper loss is not the same in all the armature inductors of a synchronous converter, but varies with the position of an

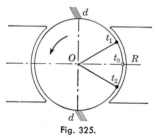

Fig. 325.

inductor with respect to the taps. The difference between the copper loss in the hottest inductor and the copper loss in the coldest inductor depends upon the number of phases for which the converter is tapped and upon the power factor at which it operates. This difference in copper loss decreases as the number of phases is increased and as the power factor is raised.

Inductor Heating. Let Fig. 325 represent the armature of a two-pole synchronous converter. The d-c brushes are dd and t_1 and t_2 are two tap inductors. t_0 is the inductor midway between the two tap inductors t_1 and t_2.

The voltage induced in the phase between t_1 and t_2 is a maximum when the axis of the field bisects the angle between the tap inductors t_1 and t_2. This occurs when t_0 is on the field axis R.

The alternating current in all inductors between t_1 and t_2 is the same at any instant, but varies as the armature rotates. The phase of the voltage generated in the winding between inductors t_1 and t_2 is the same as the phase of the voltage generated in the inductor t_0, which is midway between the two tap inductors t_1 and t_2. For unity power factor with respect to the voltage induced in the winding between t_1 and

t_2, the current is a maximum when t_0 is on the field axis R. The alternating current in all inductors between t_1 and t_2 is zero when the current in t_0 is zero. At unity power factor this occurs when t_0 is under a d-c brush.

The direct current in all inductors on the armature is the same in magnitude, but it reverses in direction in each induc-

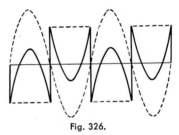

Fig. 326.

tor as it passes under a d-c brush. At unity power factor, the direct and alternating currents in the inductor t_0 reverse at the same instant. The two currents must be opposite in phase since one represents motor action and the other generator action. Neglecting the effect of the coils short-circuited by the d-c brushes, the d-c wave must be rectangular. The dotted lines in Fig. 326 show the d-c and a-c components carried by inductor t_0 when the power factor with respect to the induced voltage is unity. The full line shows the resultant current.

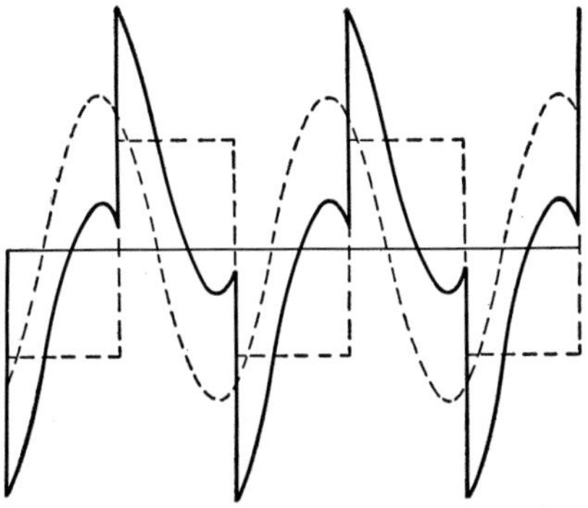

The direct current in inductor t_1 reverses when t_1 passes under a d-c brush, but the alternating current, assuming unity power factor, does not reverse until t_0 passes under the d-c brush, which for a three-phase converter occurs 60 electrical degrees later.

Figure 327 shows the resultant and component currents carried by t_1 at unity power factor in a three-phase converter. It is evident from Figs. 326 and 327 that the rms currents in inductors t_0 and t_1 are not the same.

Fig. 327.

If the current lags behind the induced voltage, the alternating current does not reverse until after t_0 has passed under a d-c brush. Considering inductor t_0, the alternating current reverses later than the direct current and the angle of lag between the reversal of the two currents is the same as the angle of lag between the alternating current and the alternating voltage in inductor t_0. If the angle of lag is 60 deg, the current relations for t_0 are the same as those for t_1 in Fig. 327, i.e., they are the same as those existing at unity power factor in an inductor 60 electrical degrees ahead of t_0.   The current relations produced in any inductor by a lagging current are the same as those which exist at unity power factor in an inductor which is ahead of the one considered by an angle equal to the angle of lag of the current behind the induced voltage.   For leading current, they are the same as those in an inductor behind the one considered by an angle equal to the angle of lead between the current and the voltage.

Inductor Heating of an n-Phase Synchronous Converter with a Uniformly Distributed Armature Winding.   Referring to Fig. 328, t_1 and t_2 are taps. c_0 is a point on the armature midway between the two taps t_1

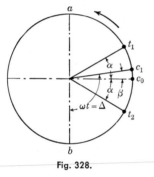

Fig. 328.

and t_2. 2α is the phase spread and is equal to $2(\pi/n)$ where n is the number of taps. When there are more than two poles, n is the number of taps per pair of poles.

The resultant current in any indicator such as c_1 is

$$\sqrt{2}\, I'_{ac} \sin (\Delta - \beta - \theta) - \frac{I'_{dc}}{2} \quad (304)$$

where I'_{ac} is the coil value of the alternating current, I'_{dc} is the direct current delivered per brush, and θ is the angle of lag between the alternating current and the induced voltage in the coil c_0.   Each path between any pair of brushes carries $1/p$ of the total direct current or one-half the direct current delivered per brush, where p is the number of poles.

The average heating in the inductor c_1 during a cycle is proportional to the mean-square current or to

$$I_c^2 = \frac{1}{\pi} \int_{\Delta=0}^{\Delta=\pi} \left[\sqrt{2}\, I'_{ac} \sin (\Delta - \beta - \theta) - \frac{I'_{dc}}{2} \right]^2 d\Delta \quad (305)$$

Replace I'_{ac} by its value in terms of I'_{dc} from Eq. (301), page 548, remembering that the current I'_{dc} per brush is equal to the total direct current I_{dc} divided by the number of pairs of poles.   Then

$$I'_{ac} = I'_{dc} \frac{1}{(\text{pf})\eta n} \frac{\sqrt{2}}{\sin \dfrac{\pi}{n}}$$

$$I_c^2 = \frac{I'^2_{dc}}{4\pi} \int_{\Delta=0}^{\Delta=\pi} \left[\frac{4 \sin (\Delta - \beta - \theta)}{(\text{pf})\eta\, n \sin \dfrac{\pi}{n}} - 1 \right]^2 d\Delta$$

$$= \frac{I'^2_{dc}}{4} \left[\frac{8}{(\text{pf})^2 \eta^2 n^2 \sin^2 \dfrac{\pi}{n}} + 1 - \frac{16 \cos (\beta + \theta)}{(\text{pf})\eta \pi n \sin \dfrac{\pi}{n}} \right] \quad (306)$$

Since the first term in Eq. (306) is constant, the average copper loss in an inductor, such as c_1, Fig. 328, has a maximum value when the last term has either its minimum positive or its maximum negative value. This term is negative when $(\beta + \theta)$ is greater than either ± 90 deg. It is evident that under ordinary conditions the maximum average copper loss occurs at one of the tap inductors of each phase. At unity power factor the copper loss in all tap inductors is the same. Under this condition, the minimum copper loss occurs in inductors midway between taps. Except in single-phase converters, which are used in practice only in small sizes, the last term of Eq. (306) is not likely to be negative under commercial operating conditions since converters are never operated at low power factor. The power factor of a converter under load conditions is seldom allowed to get below 0.9.

TABLE 10

Number of phases	Ratio of maximum to minimum inductor heating		
	Power factor = 1	Lagging current, power factor = 0.9	Leading current, power factor = 0.9
1	6.6	7.4	7.4
3	5.3	8.1	8.1
4	3.6	6.8	6.8
6	2.2	4.9	4.9
12	1.3	2.8	2.8

The ratio of the maximum to the minimum inductor heating in 3-phase, 4-phase, 6-phase, and 12-phase converters for unity power factor and for 0.9 power factor, for both lagging and leading currents, is given in Table 10. An efficiency of conversion of 100 per cent is assumed.

The ratio of the temperatures of the hottest and coldest inductors is much less than the ratio of the copper losses in Table 10 on account of the tendency for the temperature of the inductors to become equalized by heat conduction through the end connections and across the armature teeth.

The copper loss in inductors at different points on the armature of a converter is plotted in Fig. 329. All four curves are for the same converter operated at a fixed total armature copper loss. The efficiencies and relative outputs for the conditions shown are given on the plots.

Relative Outputs of a Synchronous Converter Operated as a Synchronous Converter and as a Generator. The ratio of the copper loss in the armature of an n-phase synchronous converter to the copper loss in the same machine when operated as a d-c generator is given by the ratio of average mean-square current carried by an armature inductor under the two conditions for the same d-c output. This statement assumes that the armature resistance of a converter is the same when the converter operates as a d-c generator and as a converter. Because of the badly distorted current carried by the armature when the machine

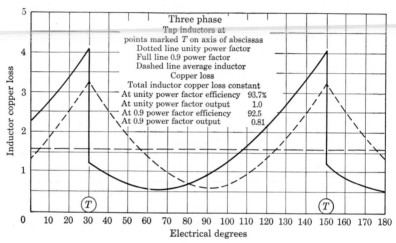

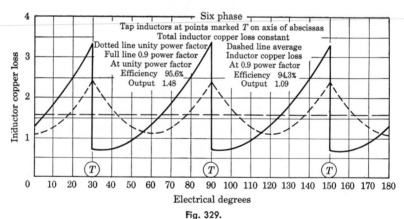

Fig. 329.

operates as a converter, the effective armature resistance must be greater under this condition than when the armature carries only direct current. The difference in the armature resistance under the two conditions is neglected in the following discussion. For this reason the ratio of outputs is actually not as great as given by Eq. (309) and in Table 11.

Assuming that the armature resistance is the same when the machine operates as a converter and as a d-c generator, the ratio of the copper losses under the two conditions is given by Eq. (307). In deriving this equation it must be remembered that the current per inductor of the d-c generator is one-half the current per terminal.

$$H = \frac{1}{\pi}\frac{n}{2}\int_{-\pi/n}^{+\pi/n}\left[\frac{8}{(\text{pf})^2\eta^2 n^2 \sin^2\frac{\pi}{n}} + 1 - \frac{16\cos(\beta+\theta)}{(\text{pf})\,\eta\pi n\,\sin\frac{\pi}{n}}\right]d\beta \qquad (307)$$

$$H = \frac{8}{(\text{pf})^2\eta^2 n^2 \sin^2\frac{\pi}{n}} + 1 - \frac{8}{(\text{pf})\eta\pi^2\sin\frac{\pi}{n}}\left[\sin\left(\frac{\pi}{n}+\theta\right) + \sin\left(\frac{\pi}{n}-\theta\right)\right]$$

$$= \frac{8}{(\text{pf})^2\eta^2 n^2 \sin^2\frac{\pi}{n}} + 1 - \frac{16}{\pi^2\eta} \qquad (308)$$

The ratio of the outputs for the same copper loss in the armature is the reciprocal of the square root of the ratio of the copper losses for the same output. Therefore

$$\frac{\text{Output of } n\text{-phase synchronous converter}}{\text{Output of d-c generator}} = \frac{1}{\sqrt{H}} \qquad (309)$$

Neglecting the difference between the effective armature resistance as a converter and as a d-c generator, the output of a converter compared with the output of the same machine as a d-c generator is given by Table 11.

TABLE 11

Number of phases	Ratio of outputs as converter and as d-c generator assuming 100 per cent efficiency	
	Unity power factor	0.9 power factor
1	0.85	0.74
3	1.33	1.09
4	1.65	1.28
6	1.93	1.45
12	2.18	1.58
∞	2.29	1.62

The gain in output by increasing the number of phases decreases rapidly as the power factor decreases. If converters are operated at low power factors, little is gained by increasing the number of phases.

However it is seldom that the power factor of a converter in commercial operation is as low as 0.9.

The decrease in output with power factor for 3-phase, 6-phase, and 12-phase converters is shown by Table 12.   The output of the converters as d-c generators is taken as unity.

TABLE 12

Power factor in per cent	Ratio of outputs as converter and as d-c generator assuming 100 per cent efficiency		
	3-phase	6-phase	12-phase
100	1.33	1.93	2.18
95	1.20	1.65	1.83
90	1.09	1.45	1.58
85	0.99	1.28	1.38
80	0.90	1.14	1.22

The results in Table 12 are shown plotted in Fig. 330.   The outputs at different power factors expressed in per cent of the output at unity power factor are plotted in Fig. 331.

From Fig. 331 it is evident that, for a fixed armature copper loss, the percentage decrease in output produced by a decrease in power factor increases slightly as the number of phases is increased.

The difference between the temperatures of the hottest and coldest inductors in the armature of a synchronous converter is less than the difference between the copper losses in these inductors on account of the equalization of temperature by conduction through the end connections and the armature core.   In spite of this tendency to equalization, there is still considerable difference between the temperatures of the hottest and coldest inductors under operating conditions.   This difference should be considered when determining the proper rating for a converter.   Since the difference in temperature decreases with an increasing number of phases, converters can more safely be given ratings determined by their average inductor heating as the number of phases is increased.   For this reason the actual gain in output by increasing the number of phases is greater than that indicated in Table 11.

Efficiency.   Since the output of a polyphase synchronous converter for given losses is greater than the output of the same machine operated either as a d-c or an a-c generator, it follows that the efficiency of a polyphase machine when operated as a converter is greater than when operated as a generator.

Table XIII gives the armature efficiencies and outputs at unity power factor of a synchronous converter when operated with different

numbers of phases and when operated as a d-c generator. These efficiencies and outputs neglect the increase in commutator friction loss and
commutator loss as the number of phases is increased. The difference
between the armature resistances as a converter and as a d-c generator
due to the difference between the skin effects in the armature conductors
under the two conditions is also neglected. No account is taken of the
effect of the uneven distribution of the armature copper loss. The

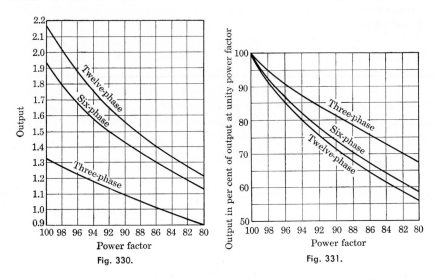

Fig. 330.   Fig. 331.

output for the d-c generator is taken as 100 per cent and the efficiency as
92 per cent.

TABLE 13

Machine	Output in per cent	Efficiency in per cent
Direct-current generator.	100	92.0
Three-phase synchronous converter.	133	93.7
Four-phase synchronous converter.	165	94.7
Six-phase synchronous converter.	193	95.6
Twelve-phase synchronous converter.	218	96.0

The increase in the efficiency of converters of the same rated output
would not be so great as the increase shown by Table 13. For example,
compare the efficiency of a 500-kw six-phase converter with a 500-kw
generator of the same speed. The converter would have an output as a
generator of (1/1.93) 500 or of about 250 kw. A 500-kw generator would
have an armature efficiency of about 95.2 per cent. The armature

efficiency of a 250-kw generator would be about 90.5 per cent.   Assuming that the 500-kw converter, when operating as a generator, has an armature efficiency of 90.5 per cent, its armature losses as a converter would be $9.5/1.93 = 4.9$ per cent.   Its armature efficiency as a converter would be 95.1 per cent or substantially the same as the efficiency of the 500-kw generator.   Although the efficiency of a converter may not be greater than the efficiency of a generator of the same rating, the efficiency of the converter is greater than the over-all efficiency of the generator and the motor required to drive it.

Whether the cost of a converter per kw rating is decreased by increasing the number of phases depends upon the relative cost of the labor and the material used in its construction.   The ratio between labor and material costs increases rapidly as the output is decreased, and below outputs of 100 or 200 kw, the cost of adding extra slip rings and increasing the over-all length to provide for these rings usually more than offsets the saving in material in other parts of the converter.

Chapter 46

Armature Reaction.   For convenience in considering the armature reaction of a synchronous converter, let the armature current be divided into four components:

1. The direct current I_{dc}
2. The component I_q of the alternating current in quadrature with the generated voltage
3. The component I_l of the alternating current opposite in phase to the induced voltage and supplying the rotational losses
4. The remainder I_e of the alternating current opposite in phase to the generated voltage and effective in producing the d-c output. I_e is the alternating current the converter would carry at unity power factor if the efficiency were 100 per cent.

Assume that the converter is lap-wound, has p poles and N uniformly distributed armature turns.   The turns per pole are N/p and the direct current is $I'_{dc} = I_{dc}/p$.   The ampere-turns per pole per elementary angle $d\varphi$ on the armature are

$$\frac{I_{dc}}{p} \frac{N}{p} \frac{d\varphi}{\pi}$$

The mmf of the turns included in the elementary angle $d\varphi$ is uniform in its space distribution and can be represented by a rectangle with a height equal to the ampere-turns.   Replace this rectangle by its equivalent Fourier series and reject the harmonics in the series as was done when considering the armature reaction of a synchronous generator on page 208.   The amplitude of the fundamental is $4/\pi$ times the height of the rectangle.   The rejected harmonics nearly cancel when the whole winding is considered.   Since the fundamental is sinusoidal in its space distribution, it can be represented by a space vector.   This vector is

$$\frac{4}{\pi} \frac{I_{dc}}{p} \frac{N}{p} \frac{d\varphi}{\pi}$$

Resolve this vector into two components, one along the axis of the resultant field and the other at right angles to this axis.   The sum of the components along the field axis taken over any pair of poles is zero. The components at right angles to the axis have the same sign and add directly. If φ is the angular displacement of any armature coil from the field axis (Fig. 332), the sum of these components is

$$\frac{4 I_{dc}}{\pi} \frac{N}{p} \frac{1}{p\pi} \int_{-\frac{\pi}{2}}^{+\frac{\pi}{2}} \cos \varphi \, d\varphi = \frac{8 I_{dc}}{\pi^2} \frac{N}{p} \frac{1}{p} \tag{310}$$

This is the magnitude of the vector that represents the d-c armature reaction per pole. It is at right angles to the resultant field provided the brushes are in the neutral plane.

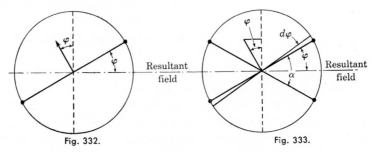

Fig. 332.   Fig. 333.

The magnitude of the vector that represents the armature reaction per pole of any one of the a-c components is

$$\frac{0.90NI}{p} \tag{311}$$

where I is the component of the coil current considered. Equation (311) gives the reaction for a concentrated winding, and to apply the equation to the distributed winding of the converter, it must be corrected for phase spread $2\pi/n$.

Let α be the phase spread in electrical radians, let the phase contain N' turns per pole, and let each turn carry a current I'. If all the turns are concentrated, the reaction per pole per phase is $0.90N'I'$. Referring to Fig. 333, it is evident that if the turns are distributed the reaction becomes

$$\int_{-\alpha/2}^{\alpha/2} \frac{0.90N'I'}{\alpha} \cos \varphi \, d\varphi = 0.90N'I' \frac{\sin (\alpha/2)}{\alpha/2}$$

The correction factor is $k_b = \dfrac{\sin (\alpha/2)}{(\alpha/2)}$ which is the breadth factor for a uniformly distributed winding. The pitch of the armature winding of a synchronous converter is so nearly unity that the pitch factor can be assumed to be unity. Applying the correction factor k_b to expression

(311) gives

$$\frac{0.90NI}{p} \frac{\sin(\pi/n)}{\pi/n} \tag{312}$$

for the magnitude of the vector that represents the armature reaction of any component, such as I, of the coil current.

The ratio of the total direct current to the coil alternating current in a converter, assuming an efficiency of 100 per cent and a power factor of unity, is

$$\frac{I'_{ac}}{I_{dc}} = \frac{2\sqrt{2}}{pn \sin \pi/n}$$

where I'_{ac} is the coil alternating current [see Eqs. (299) and (301), pages 547 and 548].

Substituting the current I'_{ac} in the expression (312) gives the armature reaction due to the component (4), the remainder of the alternating current, equal to

$$\frac{0.90Nn}{p\pi} \frac{2\sqrt{2}I_{dc}}{pn \sin \pi/n} \sin \frac{\pi}{n} = \frac{8}{\pi^2} \frac{I_{dc}}{p} \frac{N}{p} \tag{313}$$

This is equal to the d-c armature reaction and leads the resultant field by 90 deg. Therefore the armature reactions of components (1) and (4) neutralize and there remain only the reaction of the component of the alternating current supplying the rotational losses and the reaction of the quadrature component of the alternating current. This statement in regard to the neutralization of components (1) and (4) assumes that the polyphase alternating currents are balanced.

The armature reaction of the component I_l supplying the rotational losses leads the resultant field by 90 deg and produces field distortion. This distortion is small and is nearly constant since the rotational losses do not vary greatly under the usual operating conditions.

The armature reaction due to the quadrature current I_q lies along the field axis and either strengthens or weakens the field according as it lags or leads the voltage required to balance the induced voltage.

The armature reaction of a single-phase synchronous motor is pulsating. Therefore that part of the armature reaction of a single-phase converter due to the alternating current is not constant. Although the resultant of armature reactions due to 1 and 4 is on the average zero, the actual armature reaction fluctuates with double frequency between limits of plus and minus the d-c armature reaction. This is because the maximum value of I_e is equal to twice I_{dc} and is opposite to it. Owing to the presence of this fluctuating cross field, it is not possible to make the commutation of a single-phase synchronous converter so good as the commutation of a polyphase synchronous converter.

Because of the neutralization of the armature reactions caused by the direct current and the load component of the alternating current, synchronous converters can be designed with a much larger ratio of armature ampere-turns to field ampere-turns than can be used for d-c generators. Field distortion does not limit the output of a synchronous converter as it does that of a d-c generator.

Commutating Poles. All synchronous converters of any considerable size are designed with interpoles which materially improve the operation. Since there is no field distortion under steady operating conditions, the interpoles on a synchronous converter need only produce a field sufficient to take care of commutation, *i.e.*, to cause the reversal of the current in the armature coils during commutation. For this reason the interpoles required for a synchronous converter are only about 15 or 20 per cent as strong as those required on a d-c generator. A synchronous converter with interpoles should have some device for lifting the d-c brushes while the converter is being brought up to speed by alternating current. This is discussed under Methods of Starting Converters.

Because of the relatively weak commutating poles required for a converter, it is possible to use high-reluctance nonmagnetic shims between the interpoles and the frame and still get a sufficient number of turns on the interpoles to give the necessary flux for commutation. By the use of high-reluctance shims, the relation between interpole flux and interpole current can be made nearly linear over a much wider range of load than for a d-c generator where the interpoles not only have to take care of commutation but also have to neutralize armature reaction in the commutating zone. The use of interpoles with nonmagnetic shims makes it possible to design synchronous converters which can carry heavy overloads without sparking.

Methods of Starting Converters. A synchronous converter can be started by any of the following methods:

1. From its d-c side as a shunt motor

2. By means of an auxiliary motor mounted on its shaft

3. From its a-c side as a polyphase induction motor. This last assumes a polyphase converter

As a Shunt Motor. When sufficient d-c power is available, a converter can be run up to speed as a shunt motor and then synchronized on its a-c side like a synchronous generator. If a compound converter is to be started in this way, its compound field should be short-circuited to prevent weakening or even reversal of the field flux due to the starting current in the compound winding which acts differentially while the converter is operating as a d-c motor.

All switching on the a-c side of a converter is usually done on the high-tension side of its step-down transformers. Under this condition the secondary windings of the transformers are permanently connected to the slip rings of the converter and form a more or less complete short circuit on the armature when the converter is started as a shunt motor. This short circuit greatly increases the starting current and makes starting as a shunt motor difficult. Certain transformer connections in which the secondary windings are connected directly across an electrical diameter of the armature are notably bad in this respect.

By Means of an Auxiliary Motor. A small induction motor is always used for this purpose. In order to get the converter up to synchronous speed, the induction motor must have fewer poles than the converter, usually two fewer. The converter is brought up to speed by the small induction motor, synchronized and connected to the line on its a-c side. The proper instant to close the line switch must be determined by some form of synchroscope. This method of starting is seldom used.

As an Induction Motor. The shunt field is opened and from one-third to one-half normal voltage is applied to the a-c slip rings. When the converter reaches synchronous speed, its shunt field is closed and full voltage is applied to the armature. The major part of the starting torque is produced by the induction-motor action in the dampers.

To prevent puncture of the shunt-field winding by the high voltage induced in it by the armature-reaction field sweeping by the poles during starting, the shunt-field winding should be opened in several places by means of a sectionalizing switch until synchronous speed has been reached. If the converter has a series field which is shunted, the series field with its shunt forms a closed circuit about the poles. To prevent excessive current in the series field winding or its shunt by the transformer action in them during starting, either the series field or its shunt should be opened until the converter reaches synchronous speed.

The polarity of a converter brought up to speed as an induction motor cannot be predetermined. It is fixed by the direction of the armature reaction with respect to the poles at the instant of closing the field circuit. To overcome this difficulty, the field can be polarized by exciting it separately from some source of d-c power of fixed polarity. Another method of fixing the polarity is to connect a d-c voltmeter, having the zero point in the middle of the scale, across the terminals of the converter. As synchronous speed is approached, the voltmeter needle swings slowly back and forth through the zero point. The field should be closed just as the voltmeter needle starts to swing in the direction indicating the correct polarity.

A synchronous converter always sparks while coming up to speed as an induction motor since every time the armature-reaction field passes through the brush position the brushes are on active coils. As a rule, the sparking is not sufficient to cause damage unless there are interpoles, since without interpoles the reluctance for the path of the armature-reaction field is high at the instant when the field passes through the brush position and the flux is consequently low. If however interpoles are used, the reluctance of the path in the direction of the interpoles is low and bad sparking results.

If a converter with interpoles is to be brought up to speed from its a-c side, its d-c brushes must be lifted by some form of brush-lifting device to prevent short circuit of the armature coils which have voltage induced in them by the interpole flux caused by the armature reaction. One brush in each stud can be made narrow and left on the commutator to provide the necessary current for exciting the shunt field.

When a converter is brought up to speed as an induction motor it usually is pulled into synchronism by the flux produced in the poles by armature reaction. The voltage induced in the armature winding by this flux causes the converter to build up when its shunt-field circuit is closed. If the flux produced by the shunt field opposes that produced by armature reaction, there is nothing to hold the converter in synchronism and it starts to slow down. It continues to slow down until it has slipped approximately 180 deg, when the armature reaction has reversed the polarity of the converter and caused it to lock in synchronism. The converter again starts to build up but again the shunt field opposes the field owing to armature reaction and neutralizes it. This action is repeated until the shunt-field connections are reversed. Every time the converter slips 180 deg, there is bad sparking at the d-c brushes.

Chapter 47

Transformer connections.   Methods of controlling voltage.

Transformer Connections.   Since the voltage ratio of a synchronous converter is fixed by the number of taps for which it is connected, transformers are necessary in order to operate a synchronous converter from a line of standard voltage.

Most of the transformer connections given in Chap. 10 can be used, but certain of these are more common than others.   The method of connecting the primaries of the transformers is immaterial so far as the converters are concerned.

Either Δ or Y connection can be used for the secondaries for a three-phase converter.   For a six-phase converter, the diametrical connection is the one most often used when a neutral is not required on the d-c side.   This becomes the double-Y connection if the secondaries are interconnected at their mid-points.   The primaries should be connected in Δ unless a low-impedance closed path is provided by the secondaries for the third harmonics necessary in the exciting currents of transformers.

Since the two sides of a synchronous converter are in electrical connection, the neutral point of each side must also be the neutral point of the other side.   Therefore the neutral point on the d-c side for a three-wire system can be taken from the neutral point of the secondaries of the transformers, if they are connected in star.

The simple Y connection cannot be used to supply the neutral on the d-c side for a three-phase converter on account of the saturation of the cores of the transformers by the direct current in their secondary windings.   If a neutral is required on the d-c side of a three-phase synchronous converter, each of the transformers supplying it must have two identical secondaries and these must be connected in such a manner that the magnetic actions of the direct current neutralize in the two secondaries of the same transformer.   About 15 per cent more copper is required for this arrangement than for the simple Y connection.

The unbalanced direct current returning on the neutral divides equally between the secondaries connected at the neutral point.   The direction of these currents is shown in Fig. 334.   The left-hand diagram shows the simple Y connection.   The right-hand diagram shows the zigzag connection with the secondaries connected to avoid the change in the magnetic density of the core produced by the direct current with the simple Y connection.

In the right-hand figure, the two secondaries on the same transformer, as for example 1 and 1′, carry direct currents which flow in opposite directions and neutralize in so far as their magnetic effect on the core is concerned.

Any diametrical connection of secondaries with the middle points of all secondaries interconnected avoids this magnetic unbalancing.

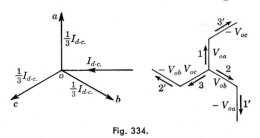

Fig. 334.

The four-phase star connection and the six-phase double-Y or double-T connection come under this class.

Methods of Controlling Voltage. The voltage ratio of any synchronous converter, except the split-pole type, is little affected by load and excitation. Any required variation in the terminal voltage must be produced externally to the converter.

If a converter is to deliver direct current, the following methods are available for controlling the d-c voltage. All of these change the voltage on the d-c side by altering the voltage impressed on the a-c slip rings. The methods are: (1) by means of a synchronous booster, (2) by means of an induction regulator, and (3) by varying the effect of the impedance drop in a series reactance by altering the power factor at which the converter operates.

A d-c booster might be used but its expense would be prohibitive. Either method 1 or method 2 can also be used to control the voltage on the a-c side of an inverted converter.

1. *Synchronous Booster.* The synchronous booster is a low-voltage synchronous generator of the revolving-armature type with its armature keyed to the shaft of the synchronous converter. The booster must be wound for the same number of phases as the converter and must have the same number of poles. It is keyed to the shaft in such a position that the axes of its poles coincide in their angular positions with the axes of the poles of the converter. The windings of the booster are connected in series with the corresponding windings of the converter. The voltages of the booster either add to or subtract from

the voltages impressed on the converter according to the direction of the field excitation of the booster. The exciting current for the booster field is often passed through an auxiliary winding on the field of the converter. If this auxiliary winding is properly adjusted, the changes in excitation produced by it, when the booster voltage is varied, maintain the power factor of the unit nearly constant. If the converter has interpoles, an auxiliary winding on them connected in series with the booster field can be made to neutralize the distorting armature reaction caused by the mechanical load put on the converter by the booster. The booster acts either as generator or motor and takes mechanical power from the converter or delivers mechanical power to it according as the booster raises or lowers the voltage.

The chief advantage of the synchronous booster is its flexibility. By its use, the control of voltage and the control of power factor are independent of each other. It is somewhat more expensive than an induction regulator and considerably more expensive than the series reactance required in the third method of controlling voltage.

Induction Regulator. The induction regulator was described on page 398. Polyphase regulators are used with converters. The regulator is placed usually between the converter and its transformers, and in this position it must be wound for the same number of phases as the converter.

Series Reactance. Controlling the voltage of a line by the use of series reactance was mentioned on page 350. When reactance is used, it is generally placed in each line between the transformers and the converter. The voltage is controlled by varying the power factor of the converter by means of the field excitation. This method of voltage control is not practical except for small changes in voltage. Any attempt to get a greater change than about 10 per cent above or below normal voltage materially lowers the output of a converter on account of the large decrease in output caused by a decrease in power factor (see Table 12, page 556). By properly compounding the converter, the voltage regulation can be made nearly automatic. If the transformer taps are adjusted to give the proper a-c voltage at unity power factor and full load, the heating at less than full load is greater than would exist if unity power factor were maintained at the lower load, but this is of little consequence and the overloads in general are of too short duration to cause serious heating. The chief advantages of the series-reactance method of controlling the voltage of the converter are its simplicity and its moderate cost. Its disadvantages are its lack of flexibility and its limited range. Neither the power factor nor the voltage can be controlled independently. Each is fixed by the other. This method of voltage control is generally employed for synchronous converters used for street railways.

Chapter 48

Inverted converter.   Double-current generator.   Motor generators versus synchronous converters.

Inverted Converter.   When a converter is operated inverted, *i.e.*, when it is used to transform from direct current to alternating current, certain difficulties arise which are not present when a converter is used to transform from alternating current to direct current.   In the latter case, the speed is fixed and any change in the excitation merely alters the power factor.   If however the converter is operating in parallel with others, a change in its field excitation may also change its load.

The conditions existing in an inverted converter are very different from those which exist in a converter delivering direct current.   An inverted converter operates as either a shunt motor or a compound motor and its a-c frequency is dependent upon the strength of its field. An inductive load weakens the field and causes the frequency of the inverted converter to increase.   This increase in frequency increases the inductive reactance of the load, increases the lag of the current, and still further increases the speed.   The action is cumulative.   If a large inductive load is thrown on an inverted converter, there is a marked tendency to race.   For this reason a converter which operates inverted must be provided with some form of speed-limiting device.   A converter not run inverted but in parallel with others or with a storage battery may under certain conditions become inverted, as for example when a short circuit occurs on the a-c line.   For this reason it is customary to use speed-limiting devices with all synchronous converters.

Certain electrical devices can be used to check the tendency of an inverted converter to speed up when an inductive load is applied.   For example, a separate shunt exciter mounted on the shaft of the converter checks this tendency, provided the exciter operates with low saturation under normal conditions in order for its voltage to be sensitive to an increase in speed.   Any tendency of the converter to race produces a rapid increase in the exciter voltage which increases the excitation of the converter and tends to check the change in speed.

The voltage ratio, heating, output, and efficiency of a converter are substantially the same whether it is operated direct or inverted.

The difficulties in the operation of inverted converters are due to instability of speed under inductive loads. Since practically all power stations now generate alternating current, there is no call for the conversion from direct current to alternating current. Inverted converters are therefore no longer used in practice.

Double-current Generator. If a synchronous converter is driven mechanically, it is capable of delivering direct current or alternating current, and it may deliver both. When equally loaded on its two sides, its total output as determined by the average copper loss in its armature, except when operated single phase, is slightly greater than its output as a d-c generator. The gain in output is only about 6.6 per cent at unity power factor even for six-phase connection.

If the voltages on the two sides of a double-current generator are to be controlled independently, some device external to the generator, such as an induction regulator, must be used for varying the a-c voltage. The d-c voltage is changed by the field excitation. This also affects the voltage on the a-c side.

Equation (308), page 555, which gives the ratio of the copper loss of a synchronous converter to the copper loss which would be produced by the d-c component of the armature current acting alone, can be made to apply to a double-current generator when the a-c output and the d-c output are equal, by putting the efficiency equal to unity, changing the sign of the last term, and dividing the right-hand member of the equation by 4. The division by 4 is necessary because the d-c output is only one-half the total output, instead of the whole output as in Eq. (306), page 552.

Motor Generators versus Synchronous Converters. In making a comparison of the relative merits of motor generators and synchronous converters, only motor generators with synchronous-motor drives are considered. Motor generators are not started under load and large starting torque is not required. Since large starting torque is not required, synchronous motors are usually preferable to induction motors on account of the high power factor at which synchronous motors can be operated. They also permit of power-factor control. A direct-connected exciter can be provided for the excitation of the synchronous motor.

The constancy of speed of the synchronous-motor drive is a slight advantage when motor generators are to be operated in parallel on their d-c sides, as it eliminates one factor which determines the division of load, *viz.*, the difference in the speed characteristics of the motors.

The reliability of a synchronous converter is not so good as the reliability of either the motor or the generator. When it is considered that the reliability of a motor generator depends upon two machines instead of upon one as in a synchronous converter, there probably is not a great deal of difference in the reliability of the two.

The factor of safety in regard to insulation is in favor of the synchronous converter, which is of necessity a low-voltage machine. The motor generator would probably be wound for full-line voltage up to about 13,200 volts. The reliability of the transformers used with a synchronous converter is so great that it requires no special consideration.

Flashover at the commutator is much more likely to occur with a synchronous converter than with a motor generator. There should be no great trouble experienced from this under ordinary conditions of operation, even at 60 cycles, provided the converters are properly designed. Insulating radial barriers placed close to the commutator on each side of the brushes are often used to prevent flashover.

One important advantage of the motor generator is the independence of the two sides of the system. The a-c and d-c voltages are independent. A variation of the power factor has no effect on the d-c voltage. A change in the frequency alters the speed of the driving motor and changes the voltage of the d-c generator. No such change in voltage takes place when the frequency impressed on a converter varies. The two sides of a converter with a series synchronous booster can be made as nearly independent as the a-c and d-c sides of a motor generator.

The efficiency of a synchronous converter is higher than the efficiency of a motor generator of corresponding speed and capacity, but when comparing the efficiencies of the two it is necessary to include the losses in the transformers and also in the series reactances, the synchronous booster, or the induction regulator in case any of these is used. Assuming that no transformers are used with a motor generator, the efficiency of a 25-cycle motor generator is 6 to 8 per cent less than the efficiency of a corresponding synchronous converter with its necessary accessories. In 60-cycle apparatus the difference is 3 to 6 per cent. The copper losses of a motor generator do not increase so rapidly with decreasing power factor as do the copper losses in a synchronous converter.

A 60-cycle synchronous converter with its transformers and other auxiliary devices usually requires somewhat less floor space than a motor generator without transformers and costs 25 to 30 per cent less.

Chapter 49

Machine. Field excitation. Efficiency.

Machine. A 1,000-kw 60-cycle 600-volt (d-c) synchronous converter is used. The data relating to this converter are:

Rating.. 1,000 kw
Direct-current voltage........................ 600 volts
Alternating-current voltage (diametrical)........ 424 volts
Direct-current output......................... 1,667 amp
Number of phases............................. 6
Frequency.................................... 60 cycles
Poles.. 12
Speed.. 600 rpm
Number of armature slots..................... 180
Inductors per slot............................ 6
Armature resistance at 25°C
   Between d-c terminals...................... 0.00589 ohm
   Between a-c diametrical terminals............ 0.00589 ohm
Shunt turns per pole......................... 864
Series turns per pole......................... 2
Resistance at 25°C of shunt field............. 39.7 ohms
Resistance at 25°C of series winding.......... 0.000610 ohm
Friction and windage loss..................... 8.1 kw

The open-circuit saturation curve and the curve of core loss are plotted in Fig. 335.

Field Excitation. The armature of a synchronous converter carries a current equal to the difference between the components due to the d-c output and the a-c input. As a result the voltage drop in the armature is relatively small and can be neglected when calculating the field excitation and efficiency.

The distorting components of the armature reaction nearly neutralize and need not be considered. The only component of the armature reaction which must be taken into account is that due to the reactive component of the alternating current. This component either strengthens or weakens the field without producing distortion. The ampere-turns corresponding to it add directly to the excitation of the shunt and series fields or subtract directly from it. In a converter delivering direct current, a reactive lagging component of the alternating current strengthens the field.   The reactive component of a leading current weakens the field.

The resultant or net ampere-turns of excitation for any terminal voltage under load conditions are approximately equal to the ampere-turns necessary to produce the required voltage when the converter is driven as a generator at no load.

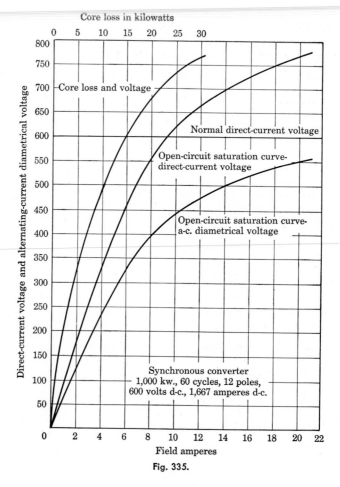

Fig. 335.

The efficiency of a synchronous converter operating at a power factor in the neighborhood of unity is always high at full load.   For a large converter it is usually 95 per cent or better.   On account of this high operating efficiency, it is usually sufficiently close to assume the efficiency to be 95 per cent as a first approximation when calculating the armature reaction caused by the reactive component of the alternating current and the armature copper loss. A second approximation can be made if necessary.

The field excitation for the 1,000-kw converter is calculated for a full d-c load and a power factor of 0.95 with a leading current.

The coil alternating current I'_{ac} can be found from Eq. (301), page 548. The coil current is the same as the inductor current.

$$I'_{ac} = I_{dc} \frac{2}{pn(\text{pf})\eta} \frac{V_{dc}}{V_{ac}}$$

$$I_{dc} = \frac{1,000 \times 1,000}{600} = 1,667 \text{ amp}$$

For an efficiency assumed to be 0.95 as a first approximation and a power factor of 0.95

$$I'_{ac} = 1,667 \frac{2}{12 \times 6 \times 0.95 \times 0.95} \frac{\sqrt{2}}{\sin \dfrac{\pi}{n}} = 145 \text{ amp}$$

The reactive component of this current is

$$I_z = 145 \sqrt{1 - (0.95)^2} = 45.3 \text{ amp}$$

The armature reaction A_x per pole for I_z can be found from the equation

$$A_x = 0.75 k_b N I_x$$

where N is the number of armature turns per pole, I_x is the reactive component of the armature current I'_{ac}, and k_b is the breadth factor. The constant 0.75 is used instead of 0.90 in finding A_x because the synchronous converter is a salient-pole machine (page 233).

The phase spread of a six-phase converter is 60 deg, or one-third of the pole pitch. The converter has 180 slots and 12 poles, or $180/(12 \times 3) = 5$ slots per phase belt.

From Table 6, page 192, the breadth factor for a spread of 60 deg and four slots per phase is 0.958. For five slots per phase it is about 0.957.

$$A_x = 0.75 \times 0.957 \times \frac{180 \times 6}{2 \times 12} \times 45.3$$

$$= 1,464 \text{ amp-turns per pole}$$

These are demagnetizing ampere-turns since a leading current is assumed. The ampere-turns per pole due to the series field are

$$1,667 \times 2 = 3,334$$

The field current required for 600 volts when the converter is driven at no load as a d-c generator is 9.25 amp (open-circuit saturation curve, Fig. 335).

This corresponds to

$$9.25 \times 864 = 7,990 \text{ amp-turns per pole}$$

The shunt excitation required under full-load conditions at a power factor 0.95, an efficiency 95 per cent, and with a leading current is

$$7,990 + 1,464 - 3,334 = 6,120 \text{ amp-turns}$$

This corresponds to a shunt-field current of

$$\frac{6,120}{864} = 7.08 \text{ amp}$$

Efficiency.   The efficiency is

$$\eta = \frac{I_{dc}V_{dc}}{I_{dc}V_{dc} + HI_{dc}^2 r_{dc} + I_{sh}V_{dc} + I_c^2 r_c + P_c + (F + W)}$$

where I_{dc} = direct current
$\quad V_{dc}$ = d-c voltage
$\quad r_{dc}$ = armature resistance between d-c terminals
$\quad I_{sh}$ = shunt-field current
$\quad I_c$ = compound-field current
$\quad r_c$ = resistance of compound winding
$\quad P_c$ = core loss
$\quad F + W$ = friction and windage loss

The armature copper loss can be found by multiplying the copper loss corresponding to the d-c component of the armature current by the ratio of the copper loss of the converter as a converter to its copper loss at the same output as a d-c generator.   This ratio H can be found from Eq. (308), page 555.

$$H = \frac{8}{(\text{pf})^2 \eta^2 n^2 \sin^2 (\pi/n)} + 1 - \frac{16}{\pi^2 \eta}$$

For a power factor of 0.95 and an efficiency to be 95 per cent as a first approximation,

$$H = \frac{8}{(0.95)^2(0.95)^2(6)^2(0.5)^2} + 1 - \frac{16}{(3.142)^2(0.95)}$$
$$= 0.385$$

The armature resistance at 75°C between d-c terminals is

$$0.00589(1 + 50 \times 0.00385) = 0.00702 \text{ ohm}$$

The armature copper loss is

$$I_{dc}^2 \times r_a \times 0.385 = (1,667)^2 \times 0.00702 \times 0.385 = 7,510 \text{ watts}$$

The ohmic resistance is used in finding the armature copper loss. This loss is small and the error introduced by using ohmic resistance

in place of effective resistance is probably not great. Since the armature inductors of a converter carry differently shaped current waves,
the ratio of the ohmic resistance to effective resistance is not the same for
all inductors. It also changes with power factor.

The shunt-field loss including the loss in the field rheostat is equal
to the shunt-field current multiplied by the voltage across the d-c brushes.
This voltage is equal to the terminal voltage plus the drop in the series
field. The drop in the series field is neglected. The shunt-field loss is

$$7.08 \times 600 = 4{,}248 \text{ watts}$$

The resistance of the series field at 75°C is

$$0.000610 \times (1 + 50 \times 0.00385) = 0.000728 \text{ ohm}$$

The series-field loss is

$$(1{,}667)^2 0.000728 = 2{,}025 \text{ watts}$$

The core loss corresponding to a d-c voltage of 600 is 14,700 watts,
from Fig. 335.

The efficiency is

$$\eta = \frac{1{,}000}{1{,}000 + 7.5 + 4.2 + 2.0 + 14.7 + 8.1} = 0.965 \text{ or } 96.5 \text{ per cent}$$

The difference between the efficiency just found and the efficiency
which was assumed to be 95 per cent as a first approximation in calculating I_z and H is too small to make a second approximation necessary.

PART NINE

MERCURY-ARC RECTIFIERS

Chapter 50

Mercury-arc power rectifiers. Multianode rectifiers. Single-anode rectifiers. Operation of the ignitron. A typical firing circuit. Arc-drop.
Arc-back.

Mercury-arc Power Rectifiers. Mercury-arc rectifiers are electronic-
type rectifiers with mercury-pool cathodes. They are usually connected
to 60-cycle sources through transformers, and single polyphase units
often supply several thousand kilowatts at d-c voltages ranging from
100 to 5,000 volts. For such purposes the rectifying parts are enclosed
in a water-cooled steel tank which may be permanently sealed in order to
maintain a high vacuum. In the larger sizes however the vacuum is
maintained by connecting the tank to a vacuum pumping system. The
earlier types generally included 6 or 12 anodes to a single "tank," but the
single-anode rectifier is now the more common.

Multianode Rectifiers. Reference to Fig. 336 shows many of the
construction details of a multianode mercury-arc rectifier. After connection has been made to the a-c source and to the load, the rectifier is
started by applying a d-c voltage between the cathode and the ignition
rod and causing the rod to plunge into the mercury. When the rod is
withdrawn, an arc is established which initiates the ionization and
vaporization of the mercury. The resulting electron flow to the excitation and main anodes then maintains the ionized condition of the mercury
vapor.

Single-anode Rectifiers. There are two general types of single-anode
rectifiers, the ignitron and the excitron. These differ mainly in the
means used for initiating the arc. In the excitron a plunger pulled down
into the mercury cathode sends up a jet of mercury which strikes the
excitation anode causing an arc. In the ignitron an igniter is used. This
is a rod of carborundum, or some similar material of high resistivity,
partly immersed in the mercury cathode and connected to a special
source of excitation through a terminal brought out through the side of
the tank. The ignitron is the more commonly used type.

Figure 337 shows an assembly of six ignitrons mounted on a common
base. Such an assembly would be used for connection to a polyphase
source. Figure 338 shows a cross section of a single ignitron. Because
this is a large-sized ignitron, provision is made for connection to a vacuum

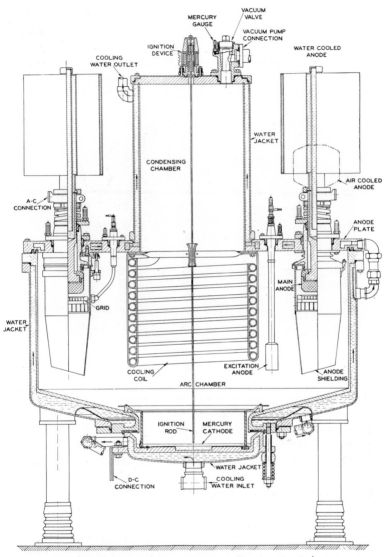

Fig. 336. *(Courtesy of Allis-Chalmers Manufacturing Co.)*

pumping system through the vacuum manifold connection. The tank has an inner and an outer shell with cooling coils between them, through which water is circulated. The cathode terminal is at the bottom of the tank. The anode terminal is a copper stud which extends through the porcelain insulator at the top of the tank. Radiating fins are attached to the top of the stud for cooling purposes and the graphite anode is attached to its bottom end. The graphite shield surrounding the anode is suspended on Mycalex insulators, and electrical connection is made to the shield through the shield entrance bushing which has a spark-plug type of construction.   The shield has holes cut in it in order to allow the electrons to pass through to the anode.   The cathode is a mercury pool enclosed in a quartz ring.   All joints are made vacuum-tight either by soldering or by the use of special gaskets.

Fig. 337.   (*Courtesy of Westinghouse Electric Corp.*)

Operation of the Ignitron.   The passage of a pulse of current through the igniter initiates the ionization of the mercury.   Whether this is due to a high potential gradient produced at the cathode surface or to some other cause is not fully understood.   It is well established, however, that the passage of several amperes through the igniter produces ionization in a few microseconds.   If this ionization takes place when the anode is positive, current flows through the "tube" from the anode to the cathode.   Because no current flows when the anode is negative, a single ignitron becomes a half-wave rectifier.   When the current even momentarily reduces to zero, the mercury vapor becomes deionized.   With an alternating voltage applied across the rectifier from the cathode to the anode terminal, a pulse current must be sent through the igniter once every cycle when the anode is positive to maintain operation.   This allows control of the "firing time" as ignition may be delayed to any point in the positive half cycle.

A Typical Firing Circuit.   Figure 339 shows a typical circuit for supplying the pulse current needed to excite the igniter of an ignitron.

This circuit may be used for supplying the excitation for two ignitrons such as R_1 and R_2 whose anode voltages are in phase opposition. Three such circuits supplied with transformer voltages differing in phase by 120 deg would be required for a six-phase rectifier. The same circuit could be

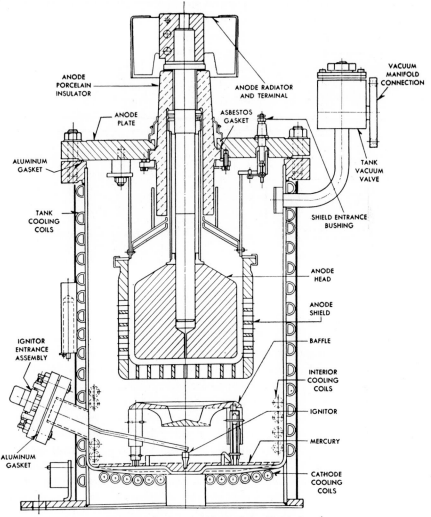

Fig. 338. (*Courtesy of Westinghouse Electric Corp.*)

used for a single ignitron. In that case the second ignitron would be replaced by a resistor. The circuit, as indicated in the figure, consists of three main parts: the firing circuit proper, a voltage-compensating network, and a phase-shifting reactor.

The capacitor in the firing circuit proper is charged through a linear reactor, *i.e.*, one in which the reactance remains constant throughout the entire range of applied voltage. The capacitor voltage is continuously applied to a saturable reactor, in which when the voltage increases to a certain point the core becomes saturated so that the effective reactance decreases. This allows the capacitor to discharge through the saturable reactor and a large peak current is set up in the igniter circuits. The arrows in the figure indicate the locations of copper-oxide or selenium rectifiers and the directions in which they allow the current to flow. When the lower line is at a positive potential, the pulse current is sent through the igniter of R_1; when the upper line becomes positive, the capacitor discharges through the igniter R_2.

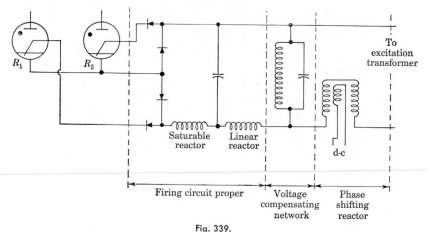

Fig. 339.

The voltage-compensating network consists of a capacitor and a saturable reactor in parallel. It ensures a nearly constant input voltage to the firing circuit proper. The capacitor and reactor of the voltage-compensating network are so proportioned as to operate close to unity power factor at normal line voltage. If the line voltage is high, the saturating reactor draws an excess of current. The resulting lagging current through the phase-shifting reactor causes a decrease in the voltage applied to the firing circuit proper. If the line voltage is low, the current drawn by the saturating reactor becomes less than that through the capacitor and the resulting leading current passing through the phase-shifting reactor causes an increase in voltage applied to the firing circuit.

The voltage drop across the phase-shifting reactor adds to the voltage supplied by the excitation transformer to give the voltage supplied to the firing circuit proper. When the voltage-compensating network is operating at unity power factor, the voltage drop across the phase-shifting reactor is nearly 90 deg out of phase with the transformer voltage. It therefore causes the voltage supplied to the firing circuit proper to differ in phase from that of the transformer by an amount depending upon this voltage drop. The d-c current in the d-c winding of the reactor changes the saturation of the iron core and so varies the effective reactance. With reactors of this type a gradual change in reactance over a 10 to 1 range may be accomplished by varying the current in the d-c winding. Because this changes the phase of the voltage applied to the firing circuit proper, the time of firing of the ignitrons may be varied over a considerable portion of a half cycle.

Other types of firing circuits make use of thyratron tubes for supplying the pulse current. Phase control is easily achieved in these circuits through the control exercised by the grids of the thyratrons. These circuits in general occupy less space than the circuit described. Their disadvantage is the necessity for tube replacement.

Arc-drop. Once a current is established in a mercury-arc rectifier it requires 3 to 5 amp to maintain ionization of the mercury vapor. With increased currents the drop in voltage across the rectifier, called the "arc-drop," remains substantially constant. This arc-drop includes a voltage of about 5 volts at the anode, 7 to 10 volts at the cathode, and a voltage drop in the electron path between the anode and cathode which is dependent on the length of the path and the introduction of baffles and shields. This results in a total voltage drop of 15 to 20 volts in the usual ignitron rectifier, while in the multianode rectifier the drop is likely to be as much as 25 or 30 volts This voltage multiplied by the d-c current represents the total loss of power within the tank.

Arc-back. Occasionally a cathode spot will appear on an anode when it is at a negative potential with respect to the cathode. When this occurs, a reversed current will flow. This is known as "arc-back." Breakers with reversed-current tripping devices are usually required in the anode circuits to protect against this. Poor vacuum or excessive temperature increases the probability of arc-backs. The frequency of arc-back is therefore a major factor in determining the ratings of mercury-arc rectifiers. Although the tolerable frequency of occurence depends on the specific application, an average of one or two arc-backs per month of continuous use may be considered acceptable for many types of service.

In the multianode rectifier, once the rectifier is started, current at all times flows from some anode to the cathode since at least one anode is always at a positive potential with respect to the cathode and ionization of the mercury vapor is thereby continuously maintained. In a single-anode rectifier such as the ignitron, however, no current flows during the negative half cycle of the voltage wave and the mercury vapor becomes deionized. Elimination of normal ionization during the period of high reversed voltage eliminates a major condition which is favorable to arc-back. This makes it possible to reduce the spacing between the anode and cathode and to reduce the shielding, thereby reducing the arc-drop.

Chapter 51

Rectifier circuits. Direct-current voltage. Anode current. Six-phase double-Y connection with interphase transformer. Overlap. Phase control or time delay. Voltage regulation. Summary.

Rectifier Circuits. In Figs. 340, 341, and 342 are shown single-anode rectifiers connected for half-wave, full-wave, and three-phase rectification, respectively. A schematic connection diagram is given at the left-hand

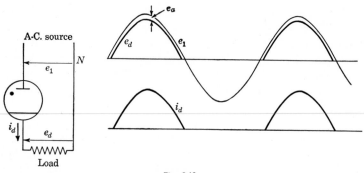

Fig. 340.

side of each figure. The voltage and current waves shown at the right-hand side are based on the following simplifying assumptions:

1. Current flows as soon as the anode becomes positive with respect to the cathode (no delay).

2. The load is a noninductive resistance.

3. The transformer impedances are negligible.

R_1, R_2, and R_3 are single-anode rectifiers. e_1, e_2, and e_3 are the voltages from neutral N to line across the transformer secondaries, e_d is the voltage across the load, and e_a is the arc-drop. i_1, i_2, and i_3 are currents in the individual rectifiers, while i_d is the current in the load.

In the circuit of Fig. 340, current flows through the circuit during the positive half cycle of the source voltage e_1. During the negative half cycle, no current can flow since the anode of the rectifier is then negative with respect to the cathode. The load voltage e_d equals e_1 minus the arc-drop while current flows and is zero at all other times.

In the circuit of Fig. 341, voltage is applied to the rectifiers through a transformer with center-tapped secondary. The voltages e_1 and e_2 are therefore displaced in phase by 180 deg. In general, current flows in rectifier R_1 when e_1 is positive and in R_2 when e_2 is positive as these are the conditions for positive anode voltages. The load voltage e_d equals e_1

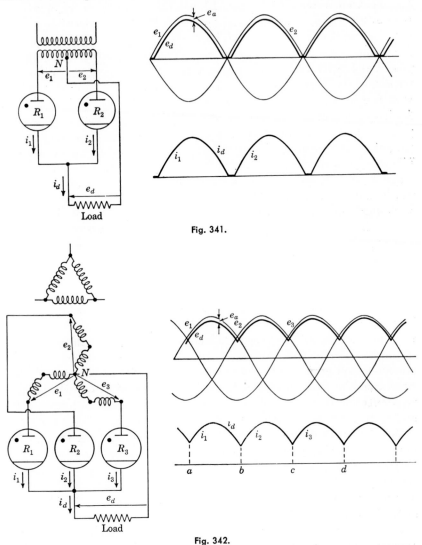

Fig. 341.

Fig. 342.

minus the arc-drop in R_1 when current is flowing in R_1 and equals e_2 minus the arc-drop in R_2 when current is flowing in R_2. This gives full-wave rectification of voltage and current. Because of the arc-drop there is a small period of zero load voltage and current between each half cycle.

In the circuit of Fig. 342, three rectifiers are connected to a three-phase source through transformers. A d-c current in the secondary winding of a transformer would shift the axis of the pulsating flux and tend to over-saturate the iron core. This is prevented by using a zigzag connection for the transformers (see page 137). The voltages e_1, e_2, and e_3 are alternating voltages displaced in phase by 120 deg. Between points a and b, current flows only in rectifier R_1 as is made clear by the following discussion. The three cathodes are connected together and are at a potential e_d with respect to the neutral N. The anode potentials with respect to the neutral are e_1, e_2, and e_3, respectively, and their potentials with respect to the cathodes are therefore $e_1 - e_d$, $e_2 - e_d$, and $e_3 - e_d$. An inspection of the voltage waves of Fig. 342 will show that, between points a and b, only R_1 has an anode potential which is positive with respect to its cathode and therefore only rectifier R_1 carries current. Likewise, between b and c, current flows only in rectifier R_2 and, between c and d, only in rectifier R_3. Starting at time a, the load current i_d is first equal to i_1, then i_2, and then i_3. This load current has the same wave shape as the d-c voltage e_d and consists of the currents i_1, i_2, and i_3 flowing in succession through R_1, R_2, and R_3.

A d-c voltmeter in the load circuit would measure the average value of the voltage wave. The difference between the instantaneous voltage and its average value is the ripple voltage. In general, increasing the number of phases reduces the magnitude of the ripple voltage and increases its frequency. For some applications the ripple is unimportant. In many circuits however the ripple produces undesirable effects and must be kept small. Increasing the number of phases reduces the ripple in both voltage and current waves. A series-connected inductive reactance will appreciably reduce the ripple in the current wave. This ripple may be still further reduced by the use of filters consisting of resonant shunts each composed of a reactor in series with a capacitor and each tuned to the frequency of the harmonic which it is to suppress.

Direct-current Voltage. The term mercury-arc rectifier is often used in referring to either a multianode rectifier or an integral assembly of one or more single-anode rectifiers. An assembly of six ignitrons as illustrated in Fig. 337, page 578, would thus constitute a mercury-arc rectifier. The output voltage of a mercury-arc rectifier depends on the a-c voltage applied to the rectifier circuit and the number of anodes involved. If there is no delay in the time of firing, i.e., each anode starts to carry current as soon as it has a positive potential with respect to its cathode and no overlap, i.e., the d-c current is carried completely by one anode at a time, then the d-c voltage will be as indicated in Figs. 340, 341, and 342, pages 582 and 583. If, too, the arc-drop is neglected, the instantaneous d-c voltage will equal the transformer voltage to neutral of the phase having the greatest positive value. This d-c voltage is shown by the full lines in Fig. 343a. The average value of this d-c voltage, sometimes called the theoretical d-c voltage, is

$$E_{d0} = \frac{1}{2\pi/p} \int_{-\pi/p}^{\pi/p} E_m \cos \theta \, d\theta$$
$$= \frac{E_m \sin \pi/p}{\pi/p} \qquad (314)$$

where E_m is the maximum voltage to neutral of the transformer secondaries and p is the number of anodes included in the mercury-arc rectifier with the exception of the single-anode half-wave rectifier. p is also the

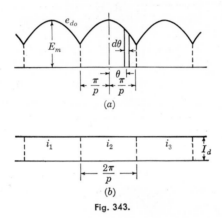

(a)

(b)

Fig. 343.

number of transformer secondary phases for all cases except single phase (both full and half wave). For the single-anode half-wave rectifier the average d-c voltage is one-half that of the full-wave rectifier and is one-half the voltage given by formula (314) if p is set equal to 2. The harmonic of greatest magnitude in the ripple voltage has a frequency of p times the frequency of the a-c source except for the single anode rectifier.

Anode Current. In many circuits supplied by mercury-arc rectifiers the ripples in the d-c load current are small because of the large amount of inductance in the load circuit. In such circuits the ripple in the current wave is often neglected and the d-c current wave is assumed to be a straight line as indicated by the full line in Fig. 343b. If there is no overlap, the load current is supplied in succession by rectifier R_1 between a and b, by R_2 between b and c, and by R_3 between c and d, and this is repeated in succeeding cycles. The current supplied to a single anode by the transformer secondaries has a rectangular wave form and persists for $2\pi/p$ radians of each voltage cycle. The rms value of this anode current is

$$I_s = \sqrt{\frac{I_d{}^2 \times 2\pi/p}{2\pi}} = \frac{I_d}{\sqrt{p}} \tag{315}$$

where I_d is the constant value of the d-c load current.

Six-phase Double-Y Connection with Interphase Transformer. Three transformers with their secondaries connected in double-Y (see page 134) may be used to supply six-phase power to a six-anode mercury-arc

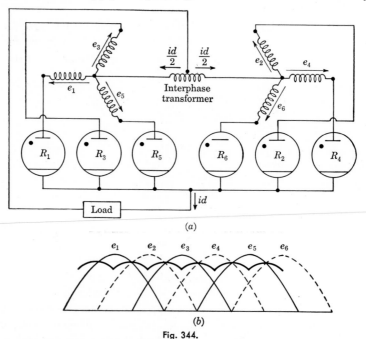

(a)

(b)

Fig. 344.

power rectifier. The most generally used circuit for mercury-arc power rectifiers however is a modification of the double-Y connection and is the circuit shown in Fig. 344a. In this circuit the three primary windings are commonly connected in delta and the neutrals of the two Y-connected secondaries are connected through a center-tapped interphase transformer. The interphase transformer is an iron-core transformer, its only winding being the center-tapped winding shown in the figure. It causes the load current to divide equally between the two Y-connected secondaries. This connection is therefore sometimes called a "double three-phase" rather than a six-phase circuit.

Figure 344b shows the positive halves of the voltage waves from neutral to line on the six transformer secondary windings, one group of

Y voltages being shown by full lines, the other group by dotted lines. If current should build up faster in one-half of the interphase transformer winding than in the other half, a voltage is produced by the resulting increasing flux; this voltage opposes the larger current and aids the smaller. This tends to equalize the voltages of the two anodes having the highest positive potentials so that both anodes conduct current. Neglecting the arc-drop, then at each instant of time the load voltage becomes the average of the transformer voltages supplied to the two current-conducting anodes and is shown by the full line of Fig. 344b.

Because each half of the d-c current is split among three anodes, the rms value of the anode current equals $I_d/2\sqrt{3}$ for the conditions specified for Eq. (315). The average d-c voltage likewise is that of a three-phase rectifier and under the conditions specified for Eq. (314) equals $0.828\ E_m$. The ripple frequency however is six times the frequency of the source voltage as may be seen by reference to Fig. 344b. This corresponds to the ripple frequency of a six-phase rectifier.

With a very light load connected across the rectifier terminals, the interphase transformer becomes inoperative and the average voltage appearing at the d-c terminals is that of a six-phase rectifier. This voltage under the conditions of Eq. (314) equals $0.956\ E_m$. As the load current is increased to about 1 per cent of its full load value, the interphase transformer becomes fully magnetized causing the voltage to drop to that of a three-phase rectifier which has an average value of $0.828E_m$ as noted above.

Overlap. The d-c voltage of a mercury-arc rectifier is given by Eq. (314), page 585, if the arc-drop is neglected and conditions are such that there is no overlap and no time delay. Because of the reactance of the windings of the supply transformers however the current supplied to any one anode is not immediately interrupted when the voltage of the next anode becomes slightly higher. The inductance in the transformer circuits causes the currents in one anode to decrease gradually to zero while the current of the succeeding anode gradually builds up to a value equal to the d-c load current as indicated in Fig. 345b. The time required for the current to transfer completely from one anode to the next is called the angle of overlap or the commutation angle and is a function of the d-c load current and the inductance in the transformer circuits. The effect of this overlap is to decrease the rms value of the anode current and to decrease the d-c output voltage of the rectifier. The full lines of Fig. 345a show the resulting d-c voltage when the commutation angle is taken into account. A comparison of Figs. 343 and 345 shows that the d-c voltage of Eq. (314), page 585, is reduced because of overlap by an amount E_x equal to one-half the shaded area of Fig. 345a divided by the

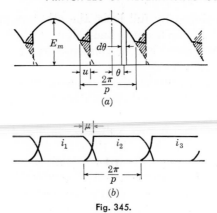

Fig. 345.

base $2\pi/p$ or

$$
\begin{aligned}
E_x &= \frac{\dfrac{1}{2} \displaystyle\int_{-\pi/p}^{\mu-\pi/p} \left[E_m \cos \theta - E_m \cos \left(\theta + \frac{2\pi}{p} \right) \right] d\theta}{2\pi/p} \\
&= \frac{E_m \sin \pi/p}{2\pi/p} (1 - \cos \mu) \\
&= E_{do} \frac{1 - \cos \mu}{2} = E_{do} \sin^2 \frac{\mu}{2}
\end{aligned}
\tag{316}
$$

Phase Control or Time Delay. The beginning of the transfer of current from one anode to the next succeeding anode can be delayed beyond the point where the voltages of the two anodes are equal. This is accomplished in multianode rectifiers by grid control and in ignitron

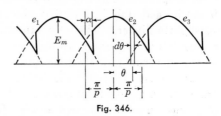

Fig. 346.

rectifiers by delaying the time of supplying the pulse currents to the ignitors. In either case this is referred to as phase control or time delay. The angle by which the beginning of current transfer lags the point of equal voltages is the angle of retard and is shown in Fig. 346 as the angle α. The full lines of this figure show the d-c voltage when time delay is taken into account but the arc-drop and commutation angle are neglected. The average value of this d-c voltage is

$$E'_{d0} = \frac{1}{2\pi/p} \left[\int_{-\pi/p}^{\alpha-\pi/p} e_1 \, d\theta + \int_{\alpha-\pi/p}^{\pi/p} e_2 \, d\theta \right]$$

$$= \frac{1}{2\pi/p} \left[\int_{-\pi/p}^{\alpha-\pi/p} E_m \cos \left(\theta + \frac{2\pi}{p} \right) d\theta + \int_{\alpha-\pi/p}^{\pi/p} E_m \cos \theta \, d\theta \right]$$

$$= \frac{E_m \sin \pi/p}{\pi/p} \cos \alpha = E_{d0} \cos \alpha \qquad (317)$$

Voltage Regulation. It has been shown that the d-c voltage of a loaded mercury-arc rectifier is less than the theoretical value given by Eq. (314) because of the arc-drop, overlap, and phase control. The d-c voltage is further reduced because of the voltage drops in the resistances of the windings of the transformers. This further decrease in voltage equals

$$E_r = \frac{P_r}{I_d} \qquad (318)$$

where P_r is the resistance loss in the transformers and I_d is the direct current in the output circuit.

The net d-c voltage taking into account all the above factors is

$$E_d = E_{d0} \cos \alpha - E_a - E_x - E_r \qquad (319)$$

The effects of phase control and arc-drop are nearly independent of the load current, but E_r and E_x are nearly proportional to this current. The voltage regulation curve of a mercury-arc rectifier is therefore a sloping straight line, showing a decrease in voltage as the load is increased.

Although phase control reduces the d-c voltage, it is often used as a means of maintaining a more nearly constant voltage at the load. If the angle of retard is decreased when the load is increased, the decrease in voltage due to overlap and resistance drop may be compensated for.

Summary. The only moving parts in a mercury-arc rectifier installation are the pumps required for circulating water for cooling and, in the case of the pumped-type rectifiers, the vacuum pumps. The vacuum pumping system most commonly employed consists of a mercury-vapor diffusion pump without moving parts and a rotary vacuum pump.

The efficiency of a mercury-arc rectifier unit is the ratio of the d-c power output to the power input to the supplying transformers. The losses of the unit therefore consist of the transformer losses, the arc-drop loss, and the power required for operation of the auxiliaries. Because the arc-drop loss is nearly independent of the rectifier voltage, improved efficiency is obtained as the operating voltage is increased. For 600-volt d-c operation, the efficiency of the ignitron-type mercury-arc rectifier is generally higher than that of a synchronous converter throughout the normal load range and the comparison is particularly favorable to the mercury-arc rectifier at light loads.

There are several rectifier circuits other than those discussed in this chapter which are of considerable importance. Included in these are circuits for 12-phase operation and double-way circuits. In the double way circuits, each transformer secondary terminal is connected to two rectifying elements. It is connected to the anode terminal of one element and to the cathode terminal of the other. This nearly doubles the output voltage obtained from a given transformer connection and is most advantageous in the higher voltage ratings.

PROBLEMS

Transformers—Chapter 2

2-1. A 2,300- to 230-volt 60-cycle single-phase transformer has a laminated core of silicon steel with a stacking factor (ratio of net to gross area of iron) of 0.90. The gross cross-section area of the core is 35 sq in. If the transformer is to be operated at rated voltage and frequency and the maximum flux density in the steel is to be 0.00070 weber per sq in., how many turns must each winding contain?

2-2. The magnetic core of a 50-kva 60-cycle transformer has a net average cross-section area of 37.0 sq in. The high-voltage winding has 5,150 turns. If this winding is connected to a 33,000-volt source, what will be the maximum flux density (a) in webers per square inch? (b) in webers per square meter? (c) in lines per square inch?

Transformers—Chapter 3

3-1. The voltage applied to the primary winding of a transformer is such as to give a maximum flux density in the core of 64,500 lines per sq in. The primary winding has 100 turns, and the mean length of the flux path through the core is 100 in. If the upper half of the hysteresis curve for the grade of steel used is given by the following data, plot for one full cycle (a) the flux density versus ωt and (b) the exciting current versus ωt. Assume the impressed voltage to be sinusoidal, and neglect the effect of eddy currents and of any joints in the magnetic circuit.

Hysteresis Curve

B, lines per sq in.	H, amp-turns per in.
0	1.35
30,000	1.80
40,000	2.15
50,000	2.75
60,000	3.85
64,500	4.50
62,500	2.50
55,000	0.25
53,000	0
50,000	−0.35
40,000	−0.85
20,000	−1.15
0	−1.35

3-2. If a voltage of the wave form $v = V_1 \sin \omega t - 0.3 V_1 \sin 3\omega t$ is impressed across the primary winding of the transformer of Prob. 3-1, plot for one complete cycle the resulting curves of (a) flux density versus ωt and (b) exciting current versus ωt. Assume the maximum flux density to be 64,500 lines per sq in. and neglect the effects of eddy currents and joints.

3-3. A resistance of 10 ohms is connected in series with the primary winding of a transformer. The transformer is then connected to a 120-volt 60-cycle source. If the voltage drop across the source is given by the equation $(-v) = \sqrt{2} \times 120 \cos \omega t$ and the current drawn by the transformer is

$$i = \sqrt{2}(0.5 \cos \omega t + 2.0 \sin \omega t - 0.8 \sin 3\omega t)$$

determine (a) the equation for the voltage drop across the primary winding of the transformer, (b) the rms value of this voltage, and (c) the ratio of the third harmonic and fundamental components of this voltage.

Transformers—Chapter 4

4-1. A certain air-core transformer draws a current of 24 amp when connected to a 120-volt 60-cycle source with its secondary open-circuited. A voltmeter is then connected across the secondary and reads 16 volts. The ratio of turns $N_1/N_2 = 5$. Assume the resistance of the primary winding to be negligible.

(a) What is the self-inductance of the primary winding?

(b) What is the leakage-inductance of the primary winding?

(c) What is the mutual inductance of the transformer?

4-2. An air-core transformer has the following constants at 1,000 cycles:

$$\omega L_1 = 200 \text{ ohms} \qquad \omega L_2 = 60 \text{ ohms} \qquad \text{and} \qquad \omega M = 50 \text{ ohms}$$

Assume the resistances of the windings to be negligible. If a noninductive resistance of 60 ohms is connected across the secondary terminals and 100 volts at 1,000 cycles is impressed across the primary, find the primary and secondary currents of the transformer.

4-3. The magnetic circuit of a 400-kva 60-cycle 2,400:600-volt transformer has a mean length of 123 in. and a net average cross-section area of 54.5 sq in. If the transformer is to operate with a maximum flux density of 85 kilolines per sq in., how many turns should there be on the high- and low-tension windings? What is the no-load current and power factor? Use the magnetic data given below and allow 200 additional ampere-turns for the joints in the magnetic circuit.

Magnetic Data

Maximum flux density, kilolines per sq in.	Magnetizing force, rms amp-turns per in.	Core loss at 60 cycles, watts per cu in.
61.3	2.0	0.550
78.0	4.0	0.840
81.0	5.0	0.910
83.0	6.0	0.950
84.7	7.0	1.000
86.2	8.0	1.020
87.0	9.0	1.050
87.5	10.0	1.070

4-4. The magnetic circuit of a certain 10-kva 60-cycle 600:120-volt transformer has a mean length of 26.5 in. and a net average cross-section area of 14 sq in. There are 50 turns on the low-tension winding. Find the core loss and the no-load current and power factor when the high-tension winding is connected to a 600-volt 60-cycle source. Use the magnetic data given below. Allow 75 additional amp-turns for the joints and assume the density of the steel to be 0.273 lb per cu in.

Magnetic Data

Maximum flux density, webers per sq m	Magnetizing force, rms amp-turns per m	Core loss at 60 cycles, watts per lb
0.4	47.8	0.114
0.8	111.4	0.350
1.0	191.5	0.520
1.2	398	0.770
1.3	670	0.920
1.4	1590	1.120

4-5. A 100-kva 2,400:600-volt 60-cycle transformer has the following constants:

$$r_1 = 0.384 \text{ ohm} \qquad r_2 = 0.0240 \text{ ohm}$$
$$x_1 = 1.360 \text{ ohms} \qquad x_2 = 0.0850 \text{ ohm}$$

The core loss and the no-load current when rated voltage at 60 cycles is impressed on the high-voltage winding are 920 watts and 1.14 amp, respectively.

(a) If this transformer is to supply rated kva at 0.8 power factor (lagging current), what voltage must be applied to the high-voltage winding in order to give rated voltage across the secondary terminals?

(b) What is the voltage regulation of the transformer for this load?

Transformers—Chapter 5

5-1. A 15-kva 2,400:240-volt 60-cycle transformer has the following constants:

$$r_1 = \quad 2.48 \text{ ohms} \qquad r_2 = 0.0212 \text{ ohm}$$
$$x_1 = 10.5 \quad \text{ohms} \qquad x_2 = 0.096 \text{ ohm}$$

Neglect the no-load component of the current. Find the voltage regulation of this transformer for a load of rated kva with rated secondary voltage at (a) unity power factor, (b) 0.8 power factor (lagging current), and (c) 0.8 power factor (leading current).

5-2. Solve Prob. 4-5 with the exciting current neglected.

5-3. The resistances of the primary and secondary windings of a 50-kva 30,000:440-volt 60-cycle transformer are 120 ohms and 0.0258 ohm, respectively, at 75°C. The transformer has a per cent reactance of 3.46. Neglect the exciting current. What is the voltage regulation of this transformer for an inductive load of 40 kw at 0.85 power factor and rated voltage?

5-4. The constants of a 1,000-kva 66,000:6,600-volt transformer have the following per-unit values based on rated kva:

$$r_1 = 0.00388 \qquad r_2 = 0.00360$$
$$x_1 = 0.0264 \qquad x_2 = 0.0218$$

Neglect the exciting current.

(a) What are the ohmic values of the primary and secondary resistances and reactances?

(b) What is the voltage regulation of this transformer for an inductive load of rated kva at 0.8 power factor with rated secondary voltage?

5-5. A 50-kva 60-cycle 4,160:240-volt transformer has the following constants:

$$r_1 = 1.48 \text{ ohms} \qquad r_2 = 0.0049 \text{ ohm}$$
$$x_1 = 5.20 \text{ ohms} \qquad x_2 = 0.0173 \text{ ohm}$$

If 4,160 volts are impressed on the high-voltage winding and the transformer supplies 40 kw at 0.8 lagging power factor, what would be the secondary voltage? Neglect the exciting current.

5-6. A transformer with taps at 4,360, 4,260, 4,160, 4,055, and 3,950-volt points on the high-voltage winding has a low-voltage winding rated at 600 volts. It is found to require 4,270 volts on the 4,160-volt tap of the high-voltage winding to produce rated voltage on the secondary when the transformer is delivering rated kva at 0.8 power factor. If the source voltage is exactly 4,160 volts, which tap will give the closest to rated secondary voltage for the above load?

Transformers—Chapter 6

6-1. In a certain transformer the ratio of hysteresis loss to eddy-current loss at rated voltage and frequency is 4:1. If P_c represents the total core loss under this condition, find the core loss if (a) the applied voltage is increased by 5 per cent, (b) the frequency is increased by 5 per cent, and (c) the voltage and the frequency are each increased by 5 per cent. *Note:* Assume the hysteresis exponent to be 1.6.

6-2. The transformer of Prob. 5-3 has a core loss at rated voltage and frequency of 645 watts. Find the efficiency of this transformer at 0.8 power factor for loads of (a) $\frac{1}{4}$, (b) $\frac{1}{2}$, (c) $\frac{3}{4}$, (d) 1, and (e) $1\frac{1}{4}$ times rated kva.

6-3. The transformer of Prob. 5-4 has a core loss at rated voltage and frequency whose per-unit value based on rated kva is 0.0092. What is the efficiency of this transformer for a load of rated kva at 0.85 power factor?

6-4. An iron-core transformer is designed to operate with a primary voltage of 440 volts at 50 cycles. With a sinusoidal voltage of rated value and frequency, the maximum flux density in the core is 1.1 webers per sq m and the core loss is 641 watts, of which 170 watts is due to eddy currents. Assume the Steinmetz exponent to be 2.0.

(a) If a voltage $v = 595 \sin 314t + 178.5 \sin 942t$ is impressed on the primary winding, what will be the core loss?

(b) If the gross cross-section area of the core is 35 sq in. and the stacking factor (ratio of net to gross area) is 0.9, how many turns are there on the primary winding?

6-5. A 100-kva transformer is connected to the supply lines for 24 hr a day. For 6 hr it delivers 90 kw at 0.9 power factor, for 4 hr it supplies 25 kw at 0.5 power factor, and for the rest of the day it operates without load. Its core loss at rated voltage is 1,000 watts, and its copper loss with full-load current is 1,680 watts. What is the all-day efficiency?

6-6. A given transformer has an efficiency at one-half load which is equal to its efficiency at full load. Find the ratio of its full-load copper loss to its core loss.

Transformers—Chapter 7

7-1. A transformer core has a mean length of 80 in. and a uniform gross cross section of 35 sq in. The stacking factor is 0.9. The core laminations are No. 29 gauge of 4.25 per cent silicon steel (see Table 2, p. 68). The density of the steel is 0.272 lb per cu in. The transformer is designed to operate from a 440-volt 60-cycle circuit.

(a) If the maximum flux density is to be 70,000 lines per sq in., how many turns should be on the 440-volt winding?

(b) What is the core loss at rated voltage and frequency?

7-2. The transformer core of Prob. 7-1 is to be used for a transformer to be rated 440:220 volts, 50 cycles.

(a) If the maximum flux density is again to be 70,000 lines per sq in., how many turns should be used on each winding?

(b) What would be the core loss at rated voltage and frequency?

7-3. The core loss of a certain transformer is found to be 60 watts when 220 volts at 60 cycles is impressed across its primary winding. When the voltage and frequency are both halved, the core loss is found to be 24.4 watts. Find (a) the hysteresis loss and (b) the eddy-current loss at 220 volts and 60 cycles.

7-4. When 2,400 volts at 60 cycles is impressed across the primary winding of a certain transformer, the power input with the secondary open is found to be 900 watts. With 2,000 volts at 50 cycles, the same transformer takes 719 watts. Find (a) the hysteresis loss and (b) the eddy-current loss at 2,400 volts and 60 cycles.

7-5. The primary winding of a transformer consists of two coils to be connected either in series or in parallel. With the secondary open and the primary coils connected in series across a 220-volt 60-cycle source, the input to the transformer is 90 watts and the primary current is 0.2 amp. What will be the power input and current if the windings are connected in parallel and to a 110-volt 60-cycle source?

7-6. The core loss with 29 gauge U.S.S. Transformer 58 grade steel is 0.123 watt per lb at a flux density of 8,000 gausses and a frequency of 25 cycles. For the same flux density the core loss is 0.368 watt per lb at 60 cycles. What would be the core loss at 8,000 gausses at 40 cycles?

7-7. When 600 volts at 60 cycles is impressed across the primary winding of a certain transformer, the power input with the secondary open is found to be 327 watts. With 300 volts at 60 cycles, the same transformer takes 100 watts. Find (a) the hysteresis loss and (b) the eddy-current loss at 600 volts and 60 cycles.

Transformer Data

Num-ber	Rating			No-load test		Short-circuit test		
	Kva	Voltages	Fre-quency	Volt-amperes	Watts	Volts	Am-peres	Watts
A	3	600:120	60	130	35	38	5.00	120
B	25	4,800:240	25	3,500	314	400	5.20	450
C	100	13,200:480	60	3,720	931	372	7.58	990
D	500	66,000:600	60	7,630	2,930	2,500	7.58	3,340
E	1,000	66,000:13,200	60	59,400	9,210	3,210	15.15	7,420

7-8. Calculate (a) the voltage regulation and (b) the efficiency of transformer A for a load of rated kva at 0.8 lagging power factor and rated secondary voltage.

7-9. Calculate (a) the voltage regulation and (b) the efficiency of transformer B for an inductive load of 20 kw at 0.85 power factor and rated secondary voltage.

7-10. Calculate (a) the voltage regulation and (b) the efficiency of transformer C for a capacitive load of 100 kw at 0.8 power factor and rated secondary voltage.

7-11. Calculate (a) the voltage regulation and (b) the efficiency of transformer D for a unity power-factor load of rated kva and rated secondary voltage.

7-12. Calculate (a) the voltage regulation and (b) the efficiency of transformer E for an inductive load of 800 kw at 0.8 power factor and rated secondary voltage.

7-13. For transformer D, determine the per-unit values of (a) equivalent resistance, (b) equivalent reactance, (c) core loss, and (d) full-load copper loss using the rated kva as a base.

7-14. A 100-kva 2,400:240-volt 60-cycle transformer has a per-unit total resistance of 0.0120 and a per-unit total reactance of 0.0370. The per-unit value of the core loss is 0.00690. Rated kva is taken as the base for all per-unit values. Determine (a) the voltage regulation and (b) the efficiency for an inductive load of rated kva at 0.8 power factor and rated secondary voltage.

Transformers—Chapter 8

8-1. A single-phase line delivers power to an inductive load. A voltmeter, ammeter, and wattmeter are connected into the circuit through instrument transformers and read 115 volts, 4.0 amp, and 230 watts, respectively. The potential transformer has a transformation ratio of 20 and the ratio for the current transformer is 10. Assume any ratio errors to be negligible, but the transformers have the following phase-angle errors for this instrument burden: angle of lag of secondary voltage of potential transformer with respect to primary voltage, 0.667 deg; angle of lead of secondary current of current transformer with respect to reversed primary current, 0.333 deg.

(a) Draw a diagram to show the connection of instruments and transformers.

(b) Determine the power delivered to the load.

8-2. The name plate of a certain two-winding transformer gives the following information: 100 kva, 480:240 volts, 60 cycles. If this transformer is connected as a 720:480-volt autotransformer, what should be its kva rating as an autotransformer?

8-3. The transformer of Prob. 8-2, when used as a two-winding transformer, has a core loss of 600 watts at rated voltage. The resistance and leakage reactance of the high-voltage winding are 0.0149 and 0.0576 ohm, respectively. The resistance of the low-tension winding is 0.00372 ohm, and the leakage reactance is 0.0144 ohm.

(a) What is its efficiency and voltage regulation at rated load, 0.8 power factor (lagging current), and rated secondary voltage?

(b) What is its efficiency when used as an autotransformer to supply a kva load of the value found in Prob. 8-2 at 0.8 power factor (lagging current) and at 480 volts?

(c) What primary voltage must be supplied to the autotransformer to give 480 volts on the secondary when transformer is loaded as in part (b)?

(d) What is the voltage regulation for the condition of part (b)?

8-4. An autotransformer has a ratio of transformation of 2, i.e., the tap b is located halfway between terminals a and c (see Fig. 59). The resistance of the entire winding ac is 0.108 ohm, and that of the common portion bc is 0.0525 ohm. The leakage reactance of the common portion is 0.159 ohm, and that of the remaining portion ab is 0.162 ohm. If an emf of 18 volts at the rated frequency is impressed across the high-tension terminals and the low-tension terminals are short-circuited, what will be the current and power taken from the line?

8-5. An autotransformer is rated 100 kva, 600:480 volts, 60 cycles. The resistance and leakage reactance of the common portion of the winding are 0.0852 and 0.144 ohm, respectively. The resistance and leakage reactance of the series portion of the winding are 0.00533 and 0.00900 ohm, respectively. If the high-tension terminals are short-circuited, what potential should be impressed across the low-tension terminals in order that there will be full-load current in the windings? What then would be the power input?

8-6. The autotransformer of Prob. 8-5 supplies 100 kva to an inductive load at 0.8 power factor and 600 volts. What voltage should be impressed across its low-tension terminals?

Transformers—Chapter 9

9-1. A certain transformer has three windings A, B, and C rated 2,400, 600, and 240 volts, respectively. Short-circuit tests give the following results:

Winding C open, winding B short-circuited:

$$I_A = 31.3 \text{ amp} \qquad V_A = 120 \text{ volts} \qquad P_A = 750 \text{ watts}$$

Winding B open, winding C short-circuited:

$$I_A = 31.3 \text{ amp} \qquad V_A = 135 \text{ volts} \qquad P_A = 810 \text{ watts}$$

Winding A open, winding C short-circuited:

$$I_B = 125.0 \text{ amp} \qquad V_B = 30 \text{ volts} \qquad P_B = 815 \text{ watts}$$

Determine the constants of the equivalent circuit for this transformer. Refer all values to the 2,400-volt winding. Neglect the exciting current.

9-2. The three-winding transformer of Prob. 9-1 supplies the following loads: (1) A load which draws 50 amp at 0.8 power factor (lagging) connected to the 600-volt terminals. (2) A load drawing 100 amp. at unity power factor connected to the 240-volt terminals.

(a) What voltage must be applied to the 2,400-volt terminals in order to give 600 volts across load 1?

(b) What then will be the voltage across load 2?

9-3. A 3-kva 600 to 240/120-volt 60-cycle transformer supplies a power load requiring 6 amp at 0.8 lagging power factor and 240 volts and a lighting load requiring 6 amp at unity power factor. The lighting load is connected across one side of the secondary winding. The transformer may be dealt with as a three-winding transformer. If the primary winding, the half of the secondary winding supplying the lighting load, and the other half of the secondary winding are designated as windings 1, 2, and 3, respectively, the equivalent circuit impedances are found to be

$$Z_1 = 1.5 + j3.0 \text{ ohms} \qquad Z_2 = 3.0 + j6.0 \text{ ohms} \qquad \text{and} \qquad Z_3 = 3.0 + j6.0 \text{ ohms}$$

These values are all referred to the primary winding.

(a) What voltage must be supplied to the primary winding in order to give 120 volts across the lighting load?

(b) What then will be the voltage across the power load?

9-4. A 3-kva 600 to 240/120-volt 60-cycle transformer supplies a power load requiring 9 amp at 0.8 lagging power factor and 240 volts and a lighting load requiring 3 amp at unity power factor. The lighting load is connected across one side of the secondary winding. This transformer may be dealt with as a three-winding transformer. If the primary winding, the half of the secondary winding supplying the lighting load, and the other half of the secondary winding are designated as windings 1, 2, and 3, respectively, the equivalent circuit impedances are found to be

$$Z_1 = 1.5 + j3.0 \text{ ohms} \qquad Z_2 = 3.0 + j6.0 \text{ ohms} \qquad \text{and} \qquad Z_3 = 3.0 + j6.0 \text{ ohms}$$

These values are all referred to the primary side.

(a) What voltage must be supplied to the primary winding in order to give 120 volts across the lighting load?

(b) What then will be the voltage across the power load?

9-5. Two 75-kva 2,400:480-volt 60-cycle transformers are to be connected in parallel on both primary and secondary sides. The per-unit equivalent resistance of transformer A is 0.0122, and the per-unit equivalent reactance is 0.0316. The corresponding values for transformer B are 0.0129 and 0.0334. These values are calculated on a 75-kva base.

(a) If a load of 150 kva at 0.8 power factor with rated secondary voltage is connected across the secondaries, what is the secondary current of each transformer?

(b) A reactor of negligible resistance is now added to the low-tension sides of transformer A. The reactor is of such a value as to cause equal currents to flow in the two transformers. What is the ohmic value of this reactance?

(c) What value of reactance would be needed if it were added to the high-voltage side?

(d) With the reactor of either part (b) or part (c) added, what current will flow in the secondaries of each transformer?

9-6. The following short-circuit data are given on two 33,000:2,300-volt transformers:

Kva rating	Amperes	Volts	Watts
500	217	105	4,320
1,000	434	153	6,540

These transformers are connected in parallel on both primary and secondary sides. What is the greatest kva load these transformers can deliver without overloading either transformer?

Transformers—Chapter 10

10-1. A 13,200-volt 3-phase generator delivers 10,000 kva to a 3-phase 66,000-volt transmission line through step-up transformers. Determine the kva, voltage, and current ratings of each of the single-phase transformers needed if they are connected (a) Δ-Δ, (b) Y-Y, (c) Y-Δ, (d) Δ-Y.

10-2. A 600-volt 3-phase load is to be supplied with 100 kw at 0.8 power factor from a 2,300-volt transmission line through step-down transformers. Determine the kva, voltage, and current ratings of each of the single-phase transformers needed if they are connected (a) Δ-Δ, (b) Y-Y, (c) Y-Δ, (d) Δ-Y, (e) V-V, (f) T-T.

10-3. Three similar 10-kva single-phase transformers are Δ-connected on both primary and secondary sides. The primaries are connected to a three-phase source. A single-phase load is connected across a pair of secondary terminals. What is the greatest single-phase load so connected that can be supplied without overloading any of the transformers?

10-4. Three similar single-phase transformers are Y-connected on both primary and secondary sides. The primary line-to-line voltage is sinusoidal with an rms value of 346 volts. The transformers are used to step down the voltage and have a ratio of transformation of 2.

(a) If the secondary voltage from line to neutral is found to have an rms value of 112 volts, what is the rms value of the third harmonic voltage produced in each transformer secondary? Neglect any higher harmonics.

(b) What is the greatest possible maximum value of the line to neutral secondary voltage?

10-5. Three single-phase transformers, each rated 1,200:120 volts and each with a tertiary winding of the same number of turns as the secondary, are Y-connected on both primary and secondary sides with their tertiary windings in Δ. The primaries are connected to a 2,080-volt 3-phase source. The primary neutral is isolated. A single-phase load drawing 30 amp is connected from line to neutral across one transformer secondary. Determine the currents in each winding of each of the three transformers. Determine also the secondary phase and line voltages.

10-6. Assume the single-phase load of Prob. 10-5 to have a power factor of unity. An additional balanced three-phase load is now connected across the line terminals of the secondaries. This load draws a line current of 50 amp at 0.8 power factor. Determine the currents in each winding of each transformer.

10-7. It is desired to transform 500 kva from three-phase to two-phase by Scott-connected transformers. The three-phase line potential is 2,400 volts, and the two-phase line potential is 600 volts. What should be the current and potential ratings of both primary and secondary and the ratio of transformation of each transformer?

Transformers—Chapter 11

11-1. Three single-phase autotransformers connected in Y are used to step up a three-phase line voltage of 2,200 volts to a voltage of 2,400 volts between lines. If a balanced three-phase load of 100 kva is supplied from the high-tension side of these transformers, what will be the current in (a) the series portion and (b) the common portion of the winding of each transformer? What line current will be supplied to the transformers by the 2,200-volt source?

11-2. Three single-phase 300:200-volt autotransformers are connected in Δ and to a 300-volt three-phase source. A balanced 3-phase 50-kva load is supplied by three lines connected to the 200-volt taps.

(a) What is the line voltage supplied to the load?

(b) What are the line currents on the primary and secondary sides?

(c) What are the currents in each portion of the transformer windings?

(d) What is the minimum phase angle between the primary and secondary voltages?

11-3. Three single-phase autotransformers rated 550:440 volts are connected in extended Δ for the purpose of raising the voltage of a 440-volt three-phase system to a higher voltage.

(a) If a balanced three-phase voltage of 440 volts is applied to the low-voltage side of the autotransformers, what will be the line voltage of the high-tension side?

(b) With a balanced 3-phase 50-kva load connected to the high-tension sides of the transformers, what will be the line currents on the input and output sides?

(c) With the load of part (b) what will be the current in each portion of the transformer windings?

Transformers—Chapter 12

12-1. Two banks of Δ-Δ connected transformers are operating in parallel. The six transformers are each rated 100 kva, are all the same design, and have equal values of equivalent resistance and equivalent reactance.

(a) What is the maximum kva that can be delivered to a balanced load without overloading any of the transformers?

(b) If it becomes necessary to remove one of these transformers, what is then the maximum balanced kva load that can be delivered without overloading any transformer?

12-2. A Δ-Δ bank of transformers is paralleled on both primary and secondary sides with a Y-Y bank. Each phase of the Δ-Δ bank has a per-unit equivalent impedance of 0.04 and negligible resistance. With the same kva base, each phase of the Y-Y bank has a per-unit equivalent impedance of 0.02 and negligible resistance. A balanced 100-kva load is connected across the transformer secondaries. What kva is delivered by each of the two transformer banks?

Synchronous Generators—Chapter 15

15-1. A three-phase synchronous generator has 18 armature slots per pole, 8 turns per coil, 2 coil sides per slot, 4 poles, and a coil pitch of $\frac{2}{3}$. With the generator operating at 60 cycles and with a sinusoidally distributed air-gap flux of 0.005 weber per pole, determine its rms voltage per phase.

15-2. The emf generated in a single inductor of a 48-pole 3-phase synchronous generator, which has two armature slots per pole per phase and two inductors in series per slot, is

$$e = 10 \sin \omega t + 2.0 \sin (3\omega t + 30°) + 1.0 \sin (5\omega t - 30°)$$

The coil pitch is $\frac{5}{6}$. The speed is 150 rpm.

(a) What is the rms value of the voltage per phase?

(b) What is the rms value of the line voltage if the generator is Y-connected?

15-3. The field current of a synchronous generator is adjusted to give an air-gap flux of 10^{-2} weber per pole. The flux is sinusoidally distributed along the air gap. The armature winding is Y-connected. There are five slots per phase per pole, and the coil pitch is $1\frac{2}{15}$. The frequency is 60 cycles at 1,200 rpm.

(a) Calculate the emf per turn.

(b) If there are 10 turns per coil and two coil sides per slot, calculate the emf per phase.

(c) Calculate the emf between line terminals.

15-4. Solve Prob. 15-3 for an air-gap flux of 10^{-2} weber but with a flux-density distribution given by the equation $B = B_1 \sin x + B_3 \sin 3x$, where x is the electrical angle measured from a point midway between a pair of poles. Assume $B_3 = 0.3B_1$.

15-5. Solve Prob. 15-3 for an air-gap flux of 10^{-2} weber but with a flux-density distribution given by the equation $B = B_1 \sin x - B_3 \sin 3x$, where x has the same meaning as in Prob. 15-4 and $B_3 = 0.3B_1$.

15-6. The flux-density distribution in the air gap of a certain generator is such that it contains a fundamental, a third harmonic, and a fifth harmonic. The flux per pole due to the fundamental distribution is 9.0×10^{-3} weber, that due to the third-harmonic distribution is 0.90×10^{-3} weber, and that due to the fifth-harmonic distribution is 0.40×10^{-3} weber. There are four armature slots per phase per pole, the connections are made for three-phase Y, and the coil pitch is $1\frac{2}{15}$. The frequency is 60 cycles. The generator has four poles.

(a) Calculate the emf per turn.

(b) If there are six turns per coil and two coil sides per slot, calculate the emf per phase.

(c) Calculate the emf between line terminals.

15-7. A 3-phase 60-cycle 32-pole synchronous generator has a sinusoidally distributed air-gap flux of 0.026 weber per pole. The armature core has 324 slots. The armature winding is Y-connected. There are two coil sides per slot and two turns per coil. The coil pitch is nine slots. Find the generated voltage between line terminals.

15-8. The air-gap flux of a 3-phase 60-cycle 24-pole synchronous generator is the same as given in Prob. 15-6. The armature core has 162 slots. The coil pitch is five slots, and the armature winding is Y-connected.

(a) Calculate the emf per turn.

(b) If there is one turn per coil and two coil sides per slot, calculate the emf per phase.

(c) Calculate the emf between line terminals.

Synchronous Generators—Chapter 16

16-1. A 3-phase Y-connected 25-cycle 13,200-volt synchronous generator is rated to deliver 7,500 kva. The field structure has 12 poles, and the armature has 180 slots. There are four series inductors per slot. The coil pitch is 12 slots. What is the armature reaction in ampere-turns per pole when the generator is delivering rated kva at rated voltage and 0.8 power factor?

16-2. A 760-kva 2,200-volt Y-connected synchronous generator delivers energy to a 3-phase 50-cycle system. The armature core has 384 slots with three inductors in series per slot. The length of the armature core parallel to the shaft is 10 in. In calculating leakage reactance, assume 6.5 leakage lines per ampere per inch of slot per inductor. The winding is full pitch. There are 64 field poles.

(a) Find the leakage reactance.

(b) Find the armature reaction at full-load current.

16-3. The armature of a three-phase Y-connected synchronous generator has two slots per pole per phase. There are two coil sides per slot with one turn per coil. The coil pitch is six slots. The generator supplies 7.07 amp to a balanced three-phase load. Assume the current in phase 2 to lag the current in phase 1 by 120.

(a) Lay out a developed view of the armature winding.

(b) Make a graphical plot of the armature reaction at the time when the current in phase 1 is a maximum. What is the maximum armature reaction in ampere-turns per pole at this instant of time?

(c) Make a graphical plot of the armature reaction at one-twelfth of a period later than in (b). What is the maximum armature reaction at this instant of time?

(d) What is the armature reaction as determined by formula (142)?

Synchronous Generators—Chapter 17

17-1. A certain 80-kva 2,200-volt 60-cycle Y-connected synchronous generator has four salient poles with 749 turns per pole. The armature has 72 slots with 32 inductors per slot. The coil pitch is 10 slots. The armature effective resistance is 1.6 ohms per phase. The leakage-reactance drop per phase with rated current is 18 per cent of the rated phase voltage. An open-circuit test gives the following data:

Field current, amp	2.0	6.0	8.0	9.0	10.0
Line voltage, volts	890	2,086	2,340	2,450	2,550

(a) Compute the field current when the generator delivers its rated kva at rated voltage and 0.8 power factor (lagging).

(b) Compute the field current when the generator delivers its rated kva at rated voltage and 0.8 power factor (leading).

17-2. Calculate, by the general method, the field current required to give rated voltage across the terminals of the synchronous generator described in the sample problem on page 234 when the generator is delivering 100,000 kva to an inductive load at 0.866 power factor.

17-3. A 16,000-kva Y-connected 11,000-volt 2-pole synchronous generator has a single-circuit armature winding. There are 36 armature slots with two coil sides per slot and one turn per coil. Assume a full-pitch winding. The d-c armature resistance at 25°C is 0.04 ohm per phase. At 25°C the effective resistance per phase is 1.6 times the d-c resistance. The armature leakage reactance is 2.6 ohms per phase. The field has a full-pitch winding of 10 coils with 10 deg between coils and 17 turns per coil.

(a) Find the effective resistance of the armature at 75°C.

(b) Find the armature reaction in ampere-turns per pole.

(c) Find the armature reaction in equivalent field amperes.

(d) Find the field current required to give rated terminal voltage when the generator delivers rated kva to an inductive load at 0.8 power factor.

(e) Find the voltage regulation for this load. The data for the open-circuit saturation curve is given below:

Field current, amp..........	100	300	400	450
Line voltage, volts..........	6,410	13,160	14,560	15,050

17-4. The following additional data are given for the synchronous generator of Prob. 17-3. With a symmetrical three-phase short circuit across the generator terminals, a field current of 187 amp is required to produce rated current in the armature windings. When delivering rated current to an inductive zero-power-factor load, a field current of 550 amp is required to give rated voltage at the generator terminals.

(a) Determine the Potier leakage reactance.

(b) Determine the unsaturated synchronous reactance.

(c) Determine the saturated synchronous reactance for an inductive load of rated kva at 0.8 power factor.

(d) Determine the field current and the voltage regulation for the load of part (c) by the synchronous-impedance method.

(e) Determine the field current and the voltage regulation for the load of part (c) by the ASA method.

17-5. Using the synchronous-impedance method, calculate the field current required to give rated voltage across the terminals of the synchronous generator whose characteristic curves are given in Fig. 154 when the generator is delivering 100,000 kva to an inductive load at 0.866 power factor. Determine also the voltage regulation and the power angle for the same load.

17-6. Solve Prob. 17-5 using the ASA method.

17-7. The following information was obtained from tests on a 1,000-kva 13,800-volt 60-cps Y-connected synchronous generator:

Effective resistance, 2.18 ohms per phase
Leakage reactance, 33.0 ohms per phase

Open-circuit Characteristic

Line voltage, volts	Field current, amp
8,800	50
15,620	110

On short circuit, a field current of 41.2 amp gives a line current of 42 amp. Using the ASA method, find the field current needed to give rated voltage when the synchronous generator is delivering 50 amp at 0.85 power factor (lagging current).

17-8. A 3-phase Y-connected 1,500-kva 60-cps 5,500-volt synchronous generator has an armature effective resistance of 0.36 ohm per phase and a synchronous reactance of 8.53 ohms per phase.

The data for the open-circuit characteristic are

Field current, amp..............	150	250	350	400
Line potential, volts...........	5,100	6,500	7,100	7,300

Find the power angle, the voltage regulation, and the field current for the rated voltage with 20 per cent overload current at 0.85 power factor (inductive load).

17-9. The direct-axis and quadrature-axis synchronous reactances for the synchronous generator of Prob. 17-8 are 8.53 and 3.00 ohms per phase, respectively. Using the two-reactance method, solve for the quantities specified in Prob. 17-8.

17-10. A 1,000-kva 2,300-volt 3-phase 60-cps water-wheel–driven synchronous generator has a direct-axis synchronous reactance of 4.8 ohms and a quadrature-axis synchronous reactance of 2.4 ohms. Calculate the open-circuit voltage and the power angle for an excitation required to give rated terminal voltage at rated kva with a power factor of 0.8 (lagging current). Neglect the armature resistance.

Synchronous Generators—Chapter 18

18-1. Determine the efficiency of the synchronous generator, whose characteristic curves are given in Fig. 154, for an inductive load of 100,000 kva and a power factor of 0.886. Assume a ventilation loss of 350 kw.

18-2. The following additional data apply to the synchronous generator of Probs. 17-3 and 17-4. The short-circuit loss at rated current is 315 kw. The friction and windage loss at rated speed is 60 kw. The field resistance at 25°C is 0.426 ohm. The ventilation loss may be neglected. Core-loss data are

Open-circuit core loss, kw...............	340	357	375
Terminal (line) voltage, volts...........	11,000	11,250	11,500

Determine the full-load efficiency for an 0.8-power factor inductive load.

Synchronous Generators—Chapter 19

19-1. Suppose a sinusoidal, 60-cycle voltage source is connected to a load through a resistanceless inductor of 2.65 henrys. Calculate the power supplied to the load when the source voltage is 110 volts, the load voltage is 100 volts, and the source voltage leads the load voltage by 30 deg.

19-2. A 3-phase Y-connected 60-cps 1,500-kva 6,600-volt synchronous generator has a direct-axis synchronous reactance of 8.00 ohms per phase and a quadrature-axis synchronous reactance of 4.00 ohms per phase. Assume the effective resistance to be negligible. With the field current adjusted so as to give an excitation voltage E_a' of $6,600/\sqrt{3}$ volts per phase and rated voltage maintained across the generator terminals, (a) plot the generator output versus the power angle and (b) determine the maximum possible generator output under these conditions.

19-3. The characteristic curves for a 93,750-kva synchronous generator are given in Fig. 154.

(a) Determine the per-unit value of the unsaturated synchronous reactance of this machine. Use rated kva as a base.

(b) A factor sometimes used in stability studies of synchronous machines is the short-circuit ratio. It is defined as the ratio of the field current required for rated voltage on open circuit to the field current required for rated armature current on short circuit. Determine the short-circuit ratio of this generator.

(c) An approximate value of saturated synchronous reactance may be found by the method for finding the unsaturated synchronous reactance if the open-circuit characteristic is used in place of the air-gap line. Using the rated voltage point on the open-circuit characteristic, find the per-unit value of this approximate synchronous reactance. Compare the reciprocal of this value with the short-circuit ratio.

(d) Using the reactance of part (c), find the voltage regulation and power angle of the generator for an inductive load of rated kva at 0.8 power factor.

Synchronous Motors—Chapter 21

21-1. A 3-phase 60-cps Y-connected synchronous motor is rated to receive 1,000 kva at 13,200 volts. Using rated kva as a base, the per-unit values of effective resistance and synchronous reactance are 0.0114 and 0.418, respectively. The data for the open-circuit characteristic are

Field current, amp..........	50	110	140	150
Line voltage, volts..........	8,800	15,600	17,250	18,900

Determine the field current and the power angle when the motor receives 1,000 kw at unity power factor from a 13,200-volt source.

21-2. The field current of the synchronous motor of Prob. 21-1 is now increased until it operates at a power factor of 0.8. If the motor still draws 1,000 kw from the 13,200-volt source, what would be its field current and power angle?

21-3. A three-phase Y-connected synchronous motor has negligible armature effective resistance and a synchronous reactance of 12.0 ohms per phase. The motor receives 12,000 kw at a line potential of 13,200 volts. If the field current is adjusted so that the excitation voltage is 9,000 volts per phase

(a) What is the power angle?

(b) At what power factor is the motor operating?

(c) What is its armature current?

21-4. A synchronous motor, whose armature windings are Y-connected, has an armature effective resistance of 0.94 ohm per phase, a direct-axis synchronous reactance of 20 ohms per phase, and a quadrature-axis synchronous reactance of 12 ohms per phase. The motor receives a line current of 50 amp at a line voltage of 12,000 volts and a power factor of 0.8.

(a) What is the power input to the motor?

(b) What is its excitation voltage E_a' if the excitation is greater than normal?

21-5. A 2,300-volt 3-phase Y-connected synchronous motor has an effective armature resistance of 0.2 ohm per phase and a synchronous reactance of 2.2 ohms per phase. It is operating at 60 per cent lagging power factor with a line current of 200 amp. Determine (a) the excitation voltage E_a' and (b) the power angle.

21-6. If the motor of Prob. 21-5 has its field current increased so that E_a' is increased by 25 per cent but the load on the motor is unchanged, calculate approximate values of (a) the power angle, (b) the armature current, and (c) the power factor.

21-7. A 3-phase 2,300-volt Y-connected 2,500-kva synchronous motor has an armature effective resistance of 0.021 ohm per phase, a synchronous reactance of 0.95 ohm per phase, and a field current of 4.50 amp when rated current is taken from the line at rated voltage and unity power factor. Find the field current required when the load is so changed that the line current is 0.9 rated value and the motor operates at a leading power factor of 0.8. Assume the open-circuit characteristic to be a straight line.

21-8. (a) For the 45-kva synchronous motor of the example given on page 316, determine its field current when connected to a source of rated voltage at rated frequency and its load is such that it receives one-half of rated kva with a leading current and a power factor of 0.8. Use the two-reactance method.

(b) Find the field current for the load of part (a) by the ASA method.

(c) Find the efficiency for the load specified in part (a).

21-9. A 3-phase Y-connected synchronous motor is connected to a 60-cps 1,732-volt 3-phase source. When it draws 100 amp at 0.8 leading power factor, the power angle is measured with a stroboscope and found to be 17.1 deg. Neglect the armature resistance.

(a) Determine the quadrature axis synchronous reactance of this motor.

(b) If the excitation voltage is now determined to be 1,544 volts per phase, determine the direct-axis synchronous reactance.

Synchronous Motors—Chapter 22

22-1. The power angle of a synchronous motor is observed with stroboscopic light. While the motor is delivering a constant power, both the applied potential and the frequency are reduced 10 per cent.

(a) If the original power angle was 30 deg, what is the power angle after the potential and frequency are reduced?

(b) What would be the power angle if the motor delivers a constant torque instead of a constant power? State any assumptions made in obtaining your solution.

22-2. A synchronous motor is operating at rated voltage. The per-unit value of its synchronous reactance is 0.577, and its effective resistance may be assumed to be negligible.

(a) If its field current is adjusted so as to give unity power factor at full load, at what per-unit value of load will the motor break down as the load is increased? Neglect hunting.

(b) What is the per-unit value of current at breakdown?

(c) What is the power factor at breakdown?

22-3. A 3-phase 1,732-volt Y-connected synchronous motor has a negligible armature effective resistance and a synchronous reactance of 10.0 ohms per phase. The friction and windage plus the core loss is 9.0 kw and may be assumed constant. The greatest excitation voltage that may be obtained is 2,500 volts per phase. The motor delivers an output power of 390 hp.

(a) What is the maximum angle by which the current can lead the voltage drop?

(b) What is the magnitude of the armature current under the conditions of part (a)?

(c) What is the smallest excitation voltage for which the motor will remain "in step"?

(d) What is the magnitude and power-factor angle of the armature current for the excitation of part (c)?

Synchronous Motors—Chapter 24

24-1. A 3-phase Y-connected 1,000-kva 13,200-volt 60-cps synchronous motor has an armature effective resistance of 2.18 ohms per phase and a synchronous reactance of 45.2 ohms per phase.

(a) From the circle diagram determine and plot the V curve of the motor for a constant electromagnetic power equal to that existing when the motor has an input equal to rated kva at unity power factor.

(b) Plot the V curve of this motor for a constant electromagnetic power equal to that existing when the input is one-half rated kva at unity power factor.

The open-circuit characteristic is as follows:

Field current, amp	50	110	140	180
Open-circuit line volts	8,800	15,600	17,250	18,900

24-2. A 3-phase Δ-connected 15-kva 230-volt 60-cps synchronous motor has an armature effective resistance of 0.302 ohm per phase and a synchronous impedance of 7.55 ohms per phase. The open-circuit characteristic is as follows:

Field current, amp	2.0	4.0	5.0	6.0	7.5
Open-circuit line volts	104	199	240	275	315

(a) From the circle diagram determine and plot the V curve for this motor for a constant electromagnetic power equal to that existing when the motor receives rated kva at unity power factor.

(b) Plot the V curve for a constant electromagnetic power equal to that existing when the motor receives one-half rated kva at unity power factor.

Synchronous Motors—Chapter 25

25-1. An induction motor load taking 2,000 kw is operating at a power factor (lagging) of 0.65. Determine the kva capacity of the synchronous condenser needed to raise the power factor to (a) unity, (b) 0.90, and (c) 0.80.

25-2. A 2,300-volt 3-phase industrial plant of 2,000-kva capacity supplies 1,800 kw to an induction motor load at 0.90 power factor (lagging current). An extension to the factory involves the installation of an additional 250-hp motor. It is proposed to install a synchronous motor whose full-load efficiency is 0.945 with its field adjusted to make the power factor of the combined load unity. Determine (a) the input rating of the synchronous motor in kva, and (b) the power factor at which it must operate.

25-3. A three-phase transmission line supplies 1,000-kw to a three-phase induction motor load at 2,300 volts line potential and 0.80 power factor. A three-phase synchronous motor of 250 kva capacity, connected in parallel with the induction motor load, takes 150 kw from the line. What is the best power factor that may be obtained for the combined load without exceeding the capacity of the synchronous motor?

Parallel Operation of Synchronous Generators—Chapter 26

26-1. Two synchronous generators of the same rating are operating in parallel. The first delivers 2,630 kw at 0.65 power factor, and the second, 5,270 kw at 0.92 power factor.

(a) If the currents in both generators are lagging their respective voltages, what adjustments should be made to have these generators operate under the best conditions?

(b) When these adjustments have been made, what power will each generator deliver, and at what power factor will it operate?

26-2. Two synchronous generators A and B are operating in parallel. Measurements show that generator A is delivering 100 amp at 0.866 lagging power factor and that generator B is delivering 142 amp at 0.707 lagging power factor.

(a) What is the power factor of the load?

(b) What percentage of the load is each supplying?

(c) In order that both may operate at the power factor of the load, in which direction does the field current of each have to be varied?

26-3. A 3-phase 5,000-kva 13,200-volt 60-cps synchronous generator is connected to a large power system and delivers 5,000 kw at rated voltage and unity power factor. Its per-unit synchronous reactance is 1.1, and its effective resistance may be neglected.

(a) Find its excitation voltage and its power angle.

(b) If now the generator's field current is increased so as to increase its excitation voltage by 10 per cent, find its power output, its armature current, and its power factor. Assume no change in line voltage.

(c) If, with its excitation adjusted as in part (a), the driving torque of its prime mover is reduced by 5 per cent, what will be its output, its armature current, and its power factor? Assume no change in line frequency. State any additional assumptions made in arriving at your answers.

26-4. A 50,000-kva 13,200-volt 60-cycle water-wheel generator is operating on a large power system and delivers 11,600 kw at unity power factor and rated voltage. The per-unit values of its direct-axis and quadrature-axis synchronous reactances are 1.15 and 0.75, respectively. Its effective resistance may be neglected.

(a) Find the excitation voltage and power angle of the generator.

(b) If now the generator loses its field excitation, so that E_a' becomes essentially zero, what will be its current and power factor? Assume the line voltage to remain constant.

26-5. A synchronous generator is driven by a steam turbine and supplies 1,000 kw at 0.8 lagging power factor and 2,300 volts to a large power system. The neutral of the generator is grounded.

(a) If the generator has a synchronous reactance of 4.66 ohms and a negligible effective resistance, what is its excitation voltage and its power angle?

(b) If now the steam is cut off from the turbine, what will be the current and the power factor of the generator? Neglect the generator and turbine losses and assume the generator field current and the system frequency to remain unchanged.

(c) If the losses of the generator plus the power required to drive the turbine equaled 100 kw, what would be the answer to part (b)?

Parallel Operation of Synchronous Generators—Chapter 27

27-1. Two three-phase synchronous generators, connected in parallel, are driven by water wheels whose speed characteristics are as follows. The speed of the first water wheel falls uniformly from 624 rpm at no load to 600 rpm with a load of 1,000 kw on the generator. The speed of the second water wheel falls uniformly from 630 to 600 rpm with a load of 1,000 kw on the generator.

(a) What will be the output of each generator when the combined load is 1,000 kw?

(b) What is the greatest load that can be delivered without exceeding 1,250 kw on either generator?

27-2. A 3-phase 1,500-kva 2,300-volt synchronous generator is connected in parallel through transformers with a 3-phase 5,000-kva 13,200-volt synchronous generator. The first generator has four poles and is driven by a turbine whose speed falls from 1,830 rpm at no load to 1,788 rpm with a load of 1,500 kw on the generator. The second generator has two poles and is driven by a turbine whose speed falls from 3,684 rpm at no load to 3,552 rpm with a load of 5,000 kw on the generator.

(a) When the combined load on the two generators is 6,000 kw, what is the load on each and what is the frequency?

(b) If the governor of the turbine driving the second generator is readjusted so as to shift its speed-load characteristic parallel to itself until the generators divide the 6,000-kw load properly, what load will each carry and at what frequency will the generators now operate?

27-3. A three-phase synchronous generator, generator A, is operating on a power system in parallel with several other generators. Its prime mover has a speed-load characteristic such that the generator frequency falls from 60 cps at no load at 57.6 cps with a load of 20,000 kw on the generator. A frequency of 60 cps is maintained on the system by adjusting the governor on the prime mover of generator A. This adjustment shifts the speed-load characteristic but always parallel to itself. With the initial adjustment, a total system load of 100,000 kw gives a frequency of 60 cps with generator A floating on the line (no output).

(a) If the system load is now increased to 110,000 kw and the prime mover of generator A is adjusted to give 60 cps, what will be the output of generator A?

(b) If with no further change in governor settings the 110,000-kw load is reduced until the output of generator A is again zero, what will be the frequency of the system?

27-4. Three 50,000-kva synchronous generators A, B, and C are driven by prime movers of such characteristics that the frequency of generator A decreases by 1.8 cps from no load to full load at unity power factor, while the frequency of generators B and C decreases by 2.2 cps from no load to full load at unity power factor.

(a) If the generators are each delivering 20,000 kw at 60 cps and generator C is disconnected from the line, what will be the load on generators A and B and what will be the frequency of the system?

(b) If the system load is now decreased to 30,000 kw, what will be the load on each generator and what will be the frequency of the system?

Polyphase Induction Motors—Chapter 30

30-1. The stator of a three-phase induction motor has three slots per pole per phase. The stator winding is Y-connected, and there are two coil sides per slot with one turn per coil. The coil pitch is nine slots. The motor receives balanced currents of 7.07 amp with the current in phase b lagging the current in phase a by 120 electrical degrees.

(a) Lay out a developed view of the stator winding.

(b) Make a plot of the distribution along the air gap of the mmf due to the stator currents at the time when the current in phase a is a maximum.

(c) Make a similar plot one-third of a cycle later than in (b).

(d) What would be the effect on the shape of the curves plotted in (b) and (c) of increasing the number of slots per pole per phase, of using a fractional-pitch winding, or of using a fractional-slot winding?

30-2. The stator for a given induction motor has 48 slots. A three-phase four-pole winding is to be placed in these slots. The winding is to have two coil sides per slot, and the coil pitch is to be nine slots. Proceeding around the air gap the top coil sides (nearest the air gap) are to be numbered in order 1, 2, 3, etc., and the bottom coil sides are to be numbered 1′, 2′, 3′, etc., such that 1′ and 1 are opposite sides of the same coil and 2′ and 2, 3′ and 3, etc., are likewise.

(a) Make a winding table for the stator winding.

(b) Draw a winding diagram.

30-3. A polyphase induction motor runs at 1,196 rpm at no load and at 1,150 rpm at full load on a 60-cps source.

(a) What is its synchronous speed?

(b) How many poles has the motor?

(c) What is its slip at no load and at full load?

(d) What is the frequency of its rotor currents at no load and at full load?

30-4. The speed of a four-pole 60-cycle induction motor at full load is 1,750 rpm. The phase voltage induced in the rotor windings at starting (standstill) is 120 volts. The rotor resistance is 6 ohms per phase, and the rotor reactance at starting is 15 ohms per phase. Assume the stator flux to remain the same at full load as at starting.

(a) Determine the slip at full load.

(b) Determine the rotor current at starting.

(c) Determine the rotor current at full load.

30-5. (a) A 3-phase 4-pole 60-cps 440-volt induction motor has a Y-connected stator winding. If the maximum flux per pole is to be 0.007 weber, how many effective turns must be wound on the stator?

(b) If the motor is rewound for eight poles using the same number of turns of the same size wire as in part (a), what should be the voltage rating of the rewound machine? Assume the maximum air-gap flux density to be the same as in part (a), and neglect any change in the pitch and breadth factors of the winding.

30-6. The stator of a certain three-phase six-pole induction motor has 54 slots with two coil sides in each slot and eight turns per coil. The coil pitch is ⅚. The resistance and leakage reactance of the stator winding are 0.114 and 0.229 ohm per phase, respectively. The rotor of this motor is of the squirrel-cage type with a single copper bar in each of the 65 rotor slots. The resistance of each rotor bar, corrected to account for the resistance of the end rings, is 4.0×10^{-4} ohm. The leakage reactance of the rotor (referred to the stator) is 0.229 ohm per phase.

(a) If the stator winding is Y-connected, how much voltage must be impressed across the stator terminals at standstill to produce a current of 40 amp in the stator winding?

(b) How much current would flow in each rotor bar?

Suggestion: It is often convenient with a squirrel-cage winding to consider the number of phases in the rotor to be equal to the number of rotor bars per pair of poles and the number of turns per phase to be equal to one-half the number of pairs of poles. Neglect the magnetizing component of the stator current.

Polyphase Induction Motors—Chapter 31

31-1. A Δ-connected induction motor draws 100 amp from each line of a three-phase source and develops 40 lb-ft torque at standstill. What will be the current and torque when connected to the same source if the machine is reconnected in Y?

31-2. A polyphase induction motor with a wound rotor breaks down at 250 per cent of full-load torque and at a slip of 0.20. What would be the maximum torque and the slip at which it occurs if (a) the rotor resistance were doubled? (b) The impressed voltage were halved? (c) The frequency of the impressed voltage were halved? Assume the stator resistance to be 0.1 of the total leakage reactance.

31-3. A certain 3-phase 60-cycle 4-pole induction motor runs at 1,750 rpm at full load.

(a) What would be its approximate speed if the load were reduced to a value requiring one-half of full-load torque?

(b) If its rotor resistance were increased to six times its original value, what would then be the approximate speed when developing full-load torque?

(c) With the rotor resistance of part (b), what would be the approximate speed when developing one-half of full-load torque?

31-4. The stator resistance and leakage reactance of a 15-hp 3-phase 240-volt 60-cps 6-pole induction motor are 0.107 and 0.320 ohm per phase, respectively. The rotor resistance and leakage reactance are 0.210 and 0.320 ohm per phase, respectively, when referred to the stator. The exciting current is 6.5 amp. When this motor is running at a speed of 1,118 rpm, what would be its internal torque in pound-feet?

31-5. A certain three-phase induction motor has a starting torque of 100 lb-ft when a balanced three-phase voltage of 240 volts is impressed across its terminals. If the line voltage became somewhat unbalanced such that the voltages across a pair of lines are 220, 240, and 240 volts, what would be the starting torque?

Polyphase Induction Motors—Chapter 33

33-1. A 15-hp 220-volt 3-phase 60-cps 4-pole 1,725-rpm induction motor has a Y-connected stator winding. The equivalent circuit constants are

$$r_1 = 0.15 \text{ ohm per phase} \qquad r_2 = 0.18 \text{ ohm per phase}$$
$$x_1 = 0.31 \text{ ohm per phase} \qquad x_2 = 0.31 \text{ ohm per phase}$$

where r_2 and x_2 are referred to the stator. The exciting current is 8.1 amp and may be assumed constant with the motor running at rated voltage. Because of the reduced flux density the exciting current may be neglected under starting conditions.

Find (a) the slip at which maximum torque occurs, (b) the maximum torque in pound-feet, (c) the starting torque at full voltage, (d) the starting current, and (e) the full-load torque.

33-2. The motor of Prob. 33-1 has, at full load, an efficiency of 0.90 and a power factor of 0.85.

(a) Find the full-load current.

(b) If the starting current is reduced by use of a compensator, what voltage applied to the motor terminals will limit the current in the supplying lines to 150 per cent of full-load current?

(c) What is the starting torque when the compensator of part (b) is used?

33-3. If the motor of Prob. 33-1 has a wound rotor, then, to limit the starting current, resistance may be added in series with the rotor circuit instead of using a compensator.

(a) How much added resistance would be needed to limit the load component of the starting current to 150 per cent of full-load current? Assume the ratio of transformation to be 1.1 to 1.0 (primary to secondary). Do not neglect the exciting current, but assume it to remain at 8.1 amp.

(b) If the rotor resistance of part (a) were added, what would be the starting torque?

33-4. A 500-kva transformer is used to step-down the voltage of a power line from 13,200 volts to 440 volts, three phase for use in a factory. This transformer has a per-unit impedance of 0.05 based on its rated kva. The secondary supplies power to a group of induction motors and also, through additional transformers, supplies the lighting circuit. Because the dip in voltage caused by the induction motor starting produces light flicker, it is decided to limit this dip to 3 per cent. The following assumptions are made:

1. The power line voltage is assumed to remain constant.

2. All circuit resistances are neglected.

3. Motor starting current is estimated to be six times full-load current when the motor is started at rated voltage.

4. The full-load efficiency and power factor of the motors is assumed to be 0.90 and 0.83, respectively.

(a) What is the maximum size of motor (in horsepower) that can be started without exceeding the 3 per cent dip in voltage if the motor is started at full voltage?

(b) If series reactors sufficient to limit the voltage at the motor terminals to 80 per cent of full voltage at starting are used, what maximum size of motor can be started?

(c) If a compensator with 80 per cent taps is used, what maximum size of motor can be started?

33-5. A 15-hp 220-volt 3-phase induction motor has, at full load, an efficiency of 0.87 and a power factor of 0.85. At standstill the motor draws six times its full-load current and develops 1.5 times full-load torque. An autotransformer or compensator is installed to reduce the starting current and to give full-load torque at starting. Neglect the exciting current. Determine (a) the voltage applied to the motor at starting, (b) the full-load line current, (c) the current drawn by the motor at starting, and (d) the line current supplied to the primary of the autotransformer when starting the motor.

33-6. (a) Determine the full-load torque of a 500-hp 2,300-volt 25-cps 12-pole 3-phase induction motor if the slip at full load is 0.02.

(b) If this motor develops three-fourths of full-load torque when started at full voltage, what would be the starting torque if the starting voltage were reduced to one-half of rated voltage?

(c) If this motor is of the wound-rotor type, and the resistance of the Y-connected rotor winding is 0.20 ohm per phase, how much resistance should be added externally to limit the starting current to rated value if (1) the external resistance is Y-connected and (2) the external resistance is Δ-connected?

(d) What starting torque would you expect when the resistance of part (c) was used?

33-7. At standstill, with 2,300 volts, three-phase impressed on the stator terminals of a three-phase wound-rotor type induction motor, the potential across the open rotor slip rings is 600 volts. With the rotor blocked and the slip rings short-circuited, the line potential, line current, and power supplied to the stator are 750 volts, 300 amp, and 120 kw, respectively. With the rotor blocked, what Y-connected resistance should be connected to the rotor slip rings so that the line current will be 300 amp when 2,300 volts is impressed on the stator? Neglect the magnetizing and core-loss components of the stator current.

33-8. A 30-hp 6-pole 60-cycle 3-phase wound-rotor induction motor is started by adding resistance to its rotor circuit. If the added resistance per phase is nineteen times the resistance per phase of the rotor winding, the starting current is found to be equal to the full-load current of the motor.

(a) What is the slip at full load?

(b) What is the speed at full load?

(c) What is the torque at full load?

(c) What is the starting torque when the added resistance is used?

33-9. A 100-hp 3-phase 60-cps 24-pole induction motor has a Y-connected rotor winding whose resistance is 0.105 ohm per phase. With the slip rings short-circuited, the motor runs at 294 rpm at full load. The motor drives a fan which requires 100 hp at 294 rpm. What Y-connected resistance should be connected to the rotor slip rings in order that the fan will be driven at 150 rpm? Assume that the slip-torque curve for the motor is a straight line from no load to full load and that the torque required to drive the fan varies as the square of the speed.

33-10. Two wound rotor-type motors A and B are wound for 10 and 12 poles, respectively. The ratio of the effective stator to rotor turns is the same for both motors. The two motors are mechanically coupled. The frequency of the supply voltage is 50 cps.

(a) Calculate the approximate no-load speed when motor A only is connected to the supply and its rotor is short-circuited.

(b) Calculate the approximate no-load speed when motor B only is connected to the supply and its rotor is short-circuited.

(c) The stator of motor A is connected to the supply and the rotor of A is connected to the rotor of B. It is assumed that the connection is so made that the torques of the two motors are aiding. The stator of B is shorted. Calculate the approximate no-load speed of the set.

33-11. The stator of an induction motor is to have 12 slots.

(a) Lay out, in developed view, a single phase of a three-phase two-pole winding for this machine using two coil sides per slot and a coil pitch of three slots. Locate the position of the north and south poles when this winding only is carrying current.

(b) Show on your drawing the minimum changes required to make your winding a four-pole winding. Locate the position of the north and south poles when the changes in connections have been made and this winding alone is carrying current.

Polyphase Induction Motors—Chapter 34

34-1. A 3-hp 3-phase 60-cps 440-volt induction motor has a Y-connected stator winding. At a given load its input is 2,652 watts with a line current of 4.19 amp and its speed is 1,739 rpm. The stator d-c resistance measured between two terminals is found to be 5.79 ohms at 75°C (the assumed operating temperature). The stray load loss is 34 watts. A no-load test gives the following data at a temperature of 25°C:

$$P = 188 \text{ watts} \qquad I = 2.0 \text{ amp} \qquad V = 440 \text{ volts}$$

Determine (a) the horsepower output, (b) the efficiency, and (c) the power factor for the load specified above. (d) If the stator winding had been Δ-connected and the above data still applied, what change would this make in the answers to (a), (b), and (c)?

34-2. A 5-hp 4-pole 3-phase 60-cps 440-volt induction motor has a Y-connected stator winding. At a given load its input is 4,410 watts with a line current of 6.92 amp and its speed is 1,751 rpm. The stator d-c resistance measured between two terminals is 3.78 ohms at 75°C. The stray load loss is 60 watts. A no-load test gives the following data at a temperature of 25°C:

$$P = 296 \text{ watts} \qquad I = 2.94 \text{ amp} \qquad V = 440 \text{ volts}$$

Determine (a) the horsepower output, (b) the shaft torque, and (c) the efficiency for the load specified above.

34-3. The following test data were obtained on a 350-hp 4,000-volt 60-amp 3-phase 60-cps 26-pole class A induction motor.

The d-c resistance of stator measured between terminals is 1.360 ohms. With motor running at no load the following values were recorded:

$$P = 15,050 \text{ watts} \qquad I = 37.6 \text{ amp} \qquad V = 4,000 \text{ volts}$$

With the rotor blocked and the stator connected to a 15-cps supply,

$$P = 23,300 \text{ watts} \qquad I = 59.9 \text{ amp} \qquad V = 341 \text{ volts}$$

All test data were taken at a temperature of 25°C. Assume an operating temperature of 75°C. The stray-load loss was found to be 5,000 watts at rated current. Find (a) the horsepower output, (b) the efficiency, and (c) the power factor when this motor is operating at a slip of 0.02.

34-4. Repeat Prob. 34-3 for a slip of 0.04.

34-5. For the motor of Prob. 34-3, determine the maximim internal torque in pound-feet and the slip at which it occurs.

34-6. If the rotor of the motor of Prob. 34-2 is blocked with full voltage at rated frequency applied to the stator, the motor takes 18,800 watts with a line current of 41.4 amp at a temperature of 25°C. Assuming the starting temperature to be 25°C, determine the internal torque developed at starting with full voltage applied.

Polyphase Induction Motors—Chapter 35

35-1. An induction generator and a synchronous generator are operated in parallel and supply an inductive load of 600 kw at 0.8 power factor. If the induction generator supplies 100 kw at a power factor of 0.707, at what power factor does the synchronous generator operate?

35-2. A six-pole induction generator and a six-pole synchronous generator are operated in parallel. The prime movers of both generators have identical speed-load characteristics. The speed of each falls from 1,250 rpm with no load on the generator to 1,180 rpm when its generator delivers 100 kw. The output of the induction generator is proportional to its slip, and the slip for an output of 100 kw is 0.02 (negative). When the frequency is 60 cycles, what is the output of each generator?

Single-phase Induction Motors—Chapter 36

36-1. Calculate the horsepower output, the efficiency, and the power factor of the motor used in the example given on page 491 for a slip of 0.01.

36-2. The following test data were obtained on a 0.5-hp 115-volt 60-cps 4-pole single-phase induction motor:

At no load,

$$V = 115 \text{ volts} \qquad I = 4.81 \text{ amp} \qquad P = 105.0 \text{ watts}$$

Rotor blocked, with starting winding disconnected

$$V = 115 \text{ volts} \qquad I = 27.2 \text{ amp} \qquad P = 1{,}308 \text{ watts}$$

Friction and windage loss = 20 watts.

Ohmic resistance of stator between terminals = 0.830 ohm.

Ratio of effective resistance to ohmic resistance of the rotor at 60 cps is 1.2 and at 120 cps is 1.6.

Neglect any difference between the ohmic and 60-cps resistance of the stator. Assume an operating temperature of 65°C and that the temperature during all the above tests was 25°C. Calculate the horsepower output, the efficiency, and the power factor of this motor when it is running with a slip of 0.02.

36-3. For the motor of Prob. 36-2 when operating at a slip of 0.02 find (a) the positive-sequence and the negative-sequence components of the phase voltages, (b) the positive-sequence and the negative-sequence components of the line voltages, and (c) the ratio of the revolving fluxes due to the positive-sequence and the negative-sequence components of the currents.

Single-phase Induction Motors—Chapter 38

38-1. A 1.0-hp 220-volt 2-phase 60-cps squirrel-cage induction motor is to be operated from a 220-volt single-phase line. When the rotor is stationary and either winding is connected across the line, the current is 30 amp at 0.20 power factor.

(a) What capacitance must be connected in series with one of these windings to produce starting torque and to give 30 amp in each winding at start?

(b) Under the conditions of part (a), what will be the starting current drawn from the single-phase line?

38-2. The following formula is sometimes used for calculating the starting torque of capacitor and split-phase types of single-phase induction motors:

$$T_{st} = \frac{p}{2\pi f} \times \frac{550}{746} \times r_2 I_m I_a \times \frac{N_a}{N_m} \times \sin \alpha$$

I_m and I_a are the currents in the main and auxiliary windings, respectively, N_m and N_a are the effective turns in each winding, r_2 is the normal frequency rotor resistance referred to the main winding, and α is the phase angle between the two currents. Prove this formula. *Hint:* Use symmetrical components as applied to two-phase machine and formula (243), page 408. Neglect the exciting currents.

38-3. The following data apply to a ¼-hp 6-pole 60-cps 1,140-rpm 115-volt split-phase type single-phase induction motor.

Blocked-rotor test at 60 cps with auxiliary winding disconnected:

$$V = 115 \text{ volts} \qquad I = 14.9 \text{ amp} \qquad P = 1,225 \text{ watts}$$

Blocked-rotor test at 60 cps with main winding disconnected and auxiliary winding connected to the source:

$$V = 115 \text{ volts} \qquad I = 9.35 \text{ amp} \qquad P = 1,012 \text{ watts}$$

The ratio $N_m/N_a = 1.13$, and the rotor resistance referred to the main winding equals 3.56 ohms.

(a) Find the full-load torque of this motor.

(b) Using the formula of Prob. 38-2, find the starting torque when both windings are connected to a rated-voltage source.

(c) Find the starting current drawn from the source.

38-4. The main and auxiliary windings of a single-phase induction motor are displaced by 90 electrical degrees on the stator. With the rotor blocked and the main winding only connected to a 60-cps source, the following readings were obtained:

$$V = 115 \text{ volts} \qquad I = 15.2 \text{ amp} \qquad P = 995 \text{ watts}$$

With the rotor blocked and the auxiliary winding only connected to the 60-cps source:

$$V = 115 \text{ volts} \qquad I = 5.20 \text{ amp} \qquad P = 527 \text{ watts}$$

The ratio $N_a/N_m = 1.30$ and the rotor resistance referred to the main winding is 4.12 ohms. The name plate gives the following data: ¼-hp 6-pole 60-cps 1,140-rpm 115-volt single-phase induction motor.

(a) Find the full-load torque of this motor.

(b) Using the formula of Prob. 38-2, calculate the starting torque.

(c) Find the starting current.

38-5. An 8-μf capacitor is connected in series with the auxiliary winding of the motor of Prob. 38-4. Neglect any resistance in the capacitor. Find (a) the starting torque, (b) the starting current, and (c) the voltage across the capacitor.

38-6. A 166-μf capacitor is connected in series with the auxiliary winding of the motor of Prob. 38-4. Assume the resistance of the capacitor to be negligible. Find (a) the starting torque, (b) the starting current, and (c) the voltage across the capacitor.

38-7. The following data apply to a $\frac{1}{2}$-hp 4-pole 60-cps 1,725-rpm 115-volt single-phase induction motor:

Blocked-rotor test at 60 cps with auxiliary winding disconnected:

$$V = 115 \text{ volts} \qquad I = 28.5 \text{ amp} \qquad P = 2,320 \text{ watts}$$

Blocked-rotor test at 60 cps with main winding disconnected:

$$V = 115 \text{ volts} \qquad I = 20.2 \text{ amp} \qquad P = 1,985 \text{ watts}$$

The ratio $N_a/N_m = 1.02$ and the rotor resistance referred to the main stator winding is 2.06 ohms. Find (a) the full-load torque, (b) the starting torque, and (c) the starting current.

38-8. (a) For the motor of Prob. 38-7, find the size capacitor required to cause the current in the auxiliary winding to be in quadrature with the current in the main winding at starting.

(b) With this capacitor inserted, what is the starting torque and the starting current?

Synchronous Converters—Chapter 47

47-1. A 6-phase 1,500-kw synchronous converter receives power through three transformers from a 3-phase 13,200-volt source and delivers power at 600 volts (d-c). The transformers are Y-connected on the primary side and have a double-Δ connection on the secondary side. If the efficiency and power factor of the converter are 0.95 and 0.90, respectively, determine the current and potential ratings of each of the three transformers.

47-2. (a) Solve Prob. 47-1 if the transformers are Δ-connected on the primary side and are connected double-Y on the secondary side.

(b) Solve Prob. 47-1 if the transformers are Δ-connected on the primary side and have a diametrical connection on the secondary side.

47-3. A 6-phase 100-kw 240-volt synchronous converter receives power through three single-phase transformers from a 3-phase 6,600-volt source. The transformer primaries are Δ-connected, and the secondaries have a diametrical connection. If the efficiency of the converter is 0.90 when it is delivering the rated load and operating at 0.85 power factor, determine the current and voltage ratings of the transformers required.

Synchronous Converters—Chapter 49

49-1. A 500-kw 6-phase 8-pole 600-volt 25-cps synchronous converter draws a line current of 415 amp at 212 volts per phase when operating at full load and unity power factor. The field current required for this condition is 2.8 amp. The armature has 144 slots with six inductors per slot. The field is wound with 1,200 turns on each pole. The resistance of the field winding without rheostat is 154 ohms. With an output of 400 kw at 600 volts, what is the least power factor at which the converter can be operated when overexcited? Neglect any change in efficiency.

49-2. A 1,500-kw 6-phase 8-pole 600-volt 25-cps synchronous converter has an armature resistance, measured between d-c terminals, of 0.00516 ohm at 25°C. At full load and when adjusted to unity power factor its field current is 8.5 amp. If core loss at normal voltage is 17.8 kw and its friction and windage loss is 11.0 kw, find the efficiency of this converter when operating at full load and unity power factor. Assume an operating temperature of 75°C.

Mercury-arc Rectifiers—Chapter 51

51-1. A full-wave single-phase mercury-arc rectifier supplies 50 amp at 120 volts to a d-c load. The arc drop is 18 volts. The rectifier is connected through a transformer to an a-c source. If the rectifier is operated without phase control and overlap is neglected, find the voltage and rms current which must be supplied to the rectifier by the transformer.

51-2. A full-wave single-phase mercury-arc rectifier receives power through a transformer connected to a 600-volt a-c source. The ratio of the primary turns on the transformer to the turns on the full secondary winding is 2.0. The arc drop is 18 volts. (a) When the rectifier is delivering a light load of 5 amp, find the voltage across its d-c terminals. (b) When the load is increased to 50 amp, the load loss of the transformer is 50 watts and the angle of overlap is 20 deg. Find the voltage across the load under this condition. (c) The firing angle is now retarded by 10 deg. If the angle of overlap remains unchanged and the load still draws a current of 50 amp, find the voltage across the load.

51-3. A three-phase mercury-arc rectifier supplies 100 amp at 240 volts to a d-c load. The transformers supplying the a-c voltage have Δ-connected primaries and zigzag-connected secondaries. The arc drop is 20 volts. If the rectifier is operated without phase control and overlap is neglected, find (a) the voltage to neutral supplied by the transformer secondaries, (b) the voltage across each winding of the transformer secondaries, and (c) the rms value of the anode current.

51-4. A three-phase mercury-arc rectifier receives power from a 600-volt a-c source through transformers with Δ-connected primaries and zigzag-connected secondaries. For each transformer, the ratio of primary turns to turns on one secondary winding equals 4.7. The arc drop is 20 volts. (a) When the rectifier is delivering a light load of 5 amp, find the voltage across its d-c terminals. (b) When the load is increased to 100 amps, the load loss in the three transformers is 200 watts and the angle of overlap is 22 deg. Find the voltage across the load under this condition. (c) The firing angle is now retarded by 30 deg. If the angle of overlap reduces to 12 deg and the load still draws a current of 100 amp, find the voltage across the load.

51-5. Six ignitions are supplied with power from a three-phase source through transformers having Δ-connected primaries and secondaries connected double-Y with an interphase transformer. The arc drop is 17 volts. The rectifier unit supplies 1,500 kw at 600 volts to a d-c load. If the rectifier is operated without phase control and overlap is neglected, find (a) the voltage to neutral across the transformer secondaries and (b) the rms value of the anode current of each ignition. (c) What voltage to neutral would be required and what would be the rms anode current if the interphase transformer were omitted?

51-6. A mercury-arc rectifier consists of six ignitrons receiving power from a 2,300-volt 3-phase source through transformers having Δ-connected primaries and secondaries connected double-Y with an interphase transformer. If the transformers are disconnected from the rectifier, the voltage to neutral measured across the open-circuited transformer secondaries is 530 volts. The arc drop is 17 volts and may be assumed to remain constant.

(a) When the load is so small that the interphase transformer is inoperative, what d-c voltage is supplied by the rectifier?

(b) When the rectifier is supplying a light load sufficient fully to magnetize the interphase transformer (approximately $\frac{9}{10}$ of 1 per cent of full-load current) find the d-c voltage supplied by the rectifier.

(c) With a load of 2,500 amp, the load loss of the transformers is 12,000 watts and the angle of overlap is 22 deg. Find the d-c output voltage of the rectifier at this load.

(d) The firing angle is now retarded by 30 deg. If the angle of overlap is reduced to 12 deg and the load current remains at 2,500 amp, find the d-c voltage of the rectifier.

51-7. A 1,500-kw 600-volt mercury-arc rectifier is composed of six ignitrons receiving power from a three-phase source through transformers having Δ-connected primaries and double-Y-connected secondaries with interphase transformer. The arc drop is 18 volts and may be assumed to remain constant. When the rectifier is fully loaded, the load loss of the transformers equals 18,000 watts. This loss is nearly proportional to the square of the d-c current of the rectifier. The core loss of the transformers is 7,000 watts, and the power used for the operation of all auxiliaries is 9.0 kw. Find the efficiency of the rectifier unit at full load and also at 50, 75, and 125 per cent of full load. Assume the d-c voltage to be maintained at 600 volts.

51-8. A 1,500-kw 250-volt mercury-arc rectifier is composed of six ignitrons and receives power from a three-phase source through transformers having Δ-connected primaries and double-Y-connected secondaries with interphase transformer. The arc drop is 18 volts. The full-load transformer losses are exactly equal to the similar losses in Prob. 51-7. Find the efficiency of this rectifier unit at full load and also at 50, 75, and 125 per cent of full load. Assume the d-c voltage to be maintained at 250 volts.

INDEX